HYDROPROCESSING FOR CLEAN ENERGY

HYDROPROCESSING FOR CLEAN ENERGY

Design, Operation, and Optimization

FRANK (XIN X.) ZHU
RICHARD HOEHN
VASANT THAKKAR
EDWIN YUH

WILEY

A Joint Publication of the American Institute of Chemical Engineers and John Wiley & Sons, Inc.

Published by John Wiley & Sons, Inc., Hoboken, New Jersey.
Published simultaneously in Canada.

Library of Congress Cataloging-in-Publication Data:

Names: Zhu, Frank Xin X., author. | Hoehn, Richard, 1950- author. | Thakkar, Vasant, author. | Yuh, Edwin, 1954- author.
Title: Hydroprocessing for clean energy : design, operation and optimization / Frank (Xin X.) Zhu, Richard Hoehn, Vasant Thakkar, Edwin Yuh.
Description: Hoboken, New Jersey : John Wiley & Sons, Inc., [2017] | Includes bibliographical references and index.
Identifiers: LCCN 2016038243| ISBN 9781118921357 (cloth) | ISBN 9781119328254 (epub) | ISBN 9781119328247 (Adobe PDF)
Subjects: LCSH: Hydrocracking. | Petroleum–Refining. | Green chemistry.
Classification: LCC TP690.4 .Z49 2017 | DDC 660–dc23 LC record available at https://lccn.loc.gov/2016038243

Cover image courtesy: UOP Unicracking Process Unit of the Bangchak Petroleum Public Company Limited, Bangkok, Thailand

Typeset in 10/12pt TimesLTStd by SPi Global, Chennai, India

Printed in the United States of America.

10 9 8 7 6 5 4 3

CONTENTS

PREFACE

It all started during a conversation between Frank Zhu and Dick Hoehn over a beer while watching the big ships wind their way through the Bosphorus Strait during a trip to Istanbul for a customer meeting in 2009. The conversation centered on how to pass on some of the things that we have learned over the years, and in doing so, pay homage to those who were willing to share their knowledge with us along the way. We decided that a book would be a good medium to do this, and thus the seed was planted.

We eventually settled on a topic currently relevant to refiners: clean energy with a focus on the production of ultra-low-sulfur diesel (ULSD) in particular. The selection of this topic came from realizing that a paradox exists in the world: people want to enjoy life fueled with a sufficient and affordable energy supply and, at the same time, live in a clean environment. There is no magic formula for achieving this, but with a knowledge of fundamentals and appropriate application of technology, the goal can be realized.

ULSD is an important part of the clean energy mix. It is made by hydroprocessing of certain fractions of petroleum crude oil. It is used in cars, trucks, trains, boats, buses, heavy machinery, and off-road vehicles. The bad news is that without adequate processing to produce clean diesel fuel and upgraded engine technology, diesel engines emit sulfur dioxide and particulates. The impact of fuel sulfur on air quality is widely understood and known to be significant.

There are challenges in producing ULSD in an economical and reliable manner. Over the years, a great deal of effort has been poured into developing the catalysts and process technology to accomplish this. It is intended that this book will be a resource for hydroprocessing technology as it relates to hydroprocessing in general and ULSD production in particular and that it will be a useful reference for plant managers, hydroprocessing unit engineers, operators, and entry-level design engineers.

We believe that there is currently no book available to provide relevant knowledge and tools for the process design and operation of facilities to produce ULSD, particularly considering the fact that these guidelines and methods have evolved over time to address the issues with the efficient production of ULSD. To this end, we decided that the book should cover four themes: fundamentals, design, assessment, and troubleshooting. That was the reason why the current team of authors was formed to create this book. The four themes correspond with each individual author's experience and expertise. An R&D specialist, Vasant, has an extensive background in the fundamentals of hydroprocessing catalysis (Chapters 3 and 4); Dick has many years of experience in the field of engineering design and development of hydroprocessing technology (Chapters 5–7); Edwin, a technical service specialist, brings a wealth of knowledge about operations and troubleshooting (Chapters 19 and 20); and Frank has both academic and practical background in process energy efficiency, process integration, and assessment methods (all other 13 chapters). The four authors represent a sum total of over 100 years of experience in the field of hydroprocessing.

The purpose of this book is to bridge the gap between hydroprocessing technology developers and the engineers who design and operate the processes. To accomplish this, 6 parts with 20 chapters in total are provided in this book. Part 1 provides an overview of the refining processes including the feeds and products together with their specifications, in particular, the fundamental aspects for hydroprocessing are discussed in detail. Part 2, mainly discusses on process design aspects for both diesel hydrotreating and hydrocracking processes. The focus of Part 3 is on process and heat integration methods for achieving high energy efficiency in design. In Part 4, the basics and operation assessment for major process equipment are discussed. In contrast, Part 5 focuses on process system optimization for achieving higher energy efficiency and economic margin. Last but not least, Part 6 deals with operation, in which operation guidelines are provided and troubleshooting cases are discussed.

Clearly, it was no small effort to write this book; but it was the desire to provide practical methods for helping people understand the issues involved in improving operations and designing for better energy efficiency and lower capital cost, which motivated us. In this endeavor, we owe an enormous debt of gratitude to many of our colleagues at UOP and Honeywell for their generous support in this effort. First of all, we would like to mention Geoff Miller, former vice president of UOP and now vice president of Honeywell, who has provided encouragement in the beginning of this journey for writing this book. We are very grateful to many colleagues for constructive suggestions and comments on the materials contain in this book. We would especially like to thank the following people for their valuable comments and suggestions: Bettina Marie Patena for Chapters 5 through 7, Zhanping (Ping) Xu for Chapter 10, Darren Le Geyt for Chapter 11, Bruce Lieberthal for Chapters 12 and 13, and Phil Daly for Chapter 14. Our sincere gratitude also goes to Charles Griswold, Mark James, and Rich Rossi for their constructive comments. Jane Shao produced beautiful drawings for many figures in the book. The contributions to this book from people mentioned above are deeply appreciated. I would also like to thank our co-publishers, AIChE and John Wiley for their help. Special thanks go to Steve Smith AIChE and Michael Leventhal for their guidance. The copyediting and typesetting by Vishnu Priya and her team at John Wiley is excellent. Finally, we would like to point out that this book reflects our own opinions but not those of UOP or Honeywell.

FRANK (XIN X.) ZHU
RICHARD HOEHN
VASANT THAKKAR
EDWIN YUH

Des Plaines, Illinois USA
June 1, 2016

PART 1

FUNDAMENTALS

1

OVERVIEW OF THIS BOOK

1.1 ENERGY SUSTAINABILITY

There is a paradox in this world: people want to enjoy life fueled with sufficient and affordable energy supply. At the same time, people wish to live in a clean environment. This paradox defines the objective of clean energy: provide affordable energy with minimum climate impact. This is a huge challenge technically, economically, geographically, and politically. There is no silver bullet for solving this paradox and the practical path forward is to determine a good mix of different kinds of energy sources. The proportions of this mix depend on the availability of these energy sources and costs of converting them to useful forms in geographic regions.

Energy demand has been increasing significantly over recent years due to the fact that people in emerging regions wish to improve their living standard and enjoy the benefit that energy can bring. Therefore, in the short and middle term, there is more oil and natural gas production to satisfy increased energy demand. To reduce the climate impact, sulfur content for the fossil fuels must be reduced – in particular, ultra-low-sulfur diesel (ULSD) is the focus in the present time. As far as energy efficiency is concerned, cars and trucks have become more fuel efficient and will continue to improve mileage per gallon. Furthermore, electrical and hybrid vehicles will improve energy efficiency even further. On the renewable energy side, the percentage of renewable energy, such as ethanol for gasoline and biodiesel blended into diesel fuel, will gradually increase over time through governmental regulation. Further technology development will make renewable energy such as wind, solar, and biofuels more cost-effective and hence these energy sources will become a sustainable part of the energy mix. These trends will coexist to achieve a balance between

Hydroprocessing for Clean Energy: Design, Operation, and Optimization, First Edition.
Frank (Xin X.) Zhu, Richard Hoehn, Vasant Thakkar, and Edwin Yuh.

increased energy demand and a cleaner environment, and at the same time, less dependence on foreign oil imports. In the long term, the goal is to increase the proportion of alternative energy in the energy mix to reduce gradually the demand for fossil fuels.

In summary, clean energy is the pathway for meeting the increased energy demand with a sustainable environment and the best future for clean energy is to capitalize on all the options: renewable energy, fossil fuels, increased efficiency, and reduced consumption. When these multiple trends and driving forces work together, the transformation becomes more economical and reliable. Technology developments in clean energy will join forces with regulations and market dynamics in the coming decades and beyond.

1.2 ULSD – IMPORTANT PART OF THE ENERGY MIX

ULSD is an important part of clean energy mix. Diesel fuel is made from hydroprocessing of certain fractions of petroleum crude. It is used in cars, trucks, trains, boats, buses, heavy machinery, and off-road vehicles. The bad news is that most diesel engines emit nitrogen oxides that can form ground-level ozone and contribute to acid rain. Diesel engines are also a source of fine particle air pollution. The impact of sulfur on particulate emissions is widely understood and known to be significant. In the European Auto Oil program, detailed study of lower effect on particulate matter (PM) was studied. This study suggests significant benefit from sulfur reductions for heavy-duty trucks. Reductions in fuel sulfur will also provide particulate emission reductions in all engines.

Testing performed on heavy-duty vehicles using the Japanese diesel 13 mode cycle have shown significant PM emission reductions that can be achieved with both catalyst and noncatalyst equipped vehicles. The testing showed that PM emissions from a noncatalyst equipped truck running on 400 ppm sulfur fuel were about double the emissions when operating on 2 ppm fuel (Worldwide Fuel Charter, Sept. 2013).

When sulfur is oxidized during combustion, it forms SO_2, which is the primary sulfur compound emitted from the engine. Some of the SO_2 is further oxidized – in the engine, exhaust, catalyst, or atmosphere to sulfate (SO_4). The sulfate and nearby water molecules often coalesce to form aerosols or engulf nearby carbon to form heavier particulates that have a significant influence on both fine and total PM. Without oxidation catalyst systems, the conversion rate from sulfur to sulfate is very low, typically around 1%, so the historical sulfate contribution to engine-out PM has been negligible. However, oxidation catalysts dramatically increase the conversion rate to as much as 100%, depending on catalyst efficiency. Therefore, for modern vehicle systems, most of which include oxidation catalysts, a large proportion of the engine-out SO_2 will be oxidized to SO_4, increasing the amount of PM emitted from the vehicle. Thus, fuel sulfur will have a significant impact on fine particulate emissions in direct proportion to the amount of sulfur in the fuel.

In the past, diesel fuel contained higher quantities of sulfur. European emission standards and preferential taxation have forced oil refineries to dramatically reduce the level of sulfur in diesel fuels. Automotive diesel fuel is covered in the European Union by standard EN 590, and the sulfur content has dramatically reduced during the last 20 years. In the 1990s, specifications allowed a content of 2000 ppm maximum of sulfur. Germany introduced 10 ppm sulfur limit for diesel from January 2003. Other European Union countries and Japan introduced diesel fuel with 10 ppm to the market from the year 2008.

In the United States, the acceptable level of sulfur in the highway diesel was first reduced from 2000 to 500 ppm by the Clean Air Act (CAA) amendments in the 1990s, then to 350, 50, and 15 ppm in the years 2000, 2005, and 2006, respectively. The major changeover process began in June 2006, when the EPA enacted a mandate requiring 80% of the highway diesel fuel produced or imported in order to meet the 15 ppm standard. The new ULSD fuel went on sale at most stations nationwide in mid-October 2006 with the goal of a gradual phase out of 500 ppm diesel.

In 2004, the US EPA also issued the clean air-nonroad-Tier 4 final rule, which mandated that starting in 2007, fuel sulfur levels in nonroad diesel fuel should be reduced from 3000 to 500 ppm. This includes fuels used in locomotive and marine applications, with the exception of marine residual fuel used by very large engines on ocean-going vessels. In 2010, fuel sulfur levels in most nonroad diesel fuel were reduced to 15 ppm, although exemptions for small refiners allowed for some 500 ppm diesel to remain in the system until 2014. After 1 December 2014, all highway, non-road, locomotive, and marine diesel fuel produced and imported has been ULSD.

The allowable sulfur content for ULSD (15 ppm) is much lower than the previous US on-highway standard for low-sulfur diesel (LSD, 500 ppm), which allows advanced emission control systems to be fitted that would otherwise be poisoned by these compounds. EPA, the California Air Resources Board, engine manufacturers, and others have completed tests and demonstration programs showing that using the advanced emissions control devices enabled by the use of ULSD fuel reduces emissions of hydrocarbons and oxides of nitrogen (precursors of ozone), as well as particular matter to near-zero levels. According to EPA estimates, with the implementation of the new fuel standards for diesel, nitrogen oxide emissions will be reduced by 2.6 million tons each year and soot or particulate matter will be reduced by 110,000 tons a year. EPA studies conclude that ozone and particulate matter cause a range of health problems, including those related to breathing, with children and the elderly those most at risk, and therefore estimates that there are significant health benefits associated with this program.

ULSD fuel will work in concert with a new generation of diesel engines to enable the new generation of diesel vehicles to meet the same strict emission standards as gasoline-powered vehicles. The new engines will utilize an emissions-reducing device called a particulate filter. The process is similar to a self-cleaning oven's cycle: a filter traps the tiny particles of soot in the exhaust fumes. The filter uses a sensor that measures back pressure and indicates the force required to push the exhaust gases out of the engine and through to the tailpipes. As the soot particles in the particulate filter accumulate, the back pressure in the exhaust system increases. When the pressure builds to a certain point, the sensor tells the engine management computer to inject more fuel into the engine. This causes heat to build up in the front of the filter, which burns up the accumulated soot particles. The entire cycle occurs within a few minutes and is undetectable by the vehicle's driver.

Diesel-powered engines and vehicles for 2007 and later model year vehicles are designed to operate only with ULSD fuel. Improper fuel use will reduce the efficiency and durability of engines, permanently damage many advanced emissions control systems, reduce fuel economy, and possibly prevent the vehicles from running at all. Manufacturer warranties are likely to be voided by improper fuel use. In addition, burning LSD fuel in 2006 and later model year diesel-powered cars, trucks, and buses is illegal and punishable with civil penalties.

The specifications proposed for clean diesel by Worldwide Fuel Charter (WWFC), which reflects the view of the automobile/engine manufactures concerning the fuel qualities for engines in use and for those yet to be developed, require increased cetane index, significant reduction of polynuclear aromatics (PNA), and lower T95 distillation temperature (i.e., the temperature at which 95% of a sample vaporizes) in addition to ultra-low sulfur levels. Automotive manufactures have concluded that substantial reductions in both gasoline and diesel fuel sulfur levels to quasi sulfur-free levels are essential to enable future vehicle technologies to meet the stringent vehicle emissions control requirements and reduce fuel consumption.

As a summary, to meet emission standards, engine manufactures will be required to produce new engines with advanced emission control technologies similar to those already expected for on-road (highway) heavy trucks and buses. Refiners will be producing and supplying ULSD for both highway and nonhighway diesel vehicles and equipment. Although there are still challenges to overcome, the benefits are clear: ULSD and the new emissions-reducing technology that it facilitates will help make the air cleaner and healthier for everyone.

In parallel, alternative technology such as electrical and hybrid-electric cars as well as biofuels for transportation is sought to address climate change issues and seek less dependence on fossil oil. The main driver for use of electrical and hybrid-electric cars is higher energy efficiency and lower greenhouse emissions; but electrical and hybrid-electric models are more expensive than conventional ones. On the other hand, biodiesel, made mainly from recycled cooking oil, soybean oil, other plant oils, and animal fats, has started to be used as blending stock for diesel. Biodiesel can be blended and used in many different concentrations. The most common are B100 (pure biodiesel), B20 (20% biodiesel, 80% petroleum diesel), B5 (5% biodiesel, 95% petroleum diesel), and B2 (2% biodiesel, 98% petroleum diesel). B20 is the most common biodiesel blend in the United States. B20 is popular because it represents a good balance of cost, emissions, cold-weather performance, materials compatibility, and ability to act as a solvent. Most biodiesel users purchase B20 or lower blends from their normal fuel distributors or from biodiesel marketers. However, not all diesel engine manufacturers cover biodiesel use in their warranties. Users should always consult their vehicle and engine warranty statements before using biodiesel.

There are two challenges to overcome in the use of biodiesel. One is the availability of feedstock and the other is the cost. Government subsidies for biofuels are currently being used to encourage expansion of production capacity. Although the social, economic, and regulatory issues associated with expanded production of biodiesel are outside the scope of this book, it is crucial that future commercialization efforts focus on sustainable and cost-effective methods of producing feedstock. Current and future producers are targeting sustainable production scenarios that, in addition to minimizing impact on land-use change and food and water resources, provide an energy alternative that is economically competitive with current petroleum-based fuels. Future growth will require a coordinated effort between feedstock producers, refiners, and industry regulators to ensure environmental impacts are minimized.

If done responsibly, increasing biofuel usage in the transportation sector can significantly reduce greenhouse-gas emissions as well as diversify energy sources, enhance energy security, and stimulate the rural agricultural economy.

1.3 TECHNICAL CHALLENGES FOR MAKING ULSD

ULSD is mainly produced from hydrocracking and diesel hydrotreating processes with crude oil as the raw feed in the refinery. Technical solutions for ULSD production can be summarized as follows (Stanislaus et al., 2010):

- Use of highly active catalysts
- Increase of operating severity (e.g., increased temperature, increase in hydrogen pressure, lower LHSV)
- Increase catalyst volume (by using additional reactor, dense loading, etc.)
- Removal of H_2S from recycle gas
- Improve feed distribution in the reactor by using high-efficiency vapor/liquid distribution trays
- Use of easier feeds; reduce feedstock end boiling point
- Use of two-stage reaction system design for hydrocrackers

A combination of the above options may be necessary to achieve the target sulfur level cost-effectively. Selection of the most appropriate option or a combination of those is specific for each refinery depending on its configuration, existing process design, feedstock quality, product slate, hydrogen availability, and so on.

Clearly, there are many design parameters to consider during process design. As an example, consider the design choice for the use of one or two reaction stages in a hydrocracking unit. In single-stage hydrocrackers, all catalysts are contained in a single stage (in one or more series or parallel reactors). A single catalyst type might be employed or a stacked-bed arrangement of two different catalysts might be used. In single-stage hydrocracking, all catalysts are exposed to the high levels of H_2S and NH_3 that are generated during removal of organic sulfur and nitrogen from the feed. Ammonia inhibits the hydrocracking catalyst activity, requiring higher operating temperatures to achieve target conversion, but this generally results in somewhat better liquid yields than would be the case if no ammonia were present. There is no interstage product separation in single-stage or series-flow operation.

However, two-stage hydrocrackers employ interstage separation that removes the H_2S and NH_3 produced in the first stage. As a consequence, the second-stage hydrocracking catalyst is exposed to lower levels of these gases, especially NH_3. Some two-stage hydrocracker designs do result in very high H_2S levels in the second stage. Frequently, unconverted product is separated and recycled back to either the pretreat or the cracking reactors.

Understanding of these fundamentals will be paramount and the related design considerations will be provided in details in this book. Apart from discussions of fundamental aspects of design, the book also provides explanation on how to design hydrocracking and distillate hydrotreating units by applying applicable theory and design considerations in order to obtain a practical and economic design with the least capital cost and energy use possible. During operation, the primary goal is to achieve safe, reliable, and economic production. Achieving the operation objectives is another focus of discussions in this book.

1.4 WHAT IS THE BOOK WRITTEN FOR

The purpose of this book is to bridge the gap between hydroprocessing technology developers and the engineers who design and operate the processes. To accomplish this, 6 parts with 20 chapters in total are provided in this book. The first part provides an overview of the refining processes including the feeds and products together with their specifications, while the second part mainly discusses process design aspects for both diesel hydrotreating and hydrocracking processes. Part 3 focuses on process and heat integration methods for achieving high energy efficiency in design. With Part 4, the basics and operation assessment for major process equipment are discussed. In contrast, Part 5 focuses on process system optimization for achieving higher energy efficiency and economic margin. Last but not least, in Part 6, operation guidelines are provided and troubleshooting cases are discussed.

REFERENCES

Stanislaus A, Marafi A, Rana M (2010) Recent advances in the science and technology of ultra-low sulfur diesel (ULSD) production, *Catalysis Today*, **153** (1), 1–68.

5th Worldwide Fuel Charter (2013) *ACEA*, European Automobile Manufacturers Association, Brussels, Belgium.

2

REFINERY FEEDS, PRODUCTS, AND PROCESSES

2.1 INTRODUCTION

Crude oils are feedstock for producing transportation fuels and petrochemical products. Many different kinds of crude oils are available in the market place with different properties and product yields and hence price. The crude price is largely based on its density (or °API gravity), sulfur content, and metals content. For example, light crude oils command a higher price as they contain a larger portion of gasoline, jet fuel, and diesel, which can be sold at higher prices and also requires less processing. Sulfur also impacts crude price. Low-sulfur crude oils command higher prices as they require less hydroprocessing.

The method to determine the properties of crude oils is called crude characterization, which provides the basis for design of new processes, or upgrading existing processes, or predicting how the refinery will need to operate to make the desired products with the required product quality. First, crude characterization can help a refiner to know if the process technology in the refinery can handle a certain crude feed in order to make desirable products with acceptable quality. Second, it provides insights into the compatibility of different crude oils being mixed together. Third, it sets the basis for developing operational guidelines for achieving predicted yields.

The ASTM (American Society for Testing Materials) methods for crude characterization are the most well known, which will be discussed in detail here. At the same time, major refining processes will be briefly explained.

Hydroprocessing for Clean Energy: Design, Operation, and Optimization, First Edition.
Frank (Xin X.) Zhu, Richard Hoehn, Vasant Thakkar, and Edwin Yuh.

2.2 ASTM STANDARD FOR CRUDE CHARACTERIZATION

Because crude oils are a mixture of many different chemical compounds, they cannot be evaluated based on chemical analysis alone. In order to characterize any crude oil and refining products, the petroleum industry has developed a number of shorthand methods for describing hydrocarbon compounds by the number of carbon atoms and unsaturated bonds in the molecule and using distillation temperatures and other easily obtained properties to specify crude and products.

In characterization of certain crude, the test methods, known as ASTM methods, are conducted in laboratory. Through distillation, crude is then cut into several products so that product yields and properties can then be predicted. Thus, crude characterization is about crude distillation, product fractions, and properties. This information can be used for simulation of the process streams in question. There are three types of ASTM crude characterization methods. The first one is the simplest using single stage of distillation, which include D-86 and D-1160. The second one is called TBP (true boiling point) distillation based on multiple theoretical stages, and D-2892 is an example of the TBP distillation. The third one is SD (simulated distillation) based on gas chromatography, which is the most consistent method. D-2287 and D-3710 are the examples of gas chromatography. The distillation data from these three methods can be intercorrelated. These ASTM methods are explained as follows.

2.2.1 ASTM D-86 Distillation

The D-86 distillation method was developed for characterization of the product fraction of the crude that can be obtained via atmospheric distillation. It can be used for characterizing crude oil and petroleum fractions with IBPs (initial boiling points) slightly above room temperature up to final boiling point above 750 °F.

D-86 is the most common refinery distillation because it is easy and quick to obtain and does not require expensive equipment. The products from atmospheric distillation include naphtha, kerosene, diesel, and atmospheric gas oil. D-86 distillation indicates the distillation temperature that corresponds to the volume percent vaporized for each product.

In determining D-86 distillation, the sample feed is put into a device whose dimensions are defined by the test specification and the sample is heated. As the sample gets hotter, more of the sample vaporizes. The vaporized material is collected and the temperature corresponding to a certain volume percent of the original sample vaporized (5%, 10%, … , 95%) is recorded.

It must be pointed out that D-86 is not a true distillation because there is only one stage. Consequently, the D-86 tends to have a much higher IBP temperature than the actual one due to entrainment as molecular interactions hold lighter molecules in the mixture. At the tail end of the D-86 distillation, due to entrainment, the heavier molecules can flash off readily, resulting in a lower EP (end-point) than the actual distillation. Therefore, D-86 has a higher IBP and a lower EP than the actual atmospheric distillation in a refinery. However, the deviations in the front and tail ends in

D-86 distillation can be overcome by TBP distillation, which is discussed below. It is important to know that the D-86 distillation can be correlated with TBP distillation.

2.2.2 ASTM D-1160 Distillation

This test is similar to D-86 using single theoretical stage but performed at vacuum conditions. It is used for determining the distillation of VGO (vacuum gas oil) and heavier materials. The reason for vacuum distillation comes from the fact that temperatures above 700–780 °F would be required for heavier oils to vaporize at the atmospheric pressure. Under these temperatures, the oil would begin to thermally crack into lighter components. Imposing a vacuum enables the heavy oils to vaporize before reaching the cracking temperature. Thus, there is a limit that is closely monitored and controlled in operation for the charge heater outlet temperature before an atmospheric distillation column. The heavy gas oil has to be recovered in a vacuum distillation column.

2.2.3 ASTM D-2892 Distillation

Instead of using flash (or single stage of distillation) in D-86, D-2892 distillation is obtained via a higher level of separation, that is, 14–18 theoretical stages. Thus, TBP distillation reasonably represents the actual atmospheric distillation. This fraction has a final boiling point below 750 °F. The following example for heavy naphtha (Table 2.1) is used to show how the shortcomings in the D-86 flash distillation, that is, the higher IBP and lower EP, can be overcome by TBP distillation.

2.2.4 ASTM D-2287 Distillation

This ASTM method also includes gas chromatography, which is regarded as most consistent method to describe the boiling range of a hydrocarbon fraction unanimously. This method can be applied to any hydrocarbon fractions with a final boiling point of 1000 °F or less under atmospheric pressure. This method is also limited to hydrocarbon fractions having an initial boiling point of 100 °F.

2.2.5 ASTM D-3710 Distillation

This method is used to determine the boiling points of gasolines below final boiling point of 500 °F at atmospheric pressure. It is also based on gas chromatography.

TABLE 2.1. Use of TBP to Overcome the Shortcomings of D-86 (Sample Heavy Naphtha)

Unit: °F	IBP	50%	EP
TBP	160	250	400
D-86	210	260	340

2.3 IMPORTANT TERMINOLOGIES IN CRUDE CHARACTERIZATION

2.3.1 TBP Cut

TBP cut defines a specific segment of the TBP distillation curve and it is usually a reference to breaking up crude. A TBP curve is indicative of the true nature of the hydrocarbons; it represents all of the material in a certain boiling range from the crude assay. For example, a 200–400 °F cut contains all components within this boiling range, but no 190 °F or 410 °F boiling range material.

TBP cut information is provided in the crude assay, often called nominal TBP cut, as it represents the fractions (or cuts) obtainable using the TBP distillation method based on infinite number of distillation trays; thus, it refers to theoretical yields. However, commercial distillation columns cannot achieve TBP cuts due to the limited number of column trays. As a result, some light material will be present in the heavier cut and some heavier material in the lighter cut. Thus, it is necessary to adjust TBP cuts provided in a crude assay to obtain a realistic product fractions, distillations cuts, and properties from commercial distillation columns.

2.3.2 Adjusted TBP Cut

In order to have a realistic estimate of the amount of material that can be obtained for a given cut, we compensate for the slumped material by making the boiling range of a corresponding TBP cut a little larger. Typical levels of slumping in the crude unit are 1% naphtha to kerosene, 3% kerosene to diesel, and 6% diesel to VGO.

For example, to compensate for the 1% of naphtha slumping, we make the naphtha cut 1% larger. This could increase the selected assay TBP end point by a few degrees.

2.3.3 Crude Assay

When a crude oil is offered for sale, a report of the physical and chemical properties for the crude and products is prepared and the report is known as "crude assay" in the industry. Crude assays are prepared from a laboratory fractionation column, and product fractions drawn from the laboratory fractionation are similar to those fractions to be drawn from the crude distillation unit in the refinery. Basically, a crude assay provides crude characterization and it is about various product fractions in TBP cuts and their properties as well as impurities such as sulfur, nitrogen, and metals.

2.3.4 Commercial Yields Versus Theoretical Yields

It is common that commercial yields from a refiner cannot match the theoretic yields obtained from TBP distillation in a crude assay because commercial distillation is not perfect with lighter materials slumping into heavier cuts. Typically, there are 6–10 fractionation theoretical stages between two adjacent cuts in a commercial crude distillation column in comparison with very large number of fractionation stages assumed in the TBP distillation. Not only are the product yields different but also the product properties can vary. A useful tool to verify the information in a crude assay is called crude oil breakup.

A crude breakup is performed to predict the potential gap or overlaps in product yields between crude assay and operation based on actual process conditions including flash zone temperature and pressure. The predicted product yields and properties can be used to compare with that claimed in crude assay and more importantly as guidelines for refinery operations, for example, determining distillation cuts and predicting their properties.

2.4 REFINING PROCESSES

Crude assay and crude breakup information can indicate what potential products are available in a crude oil, but it is up to refining processes to produce them. There are many processes involved in a modern oil refinery and only the major ones are briefly explained as follows.

2.4.1 Atmospheric Distillation

This is the first process where the crude oil enters the refinery. Under near atmospheric pressure, the crude oil is heated to vaporize most of the hydrocarbons boiling in diesel and lighter range. Diesel and kerosene are withdrawn as sidecuts from the atmospheric column and naphtha and lighter materials are produced as overhead products. The above materials are sent downstream for further processing, while the bottom product of the atmospheric distillation column is sent to the vacuum distillation to recover the heavy gas oil.

2.4.2 Vacuum Distillation

Heavy crude oil components have high boiling temperatures and cannot be boiled at atmospheric pressure. These components must be further fractionated under vacuum. The vacuum condition lowers the boiling temperature of the material and thereby allows distillation of the heavier fractions without use of excessive temperatures that would lead to thermal decomposition. Steam ejectors are commonly employed to create the vacuum conditions required.

2.4.3 Fluid Catalytic Cracking

Some of heavy gas oil or atmospheric residue fractions are sent to a fluid catalytic cracking (FCC) where the feed reacts with catalyst under high temperatures and is converted to lighter materials. The catalyst is then separated and regenerated, while the reaction products are fractionated into several products by distillation. The main product of FCC is gasoline and this is the reason why FCC has been the most widely used refinery conversion technique for around 60 years. FCC units also produce a highly aromatic distillate product called light cycle oil (LCO).

2.4.4 Catalytic Reforming

Reforming is a catalytic process that converts low-octane naphtha to high-octane product, which is called reformate with an octane number of 96 or above. Reformate is blended into gasoline to increase the octane number. The feed is passed over a platinum catalyst where the predominant reaction is the removal of hydrogen from naphthenes and the conversion of naphthenes to aromatics. The process also produces high purity hydrogen that can be used in hydrotreating processes.

2.4.5 Alkylation

In the alkylation process, isobutane, a low-molecular-weight material, is chemically combined with olefins, such as propylene and butylene in the presence of a catalyst such as hydrofluoric acid or sulfuric acid. The resulting product, called alkylate, is a branched chain hydrocarbon, which has a much higher octane number compared to a straight-chained material of the same carbon number. Similar to reformate, alkylate is blended into gasoline to increase the octane number, thus reduce knocking.

2.4.6 Isomerization

Light straight run (LSR) naphtha, mainly pentane–hexane fraction, is characterized by low octane, typically 60–70 RON. In the isomerization process, the octane numbers of the LSR numbers can be improved significantly up to 90 via converting normal paraffins to their isomers in the presence of catalysts containing platinum on various bases including alumina, molecular sieve, and metal oxide. The resulting product is isomerate, which is blended into gasoline pool.

2.4.7 Hydrocracking

This is a catalytic, high-pressure process that converts a wide range of hydrocarbons to lighter, more valuable products such as low-sulfur gasoline, jet fuel, and diesel. By catalytically adding hydrogen under very high pressure, the process increases the ratio of hydrogen to hydrocarbon in the feed and produces low-boiling materials, thus improving the product quality. It also removes contaminants such as sulfurs and nitrogen. Hydrocracking is especially adapted to the processing of low-value stocks, such as vacuum gas oils, and is most often used to produce high-quality distillate range products. FCC, which can process the same feed, is primarily geared toward naphtha production.

2.4.8 Hydrotreating

It is the most widely used treating process in today's refineries. This process uses the catalytic addition of hydrogen to remove sulfur compounds from naphtha and distillates (light and heavy gas oils). Removal of sulfur is essential for meeting product specifications for gasoline, jet fuel, diesel, and heavy burner fuel as well as for protecting the catalyst in subsequent processes (such as catalytic reforming). In addition to removing sulfur, it can eliminate other undesirable impurities (e.g., nitrogen and oxygen) and saturate olefins.

2.4.9 Residue Desulfurizing

With the increasing need for products produced from heavier and higher boiling components (the "bottom of the barrel"), desulfurization is used to process the residues from atmospheric and vacuum column distillation and the desulfurized residues can be used as a blending component of low-sulfur fuel oils. Residue desulfurizing unit can also be used to improve the quality of residue feedstock to a Residue FCC (RFCC) process.

2.4.10 Coking

The residual bottoms from the crude unit contribute lighter fractions (naphtha and gas oils) via a thermal cracking process called "coking," in which the feed is heated to high temperatures and routed to a drum where the material is allowed to remain for sufficient time for thermal cracking to take place. The lighter portions leave the drum and are recovered as liquid products. The heavier portions remain in the drum and eventually turn into coke, a nonvolatile carbonaceous material. Depending on the amount of sulfur, metals, and aromatic content of the coke, it may be used for boiler fuel or further processed into anodes for aluminum production. Since the recovered liquid products were produced through a thermal process, their quality will not meet current fuel standards, as the sulfur and nitrogen contents are usually high. In addition, they tend to have high olefin contents, making them slightly unstable for storage or blending with other materials. As a result, these products generally must be hydrotreated or upgraded further to produce suitable product blending components.

2.4.11 Blending

After the products are made from the above process units and other sources, they are blended to make final products meeting desirable specifications, which are discussed as follows.

2.5 PRODUCTS AND PROPERTIES

Although there are a large variety of products that can be produced, most refineries are designed to make liquefied petroleum gas (LPG), gasoline, jet fuel, diesel, heavy fuel oil, and feedstocks for petrochemical processes. All products must meet desirable specifications in the local markets. Making desirable products dictate refinery technology selection and process designs while optimizing product yields for high economic margin directs refinery operation.

2.5.1 LPG

LPG is a mixture of propane (C3) and butane (C4) and can be used for heating fuel as well as a refrigeration working fluid. Over half the propane produced goes to petrochemical processes as feed for olefins production and other chemical manufacturing. Some LPG range material is used in the alkylation process to produce a high-octane gasoline blending component.

Normal butane (*n*-C4) is frequently used as gasoline blending stock to regulate its vapor pressure due to its lower vapor pressure than isobutene (*i*-C4). *n*-C4 has Reid vapor pressure (RVP) of 52 psi compared with 71 psi RVP of *i*-C4. Generally, the lower the RVP of a gasoline blend, the more it costs. For example, in winter you can blend butane, which is relatively plentiful and cheap, with gasoline to promote better startup in cold weather. But butane with a high RVP cannot be used in summer as it would immediately boil off.

2.5.2 Gasoline

About 90% of the total gasoline produced in the United States is used as automobile fuel. Thus, demand of motor fuel has been the major driving force for oil refining processes. For most refineries in the United States, most crude oils contain only about 30% of gasoline range components. To make more gasoline, other components are converted into gasoline blending stock. As the result, around 50% or more of the crude oil is converted into gasoline. Most service stations provide three grades based on octane number that is designed to meet the requirements of the specific engine installed.

2.5.2.1 Gasoline Cut The standard distillation range for automobile gasoline is between 10% boiling point at 122 °F and end point of 437 °F. Typically, heavy naphtha and lighter materials are produced from the crude column overhead and then are routed to a gasoline stabilizer to remove butane and lighter materials. These lighter materials are sent to the saturate gas concentration unit where LPG is recovered out of refinery fuel gas. The pentane and heavier naphtha range materials go to a naphtha hydrotreater for making low-sulfur gasoline. Treated light naphtha consisting of primarily C5 and some C6 hydrocarbons can either be blended directly into gasoline or processed in an isomerization process to increase its octane. The heptane and heavier hydrocarbons usually have an octane lower than that required by current engine technology; thus, they are treated further by catalytic reforming to yield a high-octane gasoline blending component.

2.5.2.2 Gasoline Specifications Three properties can be used to describe the main features of gasoline: octane number, ease of startup, and RVP.

Octane number is the most important specification of motor gasoline and it delivers smooth burning in the engine without knocking. The octane number is an expression of the antiknocking performance of the engine using two reference fuels as the basis, namely normal *n*-heptane, defined to have an octane number of 0, and *iso*-octane, defined to have an octane number of 100. A gasoline with an octane number of 90 means the engine performance is equivalent to a mixture of 90 vol% of iso-octane and 10% *n*-heptane. In the past, the gasoline octane was increased by including olefins, aromatics, and alkyl lead components. Under current regulations, lead addition to gasoline is not permitted due to the health risk while many countries set limits on gasoline aromatics in general, benzene in particular, as well as olefin (alkene) content as these two components generate volatile organic compounds (VOCs) causing ground-level ozone pollution.

Second, the engine must start easily in cold weather. The easy startup is affected by the amount of light components in the gasoline, which is measured as the percentage that is distilled at 158 °F or lower. Obviously, a fuel requires more of this percentage in cold climate for the car engine to start quickly.

Third, the engine must not have vapor lock at high temperatures, which is measured by the vapor specification, namely RVP. RVP is defined as the vapor pressure of the gasoline at 100 °F in lb/in.^{2} absolute. The RVP limit is a function of ambient temperature. A lower RVP is required in warm climate compared to cold one.

Altitude has significant effects on gasoline properties. The main effect is the octane requirement, approximately with 5 RON (research octane number) less for a 5000 ft increase in elevation. In general, same model of engines could vary by 7–12 RON to suit different altitudes according to climatic conditions.

*2.5.2.3 **Gasoline Production*** Automobile gasoline is blended to meet the demand and local conditions. Blending components for making gasoline include straight run naphtha, catalytic reformate, FCC gasoline, hydrocracked gasoline, alkylates, isomerate, and *n*-butane. Proper blending is essential to achieve the proper antiknock properties, ease of startup, low vapor lock potential, and low engine deposits, which are the main characteristics that a good gasoline product must have.

LSR naphtha consists of C5-190 °F (TBP), but the final cut-point can vary from 180 °F to 200 °F (a swing cut), depending on economic conditions or local requirements. As LSR naphtha cannot be upgraded in octane in a catalytic reformer, it is processed separately from heavy naphtha. In some refineries, it is sent to isomerization units for upgrading its octane.

Heavy straight run (HSR) naphtha, the fraction of 190–370 °F, usually goes to a catalytic reformer to make high-octane reformate. Thus, a catalytic reformer is operated to give satisfactory antiknock properties within an RON of 90–100.

Historically, FCC naphtha had been blended directly into gasoline due to its inherently high octane. However, FCC naphtha contains high sulfur and it has to be hydrotreated for gasoline blending due to the gasoline sulfur limits required in regulations. Naphtha derived from hydrocracking units is usually rather low in octane and may need further upgrading via catalytic reforming if the amount available would adversely affect the overall octane rating of the gasoline pool. Alkylate is the product from the reaction between isobutene, propylene, and butylene to make a sulfur-free and high-octane gasoline blending component.

Normal butane is also used as the gasoline blending stock to adjust the RVP of the gasoline. The gasoline RVP is a compromise between a high RVP for easy startup and a low RVP for preventing vapor lock. Gasoline RVP is also subject to regulations in many regions to minimize hydrocarbon vapor emissions.

2.5.3 Jet Fuel

There are two types of jet fuels, namely naphtha and kerosene. Naphtha jet fuel, also called aviation gasoline, is made mainly for military jets. It is similar to automotive gasoline but has a narrower distillation range of 122 °F at 10% and 338 °F TBP end point. Commercial jet fuel (simply called as jet fuel in the following discussions) is in kerosene boiling range, with 401 °F at 10% and a 572 °F end point. Kerosene includes hydrocarbons boiling in the C9–C16 range; therefore, the cut can be as wide as 300–570 °F on a D-86 basis. The front end of the kerosene is limited by the kerosene flash, which is set at 100 °F.

Typically, SR (straight run) kerosene and hydrotreated or hydrocracked kerosene are blended into commercial jet fuel. Most jet fuel is SR stock and it is treated using the Merox process from which the mercaptans in the feed are converted to disulfides. Stocks containing olefins are not acceptable because they have poor thermal stability and will polymerize, forming gums that can harm jet engines. Stocks containing high quantities of straight-chain paraffins are also restricted due to unacceptable cold flow properties (i.e., freeze point) causing plugging at low temperatures.

2.5.3.1 Jet Fuel Specifications Unlike the spark ignition engines in cars, jet engines rely on continuous burn in a combustion chamber. Jet fuel needs to be mostly paraffinic as benzene and other aromatics exhibit undesirable combustion characteristics, and olefins present a gum stability risk. Thus, the main specifications for jet fuel include flash point, smoke point, freezing point, aromatics content, olefin content, and sulfur content.

Smoke Point The smoke point describes the combustion effects on mechanical integrity. It is defined as the maximum height, in millimeters, of a smokeless flame when the fuel sample is burned in a lamp of a specified design. The smoke point is related to the hydrocarbon type comprising the fuel. The more aromatic the jet fuel, the smokier the flame. The smoke point specification limits the blending percentage of cracked products that are high in aromatics. The smoke point is quantitatively related to the potential radiant heat transfer from the combustion products of the fuel. Because radiant heat transfer exerts a strong influence on the metal temperature of combustor liners and other hot parts of jet engines, the smoke point provides the basis to derive the relationship between the life of the mechanical components and the fuel characteristics.

Aromatics Aromatics in jet fuel increase with boiling range; therefore, a lighter jet fuel will have less aromatics. Some crudes are too high in aromatic content to be acceptable for jet fuel and must be cut back with other paraffinic crudes. Typical kerosene aromatics content is in the range of 20–25%.

Flash Point The flash point is an indication of the maximum temperature for fuel handling and storage without serious fire hazard. This specification provides the basis for determining the regulations and insurance requirements for jet fuel shipment, storage, and handling precautions.

In order to obtain acceptable flash point, a stripper with steam stripping is used on the kerosene sidecut from the crude column, specifically to strip out lighter molecules in order to meet the flash specification for the kerosene.

Freeze Point Freeze point is defined based on the temperature at which waxy crystals are formed as the jet fuel is cooled. Freeze point is related to the composition of the jet fuel. Higher paraffin content results in a poor freeze point because waxy crystals start to form at a high temperature. Cyclics, especially aromatics, improve freeze point. Unfortunately, as the freeze point gets better due to higher aromatics, the smoke point gets worse; therefore, there is always a trade-off between these two properties.

The freeze point specification must be sufficiently low to preclude interference with flow of fuel through filter screens to the engines at low temperatures experienced at high altitude. The fuel temperature in an aircraft tank decreases at a rate proportional to the duration of flight. Long duration flights would require lower freeze point than short duration flights. Hydrocracking is used to isomerize the paraffins and reduce the freeze point. Hydrocracking also produces jet fuel with a very low smoke point and is therefore a premium jet fuel blending component.

2.5.4 Diesel

The diesel cut is normally in the TBP range of 480–650 °F crude cut. The diesel cut typically contains C14–C22 hydrocarbons. This range can be as wide as C10–C22 if kerosene is included in diesel fuel. In the case of kerosene blended into the diesel, the diesel pour point improves and that is particularly relevant in very cold climates (e.g., Alaska, Canada, Russia) where a low pour point is critical. At the same time, the diesel IBP can be below 350 °F, but the IBP will largely be set by the flash point specification, which for a typical diesel fuel is around 125 °F. Flash point is adjusted by using a reboiled or steam stripped sidecut stripper at the crude column to strip out lighter molecules in order to meet the flash point specification.

Sulfur is a major issue for diesel produced according to current standards. To meet current standards, most diesel fuel must be hydrotreated. Achieving the required sulfur levels requires ability to remove sulfur from different hydrocarbon species, some of which are easily amenable to treatment and others are more difficult. The most difficult sulfur species is dimethyl-dibenzo-thiophene, which boils at 646 °F. This molecule is sterically hindered because the two methyl groups prevent hydrogen from getting at the sulfur in hydrotreating.

2.5.4.1 Diesel Specifications Cetane number, flash point, pour point, cloud point, and sulfur content are the most important properties of diesel fuels. Typical diesel specifications are shown in Table 2.2. The fuel volatility requirements depend on engine design and applications. For automotive diesel fuel with fluctuating speeds and loads, the more volatile fuels have advantages. For railroads, ships, and power stations, the heavier fuels are more economic due to their high heat of combustion.

Cetane Number It is the ignition performance indicator, which is similar to the octane number for gasoline. For diesel fuels, the reference fuel is cetane (n-hexadecane) with a cetane number of 100 and α-methylnaphthalene with a cetane number of 0. Cetane number is better with paraffins than with olefins or aromatics. Acceptable cetane numbers are 40 and higher at 90 is preferred for very cold climate. Hydrocracked diesel typically has a high cetane number and is a good diesel

TABLE 2.2. Typical Diesel Specifications

	US	EURO IV Property
Cetane number (Engine test)	40	51
Sulfur (wt-ppm)	15	10
Flash (°C)	52	52–60
Pour and cloud points	Seasonal by location	

blending component. Thermally cracked diesel, on the other hand, is particularly low in cetane and requires upgrading by hydrotreating before being blended into the diesel pool.

Cloud Point It indicates the suitability of the fuel for low temperature operations and it is a guide to the temperature at which it may clog filters and restrict flow as paraffinic fuel consists of precipitate as wax.

Pour Point It is another low temperature performance indicator, which defines the lowest temperature at which the fuel can be pumped. This temperature (pour point) often occurs about 8 °F below the cloud point.

Sulfur Content Sulfur has strong negative impact on the environment and there have been continuous efforts from legislation to production to reduce sulfur content in diesel. Before 2007, standard highway-use diesel fuel sold in the United States contained an average of 500 ppm (parts per million) sulfur. After 2010, nonroad diesel contained 10–15 ppm. In Europe, Germany introduced 10 ppm sulfur limit for diesel from January 2003. Other European Union countries and Japan introduced diesel fuel with 10 ppm to the market from the year 2008.

2.6 BIOFUEL

For completeness, biofuel is also briefly described and the following discussion is largely based on Wikipedia (https://en.wikipedia.org/wiki/Biofuel).

Biofuels can be derived directly from plants, or indirectly from agricultural, commercial, domestic, and/or industrial wastes. Renewable biofuels generally involve contemporary carbon fixation, such as those that occur in plants or microalgae through the process of photosynthesis. Other renewable biofuels are made through the use or conversion of biomass (referring to recently living organisms, most often referring to plants or plant-derived materials). This biomass can be converted into convenient energy containing substances in three different ways: thermal conversion, chemical conversion, and biochemical conversion. This biomass conversion can result in fuel in solid, liquid, or gas form. This new biomass can also be used directly for biofuels.

There are two major biofuels currently produced as transportation fuels, namely bioethanol for gasoline addition and biodiesel for diesel additive.

2.6.1 Bioethanol

Bioethanol is an alcohol made by fermentation, mostly from carbohydrates produced in sugar or starch crops such as corn, sugarcane, or sweet sorghum. Cellulosic biomass, derived from nonfood sources, such as trees and grasses, is also being developed as a feedstock for ethanol production. Ethanol can be used as a fuel for vehicles in its pure form, but it is usually used as a gasoline additive to increase octane and improve vehicle emissions. Bioethanol is widely used in the United States and Brazil. Current plant design does not provide for converting the lignin portion of plant raw materials to fuel components by fermentation.

2.6.2 Biodiesel

Biodiesel can be produced from oils or fats either using transesterification or hydrotreatment. Biodiesel is used as a diesel additive to reduce levels of particulates, carbon monoxide, and hydrocarbons from diesel-powered vehicles.

2.6.3 Blending of Biofuel

In 2010, worldwide biofuel production reached 105 billion liters (28 billion gallons), up 17% from 2009 and biofuels provided 2.7% of the world's fuels for road transport, a contribution largely made up of ethanol and biodiesel (Wikipedia, Biofuel cite note 2). Global ethanol fuel production reached 86 billion liters (23 billion gallons) in 2010, with the United States and Brazil as the world's top producers, accounting together for 90% of global production. The world's largest biodiesel producer is the European Union, accounting for 53% of all biodiesel production in 2010. As of 2011, mandates for blending biofuels exist in 31 countries at the national level and in 29 states or provinces (Wikipedia, Biofuel cite note 3).

The International Energy Agency has a goal for biofuels to meet more than a quarter of the world demand for transportation fuels by 2050 to reduce dependence on petroleum and coal (Wikipedia, Biofuel cite note 4). There are various social, economic, environmental, and technical issues relating to biofuel production and use, which have been debated in the popular media and scientific journals. These include the effect of moderating oil prices, the "food versus fuel" issue, poverty reduction potential, carbon emissions levels, sustainable biofuel production, deforestation and soil erosion, loss of biodiversity, impact on water resources, rural social exclusion and injustice, shantytown migration, rural unskilled unemployment, and nitrous oxide (NO_2) emissions.

3

DIESEL HYDROTREATING PROCESS

3.1 WHY DIESEL HYDROTREATING?

Diesel hydrotreating (DHT) or catalytic hydrogen treating is mainly to reduce undesirable species from straight-run diesel fraction by selectively reacting these species with hydrogen in a reactor at elevated temperatures and at moderate pressures. These objectionable materials include, but are not solely limited to, sulfur, nitrogen, olefins, and aromatics. Many of the product quality specifications are driven by environmental regulations that have become more stringent over recent time.

In the early 1900s, diesel fuel standards were first developed to ensure that diesel engine owners could buy a fuel that was compatible with the requirements of their engines. To achieve this, these early standards controlled primarily the distillation and boiling ranges, volatility, its cold flow properties, and its cetane number. Currently, these diesel standards are embodied in standards such as American Society for Testing Materials (ASTM) D975 and its equivalents under European (EN 590) and Japanese normalization organizations. In the United States, as in many other jurisdictions, several basic grades of diesel fuel are in use:

- No. 1 Diesel Fuel – A special-purpose, light distillate fuel for automotive diesel engines requiring higher volatility than that provided by Grade Low Sulfur No. 2-D
- No. 2 Diesel Fuel – A general-purpose, middle distillate fuel for automotive diesel engines, which is also suitable for use in nonautomotive applications, especially in conditions of frequently varying speed and load.

Hydroprocessing for Clean Energy: Design, Operation, and Optimization, First Edition.
Frank (Xin X.) Zhu, Richard Hoehn, Vasant Thakkar, and Edwin Yuh.

The D975 standard defines two ultra-low-sulfur diesel (ULSD) standards: Grade No. 2-D S15 (regular ULSD) and Grade No. 1-D S15 (a higher volatility fuel with a lower gelling temperature than regular ULSD). As the sulfur level in diesel is reduced, the inherent lubricity of the diesel is also reduced. So for this reason, the ASTM D975 standard also imposes lubricity requirement.

Since the 1980s, diesel fuel specifications have increasingly been tightened to meet environmental objectives, in addition to ensuring compatibility with diesel engines. The most stringent current specifications, those promulgated by the California Air Resources Board (CARB), limits sulfur, aromatics, polycyclic aromatic hydrocarbons (PAHs), and several other fuel impurities. As air quality problems persist, there is continued pressure to further reduce emissions from diesel engines, and hence to further tighten diesel specifications. By 2010, on-road diesel fuel sulfur levels was reduced to 15 parts per million (ppm) (in the United States) or even 10 ppm [in the European Union (EU)]. Recently proposed regulations will extend these specifications to virtually all diesel fuel used in engines.

In some jurisdictions, aromatics and PAH content in diesel fuel, which is strongly correlated with soot production, are also under pressure. CARB and the EU will have limits at 10% and 14%, respectively, while the US federal specifications limit aromatics to 35%.

Significant reductions in oxides of nitrogen (NO_x) and particulate matter (PM) will be required in almost all classes of diesel engines, and, in most cases, requiring significant changes in powertrain technology. The proposed rules may require electronic engine controls and exhaust after treatment system such as exhaust gas recirculation (EGR) and diesel particulate filter (DPF). Selective catalytic reduction (SCR) is frequently employed to reduce NO_x emissions.

With the advent of ultra-low-sulfur fuel regulations ushering in the first decade of the twenty-first century, however, it was required for hydrotreating (HDT) research and development to deliver quantum improvements in catalyst performance and process technology. This was accomplished in the form of so-called Type II supported transition metal sulfide (TMS) catalysts, unsupported/bulk TMS catalysts, improved bed grading catalysts and stacking strategies, advanced catalyst loading techniques, improved trickle-flow reactor internals designs, and more effective catalyst activation methodologies.

Key changes in diesel quality across various regions of the world sulfur and aromatic/cetane are summarized in Worldwide Refinery Process Review (2012) as shown in Table 3.1.

TABLE 3.1. Worldwide Diesel Fuel Specifications

Region	Sulfur (ppm)	Aromatics (vol%)
The United States	10–15	35
Canada	15	30
Latin America	2000 to as low as 10–50 in some area	NA
Western Europe	10	10
Central/Eastern Europe	50	NA
Middle East/Africa	50–5000	NA
Asia Pacific	10–350	10–35

ULSD was developed to enable the use of improved pollution control devices that reduce diesel emissions more effectively since these devices can be damaged by sulfur. ULSD is also safe to use with older diesels.

3.2 BASIC PROCESS FLOWSHEETING

Due to ULSD legislation, distillate HDT has become a key process in the refining industry. Rate limiting parameters that had previously been overlooked, such as aromatic inhibition, may need to be considered for the production of ULSD (<10 wppm S). As hydrodesulfurization (HDS) and hydroaromatics (HDA) saturation reactions occur in parallel in a shared environment, they compete for the necessary resources, namely hydrogen and catalyst sites. For this reason, the effectiveness factors for HDS and HDA decrease in the presence of increased aromatics concentrations. Furthermore, the data indicate that the presence of polyaromatics has a more significant impact on the slower HDS reactions (sterically hindered dibenzothiophenes, DBTs).

In order to successfully produce ULSD, essentially all of the organo-sulfur species must be removed including the substituted DBTs and other refractory sulfur species. The mechanisms and kinetics for deep desulfurization are more complicated than the simple first-order reactions that hold for the less refractory sulfur species, for example, thiols, sulfides, and disulfides. These more refractory sulfur species remain present at 500 and 50 ppm levels of sulfur. It is therefore critical to have a thorough understanding of desulfurization chemistry in order to successfully accomplish the deep desulfurization required.

Multiple reactions occur in parallel on the HDT catalyst surface including HDS, hydrodenitrogenation (HDN), and aromatic saturation/hydrogenation (HDA). In the development of the kinetic model, the key parameters influencing the HDS rate need careful investigation.

In modeling HDS, the sulfur species are generically characterized as "easy" or "difficult" sulfurs. The sulfur species are grouped based on the relative difficulty of sulfur extraction from the compound. In general, "easy" sulfur species (boiling range <610 °F) experience little steric hindrance and are converted quickly. Conversion of the "difficult" sulfur species (boiling range >610 °F) is much slower than it is with the "easy" sulfurs due to steric hindrance and more complex reaction pathways. Several species have been identified as being particularly difficult to convert in HDS, mostly substituted DBTs. The HDS chemistry for easy sulfur is reasonably represented by an irreversible first-order reaction. It is widely accepted that there are two primary desulfurization paths for difficult sulfur species: direct desulfurization (also known as sulfur extraction) and hydrogenation (also known as saturation). Direct desulfurization follows the same chemistry as the easy sulfur.

There are many ways to achieve sulfur removal of the more difficult molecules. Some of the options include the following:

- Use of more active catalyst
- Operating at higher temperature
- Reduction of feed endpoint
- Higher purity hydrogen

- Adding additional reactor volume
- H_2S removal from the recycle/treat gas
- Improving the feed distribution to the trickle-bed (conventional) reactor.

It is therefore goal of the refineries to find the most effective combination of these options to most economically produce ULSD in their plants. Depending on the needs of the refinery, some of these alternatives may be economically viable to consider either individually or in combination. Most of the HDS reactions are considered to be irreversible, while aromatic saturation is a reversible reaction that is controlled by equilibrium.

Typical HDT units designed have trickle-bed reactors. Reactors are operated with large quantities of hydrogen circulating over the catalyst bed, up to 10 times the quantity of hydrogen consumed by the chemical reaction. The vapor and liquid are mixed and passed through a distributor; as a result, they are in equilibrium as they enter the catalyst bed. As the reaction occurs at the catalyst surface between the dissolved hydrogen and the reactive species in the feed, hydrogen is depleted from the liquid and must be replenished from the vapor phase. Insufficient replacement of the hydrogen can lead to accelerated deactivation of the catalyst and reduced performance.

Figures 3.1 and 3.2 show basic DHT schemes used in the industry.

Diesel feed with recycle gas is introduced to the multibed reactor after exchanging heat with reactor effluent. Reactor feed is further heated in the charge heater to achieve reactor inlet temperature.

HDS and HDA reactions are exothermic; thus, reactor interbed quench is required to manage temperature and performance. The amount of quench and the number of beds depend on the aromatic and olefin content of feed. Presence of the cracked stocks, for example, fluid catalytic cracking (FCC) light cycle oil (LCO) and light coker gas oil (LCGO) require more quench and more beds to manage heat release.

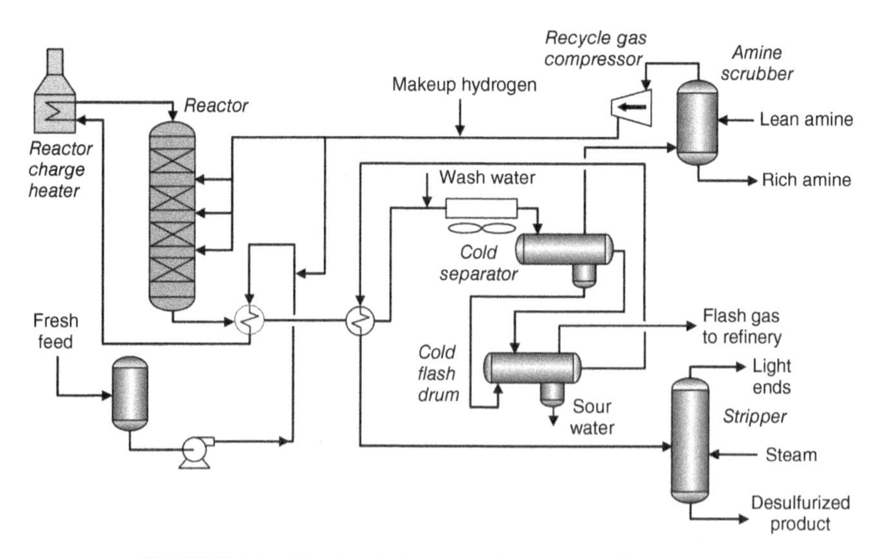

FIGURE 3.1. Diesel hydrotreating cold separator flow scheme.

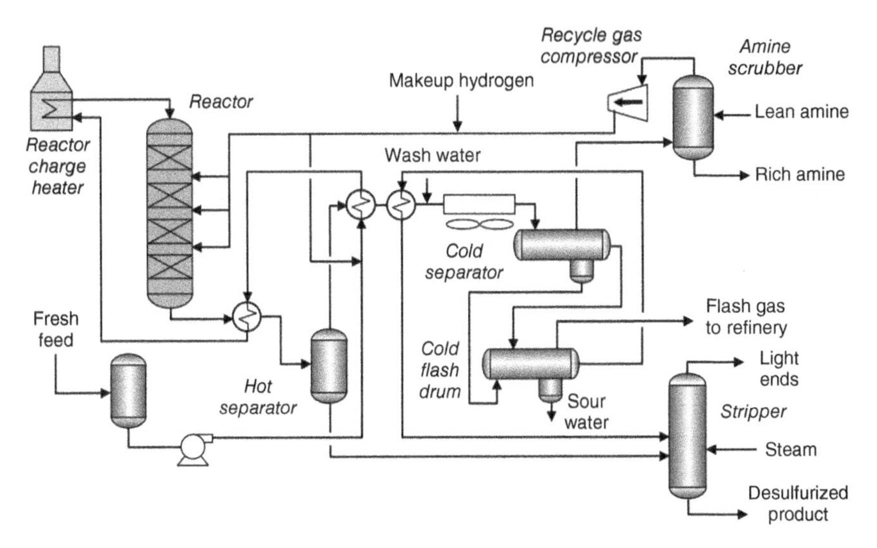

FIGURE 3.2. Diesel hydrotreating hot separator flow scheme.

Reactor effluent heats the combined feed and flows to a hot, high-pressure separator (HHPS) in some cases or to cold high-pressure separator (CHPS) separating vapor and liquid. Depending upon the flow scheme used, there may be different heat exchanger network possible for heat recovery.

Wash water is injected upstream of the product's condenser to remove ammonium hydrosulfide, which can foul heat exchanger tubes and cause plugging. HHPS liquid combines with heated CHPS liquid and flows to the product stripper.

Vapor from the CHPS contacts with amine in a scrubber for H_2S removal and flows to the recycle compressor suction drum. Makeup hydrogen is compressed and combines with recycle gas in the suction drum. The compressor suction drum can purge some recycle gas to improve recycle gas hydrogen purity. The quantity depends on the reactor's hydrogen-partial-pressure requirements and makeup hydrogen purity available.

In the product stripper, superheated steam feeds the tower's bottom and helps remove H_2S. Stripper overhead vapors condense and flow to the stripper accumulator. Accumulator vapor and liquid, known as wild naphtha, are processed in offsite facilities.

Stripped ULSD product supplies heat to the feed stream and then flows to drying facilities, which can be a coalescer-salt dryer or vacuum drying system.

Design is optimized for reactor space velocity, hydrogen treat gas quantity, hydrogen partial pressure, and reactor temperature for a given cycle length and treating severity.

Revamps are more challenging for existing equipment, and reactor loop piping will limit the hydrogen partial pressure due to hydraulic limitations. A higher treat gas rate can increase the hydrogen partial pressure; this is usually limited due to an associated increase in the reactor loop pressure drop and the corresponding maximum operating pressure of system components. Use of higher purity makeup gas can also help by reducing the quantity of light hydrocarbon gases in the recycle gas loop.

Lower-purity makeup hydrogen requires higher hydrogen circulation rates to maintain a target hydrogen partial pressure; it may even require a purge stream from the cold separator. If makeup hydrogen purity is too low, there is no combination of recycle rate and purge that will achieve the target reactor outlet partial pressure.

Catalyst cycle lengths of 24–36 months are typical for new designs. This is because, at some point, factors other than catalyst activity (such as reactor pressure drop) will limit the cycle. For a fixed space velocity, cycle length increases with a higher hydrogen partial pressure.

3.3 FEEDS

Feeds for DHT unit typically have a nominal distillation range of 300–700 °F. These feeds in the refinery are referred to as various names such as diesel, gas oil, or atmospheric gas oil (AGO). Generally speaking, streams derived from crude oil distillation are also referred to as straight run. DHT feed can also come from other process units such as FCC (LCO), Coker (LCGO), thermal cracker such as visbreaking unit (light visbroken gas oil), or low-pressure mild hydrocracking unit. Diesel range material can be produced in a synthetic fuel upgrading unit, such as those operating in Canada or Venezuela.

DHT feed distillation front end is limited by diesel flash point, typically 55–60 °C. Front end can be adjusted posthydrotreating by stripper operation. Back end of the distillation is limited by cold flow properties such as pour point, cold flow filtering point (CFPP), or cloud in addition to 95% distillation point (T95). Typical T95 is around 360 °C. Typical properties of some of the feed stream are shown in Table 3.2.

Challenges associated with gas oil produced in upstream processing unit are given, in order of importance:

- FCC gas oil (LCO): It has a high S content and low cetane number and may be hydrotreated after blending (e.g., ~30%) with straight-run gas oil, but the high aromaticity makes it difficult to improve the cetane number of the LCO (6–15 points only depending on conditions).
- Steam cracking diesel fuels: These have a very high content of aromatics especially dealkylated polyaromatics.
- Coker gas oil: These have high olefin content and higher aromatics. Coking unit uses additive that contain silicone to control foaming. These silicone compounds decompose and distribute in coker products. These compounds typically

TABLE 3.2. Typical Properties of Feed Examples

Source	*S* (ppm)	Aromatics (vol%)	Cetane	Cloud Point
GO Str. run	$3–20 \times 10^3$	20–40	42–54	−10 to +5
GO pyrolysis	$>20 \times 10^3$	30–60	28–45	−4 to −8
LCO FCC	14×10^3	60–85	18–27	−10

decompose upon encountering a hot surface (i.e., catalyst) and deposit there. Over time, this leads to catalyst deactivation. Silicone needs to be handled to avoid premature deactivation of catalyst.

The typical aromatics content of most diesel fuels is more in the 30–35% range, and although there are no particular specifications for the total aromatics content, this content drops across the HDT unit to produce ULSD. Since aromatics have low cetane number, lower aromatics feeds are desirable, particularly when attempting production of EURO diesel (EN 590), which requires a cetane number of 51.

Johnson et al. (2007) have compared various diesel range hydrocarbon composition, saturates (paraffin, naphthene, and olefins), and aromatic distribution as shown in Figure 3.3. Aromatics are identified as single- and multiring aromatics. As discussed, LCO is usually rich in aromatics and contains multiring aromatics. LCGO contains aromatics but relatively lower amount compared to LCO but has significantly higher amounts of saturates. Light gas oil (LGO) is typically composed of saturates and single-ring aromatics with relatively small quantities of 2- and 3-ring aromatics.

The relative difficulty of processing distillate range feed streams for deep HDS and/or HDA is highly dependent on their aromaticity and multiring aromatic content.

A typical commercial HDT unit processes blends of distillate streams from various sources to manage processing difficulty while achieving target sulfur and cetane number.

Distillate feed streams can also have cold flow property limitations. Cold flow characteristics are not improved by process in a distillate HDT unit but may get worst depending upon the degree of saturation. Cold flow is typically controlled by feed

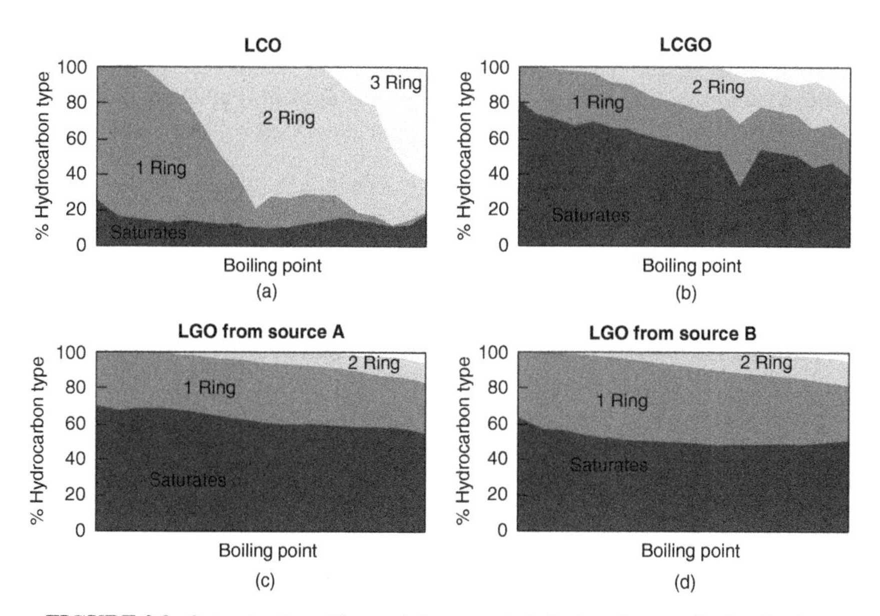

FIGURE 3.3. Saturates (paraffin, naphthene, and olefins) and aromatic distribution.

distillation unless dewaxing catalyst is utilized. Dewaxing catalysts generally fall into two categories – those that dewax the feed by cracking straight-chain paraffins and those that isomerize straight-chain paraffins to isoparaffins. The second type of catalyst is generally preferred because it retains most of the distillate range paraffins, while the first type will crack some of the paraffins to lighter products, thus reducing distillate yields.

3.4 PRODUCTS

Diesel fuel has become important transportation fuel in the last 20 years. Diesel is used for from consumer goods moved cross-country, to the generation of electric power, to increased efficiency on the nation's farms. Thus, diesel fuel plays a vital role in the nation's economy and standard of living. The major uses of diesel fuel are as follows:

- On-road for transportation
- Off-road (mainly mining, construction, and logging)
- Military transportation
- Farming
- Electric power generation
- Rail transportation
- Marine shipping.

3.4.1 Sulfur Specification

Reducing sulfur in diesel has been the focus of many countries, starting with the European Union and the United States. US Highway Diesel Program requires 15 ppm sulfur specification, known as ULSD; ULSD was phased-in for highway diesel fuel from 2006 to 2010. Off-road diesel was phased-in to use of ULSD (15 ppm) fuel from 2007 to 2014. Before EPA began regulated sulfur in diesel, diesel fuel could have contained as much as 5000 ppm of sulfur. The EU introduced regulations requiring 10 ppm sulfur in on-road in 2005, and the United States EPA began regulating diesel fuel sulfur levels in 1993. Beginning in 2006, EPA began to phase-in more stringent regulations to lower the amount of sulfur to 15 ppm in 2006. After 2010, all highway diesel fuel supplied to the market has been ULSD and all highway diesel vehicles are required to use ULSD. After 2014, all nonroad, locomotive, and marine (NRLM) diesel fuel must be ULSD, and all NRLM engines and equipment must use this fuel. More recently, China and India have announced regulations aimed at reducing the sulfur content of diesel fuel in use to 10 ppm by not later than 2021.

ASTM D975 specification is intended permissible limits of significant fuel properties used for specifying the wide variety of commercially available diesel fuel oils. Limiting values of significant properties are prescribed for seven grades of diesel fuel oils. Transportation diesel typically used is Grade 1 and/or 2 diesel. ULSD is identified as Grade No. 1-D S15, comprises the class of very low sulfur, volatile fuel oils from kerosene to the intermediate middle distillates. Fuels within this grade are

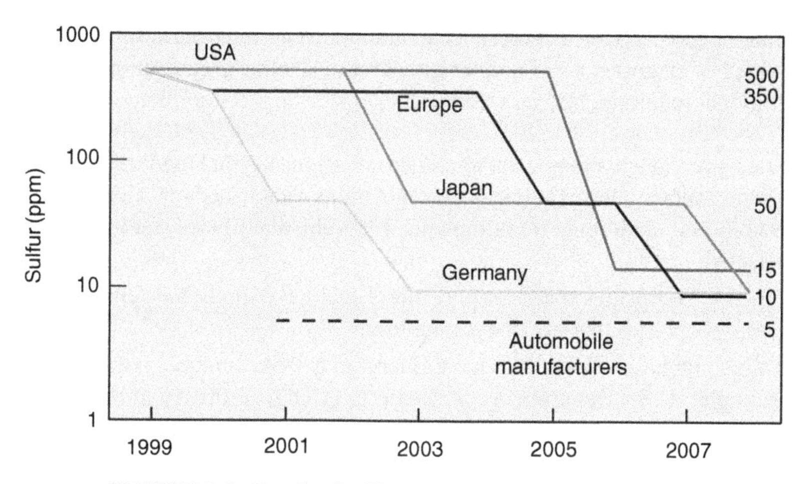

FIGURE 3.4. Trends of sulfur specification change over time.

applicable for use in (1) high-speed diesel engines and diesel engine applications that require ultra-low-sulfur fuels, (2) applications necessitating frequent and relatively wide variations in loads and speeds, and (3) applications where abnormally low operating temperatures are encountered.

ASTM Grade No. 2-D S15 includes the class of very low sulfur, middle distillate gas oils of lower volatility than Grade No. 1-D S15. These fuels are applicable for use in (1) high-speed diesel engines and diesel engine applications that require ultra-low-sulfur fuels, (2) applications necessitating relatively high loads and uniform speeds, or (3) diesel engines not requiring fuels having higher volatility or other properties specified in Grade No. 1-D S15.

Ackelson et al. (2007) have discussed how diesel sulfur specifications have changed around the world over time. In general, sulfur specifications are converging to lower sulfur over time as shown in Figure 3.4.

3.4.2 Diesel Fuel Properties

Diesel fuel needs to enable a number of engine performance characteristics that are generally recognized as important. Their relative importance depends on engine type and duty cycle (truck, passenger car, stationary generator, marine vessel, etc.). Examples of the performance criteria include starting ease, sufficient power, low noise, good fuel economy, good lubricity for low wear, low temperature operability (cold flow), long filter life (stability), and low emissions. Although engine design has the greatest impact on most of these characteristics, fuel quality impacts the designer's ability to attain the required performance.

In order to achieve the performance characteristics noted above, product from ULSD operation needs to meet certain specifications. Various properties and their relevance in meeting performance of diesel engine are given here. It is important to note that diesel in various parts of the world has its own specifications typically related to powertrain requirement environmental regulations and climate conditions, to name a few. The United States typically follows ASTM D975 specification for Grade 1 and Grade 2 Diesel. The European region uses EN 590. There also is the Worldwide

Fuel Charter (2013), which represents an attempt to harmonize diesel specification worldwide in Categories 1–5. For a variety of reasons, other organizations may establish additional requirements, for example:

- State governments may want to reduce emissions beyond the limits established by the country. One example is the regulations established in California. CARB established additional requirements for vehicular diesel fuel that became effective in 1993.
- Pipeline companies that transport diesel fuel have limits for density and pour point, properties that D975 does not limit.
- Some purchasers such as the military may have unique requirement, for example, U.S. Department of Defense (DOD) formerly purchased fuels meeting military specifications that often included special requirements in addition to the D975 requirements. Now the DOD buys commercial fuel when possible.

The following section describes several key properties discussed by Bacha et al. (1998) in a technical review of diesel fuel. Diesel fuel properties and their significance in emission and/or engine performance are discussed as follows.

Sulfur content of diesel fuel affects PM emissions because some of the sulfur in the fuel is converted to sulfate particles in the exhaust. The fraction converted to PM varies from engine to engine design, but reducing sulfur decreases PM. For this reason, the EPA limits the sulfur content of on-road diesel fuel. Also sulfur in diesel fuel is converted to SO_2 which can be converted to sulfuric acid consequently; a reduction in sulfur also reduces production of acid rain. ULSD fuel enables the use of advanced emission control devices (after treatment systems) further lowering harmful exhaust emissions.

Cetane number is measure of the ignition quality of diesel fuel based on ignition delay in an engine (readiness to spontaneously combust under the temperature and pressure conditions in the combustion chamber of the engine). It relates to the starting and warm-up characteristics of the fuel. Increasing the cetane number improves fuel combustion characteristics and reduces NO_x and PM emissions. The higher the cetane number, the shorter the ignition delay and the better the ignition quality, important for ease of ignition, better starting in cold temperature, reduced engine noise, and to control legislated emissions. Cetane number is measured by ASTM D613.

Cetane Index (CI) is an approximation of cetane number based on an empirical relationship with density and volatility parameters. CI can be calculated by ASTM D976 or D4737.

Density: Changes in fuel density affect the energy content of the fuel brought into the engine at a given injector setting. European studies have indicated that reducing fuel density tends to decrease NO_x emissions.

Aromatics: Reduction of diesel fuel reduces NO_x and PM10 in some engines. Recent European studies indicate that polynuclear aromatics content is key to the reduction and that the concentration of single-ring aromatics is not a significant factor.

Flash point is diesel property usually specified that is not directly related to engine performance. It is, however, of importance for safety precautions involved in fuel handling and storage and is normally specified to meet insurance and fire regulations.

Volatility: T95 is the temperature at which 95% of a particular diesel fuel distills in a standardized distillation test (ASTM D86). Reducing T95 decreases NO_x emissions slightly but increases hydrocarbon and CO emissions. PM10 emissions are unaffected.

Viscosity is primarily related to molecular weight and not so much to hydrocarbon class. For a given carbon number, naphthenes generally have slightly higher viscosity compared to paraffins or aromatics. Normal paraffins have excellent cetane numbers but very poor cold flow properties and low volumetric heating values. Aromatics have very good cold flow properties and volumetric heating values but very low cetane numbers. Isoparaffins and naphthenes are intermediate, with values of these properties between those of normal paraffins and aromatics.

The diesel fuel injection is controlled volumetrically or by timing of the solenoid valve. Variations in fuel density (and viscosity) result in variations in engine power and, consequently, in engine emissions and fuel consumption. The European Programme on Emissions, Fuels and Engine Technologies (EPEFE) found that fuel density also influences injection timing of mechanically controlled injection equipment, which also affects emissions and fuel consumption. Therefore, in order to optimize engine performance and tailpipe emissions, both minimum and maximum density limits must be defined in a fairly narrow range.

Fueling and injection timing are also dependent on fuel viscosity. High viscosity can reduce fuel flow rates, resulting in inadequate fueling. A very high viscosity may actually result in pump distortion. Low viscosity, on the other hand, will increase leakage from the pumping elements, and in worst cases (low viscosity, high temperature) can result in total leakage. As viscosity is impacted by ambient temperature, it is important to minimize the range between minimum and maximum viscosity limits to allow optimization of engine performance.

Cold flow: Diesel fuel can have a high content (up to 20%) of paraffinic hydrocarbons that have a limited solubility in the fuel and, if cooled sufficiently, will come out of solution as wax, causing filters to plug. Adequate cold flow performance, therefore, is one of the most fundamental quality criteria for diesel fuels. The cold flow characteristics are primarily dictated by fuel distillation range, mainly the back-end volatility and hydrocarbon composition, that is, paraffins, naphthenes, and aromatics content (Worldwide Fuel Charter, 2013).

The diesel fuel cold flow performance can be specified by cloud point (CP), by CFPP (with maximum delta between CFPP and CP), or by low temperature flow test (LTFT) (in the United States and Canada).

- If CP (only) or LTFT is used, the maximum allowed temperature should be set no higher than the lowest expected ambient temperature.
- If CFPP is used to predict cold flow, the maximum allowed CFPP temperature should be set equal to, or lower than, the lowest expected ambient temperature. In this case, the CP should be no more than 10 °C above the CFPP specified.

Diesel cold flow properties must be specified according to the seasonal and climatic needs in the region where the fuel is to be used. The low-temperature properties of diesel fuels are therefore defined by the following specific tests:

Cloud Point, CP (ISO 3015, ASTM D2500): The temperature at which the heaviest paraffins start to precipitate and form wax crystals; the fuel becomes “cloudy.”

Cold Filter Plugging Point, CFPP (EN116): The lowest temperature at which the fuel can pass through the filter in a standardized filtration test. The CFPP test was developed from vehicle operability data and demonstrates an acceptable correlation for fuels and vehicles in the market. For North American fuels, however, CFPP is not a good predictor of cold flow operability. Diesel fuel CFPP can be improved by addition of cold flow additives.

Low Temperature Flow Test, LTFT (ASTM D4539): The LTFT was developed to predict how diesel fuels in the United States and Canada will perform at low temperatures in the diesel vehicles available in these markets. LTFT is a slow cooling test and therefore more severe than CFPP. Similar to CFPP, LTFT temperature can be improved by addition of cold flow additives.

Foam: Diesel fuel has a tendency to generate foam during tank filling, which slows the process and risks an overflow. Antifoaming agents are sometimes added to diesel fuel, often as a component of a multifunctional additive package, to help speed up or to allow more complete filling of vehicle tanks. Silicon surfactant additives are effective in suppressing the foaming tendency of diesel fuels, the choice of silicon and cosolvent depending on the characteristics of the fuel to be treated. It is important that the additive chosen should not pose any problems for the long-term durability of the emission posttreatment control systems.

Carbon residue gives a measure of the carbon depositing tendencies of a fuel oil when heated in a bulb under prescribed conditions. While not directly correlating with engine deposits, this property is considered an approximation.

Electrical conductivity of fuels is an important consideration in the safe handling characteristics of any fuel. The risk associated with explosions due to static electrical discharge depends on the amount of hydrocarbon and oxygen in the vapor space and the energy and duration of a static discharge. There are many factors that can contribute to the high risk of explosion. For ULSD fuels in particular, electrical conductivity can likely be very low before the addition of static dissipater additive (SDA). The intent of this requirement is to reduce the risk of electrostatic ignitions while filling tank trucks, barges, ship compartments, and rail cars, where flammable vapors from the past cargo can be present. Generally, it does not apply at the retail level where flammable vapors are usually absent.

Lubricity: As diesel fuel refined to remove the polluting sulfur (ULSD), it is inadvertently stripped of its lubricating properties. This important lubrication property is critical for engine component as it prevents wear in the fuel delivery system. Specifically, it lubricates pumps, high-pressure pumps, and injectors. Traditional low-sulfur diesel fuel typically contained enough lubricating ability to suffice the needs of these vital components. ULSD fuel, on the other hand, is considered to be very "dry" and incapable of lubricating. As a result, engine components are at risk of premature and even catastrophic failure when ULSD fuel is introduced to the system. As a result, all oil companies producing ULSD fuel must replace the lost lubricity with additives. All ULSD fuel purchased at retail fuel stations should be adequately treated with additives to replace this lost lubricity. In addition, many additives can offer added benefits such as cetane improver, antigel agents, and water separators (demulsifiers).

Diesel fuel and other fluids are tested for lubricating ability using a device called a "high frequency reciprocating rig" or HFRR. The HFRR is currently the internationally accepted, standardized method to evaluate fluids for lubricating ability. It uses a ball bearing that reciprocates or moves back and forth on a metal surface at a very high frequency for a duration of 90 min. The machine does this while the ball bearing and metal surface are immersed in the test fluid. At the end of the test, the ball bearing is examined with a microscope and the "wear scar" on the ball bearing is measured in microns. The larger the wear scar, the poorer the lubricating ability of the fluid. The ASTM standard for diesel fuel requires diesel fuel to produce a wear scar of no greater than 520 μm. The Engine Manufacturers Association prefers a standard of a wear scar no greater than 460 μm, typical of the pre-ULSD fuels. Most experts agree that a 520 μm standard is adequate, but also that the lower the wear scar, the better.

Corrosion: A severe and rapid corrosion has been observed in systems storing and dispensing ULSD since 2007. In addition, the corrosion is coating the majority of metallic equipment in both the wetted and unwetted portions of ULSD underground storage tanks (USTs). This phenomenon was investigated by the industrial group project (Clean Diesel Fuel Alliance, 2012).

Their conclusion was that corrosion in systems storing and dispensing ULSD is likely due to the production of acetic acid throughout USTs. The acetic acid is believed to be produced by Acetobacter feeding on low levels of ethanol contamination. The presence of Acetobacter in the tank samples suggested that ethanol was being converted into acetic acid. Dispersed into the humid vapor space by the higher vapor pressure and by disturbances during fuel deliveries, acetic acid is deposited throughout the system. The source of ethanol is unknown; however, diesel fuel is often delivered in the same trucks as ethanol-blended gasoline. Also, ULSD USTs that have been converted from a gasoline tank could have manifolded ventilation systems with gasoline tanks. Thus, Clean Diesel Fuel Alliance (2012) suggests that it is possible that there be some cross-contamination of ethanol into ULSD. This results in a cycle of wetting and drying of the equipment concentrating the acetic acid on the metallic equipment and corroding it quite severely and rapidly.

The studies have also noted that corrosion was not much of a problem in tanks that were on a regular biocide treatment program to kill tank microbes. In the past, the sulfur acted as a natural biocide, but with the development of ULSD, the sulfur is now almost gone. Since there's nothing now to prevent bacterial growth, the use of biocides to treat the fuel is recommended to prevent microbial growth that normally would occur otherwise.

Storage and thermal stability of normally produced diesel fuel has adequate stability to withstand normal storage and use without the formation of troublesome amounts of insoluble degradation products.. Fuels that are to be stored for prolonged periods or used in complex applications should be selected to avoid formation of sediments or gums, which can overload filters or plug injectors. The stability properties of middle distillates are highly dependent on the crude oil sources, severity of processing, use of additives, and whether additional refinery treatment has been carried out.

TABLE 3.3. Diesel Properties and Effect on Performance

Property	Effect of Property on Performance	Time Frame of Effect
Flash point	Safety in handling and use – not directly related to engine performance	—
Water and sediment	Affects fuel filters and injectors	Long term
Volatility	Affects ease of starting and smoke	Immediate
Viscosity	Affects fuel spray atomization and fuel system lubrication	Immediate and long term
Ash	Can damage fuel injection system and cause combustion chamber deposits	Long term
Sulfur	Affects particulate emissions, cylinder wear, and deposits	Particulates: Immediate Wear: Long term
Copper strip corrosion	Indicates potential for corrosive attack on metal parts	Long term
Cetane number	Measure of ignition quality – affects cold starting, combustion, and emissions	Immediate
Cloud point and pour point	Allow low temperature operability	Immediate
Carbon residue	Measures coking tendency of fuel, may relate to engine deposits	Long term
Heating value (energy content)	Affects fuel economy	Immediate
Density	Affects heating value	Immediate
Stability	Indicates potential to form insolubles during use and/or in storage	Long term
Lubricity	Affects fuel pump and injector wear	Long term (typically)
Water separability	Affects ability to produce dry fuel	—

Available fuel additives can improve the suitability of marginal fuels for long-term storage and thermal stability, but can be unsuccessful for fuels with markedly poor stability properties.

The relationship between diesel properties and effects on performance as summarized by Bacha et al. (1998) is presented in Table 3.3.

3.5 REACTION MECHANISMS

ULSD regulations require very low sulfur (i.e., <15 ppm sulfur). North America has implemented ULSD since 2006 and regulations requiring <10 ppm sulfur are in effect in Europe since 2009. In order to comply with these regulations, essentially all of the organo-sulfur species need to be removed including the substituted DBTs and other refractory sulfur species. The mechanisms and kinetics for deep desulfurization are more complicated than the simple first-order reactions that hold for the less refractory sulfur species, for example, thiols, sulfides, and disulfides, due to the treatment of sterically hindered sulfur species. These sulfur species remain present at 500 and 50 ppm levels of sulfur and in some cases are left untouched. For this reason, some sources have suggested that it may be as difficult to reduce the sulfur levels from 50 to 10 ppm as it was to change from 500 to 50 ppm (Jones and Kokayeff, 2004, 2005).

Multiple reactions occur in parallel on the HDT catalyst surface. Companies and academia have investigated distillate HDT reaction mechanism and kinetic in great detail as ULSD became important.

In general, the following reactions occur in parallel depending upon feed composition:

- Sulfur removal, also referred to as desulfurization or HDS, in which the organic sulfur compounds are converted into hydrogen sulfide.
- Nitrogen removal, also referred to as denitrogenation or HDN, in which the organic nitrogen compounds are converted into ammonia.
- Metals (organometallics) removal, also referred to as hydrodemetallization (HDM), in which the organometallics are converted into the respective metal sulfides.
- Oxygen removal, also referred to as hydrodeoxygenation, in which the organic oxygen compounds are converted into water.
- Olefin saturation, in which organic compounds containing double bonds are converted to their saturated homologues. Olefins are not found in petroleum but are formed when processed in thermal or catalytic units. In general, olefins are unstable and thus must be protected from contact with oxygen prior to HDT to prevent the formation of polymer gums. That is especially true of feedstocks derived from thermal cracking operations such as coking and ethylene manufacturing. Typical olefin saturation reactions are rapid and exothermic, so heat release occurs near the inlet of bed.
- Aromatic saturation, also referred to as hydrodearomatization (HDA), in which some of the aromatic compounds are converted into naphthenes.

3.5.1 Hydrodesulfurization

Sulfur removal occurs via the conversion to H_2S of the organic sulfur compounds present in the feedstock. Sulfur is found throughout the boiling range of petroleum fractions in the form of many different organic sulfur compounds. Following are six sulfur types: mercaptans, sulfides, disulfides, thiophenes, benzothiophenes, and DBTs. Typical reactions for each kind of sulfur compounds are shown in Table 3.4.

Kokayeff et al. (2015) rank six sulfur types on the basis of ease of removal; easiest to hardest to remove as

$$\text{Mercaptans} \rightarrow \text{Sulfides} \rightarrow \text{Disulfides} \rightarrow \text{Thiophenes} \rightarrow \text{Benzothiophenes} \rightarrow \text{Dibenzothiophenes}.$$

TABLE 3.4. Typical Reactions for Each Kind of Sulfur Compound

Mercaptans	$R\text{–}SH + H_2 \rightarrow R\text{–}H + H_2S$
Sulfides	$R\text{–}SH + H_2 \rightarrow R\text{–}H + H_2S$
Disulfides	$R_1\text{–}S\text{–}S\text{–}R_2 + 3H_2 \rightarrow R_1\text{–}H + R_2\text{–}H + 2H_2S$
Thiophene	Step (1) 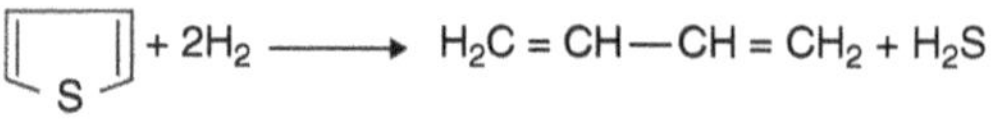 S + $2H_2 \longrightarrow H_2C = CH\text{—}CH = CH_2 + H_2S$ Step (2) $H_2C = CH\text{–}CH = CH_2 + 2H_2 \rightarrow H_3C\text{–}CH_2\text{–}CH_2\text{–}CH_3$
Benzothiophenes	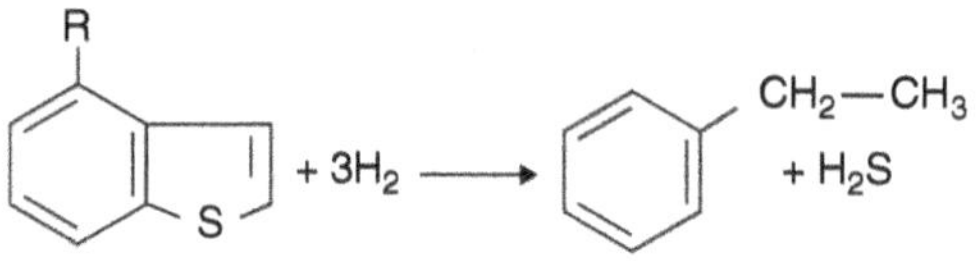 R, S + $3H_2 \longrightarrow$ $CH_2\text{—}CH_3$ + H_2S
Dibenzothiophenes	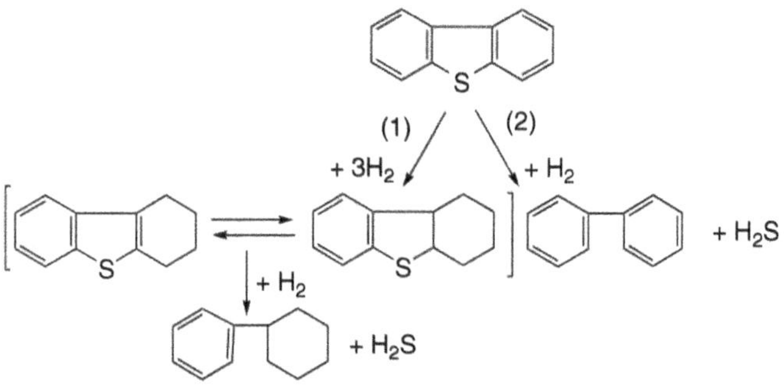 S (1) + $3H_2$; (2) + H_2 S ⇌ S; + H_2S + H_2 + H_2S

In HDS, the sulfur species are generically characterized as "easy" or "difficult" sulfurs. In general, Jones and Kokayeff (2004) defined "easy" sulfur species (present in diesel typically boiling at $<610\,°F$) experience little steric hindrance and are converted quickly. Conversion of the "difficult" sulfur species (boiling range $>610\,°F$) is much slower than the "easy" sulfurs due to steric hindrance and more complex reaction pathways. Several species have been identified as being particularly difficult to convert in HDS, mostly substituted DBTs. In order to illustrate the difference in kinetics 4,6-dimethyldibenzothiophene (4,6-DMDBT), 4-ethyl,6-methyl dibenzothiophene, and C3-DBT-A have been used as representatives of the "difficult" sulfurs. Although the specific structure of C3-DBT-A has not been identified, it has been established that it is a C3-substituted DBT.

The HDS chemistry for easy sulfur is reasonably represented by an irreversible first-order reaction with hydrogen inhibited by H_2S that competes for catalyst sites. It is widely accepted that there are two primary desulfurization paths for difficult sulfur species.

(1) Direct desulfurization (also known as sulfur extraction or hydrogenolysis) and hydrogenation (also known as saturation). Direct desulfurization follows the same chemistry as the easy sulfur.

(2) Catalyst sites are not readily accessible to difficult sulfur species such as for substituted DBTs due to the rigidity of the fused aromatic rings and steric hindrance. As a result, the second prevalent reaction route includes a reversible hydrogenation reaction preceding the sulfur extraction. The saturation of one of the aromatic rings increases the flexibility of the molecule causing the sulfur to be more accessible for desulfurization. Desulfurization of difficult sulfur species is known to be inhibited by both H_2S and organo-nitrogen.

3.5.2 Aromatic Saturation

Aromatic saturation is an important reaction in distillate HDT. The aromatic content of the feedstock varies widely depending on the source of the material; however, in all cases, the aromatic compounds undergo saturation during HDT. As aromatics can be single, di and polyaromatics, for modeling purposes Jones and Kokayeff (2004, 2005) have considered mechanism as series of reversible reactions as shown in equation (3.1):

$$A + H_2 \longleftrightarrow B + H_2 \longleftrightarrow C + H_2 \longleftrightarrow D, \quad (3.1)$$

where A, B, and C represent poly (tri+), di-, and monoaromatic compounds, respectively, and D represents cyclic compounds (saturated aromatics). When producing ULSD, hydrogen consumption increases far beyond the stoichiometric requirement for desulfurization alone due to additional aromatics saturation that takes place. Although hydrogenation is limited by the thermal equilibrium at high temperatures, the rate of aromatic saturation is much faster than HDS reactions for difficult sulfurs and also varies with the number of fused rings.

Significant disagreement exists in the literature regarding inhibition of the HDS reaction by polycyclic aromatic compounds; there is also disagreement regarding the significance and magnitude of aromatics inhibition among those who confirm that it exists. Inhibition of 4,6-DMDBT desulfurization by the addition of naphthalene was observed in studies. The conversion of 4,6-DMDBT over a CoMo catalyst decreased with increasing naphthalene concentrations present. Analyses of desulfurization product distribution results in the presence of naphthalene suggest that inhibition is stronger for the hydrogenation route than for the direct extraction pathway.

Further complicating the determination of aromatic inhibition is the variation observed in the degree of inhibition depending on the number of fused rings. Studies investigating desulfurization of DBTs in the presence of di- and tricyclic aromatic compounds indicate no inhibition in the presence of methylnaphthalene (2-ring aromatic); however, a significant decrease in HDS activity was observed in the presence of phenanthrene (3-ring aromatic). The adsorption constant increases as the number of fused rings increases. In literature review, authors have suggested that the adsorption constant may increase by as much as an order of magnitude when the number of fused rings is increased from 3 to 4.7.

The conclusions regarding the influence of aromatic compounds on the rate of HDS in the literature vary even among those reviewing HDS on a broad scale. The conflicting conclusions about the influence of aromatics, coupled with the organo-nitrogen inhibition, cause the mechanism by which aromatics influence HDS questionable. Although there is a lack of agreement in the literature as to the nature, magnitude, and significance of the influence of aromatics on HDS kinetics, it is clear that the presence of aromatics impacts HDS. Highly aromatic feedstocks, such as LCOs, are known to be difficult to treat compared to straight-run or coker material with lower aromatic concentrations. The relative difficulty is suspected to be due in part to the high aromatic concentrations. As HDS and HDA reactions occur in parallel in a shared environment, there is competition for hydrogen and catalyst sites (inhibition). The overall rate of reaction in a fixed-bed reactor is controlled by a rate-limiting step that is slow relative to the rest of the process. If the rate-limiting step is either the chemical reaction or the adsorption or desorption from the catalyst surface, the rate is kinetically controlled. Mass-transfer control can be either intraparticle or external. Intraparticle control occurs when the slowest step is the transport of reactants into the catalyst pores from the external surface. External mass-transfer control occurs when the rate-limiting step is the transfer of reactants from the bulk fluid to the catalyst surface.

Historically, HDT has been thought of as mass-transfer controlled reaction. As a result, the HDT catalysts used are small shaped extrudates, maximizing the surface area per catalyst volume. Reaction rates are modeled on the basis of the bulk fluid properties and hydrogen partial pressure, ignoring any mass transfer effects – interphase or diffusion of reactants through the pore structure of the catalyst pellet. Hydrogen is the common reactant for all reactions such as HDS, HDN, and HDA occurring in distillate HDT. If hydrogen were the limiting reagent, it is possible that the reaction environment in the catalyst pores could become hydrogen deficient, thus reducing the HDS rate. In that case, the higher the aromatics concentration, the more deficient the hydrogen supply and hence a greater impact on the rate.

In conventional trickle-bed hydrotreaters, the hydrogen available for reaction is dissolved in the liquid phase. The solubility of H_2 in a diesel feed at typical HDS condition of 650 °F and 700 psi is estimated by Jones and Kokayeff (2004): ~50 SCF/BBL or 0.13 lb mol/bbl of diesel. The authors show that the hydrogen solubility is only 5–10% of the stoichiometric requirement! Thus, a simple stoichiometric calculation shows that although excess hydrogen is available in the gas phase, the reactions are occurring in liquid that is hydrogen-deficient relative to the stoichiometric requirements for 100% conversion. The diffusion of hydrogen into the catalyst pellet is fast enough to prevent the concentration from decreasing below 65% of the bulk concentration. The stoichiometric hydrogen requirement increases with increasing aromatics concentrations. However, the supply of hydrogen, that is, the bulk concentration in the liquid phase, is constant assuming that it is continually replenished to the saturation limit for hydrogen as a result of the contact with the gas phase. The magnitude of aromatics inhibition was small relative to inhibition by organo-nitrogen; however, an inspection of the adsorption constants and the types of catalysts sites available provides possible explanations for the differences. The introduction of polyaromatic inhibition terms to the HDS kinetic model demonstrated that aromatics impact the operating conditions required for ULSD production.

3.6 HYDROTREATING CATALYSTS

HDT catalysts, in general, are high surface area materials consisting of an active component and a promoter, which are uniformly dispersed on a support. The catalyst support is normally gamma alumina (γ-Al_2O_3), doped sometimes with small amounts of silica, phosphorus, fluoride, and/or boron and prepared in such a way so as to offer a high surface area upon which to disperse the active metals and an appropriate pore structure, so that pore-plugging with coke and/or metals is sufficiently mitigated to achieve the desired operating cycle. The active component is normally molybdenum sulfide, with both cobalt (CoMo) and nickel (NiMo) as promoters. The promoter has the effect of substantially increasing (approximately 100-fold) the activity of the active metal sulfide. Commercially available catalysts (Kokayeff et al., 2015) have varying amounts of promoters and active components, depending on the desired applications, but in general they can contain up to about 25 wt% promoter and up to 25 wt% active component as oxides. HDT catalysts come in different sizes and shapes and vary depending on the manufacturer, which are shown below and in Figure 3.5.

Cylindrical	1/32–1/4 in.
Trilobe and quadrilobe	1/20–1/10 in.
Spheres	1/16–1/4 in.
Hollow rings	Up to 1/4 in.

Both Co/Mo and Ni/Mo catalysts have been employed in the hydrotreatment of distillate fuels. While in the early stages of development, the Ni/Mo catalysts were judged to be slightly more active in hydrodesulfurizing diesel fuels to ULSD (<10 wppm sulfur) specifications, more recent developments have provided Co/Mo catalysts with equivalent activity to ULSD. The observations that Ni/Mo catalysts may be more active for ULSD is in accord with the theory that for the desulfurization of the most difficult sulfur molecules, dimethyl dibenzothiophenes with substituent groups in the 4- and 6-positions, such as 4,6-dimethyldibenzothiophene, one of the rings needed to undergo saturation of one of the rings, allowing the sulfur atom to be more accessible due to the greater flexibility of the saturated ring. This pathway for desulfurization was termed the "hydrogenation" pathway and was most pronounced over Ni/Mo catalysts with their higher activity for saturation. The other pathway involved direct abstraction of the sulfur atom and formation of H_2S was termed the "direct abstraction" route and was the dominant route on Co/Mo catalysts.

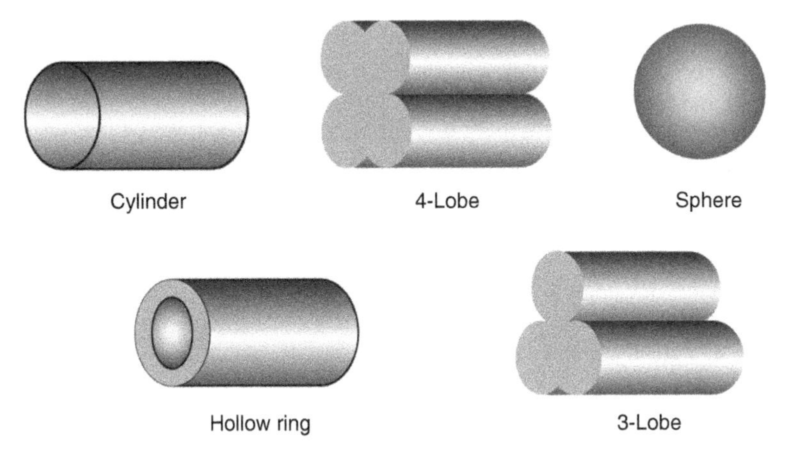

FIGURE 3.5. Hydrotreating catalysts shapes.

Other shapes include nodular beads and wagon wheels. In general, the size and shape of the catalyst pills is a compromise between the desire to minimize pore diffusion effects in the catalyst particles and pressure drop across the reactor (requiring large particle sizes). The physical characteristics of catalysts, as summarized in Handbook of Petroleum (2014), also vary from manufacturer to manufacturer and the intended use of the catalyst, but in general are as follows:

Surface area (m^2/g)	100–250
Pore volume (cc/g)	0.5–1.0
Median pore diameter (Å)	65–150
Compacted bulk density (lb_m/ft^3)	35–55
Crushing strength (lb_f/in^2)	4–20

CoMo catalysts have been designed primarily for desulfurization with minimum hydrogen addition; CoMo catalysts have the low hydrogenation activity; therefore, they have the lowest hydrogen consumption per mole of sulfur removed. CoMo catalysts have sufficient desulfurization performance at low operating pressures (<600 psig, or <~40 barg). These catalysts also have the low denitrogenation activity due to low hydrogenation capability. Because CoMo catalysts exhibit the highest sulfur removal per unit of hydrogen consumed, they are preferred for desulfurization at lower pressures and when hydrogen is in short supply.

NiMo catalysts have been designed for desulfurization, with intention to improve hydrogenation and denitrogenation activity. NiMo catalysts have higher denitrogenation activities than CoMo and are therefore preferred for cracked stocks or other applications where denitrogenation and/or saturation is as important as desulfurization. The performance of NiMo catalysts is relatively good at high pressures. NiMo catalysts show a greater response in denitrogenation and desulfurization performance to changes in H_2 partial pressure than CoMo.

3.6.1 Type I Versus Type II Hydrotreating Catalysts

The designations "Type I" and "Type II", as originally referred to the nature of the bonding between the active phase (e.g., CoMoS, NiMoS) and the support, the extent of stacking of the active phase as dispersed MoS_2 nanosheets, and the coordination of the promoter metal at the edges of the nanosheets. The promoted active phase along the edges, as Co(Ni)–Mo–S structures, for example, are believed to be responsible for the relatively greater HDT activity of Type II catalysts. Catalysts of the former type consist predominantly of an active phase that is more strongly bound to the support, as in Mo–O–Al where the support is γ-alumina, with relatively shorter stacks of MoS_2 nanosheets, and wherein a relatively smaller portion of the promoter metal is coordinated into the edges of MoS_2 nanosheets. Catalysts of the latter type consist predominantly of an active phase that is more weakly bound to the support via dipole–dipole (van der Waals) interactions, with relatively taller stacks of MoS_2 nanosheets, and wherein a relatively greater portion of the promoter metal is coordinated into the edges of MoS_2 nanosheets. The active phase is, in general, fully sulfidable.

3.6.2 Unsupported, or Bulk Transition Metal Sulfide Catalysts

The step change in improvement in HDT catalyst performance can be observed in the activity of unsupported TMS catalysts. In principle, today's unsupported catalysts as described in Handbook of Petroleum (2014) are comprised only of active phase, representing a much higher population density of active phase per unit volume than Type I and Type II supported catalysts, without the limitation of active phase-support interactions. Such catalysts are comprised almost entirely of metals, as in NiMoW. Their HDS and HDN HDT relative volumetric activities (RVAs) can exceed those of the traditional Type II supported catalyst by 50–100%. It is for this reason that unsupported catalysts are typically loaded in addition to supported catalysts, since the hydrogen consumption and heat release of a full reload with an unsupported catalyst could exceed process design allowances. In general, these catalysts are prepared via precipitation of the oxidic precursor followed by the sulfidation of the precipitate, or based on the direct precipitation of the (mixed) sulfide.

3.6.3 Catalyst Activation

HDT catalysts have to be activated in order to be catalytically useful. The activation of the catalyst is performed by conversion of metal oxides to metal sulfides, commonly called sulfiding. Other names that are used to describe catalyst activation techniques are presulfiding or presulfurizing. The metals on the catalysts are in an oxide form at the completion of the manufacturing process. The catalysts are activated by transforming the inactive metal oxides into metal sulfides. More often than not, this is typically accomplished *in situ*; however, more refiners have started to use catalyst that had the sulfiding compound loaded onto the catalyst outside the unit (*ex situ* presulfidation). The trend is that more and more refiners are opting to receive the catalyst at the refinery site in presulfided state to accelerate the startup of the unit, and because it is more environmentally friendly. There is a difference between presulfidation and *ex situ* presulfiding. The former adds the sulfiding compound to the catalyst and the latter actually provides with the oxide converted to a metal sulfide making it active; it will require special handling.

In situ sulfiding can be accomplished either in vapor or liquid phase. In vapor phase sulfiding, the activation of the catalyst is accomplished by injecting a chemical into the recycle gas stream, which decomposes easily to H_2S, such as dimethyldisulfide (DMDS) or dimethylsulfide (DMS). Liquid phase sulfiding is desirable because the liquid phase provides a heat sink for the exothermic sulfiding reactions that helps prevent high catalyst temperatures and temperature excursions that can damage the catalyst. It also enables quicker sulfiding due to the innate ability of the liquid to impart heat to the reactor walls and allows them to more rapidly reach minimum pressurization temperature (MPT). The active phase in Type II HDT catalysts, orders of magnitude more active than Type I at start of run, is especially susceptible to such damages during the sulfiding process. Another advantage of liquid phase over gas phase sulfiding is that by having all the catalyst particles wet from the very beginning there is very little chance of catalyst bed channeling that can occur if the catalyst particles are allowed to dry out. The *in situ* sulfiding occurs at temperatures between 450 and 650 °F (230–345 °C) regardless of the method used. Some catalyst manufacturers recommend that the sulfiding be conducted at full operating pressure, while others prefer it be done at pressures lower than the normal operating pressure. It is also very typically recommended that cracked feedstocks be introduced very gradually, over a period of 3–7 days. Cracked feedstocks are composed of relatively more olefins and aromatics, and so introducing these during the start-of-run (SOR) period would lead to excessive heat release, leading to catalyst damage via sintering of the active phase.

In the case of *ex situ* presulfurization of the catalyst, sulfur compounds are loaded onto the catalyst. The activation occurs when the catalyst, which has been loaded in the reactor, is heated up in the presence of hydrogen and the sulfur compounds decompose to H_2S. During this period, the H_2S scrubber is offline so that the H_2S can accumulate in the recycle gas circuit.

3.7 KEY PROCESS CONDITIONS

On average process conditions required for distillate HDT today are much more severe than they were historically. This is mainly due to the need to produce ever-increasing quantities of ULSD as regulations globally move to this specification. Typical process conditions are shown in Table 3.5.

The variations in process conditions can be attributed to a number of factors including feedstock contaminants (S, N, O), feedstock type, and process objective (ULSD, cold flows, and/or cetane improvement). Note that the operating conditions required to result in significant cetane improvements is generally sufficient to produce ULSD sulfur specifications.

As mentioned earlier, products derived from cokers contain silicon due to the use of antifoaming agents in the delayed coking unit. Silicon is a permanent poison for HDT catalyst. In order to protect catalyst and achieve the desired cycle length in the HDT unit when processing coker feed, it is important to load a Si guard catalyst to essentially trap the Si. Although HDT catalysts do have the innate ability to trap Si, it is more efficient to utilize catalyst specifically designed for higher Si uptake capacity and relatively low HDT activity to optimize the cost of reactor loadings. Therefore, a layer of Si trapping material is specified for units processing coker-derived feedstocks

TABLE 3.5. Typical Process Conditions for a Diesel Hydrotreating Process

Distillate Hydrotreater Typical Process Conditions			
Feed Type	Straight Run	Coker Distillate	Light Cycle Oil
LHSV (h^{-1})	1–2.5	1.0–2.0	0.75–1.5
Temperature (°F (°C))	630–660 (330–350)	640–680 (340–360)	670–710 (355–375)
Pressure (psig (barg))	600–900 (41–62)	700–900 (48–62)	750–1000 (51–70)
H_2/Oil, SCF/B (N m^3/m^3)	1500 (250)	2000 (340)	2000 (340)

at the top of the reactor to trap the bulk of the Si and to protect the higher activity bulk HDT catalyst downstream.

Process conditions typically are selected for a new DHT unit to achieve the desired product quality specifications and cycle length while optimizing unit capital and operating costs. Operating flexibility is also considered, since many refiners process a variety of crude oils. In some cases, refiners are also required to produce different grades of diesel depending on local market requirements. For example, some refiners are required to produce winter grade of diesel to meet cold flow requirements they serve.

Unit design typically involves a trade-off between various key process design parameters; Palmer et al. (2009) summarized the following key process variables:

- Space velocity (liquid hourly space velocity, LHSV)
- Total pressure
- Hydrogen partial pressure
- Make-up hydrogen purity
- Recycle gas rate
- Reactor temperature.

Cycle length can be variable but is frequently set by the owner based on overall refinery maintenance schedule.

Reactor temperature is variable up to metallurgical, and the cycle-ending temperature limits are specified based on parameters such as hydrogen and H_2S partial pressure and total pressure.

Space velocity is kinetic parameter that defines time required for desired extent of reaction and is used to calculate the volume of catalyst required to process a given volumetric feed rate. Since it specifies catalyst volume, it also directly impacts the size of the reactor required. Space velocity is defined as volumetric feed rate divided by catalyst volume as

$$\text{LHSV (h}^{-1}\text{)} = \frac{\text{Charge rate (ft}^3\text{/h)}}{\text{Catalyst volume (ft}^3\text{)}}. \quad (3.2)$$

Space velocity is typically expressed as the inverse of time unit. Reciprocal of space velocity is the residence time of feed molecules time within reactor. Knowing LHSV, one can determine SOR temperature requirement from reaction kinetics and catalyst activity. Higher activity catalyst requires less catalyst volume at same temperature for similar HDS. Catalyst life depends on end-of-run (EOR) temperature that is allowed based on metallurgical constraints and catalyst deactivation rate.

There are different types of grading material available with low HDT activity and large voidage to load in top section of reactor. The grading materials provide catalytic activity for easy reactions such as olefin saturation, metals removal, and some easy sulfur removal. One can then estimate the approximate cycle length based on SOR and EOR temperatures and deactivation rates.

Hydrogen partial pressure impacts product quality in terms of sulfur and nitrogen removal and aromatic saturation (cetane). Since aromatic saturation is equilibrium-limited reaction, lower pressure and higher temperature limit extent of saturation. Hydrogen partial pressure also affects catalyst deactivation rate. Higher deactivation rate will increase catalyst temperature at higher rate thus approaching equilibrium limitation sooner. Although higher hydrogen partial pressure is desirable, it is expensive so design pressure needs to be optimized for the product quality and catalyst life objective associated with a given feed. Although this above discussion refers to hydrogen partial pressure, hydrogen purity and total system pressure are critical considerations as they help set hydrogen partial pressure. Higher purity not only helps improve hydrogen partial pressure but also allows to maintain good recycle gas purity.

Make-up hydrogen purity is another process variable critical for design. As discussed, hydrogen partial pressure is key for product quality and catalyst life. Hydrogen partial pressure is based on recycle gas hydrogen purity and system total pressure. To maximize hydrogen partial pressure, it is desirable to maximize recycle gas purity. The recycle gas purity depends on the amount of light ends produced, make-up gas purity, and purge rate. If the recycle gas purity is too low, one needs to purge part of recycle gas. Purity can be improved to some extent by scrubbing recycle gas to remove H_2S. Selection of purge rate needs to be optimized as it will impact compressor size and make gas rate and therefore capital and operating cost.

Recycle gas rate (also referred to as treat gas rate) has impacts similar to that of hydrogen partial pressure. Higher recycle gas rate is desirable for efficient reaction and lowering deactivation rate. Higher gas rate also costs in compressor size and unit size. Unit design typically aims to maintain minimum recycle gas rate that is typically 5–8 times of consumption in order to ensure the catalyst and reactors have adequate hydrogen available. HDT reactions are exothermic and generating heat rise in the reactor. Catalyst life is limited by peak temperature and for operational control, a good design will control heat rise by dividing catalyst volume in multiple beds and inject quench. Quench is typically a portion of the recycle gas stream. The heat rise is a function of feed quality (aromatics and olefins) and extent of reactions required to meet product quality target.

H_2S formed in the reactors will reach equilibrium concentration in the recycle gas depending upon pressure, feed sulfur, and recycle gas rate. Since H_2S has a depressing effect on catalyst activity, it is desirable to remove the H_2S from the recycle gas. The removal of H_2S is performed in a scrubber where the recycle gas in contacted with an amine (generally Monoethanolamine (MEA), Diethanolamine (DEA), or Methyldiethanolamine (MDEA)) solution. In this manner, the H_2S content of the recycle gas can be reduced and hydrogen purity of recycle gas can be improved.

Reactor temperature is a critical parameter in design for several reasons. SOR temperature is set to make catalyst life target that depends on EOR temperature (peak temperature) and must be sufficient to drive the required reactions. As discussed previously, exothermic reactions cause a temperature rise across the bed. The resultant peak temperature is typically limited for either a product quality reason or mechanical reason. Typically, temperature rise per bed is limited to 50–70 °F. Catalyst temperature for adiabatic reactor is usually calculated as weighted average bed temperature (WABT). WABT is used for kinetic calculation to set LHSV from kinetics. Typically, HDT reactors are multibed design that allows injection of cold recycle gas as quench for temperature control at the inlet of subsequent bed. Too high of peak temperature can deteriorate color of diesel. Diesel color is specification in some regions such as Japan. Typically, HDT reactor temperature is limited between 730 and 760 °F.

Cycle length is typically determined by the combined effect of the previously discussed variables. Typically, units are designed for a 2–4-year cycle length although some units can have cycle lengths as short as 1 year. This can be particularly true when a unit was originally designed for a higher product sulfur and then revamped to a more stringent sulfur specification.

In summary, optimum design considers catalyst volume, hydrogen partial pressure, operating temperature requirements, feed characteristics, and recycle gas rate. A good design will meet product quality specifications for entire cycle for given feed. Since process variable selection is dependent on catalyst activity, the selection of catalyst type is also important. This is particularly relevant if there are impurities in feed that are better handled by specific catalysts designed to handle those impurities. For example, it is better to utilize silica trap catalysts to remove silica from coker-derived feed as opposed to more expensive HDT catalyst.

3.8 DIFFERENT TYPES OF PROCESS DESIGNS

Different process design and flow schemes can be employed for DHT depending on the process objectives and characteristics of the feed being processed. One typical process design in distillate or DHT unit is a simple once-through liquid flow design with circulating/recycle gas. A schematic of such a flow scheme is shown in Figure 3.6.

Units employing a once-through design can have multiple reactor vessels, each having multiple beds to accommodate high feed rates and large catalyst volume. Once-through design can also use multiple types of catalysts ranging from graded beds to address feed contaminants to different types of HDT catalysts stacked to provide specific function in desired reaction zone/section.

There is also a two stage hydrotreating concept is proposed by (Masaomi et al., 2002) for deep HDS based on concept of minimizing H_2S and NH_3 inhibition in

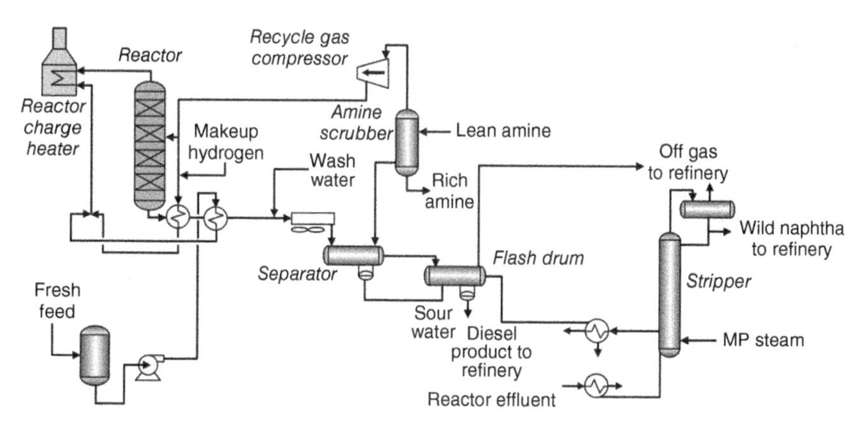

FIGURE 3.6. Simplified process flow diagram.

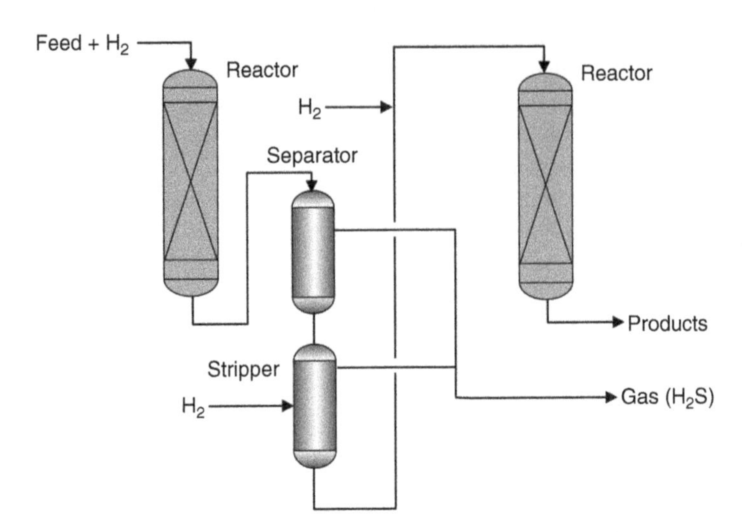

FIGURE 3.7. Description of two-stage process pilot plant.

second stage by removing H_2S between stages. A basic conceptual two stage flow scheme is depicted in Figure 3.7. A further improved two-stage process is possible by the use of gas/liquid separation in the middle of the unit. Removal of produced hydrogen sulfide and ammonia in the middle of the unit accelerates HDS in the following second stage and has the potential to produce product containing 10 ppm S or less. Significantly reducing the concentration of H_2S and NH_3 in the second stage enables the application of NiW catalyst, bringing further improvement to the performance of the process.

REFERENCES

Ackelson D, Thakkar V, Dziabala B (2007) Innovative hydrocracking applications for conversion of heavy feedstocks, Paper AM-07-47, NPRA Annual Meeting March 18–20.

Bacha J, Blondis L, Freel J, Hemighaus H, Greg K, Hogue N, Horn J, Lesnini D, McDonald C, Nikanjam M, Olsen E, Scott B, and Sztenderowicz M (1998) *Diesel Fuels Technical Review (FTR-2)*, Chevron Products Company.

Clean Diesel Fuel Alliance (2012) Battelle Memorial Institute Study No. 10001550, Final Report, Corrosion in Systems Storing and Dispensing Ultra Low Sulfur Diesel (ULSD), Hypotheses Investigation, September 5.

Johnson JA, Frey S, Thakkar V (2007) Unlocking high value xylenes from light cycle oil, Paper AM-07-40, Annual NPRA Meeting March 18–20.

Jones L, Kokayeff P (2004) Distillate hydrotreating to ULSD, the impact of aromatics, Paper No. 90a, AIChE 2004 Spring National Meeting.

Jones L, Kokayeff P (2005) Equilibrium limitations in distillate hydrotreating, Paper No. 61d, AIChE 2005 Spring National Meeting.

Kokayeff P, Zink S, Roxas P, Hydrotreating in petroleum processing, in Treese SA, Pujado PR, DSJ Jones, *Handbook of Petroleum Processing*, 2nd edition, Springer International Publishing 2015, pp. 361-434

Masaomi A, Masanari M, Ryutaro K, Yasuhito G, Manabu K, Katsuaki I, Hideo S (2002) Ultra low sulfur diesel fuel production by two-stage process with gas/liquid separation system, *Fuel Chemistry Division Preprints*, **47**(2), 460–461.

Palmer E, Polcar S, Wong A (2009) Clean diesel hydrotreating, *Petroleum Technology Quarterly*, **Q1 2009**, 91–100, www.digitalre ning.com/article/1000587.

Worldwide Fuel Charter (2013) Fifth Edition, September '13, published by the members of the Worldwide Fuel Charter Committee. http://www.autoalliance.org/auto-issues/fuels

Worldwide Refinery Process Review (2012) *Hydrotreating and Environmental Controls*, Second Quarter '12, Hydrocarbon Publishing Co.

4

DESCRIPTION OF HYDROCRACKING PROCESS

4.1 WHY HYDROCRACKING

In simple terms, hydrocracking is a process to convert larger hydrocarbon molecules into smaller molecules under high hydrogen pressure and elevated temperature. Thus, hydrocracking is a flexible catalytic refining process that upgrades a variety of petroleum fractions to higher value products. For this reason, hydrocracking is commonly applied to upgrade the heavier fractions of the crude oils to produce higher value transportation fuels. The process adds hydrogen to the hydrocarbon, removes impurities such as sulfur and nitrogen, and saturates aromatics to produce a product that meets the environmental specifications. Hydrocracking processes are designed to operate at a variety of conditions with a variety of hydrocarbon feeds. The process design will depend on many factors such as feed type, desired cycle length, and the desired product slate. Hydrocracking units generally operate at the following conditions: liquid hourly space velocity (LHSV): 0.5–2.0, H_2 circulation: 5000–10,000 standard cubic feet per barrel (SCFB), H_2 partial pressure: 1500–2500 psig (103–172 barg) and start of run (SOR) temperatures: from 650 °F (288 °C) to 750 °F (399 °C). Products from hydrocracking are high-quality saturated products that meet or exceed specifications (Bricker et al., 2015; Scherzer and Gruia, 1996).

Modern hydrocracking technology has been in use since 1960 and traces its origin to the development in the 1910s when it was used for coal conversion to secure a supply of liquid fuels from domestic coal deposits. Coal conversion to liquids was a relatively very high pressure, 3000–10,000 psig (207–690 bar), and high temperature (700–1000 °F, 371–538 °C) catalytic process. After World War II, the Middle

Hydroprocessing for Clean Energy: Design, Operation, and Optimization, First Edition.
Frank (Xin X.) Zhu, Richard Hoehn, Vasant Thakkar, and Edwin Yuh.

Eastern crudes became available, and their gas oils and cracked stocks were easily processed in fluid catalytic cracking (FCC). The FCC technology commercialized in the 1940s. Catalytic cracking processes proved to be more economical for converting heavy petroleum fraction into gasoline based on the demands for gasoline. Consequently, hydrocracking of coal became less important. However, environmental regulations and increased diesel consumption have made hydrocracking the technology of choice when conversion of heavier hydrocarbon is desired.

In the mid-1950s, the automotive industry started producing high-performance cars with engines that required higher octane gasoline. This resulted in a large expansion of catalytic cracking capacity to produce gasoline. This also increased production of low-value by-product light cycle oils (LCOs) that were difficult to convert to gasoline blendstock. The hydrocracking process can convert LCO and other difficult by-products to naphtha and, as a consequence, was adopted by some refiners to produce naphtha that can be catalytically reformed to high-octane gasoline blendstock. As industrialization grew, railroads switched from steam to diesel engines and increased commercial aviation needed more jet fuel, further motivating refiners to employ hydrocracking, making it a major refinery process.

Presently, various environmental regulations stipulate a low level of sulfur and in some cases a low level of aromatic composition in both gasoline and diesel (ultra-low-sulfur diesel, ULSD) products, which has spurred the addition of hydroprocessing complexes in global refineries. In the United States, the Tier 2 gasoline sulfur program reduced the sulfur content of gasoline by up to 90% from an uncontrolled level, phasing in an average sulfur level of 30 ppm between 2004 and 2007. The final Tier 3 gasoline sulfur program lowers the sulfur level to 10 ppm. Starting in 2017, the European Union and Japan both specify sulfur content of below 10 ppm. As gasoline and diesel sulfur specifications level approached similarly low levels, a large increase in both hydrotreating and hydrocracking capacity has taken place globally and is expected to continue to increase over the next decade. In many cases, even when a refiner does not need to achieve the stringent specification for their regions, the process is designed to achieve the stiffer requirement so that the refiner is able to export their products to any region of the world to improve their economics.

Vermeiren and Gilson (2009) have compared two catalytic cracking processes, FCC and hydrocracking process, as summarized in Table 4.1. In principle, FCC is a lower pressure carbon rejection process, while hydrocracking is a higher pressure hydrogen addition process.

The demand for hydrocracking is predicted to be a stable growth market for the foreseeable future. The installed hydrocracking capacity is estimated to be between 6 and 7 million barrels per stream day (bpsd). The 2013 OPEC world oil outlook shows that the demand for hydrocracking increases with a continuing increase in value of distillate products over gasoline products as well as a large increasing demand for desulfurization processes, which reflects the lower sulfur requirements in gasoline and diesel around the world. Due to changes in the demand growth of diesel relative to gasoline, many refiners seek to modify their existing facilities to allow flexible yields and meet market product needs. Since 2009, the worldwide capacity for hydrocracking has grown yearly about 2%, and the global demand for diesel has also grown about 2%. The demand for combined hydrotreating and hydrocracking is projected to continue to grow in the near future.

TABLE 4.1. FCC and Hydrocracking Process Key Differences

Cracking Process	FCC	Hydrocracking
Principle	Carbon rejection	Hydrogen addition
Catalytic function	Monofunctional	Bifunctional
Operating pressure	Near atmospheric	High pressure (>100 bar)
Hydrogen need	No	Lot of hydrogen needed
LPG	Highly olefinic	Nearly no olefins
Naphtha product		
Olefins (wt%)	>20	Nil
Aromatics (wt%)	>20	Low aromatics
Octane number	High	Low
Use	Gasoline pool	Reformer or steam cracker feed
Kerosene/jet fuel product		
Hydrogen (wt%)	~10–10.5	~13.5–14.0
Sulfur, ppm	>1000	~10–20
Gasoil/diesel product		
Hydrogen (wt%)	9.5–10	~13–13.5
Sulfur, ppm	>5000	~20–40
Cetane number	<30	>50

4.2 BASIC PROCESSING BLOCKS

Hydrocracking is an important unit in today's modern refinery as it is one of the three primary conversion processes available to the refiner along with FCC and Coker. FCC is considered as gasoline making machine, although it also produces olefins, and delayed coking has been the primary technology of choice for converting vacuum resid to lighter products. Hydrocracking unit is flexible – it can produce a range of products as light as liquefied petroleum gas (LPG) to heavy diesel and lube oil base stock or ethylene cracker feed. Hydrocracking provides value uplift, converting heavier low-value stream such as vacuum gas oil (VGO), coker gas oil (CGO), and LCO to a better quality and high-value product stream. Hydrocracking is a versatile process that takes many different feed streams to convert into range of desirable products. Figure 4.1 shows the typical refinery configuration with hydrocracking unit incorporated.

Hydrocracking units can be configured in a number of ways. The unit can consist of one or more reactors with either one or more type of catalysts. Bricker et al. (2015) discuss hydrocracking process that can use one or two stages and be operated in once-through mode for partial conversion or recycle mode for maximum conversion. In single-stage hydrocracking, feed is hydrocracked in single stage followed by product separation, while in two-stage hydrocracking, a portion of conversion takes place in the first stage and then the product is separated before the second stage of hydrocracking on unconverted oil (UCO) is carried out in the second stage. Rossi et al. (2007) have discussed various flow schemes available for hydrocracking process. The choice of the configuration depends on the feed properties and the specific product slate desired by the refiner. Naphtha produced from hydrocracking unit

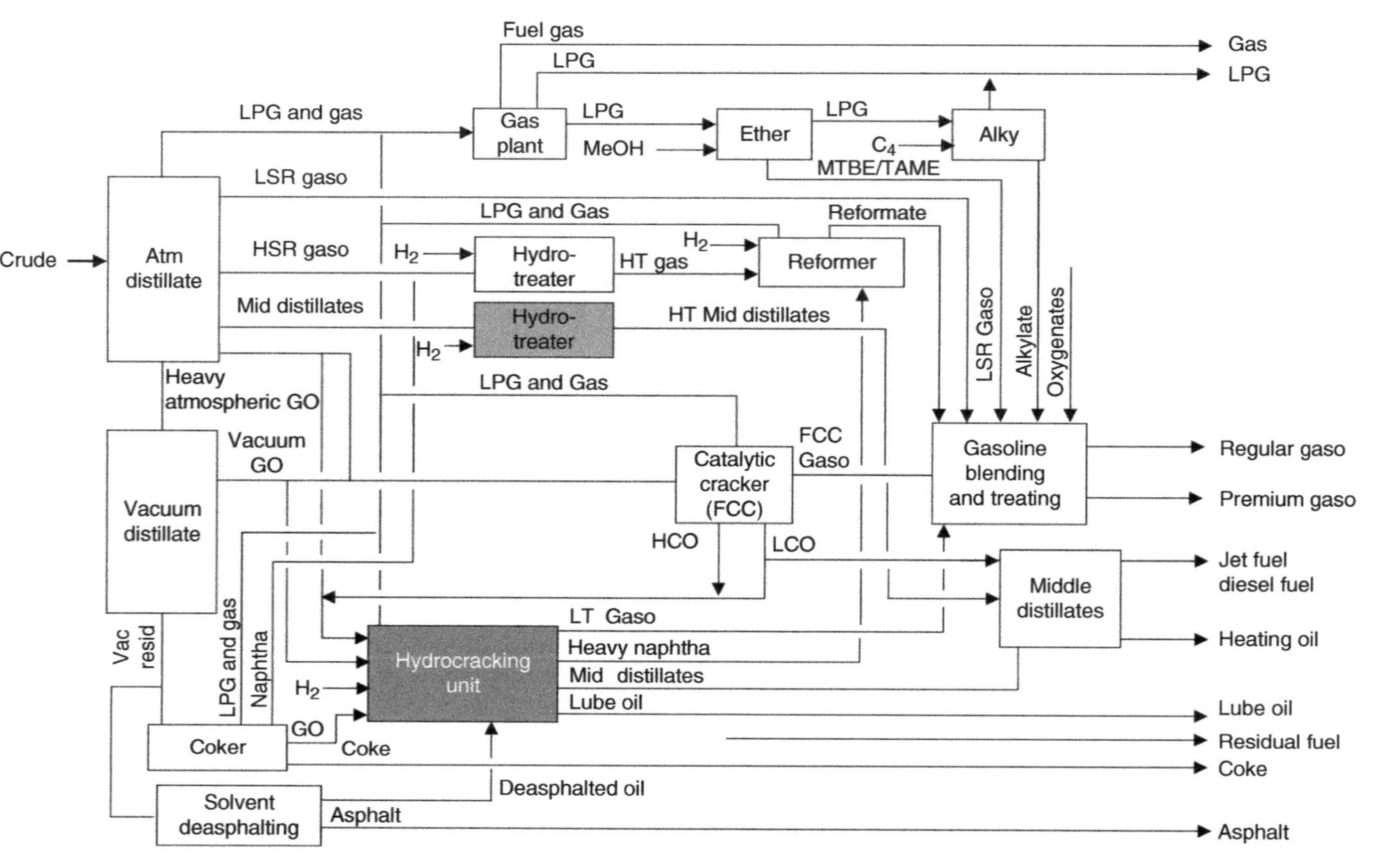

FIGURE 4.1. Where hydrocracking fits in the refinery configuration.

TABLE 4.2. Nominal Operating Conditions for Typical Hydrocracking Unit

Unit Type	Typical Conversion	Total Pressure (bar/psig)	Reactor Temperature (°C/°F)
Mild (MHC)	20–40	60–100/870–1450	350–440/662–824
Moderate pressure	40–70	100–110/1450–1600	340–435/644–815
Conventional HC	50–100	110–200/1600–2900	340–435/662–842
Residue HC	65–100	97–340/1400–3500	385–450/725–914
Slurry hydrocracking	80–97	138–241/2000–3500	426–471/800–880

usually requires reforming to boost its octane, kerosene/jet and diesel produced in a hydrocracking normally do not require additional processing, while unconverted oil (UCO) used to produce lube base oil will require additional hydroprocessing. Table 4.2 shows the nominal operating conditions for the typical hydrocracking unit. A brief description of various flow configurations is given in the following sections.

4.2.1 Once-Through Configuration

Figure 4.2 shows a schematic of a once-through hydrocracking unit, which is the simplest configuration for a hydrocracking unit. This configuration represents the simplest design and minimum cost option. The feed mixes with hydrogen and passes through the reactor(s). The reactor effluent goes through vapor–liquid separation and then to fractionation, with the UCO removed as the heaviest cut from the bottom of a fractionation column. High-pressure hydrogen-rich gas is recycled back to the reactor and makeup hydrogen is added to compensate for hydrogen consumption by the hydrocracking reactions and losses due to purge as well as hydrogen dissolved in the liquid stream. Note that there is no recycle of liquid in this configuration. This type of unit is the lowest cost hydrocracking unit and can process heavy, high-boiling feed stocks and produce high-value products including unconverted hydrocarbon material

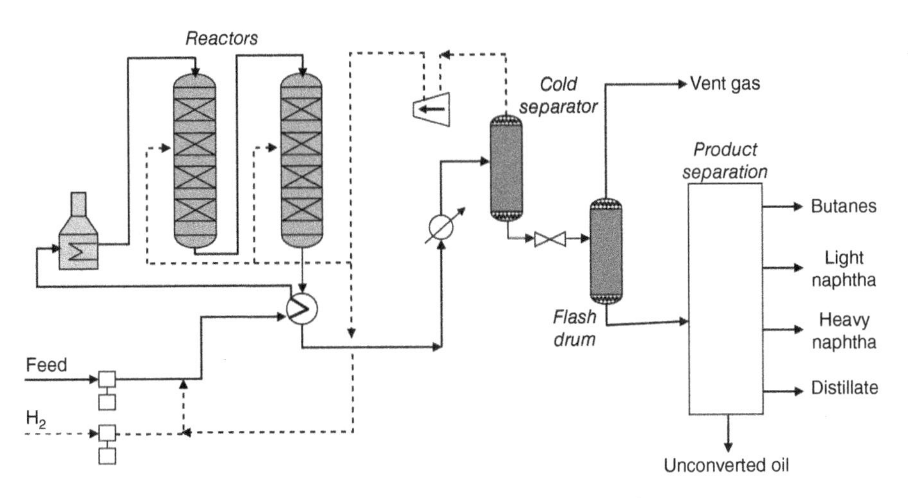

FIGURE 4.2. Once-through process flow scheme.

that can be a good feedstock for FCC units, ethylene cracking plant, or high-quality lube base oil. In general, for the once-through units, conversion of the feedstock to products can range between 30% and 90% volume.

4.2.2 Single-Stage Hydrocracking

Single-stage hydrocracking is a high conversion (close to extinction of feed, 100% conversion) design with liquid recycle. This design is most widely utilized in hydrocracking units in which the recycle oil is sent back to the reactor section to maximize conversion at a reduced conversion per pass across the reactor. Figure 4.3 depicts this type of configuration. It is the most cost-effective design for 100% conversion and is mainly used to maximize the heaviest product cut such as diesel or jet fraction. The fresh feed mixed with hydrogen is preheated to reaction temperature by exchanging heat with hot reactor effluent and then in fired heater. The reactors in single-stage design contain pretreating catalyst and followed by hydrocracking catalyst. Effluent from the reactors goes through a series of vapor–liquid separators where hydrogen-rich vapor is recovered and, together with makeup hydrogen, recycled to the reactor inlet. Liquid product is sent to a fractionation column where the final products are separated from the UCO, which is recycled. When operating in high conversion mode, careful consideration must be given to the potential production of polynuclear aromatics. Hydrocracking catalysts are acidic and when operated at high temperatures tend to produce condensed ring or polynuclear aromatic (PNA) structure that is precursor for catalyst deactivation by coking. There are options available for managing PNA, including removal of heavy polynuclear aromatic (HPNA) from recycle oil that improves hydrocracking unit performance.

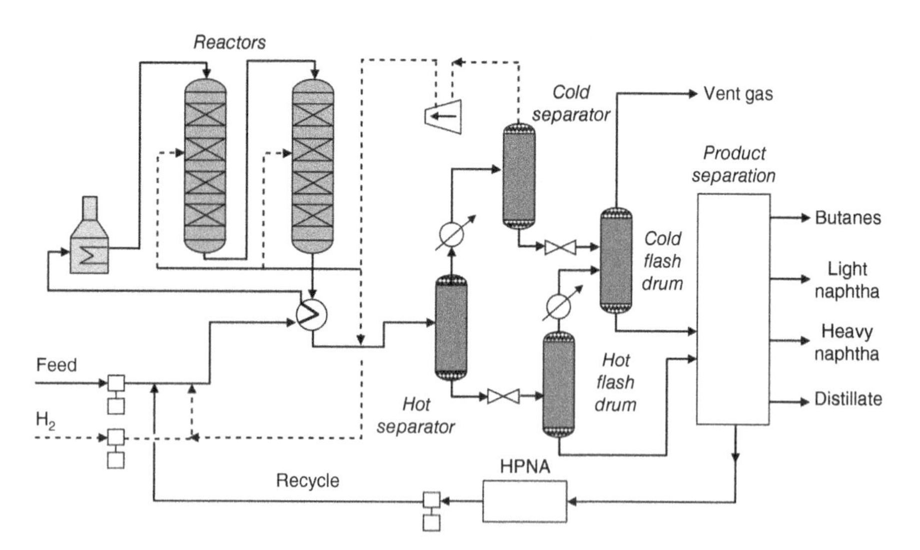

FIGURE 4.3. Single-stage process flow scheme.

4.2.3 Two-Stage Recycle Hydrocracking

The two-stage hydrocracking process configuration is also widely used, especially for high capacity and high conversion units. When the feed rates increase and reactor flux criterion may require to design two reactor trains, in such cases two-stage design offers option to maintain design in single train. Other reason one may choose two-stage design because of possibility to use two or more different types of cracking catalysts to optimize desired product selectivities. Two-stage design will have product separation between stages allowing removal of H_2S/NH_3 so that second stage can run in relatively low H_2S/NH_3 environment. In two-stage units, the hydrotreating and some cracking take place in the first stage. The effluent from the first stage goes through vapor–liquid separation then to fractionation to separate light products from UCO. The UCO is then sent to the second stage for further cracking. The reactor effluent from the second stage reaction section is typically combined with the first stage effluent for efficient vapor–liquid separation and fractionation. A simplified schematic of a two-stage hydrocracking unit is shown in Figure 4.4. In general, a two-stage design will consist of pretreating section in first stage followed by cracking catalyst. Thus, there are two stages of cracking in such designs. The cracking catalysts in the first stage are the same types as those used in the once-through or single-stage recycle configuration as it needs to perform effectively in the presence of H_2S and NH_3. The reaction environment in the second stage is usually sweet as recycle gas is scrubbed to low hydrogen sulfide levels. Ammonia levels are also low, meaning the catalyst in the second stage in such a design operates in the near absence of ammonia and hydrogen sulfide. This environment enhances cracking activity. Catalysts used in the second stage are tailored for optimal performance in a sweet reaction environment to maximize desired product selectivity.

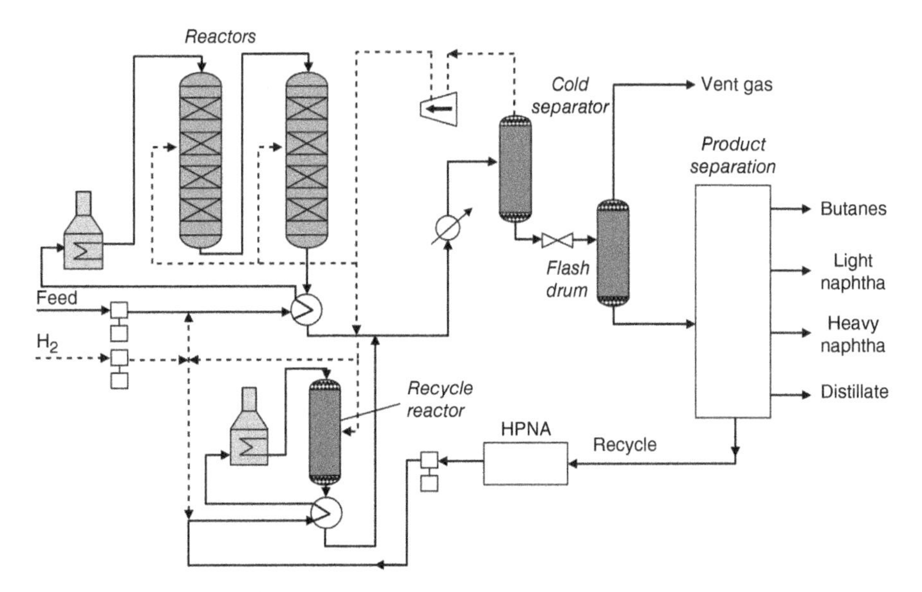

FIGURE 4.4. Two-stage process flow scheme.

4.2.4 Separate Hydrotreating Hydrocracking

A variation of the typical two-stage hydrocracking process flow scheme shown in Figure 4.4 is used in certain situations where the first stage contains only pretreating catalyst and no hydrocracking catalyst. Such flow schemes can be specified when the feed contains large amount of product range material that just need hydrotreating and/or feed contains high levels of nitrogen that make first stage hydrocracking catalyst ineffective. The separate hydrotreating scheme can use either common or separate hydrogen circulation loop. This flow scheme allows second stage with cracking catalyst to operate in sweet environment, that is, near absence of hydrogen sulfide and ammonia.

Hydrocracking configurations typically utilize both pretreating catalyst and hydrocracking catalyst. Pretreating catalyst function is to remove sulfur and nitrogen, and improve the quality of the feed. Hydrocracking catalyst provides boiling point conversion and aromatic saturation. These reactors are downflow, fixed-bed catalytic reactors and generally operate in the trickle flow regime. The reactors usually have multiple beds for managing heat release of exothermic reactions, with cold recycle gas used as the quenching medium.

In many cases, the first reactor employs a graded bed for activity grading to manage the reactions and also provides voidage for trapping feed particulate matter that would otherwise create pressure drop in tightly packed main catalyst bed. Depending on feedstock quality, specialized catalyst can be loaded in the top bed of the reactor for hydrodemetallization (HDM) of feed.

4.3 FEEDS

As previously discussed, hydrocracking can upgrade a variety of petroleum fractions. Table 4.3 shows the feedstock typically used in the hydrocracking unit and products that can be produced. Since the process adds hydrogen to the hydrocarbon molecules, it improves the hydrogen to carbon ratio of the reactor effluent and removes impurities such as sulfur and nitrogen. Hydrogen addition generally enables

TABLE 4.3. Hydrocracking Feeds and Products from the Processing

Feedstock	Products
Straight run gas oils (AGO)	LPG
Vacuum gas oils (VGO)	Motor gasoline
FCC light cycle oils (LCO)	Reformer feeds
Coker gas oils (LCGO HCGO)	Jet fuels
Thermally cracked gas oils	Diesel fuels
Deasphalted oils (DAO)	Distillate fuels
	Heating oils
Straight run and cracked naphtha	Steam cracker feedstock
Fischer Tropsch liquids	Lube plant feedstock
Tar sands-derived liquids	FCC feedstock

products to meet the specifications that relate to environmental regulations. The hydrocracking chemistry involves conversion of higher molecular weight compounds to lower molecular weight compounds through carbon–carbon bond breaking and hydrogen addition. The products have lower boiling points and are highly saturated. Hydrocracking processes are designed for, and run at, a variety of conditions. The process design will depend on many factors such as feed type, desired cycle length, and desired product slate.

These feeds are a complex mixture of different types of hydrocarbons. Feed to the hydrocracking unit comes from crude tower or other units such as Coker, FCC, or Deasphalting unit. These feed streams contain various contaminants that include sulfur, nitrogen, olefins, conradson carbon, and metals (such as Ni, V, Si, and As). If the feed fractionation is not efficient, low levels of asphaltenes can carry over and potentially accelerate catalyst deactivation or cause pressure drop. The concentration and type of contaminant present impacts the hydrocracking unit design, catalyst selection, and the design of the catalyst loading.

Although straight run feed stream do not contain many olefins, feeds from thermal processes such as coking and thermal cracking unit will contain olefins. Olefins are not a big issue in processing, as they are easy to convert, but they consume hydrogen and generates high exotherm. Diolefins are more reactive and can form gums and polymers capable of fouling heat exchanger surfaces and blocking catalyst beds. These are present in thermally cracked streams, particularly naphthas, but can be processed with little or no problem if proper storage and design procedures are followed.

Metals are permanent poison for main pretreating and cracking catalyst. Hence, main active catalyst needs to be protected. Metals are removed by demet catalyst that is designed for such services.

Feed endpoints are critical to control in order to minimize contaminant level and catalyst deactivation. Higher endpoint feeds tend to increase production of HPNA, which are precursor to coking reaction deactivating catalyst. Highly aromatic feed streams such as LCO, HCO, and HCGO contain polyaromatics in heavy back-end tail, which accelerates deactivation.

4.4 PRODUCTS

Since the feed to the hydrocracking unit can cover a wide variety of boiling range, the products from hydrocracking unit will also vary in boiling range depending upon cracking severity and mode of operation (once-through or recycle unit). The product fractions and the general application of them are summarized in Table 4.4.

Since hydrocracking is a high-pressure hydrogen addition process, products from hydrocracking are fairly saturated. The product quality depends on the desired use or application, for example, smoke point for jet while cetane for diesel is important. Naphtha product from hydrocracking unit is saturated so needs reforming to increase its octane for use as gasoline. Jet fuel and diesel property such as smoke point and cetane are good as produced. The products from hydrocracking unit are low in sulfur content meeting ULSD requirement. Table 4.5 lists some of the important product qualities and the chemical basis for better quality.

Jet-A is typically used in the United States, while most of the rest of the world uses Jet A-1. The difference in two grades is freeze point. Jet A-1 has a lower maximum

TABLE 4.4. Product Fractions and the General Use

Fraction	Product	Use
Light gases, C_4^-	C_3, C_4	Fuel gas, feed for alkylation, LPG, petrochemical feedstock
Light naphtha	C_5-80 °C/C_5-175 °F	Gasoline pool component, feed to isomerization unit
Heavy naphtha	80–150 °C/175–300 °F	Reformer feedstock to be converted in high octane gasoline and hydrogen or aromatics for petrochemicals
Jet fuel/kerosene	150–290 °C/300–550 °F	Fuel for turbine engines, or heating
Diesel fuel	290–370 °C/550–700 °F	Fuel for diesel engines
Unconverted oil	370 °C$^+$/700 °F$^+$	Recycle feed or feedstock for lube plants or ethylene plants or FCC plants

TABLE 4.5. Chemical Basis for Product Quality Measurements

Desired Product Quality	Chemical Basis
High smoke point	Low concentration of aromatics
Low pour point	Low concentration of *n*-paraffin
Low freeze point	Low concentration of *n*-paraffin
Low cloud point	Low concentration of *n*-paraffin
Low cold flow pour point (CFPP)	Low concentration of *n*-paraffin
High octane	High ratio of *i*/*n* paraffin High concentration of aromatics
Cetane number	Measure of hydrocarbon type
Cetane index	Estimate of cetane number based on distillation range and density

freezing point than that of Jet A (Jet A: −40 °C, Jet A-1: −47 °C). The lower freezing point makes Jet A-1 more suitable for long international flights, especially on polar routes during the winter.

However, the lower freezing point comes at a price. Other variables being constant, a refinery can produce a few percent more Jet-A than Jet A-1 because the higher freezing point allows the incorporation of much higher boiling components, which, in turn, permits the use of a broader distillation range. Jet fuel specifications used in the industry are ASTM D1655. Other specifications widely used are Defense Standard 91-91; the United Kingdom Ministry of Defense maintains this specification (formerly titled DERD 2494). Airlines Association and International Air Transport Association (IATA) publishes a document entitled Guidance Material for Aviation Turbine Fuels Specifications. The guidance material contains specifications for aviation turbine fuels.

Jet fuel smoke point is one of the critical properties, which is easily achieved by product from hydrocracking process. Smoke point specification usually requires either 25 or 19 mm smoke point with less than 3% naphthalene's. Jet fuel produced by a hydrocracking unit will typically meet less than 3% naphthalene's content.

Diesel fuel has become important transportation fuel in the last few decades. Diesel is used as fuel for many applications such as moving consumer or industrial goods cross-country, or to generate electric power, or to run farm equipment for increased efficiency on the farms. Thus, diesel fuel plays a vital role in a nation's economy and standard of living.

US Highway Diesel is mandated to 15 parts per million (ppm) sulfur specification, known as ULSD. Europe (<10 ppm S) and most of the developed countries have adapted ULSD specifications. Other nations are trending to achieve ULSD target. The other key specification for diesel is cetane. Most regions other than the United States require higher cetane diesel product. ASTM D975 diesel specification requires 40 cetane number. Diesel cetane can be improved by lowering aromatics content of diesel, which is what happens (aromatic saturation) in the hydrocracking unit. Transportation diesel typically used is Grade 1 and/or Grade 2 diesel. ULSD is identified as Grade No. 1-D S15, comprises the class of very low sulfur, volatile fuel oils from kerosene to the intermediate middle distillates. Fuels within this grade are applicable for use in (1) high-speed diesel engines and diesel engine applications that require ultra-low-sulfur fuels, (2) applications necessitating frequent and relatively wide variations in loads and speeds, and (3) applications where abnormally low operating temperatures are encountered. ASTM Grade No. 2-D S15 includes the class of very low sulfur, middle distillate gas oils of lower volatility than Grade No. 1-D S15. These fuels are applicable for use in (1) high-speed diesel engines and diesel engine applications that require ultra-low-sulfur fuels, (2) applications necessitating relatively high loads and uniform speeds, or (3) diesel engines not requiring fuels having higher volatility or other properties specified in Grade No. 1-D S15. Diesel qualities of significance have been discussed in detail in Chapter 3.

While trying to maximize diesel production, one can adjust cut point (IBP, initial boiling point) to the flash point limits to maximize diesel cut if diesel is the most desirable product. Diesel product front-end distillation is limited by flash point and back end is limited by T95 or cold flow property. Hydrocracking product is saturated and isomerized product, so cold flow properties are good.

Jet fuel and diesel products from hydrocracking unit are usually blending component in refinery product pool. Additives are used to improve some of the properties as discussed in the hydrotreating chapter.

4.5 REACTION MECHANISM AND CATALYSTS

The chemistry of hydrotreating and hydrocracking is commonly discussed together in terms of hydroprocessing and is similar for both sections, that is, hydrotreating and hydrocracking. Bricker et al. (2015) and Scherzer and Gruia (1996) have reviewed hydrocracking process chemistry and reaction mechanism. The function of hydrotreating is to convert sulfur- and nitrogen-containing heteroatoms to H_2S and NH_3; if oxygen is present it is converted to H_2O. Significant aromatic saturation also occurs. Modern catalysts designed for deep hydrotreating also have some acidity, thus some hydrocracking also occurs.

TABLE 4.6. Hydroprocessing Reactions

Reaction Type	Reaction
Minimal C—C bond breaking	
Hydrodesulfurization (HDS)	$R{-}S{-}R^* + 2H_2 \rightarrow RH + R^*H + H_2S$
Hydrodenitrogenation (HDN)	$R{=}N{-}R^* + 3H_2 \rightarrow RH + R^*H + NH_3$
Hydrodeoxygenation (HDO)	$R{-}O{-}R^* + 2H_2 \rightarrow RH + R^*H + H_2O$
Hydrodemetallation (HDM)	$R{-}M + \frac{1}{2}H_2 + A \rightarrow RH + MA$
Saturation of aromatics	$C_{10}H_8 + 2H_2 \rightarrow C_{10}H_{12}$
Saturation of olefins	$R{=}R^* + H_2 \rightarrow HR{-}R^*H$
Isomerization	n-RH → i-RH
Significant C—C bond breaking	
Dealkylation of aromatic rings	$\Theta{-}CH_2R + H_2 \rightarrow \Theta{-}CH_3 + HR$
Opening of naphthene rings	*cyclo*-$C_6H_{12} \rightarrow C_6H_{14}$
Hydrocracking of paraffins	$R{-}R^* + H_2 \rightarrow RH + R^*H$
Other reactions	
Coke formation	$2\Theta{-}H \rightarrow \Theta - \Theta + 2H_2$
Mercaptan formation	$R{=}R^* + H_2S \rightarrow HSR{-}R^*H$

R represents alkyl group, Θ represents aromatic, M represents metal, and A represents metals adsorbing material.

Hydrocracking catalysts convert the higher carbon number feed molecules to lower molecular weight products by cracking the side chains and by saturating the aromatics and olefins. Hydrocracking catalysts will also remove any residual sulfur and nitrogen remaining after hydrotreating. Table 4.6 is a list of the hydroprocessing reactions.

The early hydrocracking units used catalysts that were based on iron, nickel, or nickel–tungsten metal supported on alumina or amorphous silica alumina. Zeolite components were introduced to the catalysts during the rapid growth of hydrocracking technology during the 1960s. Zeolite-containing catalysts were significantly different from amorphous hydrocracking catalysts; zeolites had higher activity and better selectivity to gasoline. In the United States, hydrocracking was primarily used in the production of gasoline, but in the other parts of the world it was used primarily for the production of middle distillates. Different catalyst formulations are used to tailor the product slate desired by refiners. Amorphous hydrocracking catalysts used in the industry are not as active as zeolitic catalysts and generally are more selective toward heavier products such as diesel. Higher activity zeolitic hydrocracking catalysts can also be designed to maximize distillate selectivity. Most hydrocracking designs incorporate pretreating catalyst upstream of the hydrocracking catalyst. This is done to prepare the feed to ensure optimum performance of the cracking catalyst. Since amorphous catalyst is not as active as zeolitic catalyst, it requires higher pressure and higher temperature. It is not unusual for amorphous catalyst to perform both hydrotreating and hydrocracking functions.

4.5.1 Hydrotreating Reactions

The hydrotreating reactions are removal of sulfur, nitrogen, organometallic compounds, oxygen, and halide. Olefin and aromatic saturation will also occur as part of hydrotreating. Sulfur, nitrogen, and metals are almost always present in hydrocracking feed and the levels depend on the crude source and/or the upstream conversion unit, such as a delayed coking unit, thermal cracking, or FCC unit. Oxygen and halides may or may not be present in the feed. The hydrotreating reactions summarized proceed in the following order: metal removal, olefin saturation, sulfur removal, nitrogen removal, oxygen removal, halide removal, and aromatic saturation.

Hydrotreating Reactions

- Remove
  - Sulfur
  - Nitrogen
  - Oxygen
  - Halides
  - Metals
- Saturate
  - Olefins
  - Aromatics
  - Polycyclic aromatics (PCA).

Hydrogen is consumed in all of the treating reactions. As a guideline, the desulfurization reactions consume 100–150 SCFB/wt% S change (17–25 $N\,m^3/m^3$/wt% change) and denitrogenation reactions consume 200–350 SCFB/wt% N change (34–59 $N\,m^3/m^3$/wt% change). Typically, the heat release in the hydrotreating section is about 0.1–0.2 °F/SCFB H_2 consumed.

4.5.2 Hydrocracking Reactions

Hydrocracking reactions proceed on acidic cracking catalysts, which are designed as a dual-functional catalyst. Two distinct types of catalytic sites are required to catalyze the cracking reaction sequence. One of the functions is called the acid function, which provides cracking and isomerization activity, and the second function is metal function, which provides olefin formation and hydrogenation/saturation activity. Figure 4.5 is a representation of dual-function hydrocracking catalyst components as discussed by Scherzer and Gruia (1996), which are providing acid and metal function.

Thus, the cracking catalyst accomplishes the following class of reactions at high temperature and under hydrogen partial pressure.

Hydrocracking Reactions

- Hydrodesulfurization
- Hydrodenitrogenation
- Aromatics saturation
- Paraffin isomerization
- Paraffin cracking

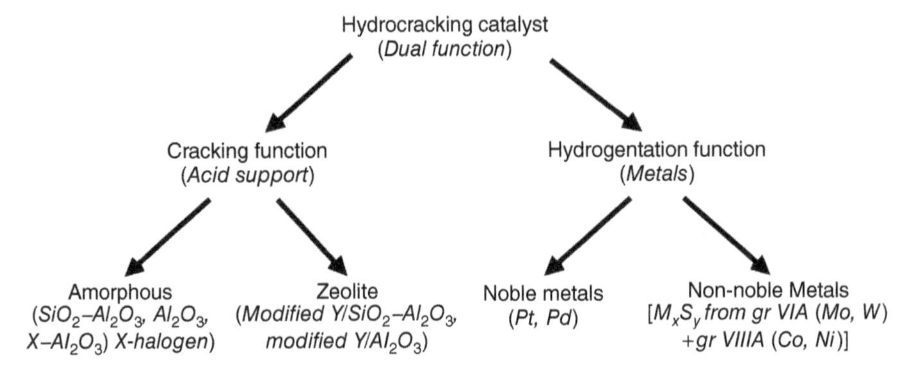

FIGURE 4.5. Dual-function hydrocracking catalyst components.

- Cyclization
- PNA formation.

Overall, there is a heat release during hydrocracking reactions; the heat release is a function of the hydrogen consumption. Generally, the hydrogen consumption in hydrocracking by Bricker et al. (2015) is estimated to be 1200–2400 SCFB (200–420 N m^3/m^3), resulting in a typical heat release of 50–100 Btu/SCFB H_2 (2.1–4.2 kcal/m^3 H_2), which translates into a temperature increase of about 0.1–0.2 °F/SCFB. This amount includes the heat release generated in the hydrotreating section.

In general, the hydrocracking reaction starts with the generation of an olefin or a cyclo-olefin on a metal site on the catalyst. Next, an acid site adds a proton to the olefin or cyclo-olefin to produce a carbenium ion. The carbenium ion cracks to a smaller carbenium ion and a smaller olefin. These are the primary hydrocracking products. These primary products can react further to produce still smaller secondary hydrocracking products. The reaction sequence can be terminated at primary products by abstracting a proton from the carbenium ion to form an olefin at an acid site and by saturating the olefin at a metal site. Next, the cracking reaction occurs at a carbon–carbon bond that is beta to the carbenium ion charge. The beta position is the second bond from the ionic charge. Carbenium ions can react with olefins to transfer charge from one fragment to the other. In this way, charge can be transferred from a smaller hydrocarbon fragment to a larger fragment that can better accommodate the charge. Finally, olefin hydrogenation completes the cracking mechanism (Bricker et al., 2015).

Since hydrotreating and hydrocracking reactions are exothermic, the fluids exiting catalyst bed have to be cooled prior to entering the next catalyst bed to control reaction rate and allow a safe and stable operation. This is accomplished by injecting cold hydrogen for quench. Cold hydrogen gas, introduced in the interbed quench zones, is used to control reactor temperature and improve hydrogen partial pressure.

Furthermore, the temperature distribution in the cooled fluid entering the next catalyst bed has to be uniform to minimize the radial temperature gradients in successive catalyst beds. Unbalanced temperatures in a catalyst bed could result in different reaction rates at different points in the same bed. This can lead to different deactivation rates of the catalyst and, in worst cases, to temperature excursions and uncontrolled reaction rates. It is also important to achieve a good mass flow distribution across the catalyst bed. Design of the unit nominally targets a mass flux of 4000 lb/h/ft^2.

During operation, the hydrotreating and hydrocracking catalysts gradually lose their activity. In order to maintain constant HDN and conversion, the average bed temperature is gradually increased. The temperature increase in many cases is targeted small, less than 2 °F/month (1 °C/month) by proper selection of operating conditions. When the average bed peak temperature reaches a value close to the design maximum, the catalyst has to be replaced or reactivated.

In hydrocracking, activity is defined as the temperature required for obtaining a fixed conversion under certain process conditions. Hydrocracking conversion is usually defined, in equation (4.1), in terms of shifting true boiling point (TBP) range of feed, commonly referred to as a cut point.

$$\%\text{Net conversion} = \left(\frac{\left(\text{EP}^{+}_{\text{feed}} - \text{EP}^{+}_{\text{product}}\right)}{\text{EP}^{+}_{\text{feed}}} \right) \times 100 \tag{4.1}$$

where EP$^+$ indicates the fraction of material in the feed or product boiling above the desired cut point.

The hydrocracking mechanism is selective for cracking of higher carbon number paraffins (Bricker et al., 2015). This selectivity is due in part to a more favorable equilibrium for the formation of higher carbon number olefins. In addition, large paraffins adsorb more strongly. The carbenium ion intermediate results in extensive isomerization of the products, especially to *a*-methyl isomers, because tertiary carbenium ions are more stable. Finally, the production of C_1–C_3 hydrocarbon is low because the production of these light gases involves the unfavorable formation of primary and secondary carbenium ions. Other molecular species such as alkyl naphthenes, alkyl aromatics, and so on react via similar mechanisms, for example, via the carbenium ion mechanism. Steps involved in hydrocracking of paraffins are shown in Figure 4.6.

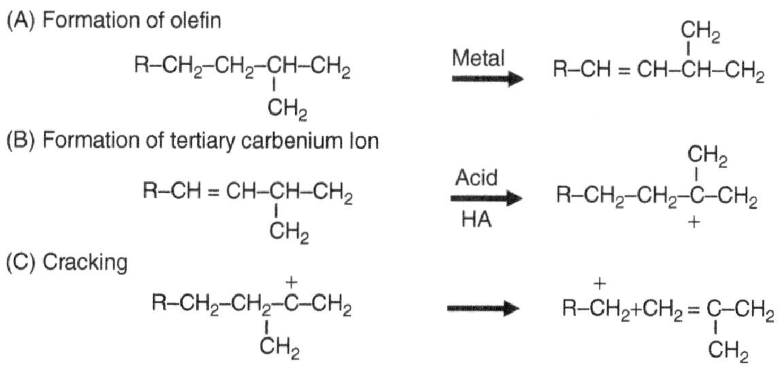

(A) Formation of olefin

$R{-}CH_2{-}CH_2{-}CH(CH_2){-}CH_2 \xrightarrow{\text{Metal}} R{-}CH{=}CH{-}CH(CH_2){-}CH_2$

(B) Formation of tertiary carbenium Ion

$R{-}CH{=}CH{-}CH(CH_2){-}CH_2 \xrightarrow[\text{HA}]{\text{Acid}} R{-}CH_2{-}CH_2{-}\overset{+}{C}(CH_2){-}CH_2$

(C) Cracking

$R{-}CH_2{-}CH_2{-}\overset{+}{C}(CH_2){-}CH_2 \longrightarrow R{-}\overset{+}{C}H_2 + CH_2{=}C(CH_2){-}CH_2$

FIGURE 4.6. Metal function.

(D) Reaction of carbenium Ion and carbon

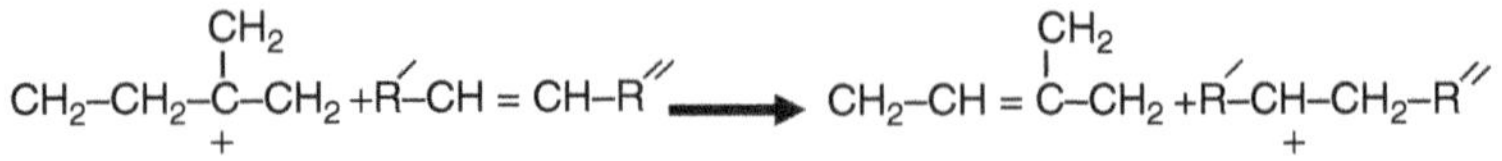

$$CH_2\text{–}CH_2\text{–}\overset{\overset{CH_2}{|}}{\underset{+}{C}}\text{–}CH_2 + R'\text{–}CH = CH\text{–}R'' \longrightarrow CH_2\text{–}CH = \overset{\overset{CH_2}{|}}{C}\text{–}CH_2 + R'\text{–}\underset{+}{CH}\text{–}CH_2\text{–}R''$$

(E) Olefin hydrogentation

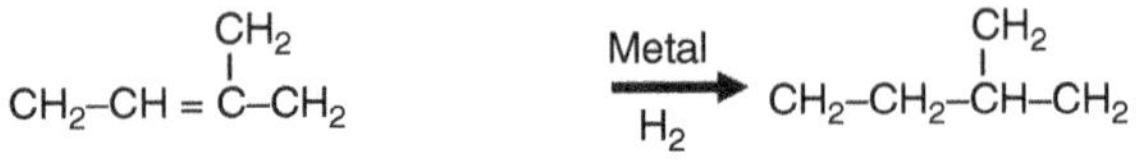

$$CH_2\text{–}CH = \overset{\overset{CH_2}{|}}{C}\text{–}CH_2 \xrightarrow[H_2]{\text{Metal}} CH_2\text{–}CH_2\text{–}\overset{\overset{CH_2}{|}}{CH}\text{–}CH_2$$

FIGURE 4.6. (*Continued*)

In summary, hydrocracking occurs through a bifunctional mechanism that involves olefin dehydrogenation–hydrogenation reactions on a metal site, carbenium ion formation on an acid site, and isomerization and cracking of the carbenium ion. The hydrocracking reactions tend to favor conversion of large molecules because the equilibrium for olefin formation is more favorable for large molecules and because the relative strength of adsorption is greater for large molecules. In hydrocracking, the products are highly isomerized, C_1 and C_3 formation is low, and single rings are relatively stable.

In addition to treating and hydrocracking, several other important reactions take place in hydrocrackers. These are aromatic saturation, PNA formation, and coke formation. Some aromatic saturation occurs in the treating section and some in the cracking section. Aromatic saturation is the only reaction in hydroprocessing that is equilibrium limited at higher temperatures that is reached toward the end of the catalyst cycle life. Because of this equilibrium limitation, complete aromatic saturation is not possible when reactor temperature is increased to make up for the activity loss due to coke formation and deposition on catalyst. The initial step, which generates an olefin or cyclo-olefins, is unfavorable under the high hydrogen partial pressure used in hydrocracking. The dehydrogenation of the smaller alkanes is most unfavorable. The concentration of olefins and cyclo-olefins is sufficiently high, and the conversion of these intermediates to carbenium ions is sufficiently fast so that the overall hydrocracking rate is not limited by the equilibrium olefin concentration.

PNA, also called polycyclic aromatics (PCAs), or polyaromatic hydrocarbons (PAHs) are compounds containing at least two benzene rings in the molecule. Normally, the feed to a hydrocracker can contain PNA with up to seven benzene rings in the molecule. The PNA formation is an important, though undesirable, reaction that occurs in hydrocrackers. Figure 4.7 shows the competing pathways for conversion of multiring aromatics. One pathway starts with metal-catalyzed ring saturation and continues with acid-catalyzed cracking reactions. The other pathway begins with an acid-catalyzed condensation reaction to form a multiring aromatic-ring compound. This molecule may undergo subsequent condensation reactions to form a large PNA.

The consequence of operating hydrocracking units in recycle mode is the creation and concentration of large PNA molecules that can contain more than seven aromatic rings. These are also called HPNA as the ring number increases to 11 or more. Understanding the formation and prevention of PNA and HPNA in hydrocracking is critical for good stable operation of a hydrocracking unit. This subject is discussed in

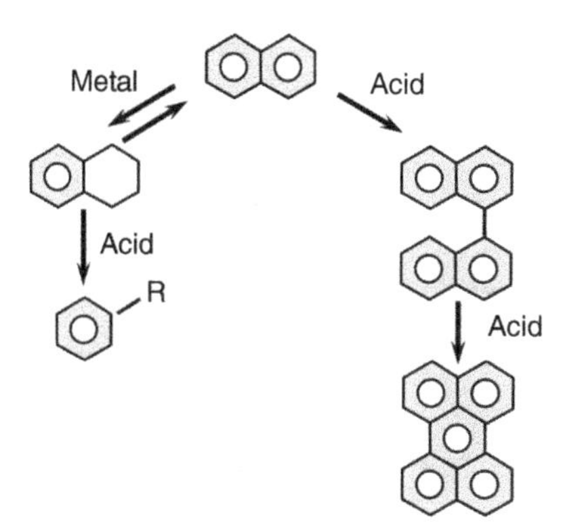

FIGURE 4.7. Competing pathways for conversion of multiring aromatics.

detail by Bricker et al. (2015) and Scherzer and Gruia (1996). The HPNA produced on the catalyst may exit the reactor and cause downstream equipment fouling or they may deposit on the catalyst when recycled and eventually form coke (catalyst deactivation). As reactor effluent is cooled, HPNA plate out on equipment resulting in fouling and plugging, leading to premature shut down. One possible way for HPNA mitigation is to allow a stream of 5–10% of the unconverted material bleed out of the hydrocracker for purging HPNA, resulting in lower-than-desired conversion of the feed. Technology also exists to adsorb HPNAs on a bed of adsorbent, thus reducing HPNA buildup while maintaining high conversion. Another option to reduce HPNA is to reduce endpoint of the feed.

4.6 CATALYSTS

As stated earlier, hydrocracking catalysts are dual functional having both an acid function and a metal function. The acid function from amorphous silica–alumina (ASA) and/or zeolitic material promotes cracking. The metal function, also sometimes referred to as hydrogenation function, is provided by one or more base metals. Base metal systems are usually composed of Ni, Mo, and W (Group VI A and VII A), and noble metal systems use Pt and/or Pd. Both metallic and acidic sites need to be present on the catalyst surface for the full suite of hydrocracking reactions to occur that include hydroisomerization, dehydrogenation, and dehydrocyclization. Base metal must be present in its sulfide form in order to be active. For optimum catalyst performances (i.e., good catalyst activity, selectivity, and stability), acid and metal functions must be in balance. Design of the catalyst using proper concentration and the type of acid and metal component result in desirable performance.

4.6.1 Acid Function of the Catalyst

Cracking and isomerization reactions take place on the acidic support. ASA provides the cracking function of amorphous catalysts and serves as support for the hydrogenation metals. Sometimes, ASA catalysts or a combination of ASA and zeolites can be used to produce high-yield distillate hydrocracking catalysts. Zeolites, particularly Y and beta, are commonly used to produce high-activity hydrocracking catalysts.

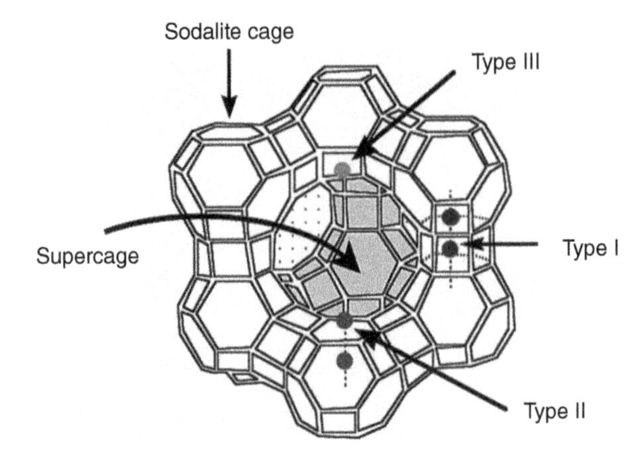

FIGURE 4.8. Commonly used Y zeolite structure.

Acid-treated clays, pillared clays, layered silicates, acid metal phosphates, and other solid acids have also been either used or investigated as acidic supports.

Advantages of ASA include its large pore, which permits access of bulky/large molecules of feedstock to the acidic sites, and moderate activity, which facilitates achieving the metal–acid balance needed for distillate selectivity.

Zeolites use became prevalent in hydrocracking catalysts because they provided high activity due to their higher acidity compared to the ASA materials. Zeolites are crystalline aluminosilicates composed of Al_2O_3 and SiO_2 tetrahedral units that form a negatively charged microporous framework structure (Scherzer and Gruia, 1996). A pictorial representation of commonly used Y zeolite structure is shown in Figure 4.8.

Hydrotreating and hydrocracking catalyst can come in cylindrical, trilobe, or quadrilobe shapes commercially (see Figure 4.9). Catalysts with multilobed cross sections have a higher surface-to-volume ratio than simple cylindrical extrudates. When used in a fixed bed, such shaped catalyst particles help reduce diffusion resistance, create a more open bed, and help reduce pressure drop.

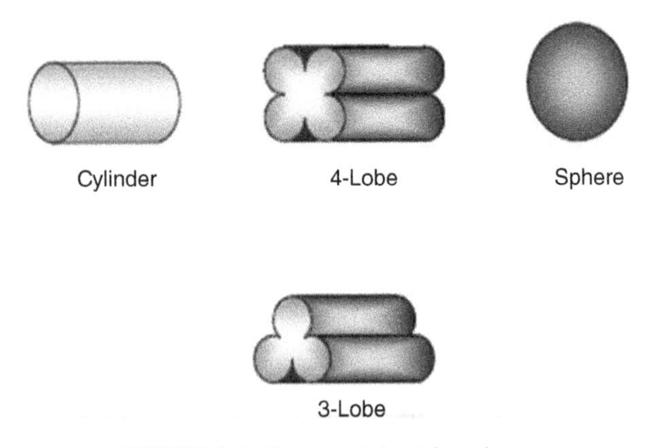

FIGURE 4.9. Commercial catalyst shapes.

4.6.2 Catalyst Activation

The metals on the catalysts as manufactured are in the oxide form. As mentioned previously, the active form of the base metals used in hydrocracking catalyst is the sulfide form. Base metal hydrocracking catalysts are activated by converting oxide form of metal to sulfide form. Several names are used for this treatment, such as sulfiding, presulfiding, presulfurizing in addition to the term activation. Ni/W is more difficult to sulfide than Ni/Mo; that is, they require higher temperatures. The catalyst presulfiding is accomplished mainly *in situ* though some refiners have started to do the activation outside the unit (*ex situ*). *In situ* sulfiding can be accomplished either in vapor or liquid phase. In vapor phase sulfiding, the activation of the catalyst is accomplished by injecting a chemical, which easily decomposes to H_2S. These chemicals may be dimethyl disulphide (DMDS) or dimethyl sulfide (DMS). *In situ* sulfiding occurs at temperatures between 450 and 600 °F (230–315 °C). Some catalyst manufacturers recommend sulfiding be conducted at full operating pressure, while others prefer it be done at pressures lower than the normal operating pressure. Ammonia injection is practiced during sulfiding of high-activity (high zeolite content) catalysts to prevent premature catalyst deactivation. Care must be taken to avoid exposing the base metal oxide to hydrogen at high temperatures in the absence of H_2S to avoid reduction to the base metal, which has minimal activity.

In the case of *ex situ* presulfurization of catalyst, sulfur compounds are loaded onto the catalyst. The activation actually occurs in the unit when the catalyst loaded in the reactor and heated up in the presence of hydrogen. The activation can be conducted either in vapor or liquid phase.

The noble metals, which are in oxide form on the catalyst, require activation by reducing metal in hydrogen. The activation of noble metal catalysts by hydrogen reduction occurs at 570–750 °F (300–400 °C).

4.6.3 Catalyst Deactivation and Regeneration

Catalyst deactivation is the gradual loss of the catalyst's activity while maintaining conversion severity. In practical terms, it is the temperature increase required to obtain a fixed conversion as the catalyst ages. As the run progresses, the catalyst loses activity, which may be due to one or more of the ways as described below.

4.6.4 Catalyst Coking

Coke deposition is a by-product of the cracking reactions. The deposition of coke on a catalyst is a function of time and temperature. The longer the catalyst is in service and/or the higher the temperature of the process, the deactivating effect will be more severe. The coking of the catalyst begins with the adsorption of high-molecular-weight hydrocarbon and low hydrogen/carbon ratio containing ring hydrocarbon. These are polynuclear aromatics (PNA and HPNA), discussed earlier, that cause coking on catalyst. The coke deposited on catalyst covers active sites and/or prevents access to these sites by physical blockage of the entrance to the pores leading to the loss of active sites. Coke is not a permanent poison. Catalyst, deactivated by coke deposition, can be easily restored to near original condition by controlled coke burning, that is, regeneration.

4.6.5 Reversible Poisoning/Catalyst Inhibition

Hydrocracking catalyst sometimes loses activity due to organic nitrogen or NH_3 poisoning. This is primarily the result of strong chemisorption on active sites. This type of poisoning is reversible, that is, when deactivating agent is removed, the deactivating effect is gradually reversed. If there is a temporary exposure to high ammonia, the catalyst will require a higher temperature to maintain its activity. However, when NH_3 concentration returns to normal value, that is, more typical value, the catalyst should return to the normal activity as the ammonia desorbs from the catalyst surface. Another example of a reversible poison is carbon monoxide, which can impair the hydrogenation activity by preferential adsorption on active metal sites.

4.6.6 Noncatalyst Metals Deposition

Some metals may come into the system via additives in the feed (i.e., decomposition product of silicon based antifoam compounds in delayed), or organometallic compounds in the feed such as Ni, V, and others Pb, Fe, As, P, Na, Ca, Mg. These compounds cause deposition at the pore entrances and near the outer surface of the catalyst. Such deposition effectively chokes off the access to the interior part of the catalyst, where most of the surface area resides thus causing deactivation. Metal deposition can damage acid sites, metal sites, or both. Deposition of metals is not reversible even with catalyst regeneration.

4.6.7 Catalyst Support Sintering

Catalyst support sintering is another reason for loss of catalyst activity and it also is irreversible. Sintering is a result of extremely high temperatures and is particularly damaging when it occurs in the presence of high water partial pressure. In this case, the catalyst support material can lose surface area from a collapse of pores.

4.6.8 Catalyst Regeneration

A coked catalyst is usually regenerated by combustion of carbon in controlled oxygen or air environment. Upon combustion, coke is converted into CO_2 and H_2O. In the absence of excess oxygen, CO may also form. Except for the noble metal catalysts, hydrocracking catalysts contain sulfur, as the metals are in the sulfide form. In the regeneration process, the sulfur will be emitted as SO_2. In general, sulfur oxide emission starts at a lower temperature than CO_2 emission. Regeneration of commercial catalysts can be done *in situ* or *ex situ*. The majority of commercial catalysts regenerations are performed *ex situ* because of environmental considerations.

4.7 KEY PROCESS CONDITIONS

Understanding of key process variables is critical for safe hydrocracking unit operation as the process is exothermic, generating heat and if not controlled it can lead to temperature excursion and runaway situation. The proper operation of the unit will depend on the understanding the interaction of process variables and controlled variation of the processing conditions. By careful monitoring of these process variables, the unit can operate safely to its full potential.

Typical operating conditions for hydrocracking unit range in LHSV of 0.5–2, conversion can be between 30% and 100%, unit pressure can range from 1400 to 3000 psig, recycle gas rate can be 8000–15,000 scfb. Combined feed ratio (CFR) for recycle unit can range from 1.2 to 2.0 CFR. The unit is generally designed for a given feed at design operating conditions with some operating flexibility. The following discussion provides an understanding of the impact of process variables changes on unit performance. Key process variables that impact catalyst performance with respect to catalyst activity, stability, product selectivity, and product quality are discussed by several authors (Bricker et al., 2015; Gary and Handwerk, 2001; Scherzer and Gruia, 1996; Ray, 2010; Parihar et al., 2012). These variables are as follows:

- Conversion
- Catalyst temperature
- Total pressure and hydrogen partial pressure (hydrogen purity)
- Recycle gas rate
- LHSV
- Feed quality.

4.7.1 Conversion

In hydrocracking, conversion is the key variable representing boiling point conversion and it is defined in several different ways. Conversion defines the severity of operation, which has impact on the catalyst temperature and catalyst stability. Conversion is defined at a specific cut point. Simplest conversion defined is gross conversion. Gross conversion depends only on percentage of UCO from the fractionator bottom (once-through or recycle operation). It does not account for the product range material, that is, cut point minus material in feed.

$$\text{Gross conversion, vol.\%} = 100 - \frac{\text{Fractionator bottom yield}}{\text{as vol.\% of fresh feed}}$$

The product fractionator sets the conversion cut point. Another more fundamental definition of conversion is net conversion, which takes into account product range material present in the feed. The net conversion is defined as the fraction of the feed cut point+ material converting into product. Thus, feed cut point+ material and unconverted cut point+ material in the product defines net conversion.

$$\%\ \text{Net conversion} = 100 - \left(\frac{(\%\ \text{Feed cut point} + -\%\text{Product cut point+})}{\%\text{Feed cut point+}}\right) * 100$$

For recycle mode operation, conversion per pass is true representation of the severity, a meaningful parameter for conversion kinetics and reactor severity measure. In recycle mode, the reactor feed is a blend of fresh feed and recycle oil. When combined feed goes through the reactor, a portion of this feed is converted while maintaining constant CFR. CFR is the volumetric ratio of fresh feed plus recycle feed to fresh feed. CFR of 1 is once-through operation and as CFR is increased in recycle operation at fixed gross conversion the conversion per pass across reactor drops.

Conversion is a measure of the reactor severity. Higher conversion requires higher temperature. For practical purposes, hydrocracking can be assumed to follow first-order kinetics. Conversion severity impacts product selectivity and product quality. As conversion increases, product selectivity and quality decline. Therefore, once-through high conversion produces a lower product selectivity compared to the recycle operation at high gross conversion. In recycle operation, conversion per pass determines selectivity, which is a function of CFR. A hydrocracking unit, when designed for recycle operation, is designed to operate at constant recycle rate (CFR). Higher conversion operation will directionally produce higher quality products due to increased saturation (assuming saturation is not equilibrium limited under more severe conversion conditions). Figure 4.10 summarizes effects of change in conversion on various aspects of hydrocracking unit performance.

4.7.2 Catalyst Temperature

The conversion level in the reactors is determined by several variables: the type of feedstock, space time or LHSV, partial pressure of hydrogen in the catalyst bed, and, most importantly, temperature of the catalyst. A high temperature in the reactor increases rate of reaction and, therefore, increases conversion. Since temperature increases as the reactor effluent passes through catalyst bed, maintaining a good

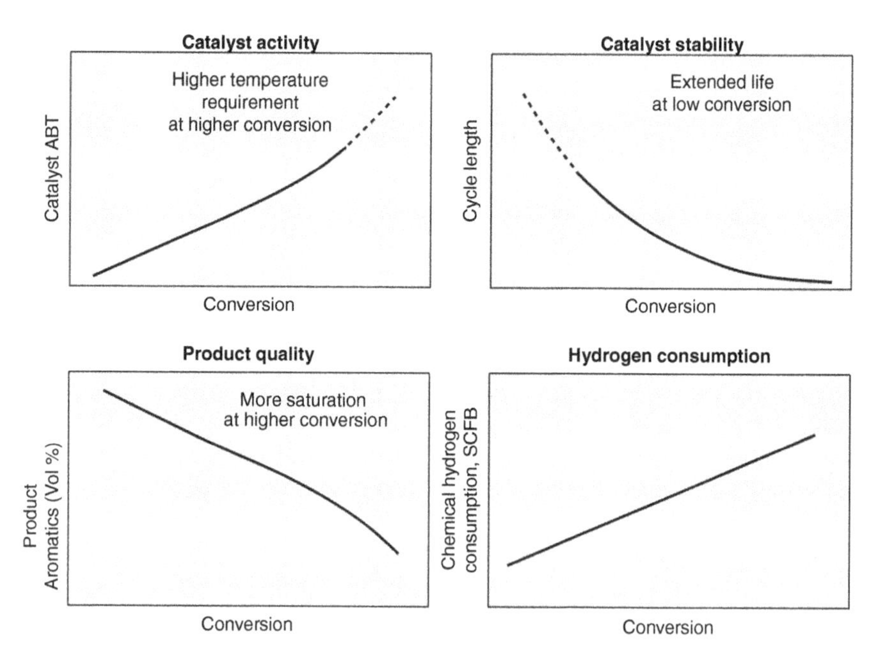

FIGURE 4.10. Effect of conversion on hydrocracking performance.

temperature control through catalyst bed is very important for stable operation. More heat is generated at higher conversion, so conversion needs to be controlled within allowable temperature rise limits and meet heat removal capability to avoid temperature excursion or a temperature runaway situation. A temperature runaway is a very serious, unsafe situation in hydrocracking unit since extremely high temperatures can be generated within a short period. These high temperatures can cause serious and potentially catastrophic damage to the catalyst and/or to the reactors. Good flow distribution devices (called reactor internals) are designed to achieve even liquid and gas flow through bed to promote mixing and thermal equilibrium between two fluids. To avoid temperature excursion, temperature guidelines have to be observed. These guidelines are dependent on the type of feedstock and type of catalyst and vary from catalyst supplier to catalyst supplier. But by and large, the temperature rise for noble metal catalyst is limited nominally to about 30 °F (17 °C). The temperature rise for high-activity base metal catalysts (for naphtha production) is limited to about 25 °F (14 °C) and for those with low-zeolite-content catalyst (such as middle distillate production) the temperature rise is limited to 40 °F (22 °C). Finally, maximum bed temperature rise of about 50 °F (28 °C) is recommended for amorphous catalysts. The maximum bed temperature rise of 50 °F (28 °C) is recommended for most pretreating catalyst bed. For optimum catalyst utilization, catalyst beds should be operated with equal catalyst bed peak temperatures.

For kinetic calculations, typically weight average bed temperature (WABT) is used as the reactor is adiabatic and has temperature profile due to exothermic reaction.

The rate of increase of the reactor WABT to maintain both hydrotreating and hydrocracking catalyst activity is referred to as the deactivation rate. The deactivation rate can be expressed in °F per barrel of feed processed per pound of catalyst (°C per m^3 of feed per kilogram of catalyst) or more simply it can be °F per day for the unit (°C per day). The loss in catalyst activity for hydrotreating catalyst is observed as increase in nitrogen level in the hydrotreating reactor effluent. For hydrocracking catalyst, a decrease in catalyst activity will be reflected in loss in conversion. To maintain catalyst performance over time, reactor WABT is gradually increased.

4.7.3 Total Pressure and Hydrogen Partial Pressure (Hydrogen Purity)

Total pressure is relevant for unit design, but for process performance hydrogen partial pressure is key parameter. Hydrogen partial pressure is based on the recycle hydrogen gas purity (H_2 partial pressure = recycle gas %H_2/100* total pressure). Recycle gas purity is dependent on feedstock, process severity (light end production), makeup H_2 purity, gas purge rate, and whether recycle gas scrubbing is employed. Feed quality and conversion severity determines how much fresh makeup gas will be entering the unit. Makeup hydrogen for hydrocracking typically can be high purity (i.e., >99%) H_2 that comes from PSA (pressure swing adsorption), but can also come from steam methane reformer, or catalytic reforming unit.

Hydrogen partial pressure impacts product quality (saturation) and catalyst stability (deactivation rate). It is always good to maintain maximum hydrogen partial pressure for maximum product quality and catalyst stability. Figure 4.11 gives a trend for relative deactivation rate change for change in pressure.

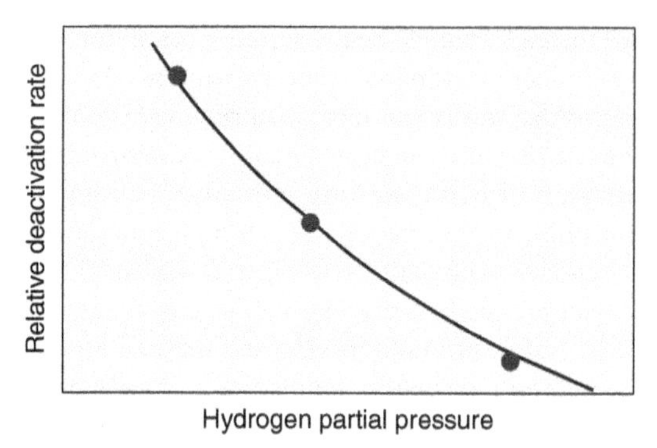

FIGURE 4.11. Relative deactivation rate change for change in pressure.

4.7.4 Recycle Gas Rate

Recycle gas rate is expressed as volumetric gas flow per volume of feed. So recycle gas rate can be expressed as standard cubic feet per barrel of fresh feed ($SCFB_{FF}$) or normal cubic meter per cubic meter of feed (nm^3/m^3). Figure 4.12 indicates effect of recycle gas rate change. Higher recycle gas rate improves catalyst activity and stability, resulting in increased catalyst life possible with increased gas rate.

4.7.5 LHSV (Fresh Feed Rate)

The amount of catalyst loaded into the reactor is based on the quantity and quality of the design feedstock and the desired conversion level. LHSV is defined as volumetric feed rate per unit volume of catalyst. Space time or the residence time is inverse of LHSV. Increasing the fresh feed rate increases severity on catalyst in terms of increased LHSV, and if recycle gas compressor capacity is limited, there may be lower recycle gas rate in SCFB. Increasing the severity will require higher WABT for the same conversion and thus will increase deactivation rate.

4.7.6 Fresh Feed Quality

The quality of the fresh feed has effect on catalyst activity (i.e., temperature required to achieve desired conversion). In general, increasing the amount of nitrogen and

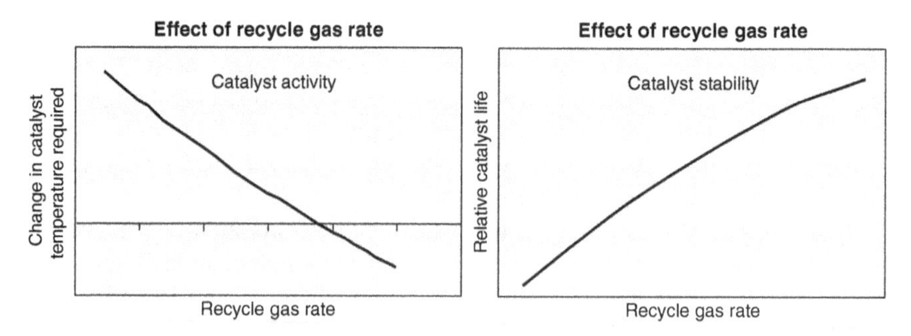

FIGURE 4.12. Effect of recycle gas rate.

sulfur content of the feed results in an increase in the severity of the operation, thus requiring an increase in temperature. The organic nitrogen compounds, which are converted to ammonia, have inhibiting effect on the hydrocracking catalyst activity thus resulting in higher temperature requirement. The amount of unsaturated compounds (such as olefins and aromatics) present in the feed has effect in increasing hydrogen consumption, thus increased heat release. In general, for a given boiling range feedstock, a reduction in API gravity (increase in specific gravity) indicates an increase in the aromaticity and, therefore, higher heats of reaction and higher hydrogen consumption. Presence of cracked feedstock such as one derived from catalytic cracking or thermal cracking can have higher contaminants such as sulfur, nitrogen, aromatics, and particulates. These components are harder to process and impact product quality negatively.

4.8 TYPICAL PROCESS DESIGNS

Hydrocracking is one of the important processes in the refinery contributing to the profitability. The versatility of the process is in its capability to use a wide variety of low-value feed stream in the refinery and converting into high-value transportation fuels. The process uses various conventional flow schemes as described earlier. There is once-through design for partial conversion and recycle design for high conversion. The high conversion recycle flow scheme can be designed as single-stage, two-stage, and separate hydrotreat flow scheme. These flow schemes are selected based on the feed quality, feed rate, and conversion level desired. The selection of catalyst type used in these flow schemes allows to maximize desired product slate.

Since hydrocracking is one of the key processes in the refinery, there is continuous development of new catalysts and flow schemes. In recent past, as distillates have become important products requiring ULSD quality, several new schemes have been developed. Effort is made to describe some of them in this section.

Once-through option is used in many cases while trying to prepare good hydrotreated feed for FCC. While FCC feed requirement is low sulfur, nitrogen and low aromatics intent is to maximize UCO quantity (typically 700 °F+/370 °C+ is FCC feed). In such case small amount of distillate is produced that typically use to be blended in refinery diesel pool (pre ULSD era). Since recent ULSD requirement there is not much room for blending if sulfur level is high, which will be the case for low conversion conventional hydrocracking unit. To overcome such a deficiency, an innovative flow is described by Thakkar et al. (2007) that assures ULSD product at low conversion. This process, developed by UOP, is called advanced partial conversion Unicracking™ process. A schematic flow diagram is shown in Figure 4.13. The concept incorporates posttreating reactor to treat only the hydrocracked product lifted by enhanced hot separator (EHS).

A similar concept is employed by Meister and Kokayeff (2012) to produce ULSD product from low-pressure partial conversion hydrocracking unit. Figure 4.14 shows a flow scheme that incorporates finishing reactor in single gas and pressure loop.

A variation of two-stage flow scheme that is available for high conversion is HyCycle Unicracking process described by Thakkar et al. (2007). This flow scheme is based on understanding of effect of conversion per pass to improve distillate selectivity.

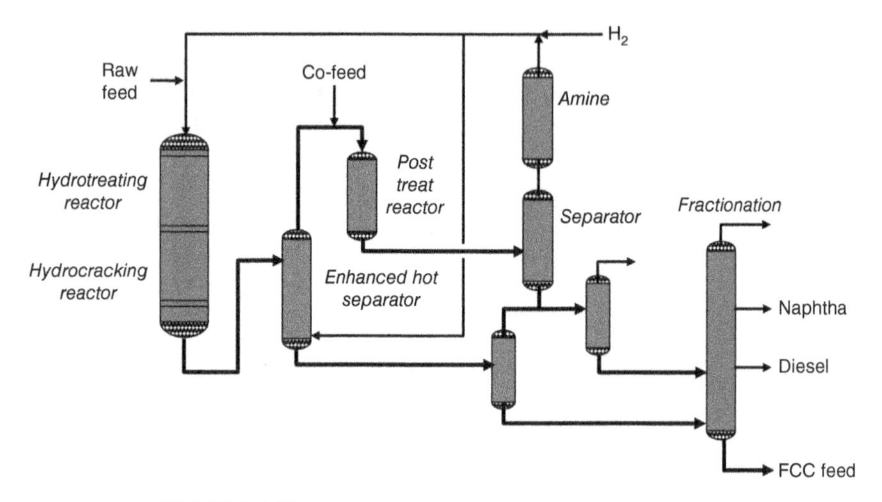

FIGURE 4.13. Advanced partial conversion hydrocracking.

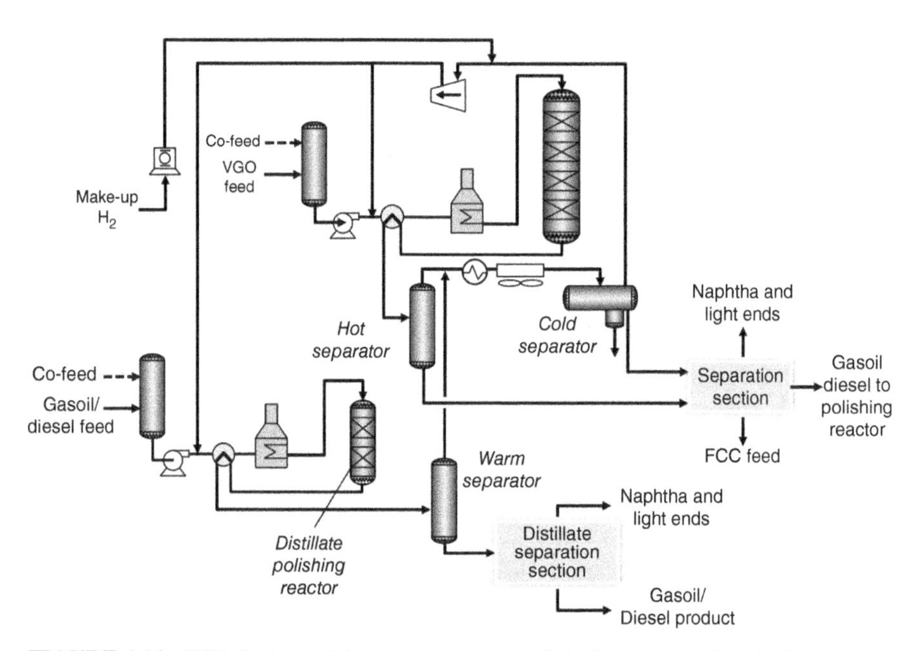

FIGURE 4.14. Mild hydrocracking – incorporates finishing reactor in single gas and pressure loop.

A higher CFR is required to obtain lower conversion per pass design, which is expensive proposition in conventional design due to increased recycle oil rate. This is overcome in HyCycle process by incorporating enhanced hot high pressure separator (EHS). Thus, keeping most of the recycle stream in high pressure loop at high temperature resulting in better efficiency than conventional route where recycle needs to be depressured and cooled and then need to be pressured up to recycle. Also, inhibiting

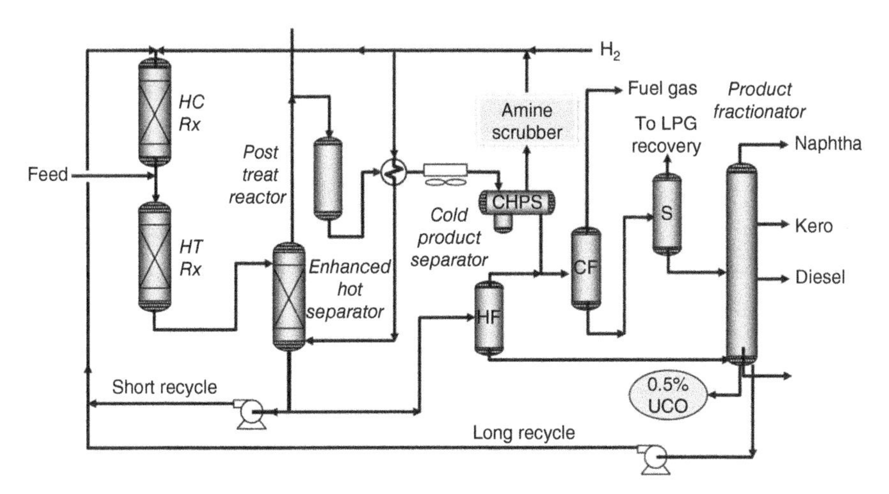

FIGURE 4.15. One pressure loop and one recycle gas loop for cost-efficient design.

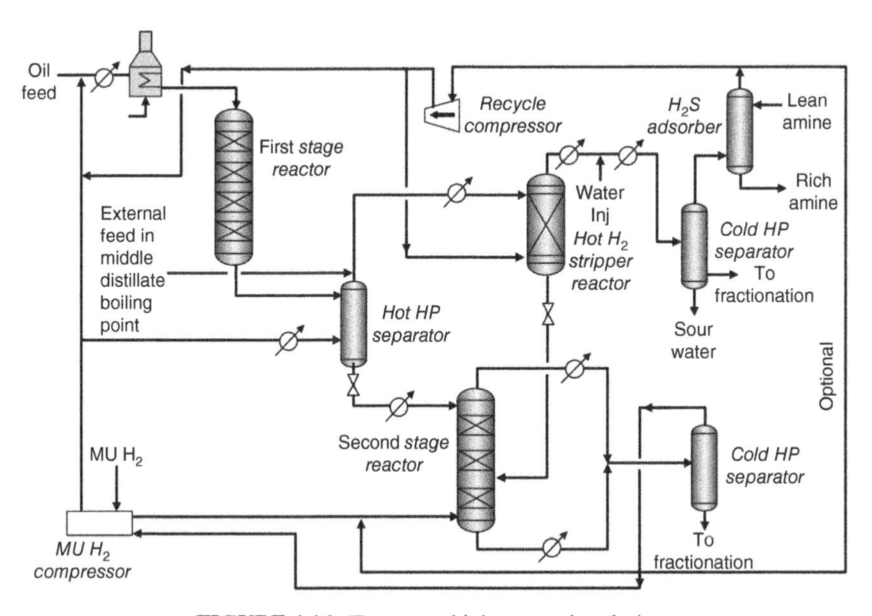

FIGURE 4.16. Two-stage high conversion design.

effects of NH_3 are removed for back-flow recycle reactor. The flow scheme in Figure 4.15 uses one pressure loop and one recycle gas loop for a cost-efficient design.

Another version of two-stage high conversion design described in US patent by Mukherjee et al. (2004) is shown in Figure 4.16. Here, second stage is at lower pressure than first stage, and for first stage product separation a hot hydrogen reactor stripper is included.

REFERENCES

Bricker ML, Thakkar V, Petri J (2015) Hydrocracking in petroleum processing, in Treese SA, Pujado PR, Jones DSJ (eds), *Handbook of Petroleum Processing*, 2nd edition, pp. 317–359, Springer International Publishing.

Gary JH, Handwerk GE (2001) *Petroleum Refining – Technology and Economics*, Marcel Dekker, Inc.

Meister J, Kokayeff P (2012) Mild Hydrocracking: New Challenges for a Mature Technology, Technology and More – 2012 Winter, UOP News letter, http://www.uop.com/mild-hydrocracking-challenges-mature-technology/?trk=profile_certification_title (accessed 14 July 2016).

Mukherjee UK, Louie WSW, Dahlberg AJ (2004) Hydrocracking process for the production of high quality distillate from heavy gas oil. US Patent 6,797,154, Sep. 28.

Parihar P, Voolapalli RK, Kumar R, Kaalva S, Saha B, Viswanathan PS (2012) Optimise hydrocracker operations for maximum distillates, *Petroleum Technology Quarterly* **Q2**, www.digitalrefining.com/article/1000369.

Ray D (2010) PetroFed Presentation, Aug. 28, https://www.yumpu.com/en/document/view/42466663/debangsu-raypdf-petrofedwinwinho (accessed 14 July 2016).

Rossi RJ, Anderle CJ, Abou Chedid NK (2007) Unicracking technology for flexibility and maximizing diesel production. Asian Refining Technology Conference 10th Annual Meeting, March 20–22, 2007, Bangkok.

Scherzer J, Gruia AJ (1996) *Hydrocracking Science and Technology*, Marcel Dekker, Inc.

Thakkar V, Ackelson D, Rossi R, Dziabala B, McGehee J (2007) Hydrocracking for heavy feedstock conversion, *Petroleum Technology Quarterly*, **Q4**, 33–40.

Vermeiren W, Gilson JP (2009) Impact of zeolites on the petroleum and petrochemical industry, *Topics in Catalysis*, **52**(9), 1131–1161.

PART 2

HYDROPROCESSING DESIGN

5

PROCESS DESIGN CONSIDERATIONS

5.1 INTRODUCTION

A safe, reliable, and economical design relies on many decisions made throughout the design process. Some of them are obvious, some are more subtle and only become apparent after some years of experience during which the better design alternative is proven in actual operation. The designer frequently encounters trade-offs between competing constraints or criteria. These must be addressed in a manner that results in the best overall design taking into account all the relevant factors. The designer must also be aware of developing technology as well as continually championing new process concepts that would offer better alternatives to achieve the design objectives.

High-quality ultra–low-sulfur diesel (ULSD) can be produced either by hydrotreating of distillate range material or by catalytic cracking of heavier gas oils. General design aspects of both processes will be covered.

5.2 REACTOR DESIGN

The reactor is arguably the most critical component in the process, in that this is where the essential reactions take place. Specific catalysts are chosen to achieve the process objectives, but efficient and effective use of the catalyst is critical if the yields, product quality, and required run length are to be achieved with a minimum of catalyst volume. Thus, the design of the reactor plays a central role in the proper functioning of a hydroprocessing unit, as well as any catalytic process. The focus of

Hydroprocessing for Clean Energy: Design, Operation, and Optimization, First Edition.
Frank (Xin X.) Zhu, Richard Hoehn, Vasant Thakkar, and Edwin Yuh.

all aspects of reactor design must be the most effective use of the catalyst. In order to accomplish this, the designer must address a number of issues related to the design of the reactor vessel, its internals, the method of catalyst loading, and other important mechanical details.

5.2.1 Flow Regime

A great many studies have been published related to flow in packed beds (Ergun, 1952; Midoux et al., 1976; Larkins and White, 1961; Gianetto et al., 1978). With the exception of lighter feeds, such as naphtha, most hydroprocessing reactions will take place in the two-phase flow regime, meaning that both liquid and vapor will be present in the catalyst bed. Similar to two-phase flow in pipes, two-phase flow in catalyst beds exhibit different flow regimes, according to the amounts of liquid and vapor present. One flow map that has been developed (Charpentier and Favier, 1975) shows various flow regions for two-phase flow in packed beds. A simplified version is shown in Figure 5.1. The parameter $(L/G)\lambda\psi$ versus G/λ is plotted. The symbols are defined as follows:

L = liquid superficial mass velocity, kg/m^2 s
G = vapor-phase superficial mass velocity, kg/m^2 s

$$\lambda = \text{flow parameter}: \quad \lambda = \left[\frac{\rho_L}{\rho_{wat}} \frac{\rho_G}{\rho_{air}}\right]^{0.5} \tag{5.1}$$

$$\psi = \text{flow parameter}: \quad \psi = \frac{\sigma_{wat}}{\sigma_L}\left[\frac{\mu_L}{\mu_{wat}}\left(\frac{\rho_{wat}}{\rho_L}\right)^2\right]^{0.33} \tag{5.2}$$

$\rho_G, \rho_L, \rho_{wat}$ = density of gas, liquid, and water
μ_G, μ_L, μ_{wat} = viscosity of gas, liquid, and water
σ_L, σ_{wat} = surface tension of liquid and water

Several data points were generated from simulations of two distillate hydrotreating units operating at about 50 and 100 bar reactor inlet pressure, as well as a hydrocracking unit operating at 170 bar reactor inlet pressure and have been placed on the flow map. As shown in Figure 5.1, the conditions that exist in the reactor are generally in the trickle flow region. This indicates that on a volumetric basis, the gas phase predominates. Thus, the design concepts discussed in this chapter will assume a trickle flow reactor.

5.2.2 Design Flux

The first critical parameter to be determined is the diameter of the reactor. The usual means to set this value is by setting the design flux. Flux is defined as the mass of material flowing over a given cross-sectional area of the catalyst bed in a fixed length of time. Therefore, units are in mass per area per time. Flux is closely related to other parameters, such as pressure drop, as well as catalyst wetting efficiency and flow regime for a two-phase system.

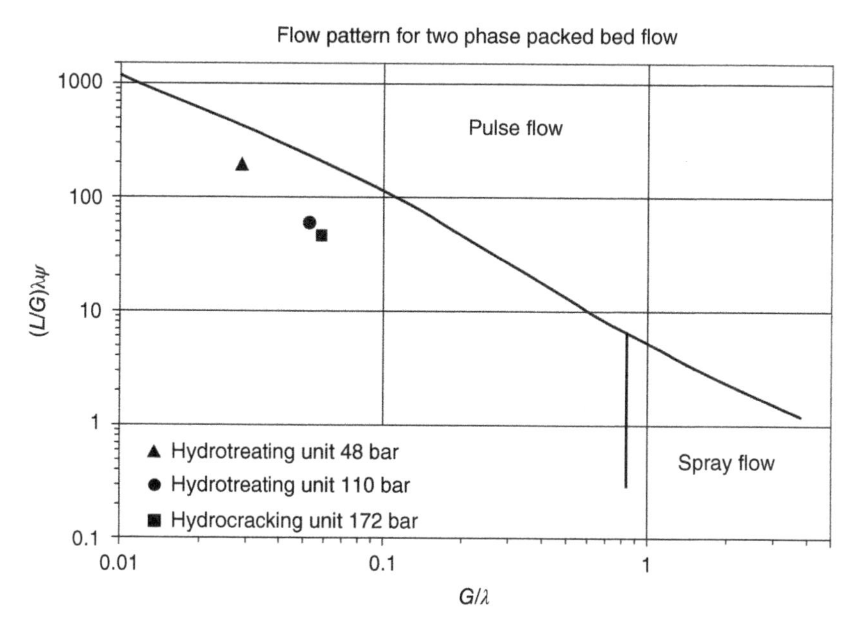

FIGURE 5.1. Flow map for two-phase packed bed reactor system.

For reactions carried out entirely in the vapor phase, the choice of flux is an economic trade-off between reactor cost and utility cost. Higher flux results in higher catalyst bed pressure drop. At the same time, higher flux results in smaller reactor diameters, which reduces the wall thickness and actually decreases the weight of the reactor for a given catalyst volume. For most feeds, the normal range of fluxes tends to be in the 3000–5500 lb/h ft^2 range (125–230 kg/h m^2).

For two-phase flow, the situation is more complicated. While the reactor diameter is still determined by a characteristic flux value, there is some variation in how the flux for a two-phase system is determined. Some designers use the total (liquid and vapor) mass flow, while others use only the liquid mass flow. Either basis has been successfully applied. The flux value chosen for design affects the distribution of vapor and liquid in the beds. Distribution refers to the degree of evenness that both the vapor and liquid material are dispersed in the catalyst bed. Poor distribution leads to ineffective use of the catalyst, with the result that run lengths will be decreased and/or product quality will suffer. In extreme cases, hot spots can be generated due to areas of the catalyst bed that are relatively stagnant due to poor distribution. The stagnant areas of the bed create regions that have a low space velocity and an inability to dissipate heat generated from the exothermic reactions. Since the reaction rate increases with increasing temperature, hot spots can sometimes lead to uncontrolled temperature excursions and the generation of coke agglomerated catalyst ("coke balls"). Either situation is undesirable. In general, low mass fluxes lead to poor distribution and therefore poor performance. Higher mass fluxes generally lead to better distribution, but at the cost of higher catalyst bed pressure drop. While a number of factors can influence distribution in the catalyst bed, it is generally acknowledged that a minimum flux must be applied particularly for a

TABLE 5.1. Flux Values for Hydroprocessing Reactors

Unit Type	Flux Range	
	lb/h ft^2	kg/h m^2
Hydrotreating (distillate, VGO)	3500–4500	146–188
Hydrotreating (residuum)	2000–3000	83–125
Hydrocracking	4000–5000	167–208

two-phase system to achieve good distribution. While no definite standards have been established, some observations can be made about the range of values that have been applied to the design of hydroprocessing units. These are included in Table 5.1.

The above values are not meant to be limiting but represent a range of values that have been applied in various applications. Lower fluxes reduce pressure drop through the reactor, and thus operating costs, but the reactor walls will be thicker compared with that of higher flux designs, resulting in a higher weight. Since the cost of the reactor is largely a function of the weight of the vessel, lower flux reactor designs will tend to cost more than higher flux designs. This effect is illustrated in Table 5.2. This is a tabulation of reactor size for a distillate hydrotreating unit of 40,000 BPSD capacity with an operating pressure of 48 bar. The values are tabulated as a function of reactor inside diameter, since most reactors and other vessels sized to the nearest 6 in. (100 mm). Therefore, there are two competing factors: reactor design flux (pressure drop) and operating costs. The designer can optimize this to a certain extent, but the focus must be on optimum reactor performance by selecting the proper flux.

For very heavy feeds, such as reduced crude and vacuum residuum, there are two complicating factors. First, the viscosity of the feed is much higher than for lighter feeds.

It is noticed from Table 5.1 that residuum hydrotreating flux values are significantly lower than other applications. The main reason for this is that a compromise must be made between the overall pressure drop and bed distribution. Residuum hydrotreaters feature both very low space velocities and very heavy viscous feeds, both of which can combine to create a rather large total catalyst bed pressure drop. Therefore, the applied flux tends to be lower as a means of mitigating this. It does, however, make such units more susceptible to coke ball formation and large bed temperature variations, particularly at values below the minimum flux value. This may be

TABLE 5.2. Effect of Flux on Reactor Size and Weight

Inside Diameter (mm)	Flux		Tangent Length (mm)	Weight (MT)	Wall Thickness (mm)
	lb/h ft^2	kg/h m^2			
3,500	5,400	225	29,900	302	85
4,000	4,150	173	23,300	316	97
4,500	3,300	137	18,200	332	108

mitigated to a certain extent by careful attention to internals design, proper loading, and efficient fines removal.

5.2.3 Pressure Drop

The most commonly used correlation for pressure drop was developed by Ergun (1952), which has the following general form:

$$\Delta P = \frac{150\,(1-\varepsilon)^2 \mu\, U}{\varepsilon^3 {D_\mathrm{p}}^2 g_\mathrm{c}} + \frac{1.75\,(1-\varepsilon)\,\rho\, U^2}{\varepsilon^3 D_\mathrm{p}\, g_\mathrm{c}}, \tag{5.3}$$

where

ΔP = pressure drop across the bed per unit length
D_p = equivalent spherical diameter of the catalyst
ρ = density of fluid
μ = viscosity of the fluid
U = superficial velocity (i.e., the velocity that the fluid would have through an empty vessel)
ε = void fraction of the bed (volume not occupied by catalyst particles)
g_c = gravitational constant

This equation works well for predicting the pressure drop of a single-phase material flowing through a packed bed of spherical particles. However, modern hydroprocessing catalysts are rarely manufactured in spherical form. Therefore, it is necessary to determine the diameter value to use for a nonspherical particle to obtain an accurate pressure drop value when using the Ergun equation. This involves estimating an equivalent particle diameter. A method has been proposed (McCabe and Smith, 1967a) for determining the equivalent diameter of a particle according to the following equation:

$$D_\mathrm{p} = \frac{6\,V_\mathrm{p}}{s_\mathrm{p}}, \tag{5.4}$$

where

V_p = volume of particle
s_p = surface area of particle

Using this relation, a 1.5 mm extrudate catalyst having a length to diameter ratio of 3:1, the equivalent diameter would be calculated as follows:

$$V_\mathrm{p} = \frac{\pi}{4}\, 1.5^2 \times 4.5, \tag{5.5}$$

$$s_\mathrm{p} = \pi 1.5\ \times 4.5, \tag{5.6}$$

$$D_\mathrm{p} = \frac{6\,V_\mathrm{p}}{S_\mathrm{p}} = 2.25\,\mathrm{mm}. \tag{5.7}$$

Note that in this equation for D_p, the length of the particle cancels out and the equivalent diameter is related only to the diameter.

The void fraction is a function of the catalyst loading method and the catalyst shape. It is a function of the amount of the actual catalyst that was loaded into a given volume of the reactor according to the following relation.

$$\varepsilon = 1 - \left(\frac{V_c}{V_r}\right), \tag{5.8}$$

where

ε = void fraction
V_c = volume of catalyst loaded
V_r = volume of reactor into which the catalyst was loaded

The actual void fraction can be as much as 0.5 for sock loaded shaped catalyst, to as low as 0.35 for dense loaded spherical catalyst. It also varies slightly for each loading, even when the same loading method is used, although dense loading usually results in the least variation in void volume.

An additional complicating factor for reactors operating in the two-phase flow region is how to account for the two phases flowing simultaneously. The Ergun equation is still the basis for calculating the pressure drop, but an adjustment must be made to account for the presence of two separate phases in the catalyst bed. A number of correlations have been developed, among them by Larkins and White (1961), Charpentier (1978), and Tosun (1984). Since reactor pressure drop is one of the critical design parameters, an example is provided below for determining the two-phase pressure drop using the Ergun–Tosun method.

Input Data:

Reactor ID, ft	$D = 16.0$
Catalyst nominal diameter, in.	$D_p = 0.0825$
Catalyst bed void fraction	$\varepsilon = 0.355$
Bed volume, ft^3	$V_c = 1250$
Gravitational constant, ft/s^2	$g_c = 32.17$

Gas Properties

Flow, lb/h	$W_g = 250{,}450$
Density, lb/ft^3	$\rho_g = 4.588$
Viscosity, cP	$\mu_g = 0.354$

Liquid Properties

Flow, lb/h	$W_l = 459{,}872$
Density, lb/ft^3	$\rho_l = 38.455$
Viscosity, cP	$\mu_l = 0.354$

Calculations:

(1) Calculate superficial gas and liquid velocities for each phase:

$$U_g = \frac{W_g}{3600\left(\frac{\pi}{4}\right)D^2\rho_g} = 0.075\,\text{ft/s}, \tag{5.9}$$

$$U_l = \frac{W_l}{3600\left(\frac{\pi}{4}\right)D^2\rho_l} = 0.017\,\text{ft/s}. \tag{5.10}$$

(2) Calculate the gas only and liquid only pressure drops using the Ergun equation.

$$\Delta P_g = \frac{150\,(1-\varepsilon)^2\mu_g\,0.000672\,U_g}{\varepsilon^3\left(\frac{D_p}{12}\right)^2 g_c} + \frac{1.75\,(1-\varepsilon)\rho_g\,{U_g}^2}{\varepsilon^3\left(\frac{D_p}{12}\right)g_c}$$

$$= 3.814\,\text{lb/ft}^2\text{/ft catalyst bed}, \tag{5.11}$$

$$\Delta P_l = \frac{150\,(1-\varepsilon)^2\mu_l\,0.000672\,U_l}{\varepsilon^3\left(\frac{D_p}{12}\right)^2 g_c} + \frac{1.75\,(1-\varepsilon)\,\rho_l\,{U_l}^2}{\varepsilon^3\left(\frac{D_p}{12}\right)g_c}$$

$$= 4.803\,\text{lb/ft}^2\text{/ft catalyst bed}. \tag{5.12}$$

(3) Next parameter χ is calculated.

$$\chi = \left(\frac{\Delta P_l}{\Delta P_g}\right)^2 = 1.122. \tag{5.13}$$

(4) The parameter Φ_l is calculated

$$\Phi_l = 1 + \frac{1}{\chi} + \frac{1.424}{\chi^{0.576}} = 3.224. \tag{5.14}$$

(5) The two-phase pressure drop is calculated as follows:

$$\Delta P_{lg} = {\Phi_l}^2\,\frac{\Delta P_l}{144} = 0.347\,\text{psi/ft of catalyst bed length}. \tag{5.15}$$

5.2.4 Number of Catalyst Beds

Once the reactor diameter has been selected, the designer must determine the number and length of catalyst beds. Since hydrotreating and hydrocracking reactions are generally exothermic, there will be a temperature rise as the feed is processed in the catalyst bed and converted into the desired products. If the heat of reaction is high enough, the resulting temperature rise will exceed a desired value. It is then necessary to cool the reactor effluent before passing the material over the remaining catalyst. The allowable temperature rise is generally a function of the catalyst type and, to some extent, the type of reactions involved. Lower activity hydrotreating catalysts performing hydrogen addition reactions can tolerate a higher temperature rise than the more active cracking catalysts. The catalyst manufacturer will designate the allowable temperature rise. Once this value is known, the number of required reactor beds is a function of the heat of reaction.

One reason for limiting the temperature rise is to minimize undesirable side reactions, such as coking of the catalyst, which is more likely to occur at elevated temperatures, with resultant decrease in catalyst life. For units processing naphtha that would be further processed in a catalytic reformer that usually requires a feed sulfur content of less than 0.5 wt ppm, it is desirable to limit the maximum catalyst bed temperature to less than about 650 °F (323 °C) to prevent recombinant mercaptan formation.

Units processing thermally cracked materials, such as FCC light cycle oil (LCO) or coker gas oil, contain large amounts of olefins, and this will translate into large temperature increases as the material is processed in the reactor. In this case, multiple catalyst beds utilizing recycle gas as the quench medium would be employed. However, for some materials, such as olefinic naphtha, the heat of reaction is such that the amount of quench gas would be uneconomical. In this case, recycling of treated product to act as a heat sink and limit bed temperature rise has been practiced. Product recycling, however, has the disadvantage of increasing the hydraulic capacity of the unit. Therefore, the designer must evaluate the economics of each available option.

For hydrotreating reactors that have multiple catalyst beds, it is important to know how nonlinear the heat of reaction curve is. The various reactions that take place all have a certain reaction rate and this determines where and how much heat is generated in the catalyst bed. For example, olefin saturation reactions take place rapidly and generate a substantial amount of heat. Therefore, processing highly unsaturated feeds will require that the catalyst beds be progressively longer.

By contrast, hydrocracking reactions are typically more linear and cracking catalyst beds tend to be of equal length.

If the reactor is to have more than one bed, it is necessary to determine the inlet temperature for each bed. In order to do this, the designer must keep in mind a parameter called the weighted average bed temperature (WABT). This is an important parameter used to judge the performance of the catalyst. Since the pilot plants used for testing catalyst tend to be small and operate under adiabatic conditions, the performance data is recorded at the temperature at which the entire bed is maintained. That, of course, is not usually achievable in the actual operation of a unit. However, since pilot plant data is used to predict catalyst performance, the catalyst supplier will usually supply the required reactor temperature as a WABT.

There are two basic ways by which the WABT is determined: the first can be stated simply as follows:

$$\mathrm{WABT} = \sum_{i=1}^{n} \frac{(T_\mathrm{i} + T_\mathrm{o})W_i}{2}, \tag{5.16}$$

where

n = number of reactor beds
T_i = reactor bed inlet temperature
T_o = reactor bed outlet temperature
W_i = weight fraction of total catalyst in reactor bed i

An alternative method for calculating WABT is as follows:

$$\mathrm{WABT} = \sum_{i=1}^{n} \left(\frac{1}{3}T_{\mathrm{i}} + \frac{2}{3}T_{\mathrm{o}}\right) W_i. \tag{5.17}$$

This second method takes into account the fact that some hydroprocessing reactions are nonlinear resulting in a somewhat rapid initial temperature rise followed by a more gradual temperature rise near the end of the catalyst bed. The particular method to be used may be customary to a given catalyst vendor or catalyst type, and it is important to determine which method to use so that the catalyst performance can be properly judged.

5.2.5 Reactor Internals

A great deal of work has been done over the years to develop reactor internals that allow for the most efficient use of the catalyst. While a number of proprietary designs have evolved, the basic principles have largely remained the same. These include the following:

(1) *Dissipate the momentum of the material entering the reactor.*

It is a device usually mounted in the inlet nozzle of the reactor, often composed of vanes or a perforated plate with outlets or openings geared toward dispersing the exiting material as widely as possible. It is important to insure that the distance between the inlet distributor device and the top support material is not too short; otherwise, the support material loaded above the catalyst bed can be displaced and the catalyst bed will be fluidized causing a large amount of catalyst fines to be generated. In time, this will lead to maldistribution as seen by a large temperature spread in the thermocouples located in the catalyst bed, as well as a large pressure drop buildup in the bed, eventually requiring a shutdown to remove the fines laden catalyst. If the reactor is operating under vapor-phase conditions, the inlet distributor usually serves to provide adequate dispersion of the reactor inlet material.

(2) *Collect the liquid and distribute it along with the vapor evenly across the reactor cross section.*

For systems operating in the two-phase region, this is important because liquid does not disperse axially in the catalyst bed nearly as evenly as vapor. It is critical that the liquid and vapor distribution across the top of the catalyst bed be as equal as possible, because the catalyst bed itself does not promote further distribution. In fact, if care is not taken in the loading of the catalyst, or if excessive fines are present, the catalyst bed itself can actually contribute to poor distribution and therefore poor catalyst performance.

A number of proprietary vapor–liquid distribution devices have been developed over the years. Of the devices that have been developed, two main types of designs have evolved. These can be broadly categorized as gravity flow (Figure 5.2) and vapor lift (Figure 5.3). Both designs rely on a collection tray on which a liquid level is first established.

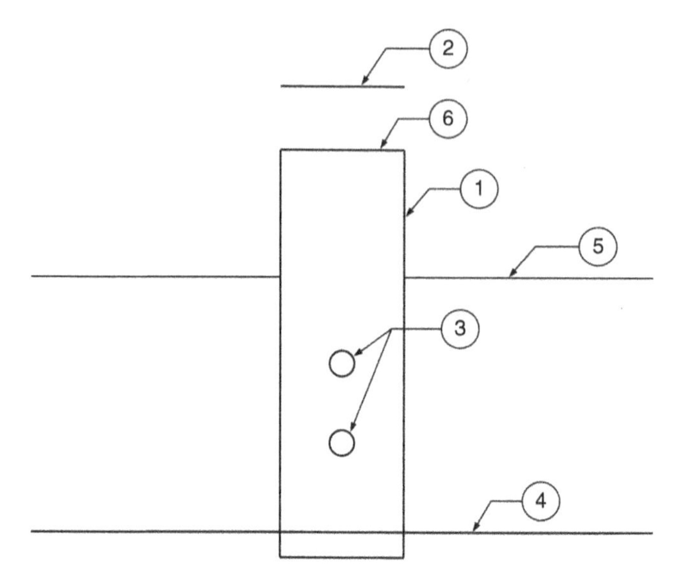

FIGURE 5.2. Example of gravity flow distributor.

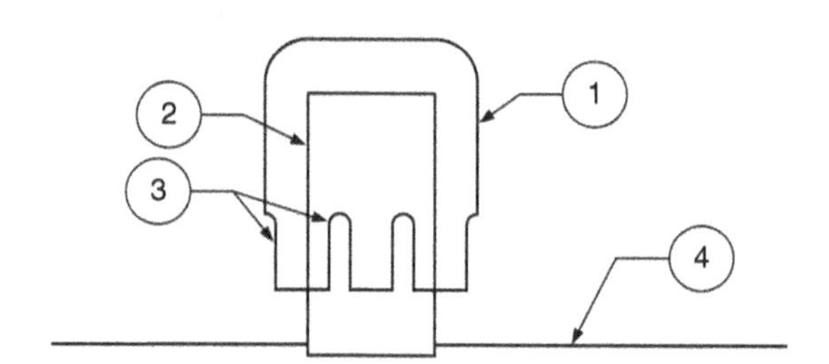

FIGURE 5.3. Example of vapor lift distributor.

Gravity flow distribution devices feature a series of pipes (1) installed in the tray (4). The pipes are provided with a cover (2) to minimize liquid entry into the vapor inlet. The pipes are usually fabricated with one or more holes (3) through which liquid passes and joins with the vapor that entered at the top of the pipe. The holes are sized so that a liquid level (5) is maintained on the tray (4) at a point above the bottom hole and below the top of the pipe for the various operations that the unit will undergo. The holes may be oriented either horizontally as shown or vertically. In the latter case, the holes provide a better ability to vary the amount of liquid entering the device due to variations in liquid rate that can occur as the reactor bed temperatures are increased from start of run to end of run. This also provides for accommodating different feed cases. The pipes sometimes project a certain distance below the tray floor (4). The liquid drips onto the catalyst bed below and the vapor and liquid flow through the catalyst bed.

Recent advances in this type of device have featured restrictions placed in the throat of the chimney to increase pressure drop and various devices placed at the bottom of the pipe, the result of which is to induce a liquid spray pattern that distributes more effectively over the cross section of the catalyst bed below.

Vapor lift distributors operate on a different principle. These devices generally consist of a cap (1) mounted over a riser pipe (2), which is inserted into a hole in the distribution tray (4). A liquid level is still established on the tray, but instead of the vapor and liquid passing through the device via separate openings, the vapor and liquid pass cocurrently through the device. The vapor entrains the liquid and both pass upward inside the annular area between the outer cap (1) and the inner pipe (2). The outer cap is constructed with a number of slots (3) that provide a variable vapor flow area to account for variations in liquid and vapor flow throughout the operation or for alternate feed cases.

For both types, a number of devices are installed in an even pattern on the tray and these provide the means by which the liquid and vapor is passed onto the catalyst bed below. Both designs aim at an equal distribution of liquid and vapor across the catalyst bed, which is critical in achieving good catalyst performance.

Similar to gravity flow devices, recent developments provide a means of providing a spray pattern at the outlet of the device, thus greatly facilitating the even distribution of liquid over the reactor cross section relative to older designs.

(3) *Provide a support for each catalyst bed.*

Each catalyst bed must be supported in a way that allows material to exit the bed without affecting the distribution in the catalyst bed and assure that the catalyst is retained in the bed. For a single bed reactor, a device consisting of perforated or slotted plate is typically mounted above the reactor outlet nozzle. It is covered by a layer of appropriately sized support material to prevent migration of catalyst, which would easily pass through the openings in the outlet collector. So as not to disrupt the flow in the catalyst bed, certain constraints are placed on the size of the outlet collector and the location of the bottom of the catalyst bed in relation to the outlet collector.

For intermediate reactor beds, a more elaborate method of support is required. This involves catalyst support grids, composed of sections of reinforced wire screen that are supported by beams mounted on rings located along the wall of the reactor. The support grid sections rest on the flange of the support beams. The support beams are sized to carry the load associated with the static (catalyst weight) and dynamic (pressure drop) load associated with the catalyst bed to the walls. Various methods have been employed to design the support grid and support beams. The design of the support grids should minimize any disturbance to flow in the catalyst bed.

(4) *Introduce quench material between reactor beds.*

If the reactor has more than one bed, it is usually because the temperature rise due to the exothermic reactions requires cooling the material after contacting it with a portion of the catalyst before contacting it with the remainder of the catalyst.

A number of proprietary designs have been developed over the years to mix the quench material with the effluent from the bed above and establish a uniform temperature before the material is distributed to the catalyst bed below.

The main principle is to introduce sufficient turbulence to allow a uniform temperature to be quickly established. Achieving thermal equilibrium before redistribution is desirable to minimize temperature variations from occurring in the next catalyst bed.

In the vast majority of cases, the quench material is unheated recycle gas. This is convenient since recycle gas, which contains a high proportion of hydrogen, can replenish hydrogen that has been consumed in the previous beds. Recycle gas is also appropriate in that it is supplied by the recycle gas compressor, thus a separate means for providing a quench medium is not required. In some situations, liquid-phase quench material has been used. The main drawback with liquid quench is that a separate pump must be provided to supply the quench and consideration must be given to the reliability of this auxiliary source relative to the reliability of the recycle gas compressor. Since liquid quench requires an additional pump, the capital cost tends to be higher, making recycle gas the usual choice for quench material.

Figure 5.4 (Aly et al., 1989) shows an example of an intrareactor quench zone for a two-phase reactor system using recycle gas as the quench medium.

The catalyst bed above the quench zone is supported by a support grid (11). The liquid and vapor from the bed above drop onto a collection tray (12). Quench gas is sprayed into the space above the collection tray via a distributor (13). Liquid collects on the tray and, along with the vapor, enters one or more devices provided in the collection tray (15). The devices have openings (17) through which the liquid and vapor pass. The material passing through the devices is deposited onto a mixing tray (16). This consists of a circular shaped tray with an opening in the center (21). The devices on the collection tray (15) are oriented so as to induce a circular motion of the material in the mixing tray. In that way, the liquid and vapor from the reactor bed above and quench zone, as well as the quench gas, have an opportunity to mix and reach a uniformity of temperature. The material exits the mixing tray and is deposited onto another

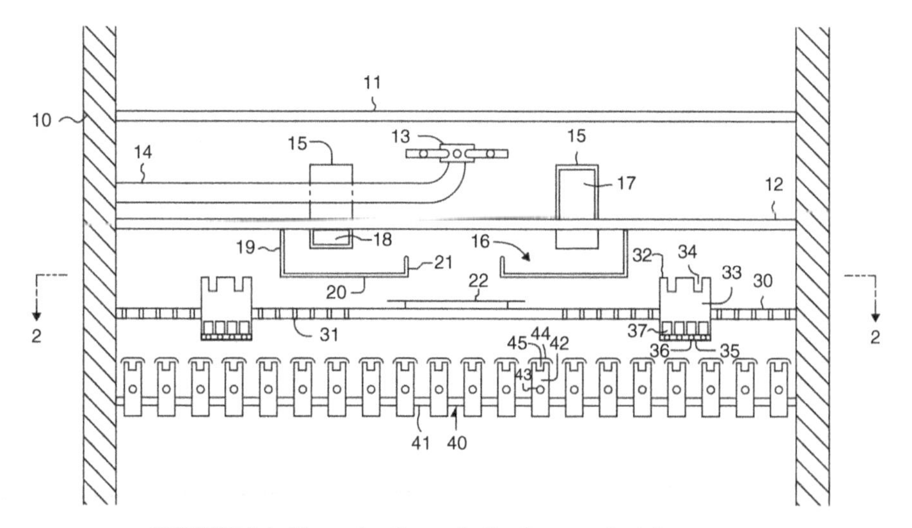

FIGURE 5.4. Example of quench distributor and mixing system.

tray (30) after being dispersed by a splash plate (22). The tray (30) is provided with a number of holes (31) through which the liquid passes and vapor chimneys (33) through which the vapor passes. The height of the vapor chimneys is such that the opening is above the expected liquid level that would exist on the tray. The liquid and vapor exiting the tray (30) are then further distributed onto the catalyst bed by the vapor liquid distribution devices (42). The type shown in this particular case is a chimney type, although other devices could be employed as well.

The quench zone described above is one of many possible designs that could be employed. All such designs, however, would be expected to employ some or all of the above steps in order to achieve effective equilibration of the temperature of the material entering the catalyst bed below the quench zone.

5.2.6 Catalyst Loading

The method used to load the catalyst has a significant impact on its performance. Early units tended to load catalyst by means of a flexible hose or sock. The sock is moved across the bed so as to minimize mounding of catalyst. Even so, it has been shown that this method results in an uneven void fraction distribution throughout the catalyst bed. Particularly for two-phase systems, this can contribute to maldistribution of liquid and vapor flow, regardless of how well the liquid and vapor are distributed at the top of the catalyst bed. Maldistribution is most clearly evident via the temperatures measured at the bottom of the reactor. Figures 5.5 and 5.6 show examples of the temperature profile at the bottom of a bed with both good and poor liquid–vapor distribution.

While distribution can be affected by a number of factors, catalyst loading technique plays a role. In general, it is desirable to achieve a uniform void fraction distribution in the bed and various mechanical devices have been developed to accomplish this. This technique is called dense loading and is practiced by nearly every catalyst vendor. Because the void fraction is more uniform, it is possible to load more catalyst into the same reactor volume. Typically, the amount of increased catalyst

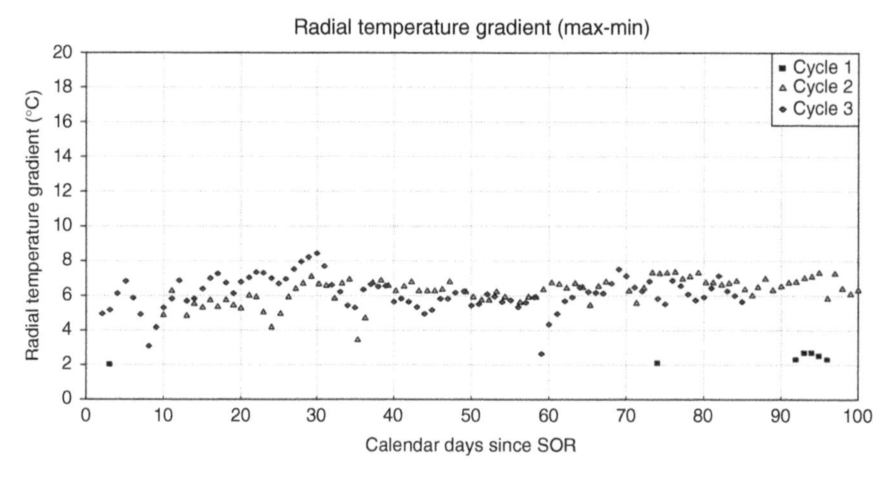

FIGURE 5.5. Example of good catalyst bed distribution.

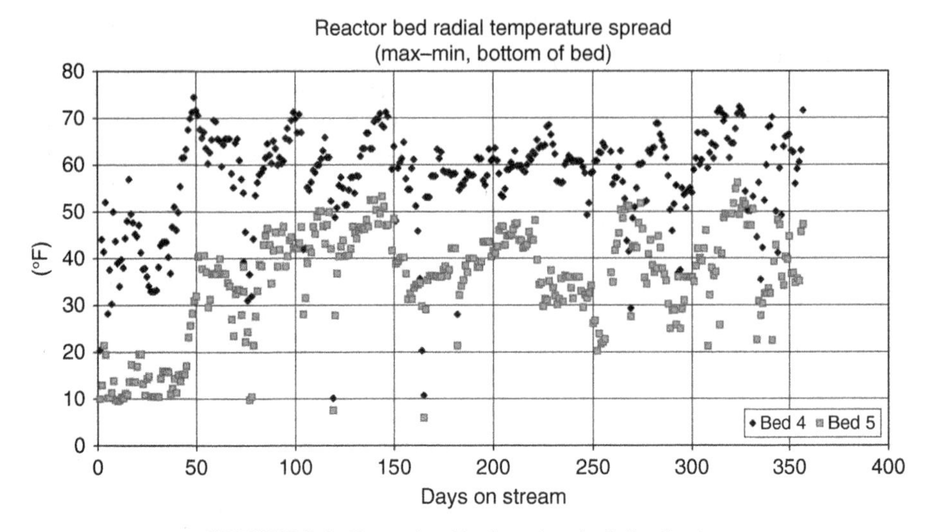

FIGURE 5.6. Example of bad catalyst bed distribution.

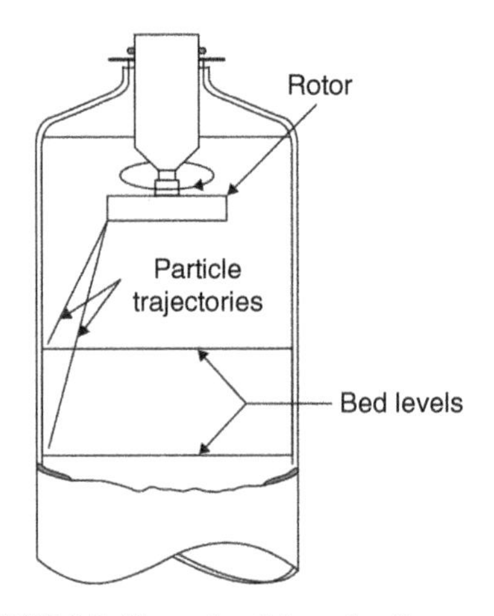

FIGURE 5.7. Example of dense loading machine.

volume that can be loaded is on the order of 10–15%, thus helping to minimize the size requirements of the reactor, which is the most expensive equipment item in the unit. An example of a machine used for dense loading catalyst is shown in Figure 5.7. Its main feature is a rotating head with slotted openings on each side.

The slot openings and rotational speed are adjusted so that the catalyst displaced by the largest radial distance via the rotating head just reaches the reactor wall at the location of the top of the loaded catalyst. Therefore, the rotational speed must be constantly monitored and adjusted to maintain the proper catalyst dispersion. Too low a speed can lead to catalyst accumulating in a mound of catalyst toward the center of the reactor, while too high a speed will cause too much catalyst to be deposited at the reactor walls, as well as create undesirable fines generation as the catalyst contacts

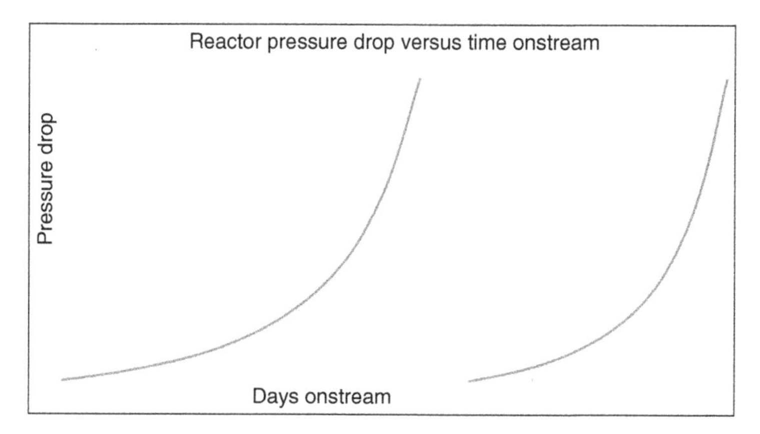

FIGURE 5.8. Example of reactor bed pressure drop buildup due to fines.

the wall at high velocity. Typically, the amount of increased catalyst volume that can be loaded is on the order of 10–15%, thus helping to minimize the size requirements of the reactor, which is the most expensive equipment item in the unit.

5.2.7 Graded Bed Design

During the course of the operation, it is often noted that the pressure drop will increase in one or more of the catalyst beds. This can occur particularly when processing material that contains particulates, which can be present in coker gas oils, FCC LCO, or reduced crude. It can also occur because of fines present in the feed due to corrosion occurring in upstream units or operating upsets that cause coking of the catalyst bed. Figure 5.8 shows an example where pressure drop increases during the course of the run.

In some cases, the pressure drop increase can begin to exceed the design load of the catalyst support components. When that occurs, it is necessary to shut down the unit and remove the catalyst that has become fouled. While the first line of defense is feed filtration, the practical limitation for filter opening and the particle size distribution of fines in the feed usually mean that even for filters in good working order, a substantial amount of fines can enter the reactor. There is a practical limit to the particle size that can be removed by feed filtration. Typically, this limit is about 10 μm (0.00039 in.). Below a certain size, some fines can pass through the reactor. However, the catalyst bed itself acts as a rather effective filter medium. As such it can accumulate solid material present in the feed.

To mitigate this, various methods and techniques have been applied over the years to prevent fines accumulation in the first catalyst bed of the reactor from causing a premature shutdown due to high pressure drop. One of the first methods was to install devices (Carson, 1972), sometimes called distributor or trash baskets shown in Figures 5.9 and 5.10. The baskets were constructed primarily of wire mesh and were located at the top of the catalyst bed such that they extended some distance into the catalyst bed itself. They were intended to provide extra capacity to absorb fines in the feed and thus mitigate the pressure drop buildup associated with feeds heavily

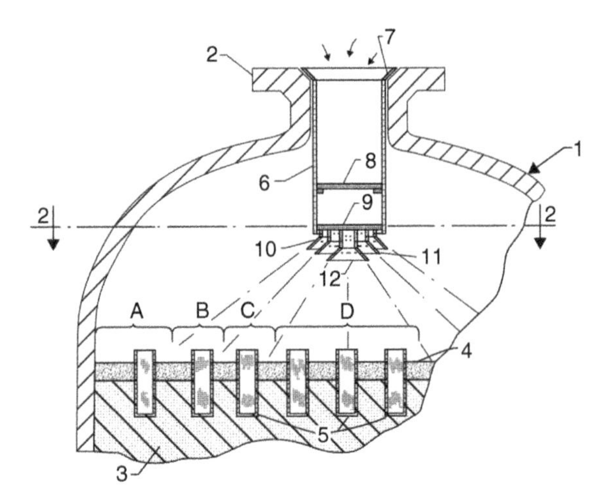

FIGURE 5.9. Distributor basket location.

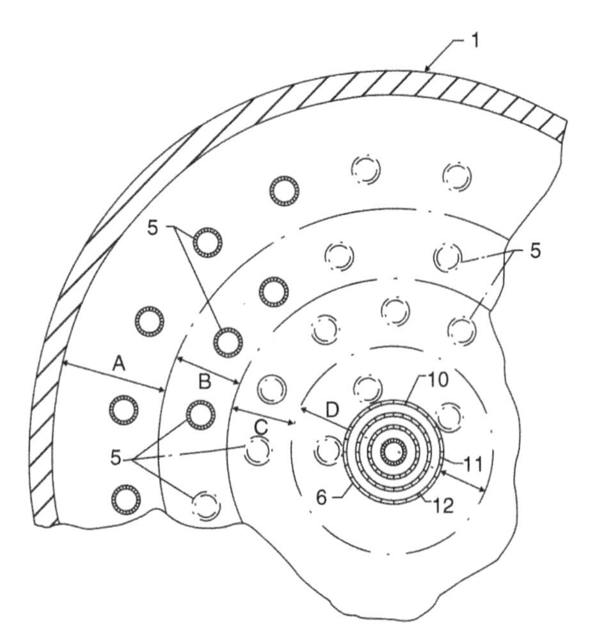

FIGURE 5.10. Distributor basket orientation.

loaded with fines. These were used particularly for heavy feeds, such as coker gas oils, reduced crude, or vacuum residuum, as these materials tended to contain more fines than other materials. These devices had an inherent disadvantage in that fines still tended to accumulate in the catalyst bed and computational fluid dynamics calculations subsequently showed that their presence tended to disrupt the initial distribution of liquid onto the catalyst bed, which is critical to effective catalyst utilization.

Distributor baskets were gradually replaced by increasingly sophisticated inert material having various shapes and sizes, all designed to accumulate fines so as to delay the onset of catalyst bed pressure drop, thus avoiding a shutdown to skim the catalyst before the end of cycle is reached. Some examples of materials used at the top of the bed to accumulate fines are shown in Figure 5.11.

These materials, when properly applied, have shown to be highly effective when used in combination with appropriate feed filtration in achieving the full intended

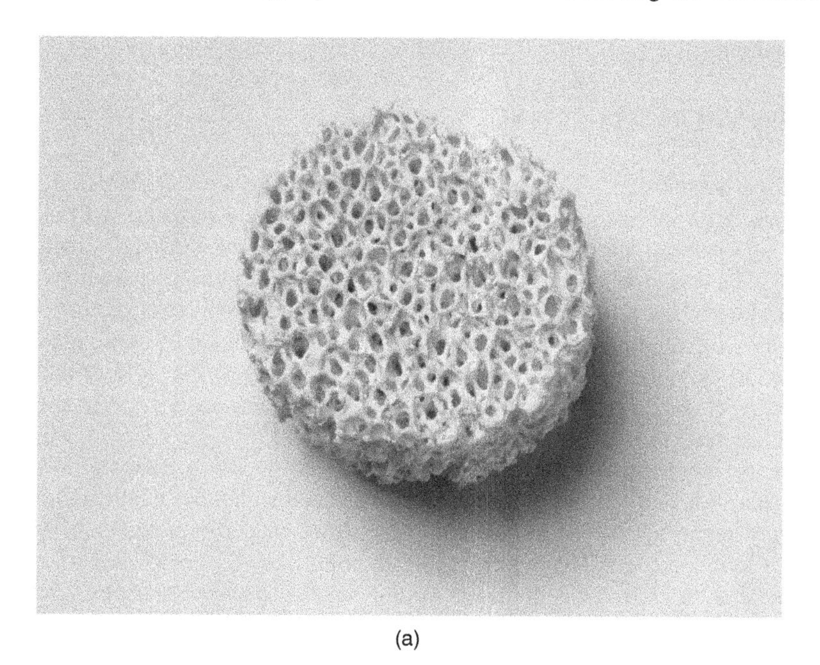

(a)

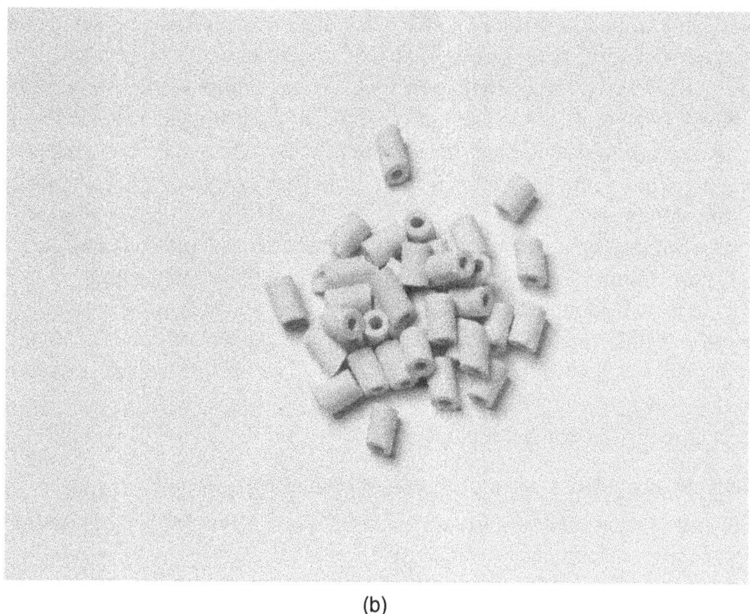

(b)

FIGURE 5.11. Examples of graded bed loading. Source: Reproduced with permission of Crystaphase.

run length without having to shut down due to premature catalyst bed pressure drop increase.

Previous sections of this chapter focused solely on design of the reactor, as this vessel is arguably the most important in terms of assuring that the catalyst performs as intended over the expected operating cycle. Other systems are also necessary for the proper functioning of the reactor section. This chapter discusses some of the more important items.

5.3 RECYCLE GAS PURITY

One of the parameters that determine the cycle length for a given operation is the hydrogen partial pressure existing in the reactor. The yield estimator will usually specify an operating pressure at the reactor or separator and recycle gas hydrogen purity. Taken together, these two parameters constitute a hydrogen partial pressure that must be maintained. For a single high-pressure separator flow scheme, the recycle gas purity will, among other things, be a function of the amount of noncondensable compounds such as methane or nitrogen in the makeup gas, the yield of methane from the reactions, and the temperature of the high-pressure separator. The yield of methane is fixed for a given feed, catalyst system, and process objective. Makeup hydrogen is produced either by a steam–methane reformer or a catalytic reformer, and the purity is determined by the systems associated with those facilities. In most new unit designs, hydrogen purification by pressure swing adsorption (PSA) will be the most common form of hydrogen purification. Purities of greater than 99.9 mol% are usually achievable. This minimizes compression cost as well as entry of impurities into the unit. However, such purity is not always available. To illustrate the effect of makeup gas purity on the recycle gas purity as well as separator temperature, a series of simulations were run for a distillate hydrotreating unit with a single high-pressure separator operating at different temperatures. The results are shown in Figure 5.12. Two different hydrogen makeup gas compositions were used: one of high purity (99 +mol%), representative of the PSA hydrogen, and one low purity (91 mol%), representative of catalytic reformer net gas. As can be seen, the purity of the makeup gas has a much bigger effect on the separator vapor hydrogen purity than the separator temperature.

A similar picture emerges with looking at the effect of makeup gas purity and separator temperature for a unit with a hot separator configuration. Similar to before, a series of simulations was run for a distillate hydrotreating unit with a hot separator configuration. Two series of data were generated: one for high purity makeup gas (99 +mol%) and one for low purity (91 mol%). The results are shown in Figure 5.13.

Several conclusions can be drawn:

(1) Recycle gas purity is strongly affected by makeup gas purity.
(2) The use of a hot separator has an effect on recycle gas purity, particularly when low purity makeup gas is used.
(3) Recycle gas purity is lower when using a hot separator flow scheme and when low purity makeup gas is used.
(4) Separator temperature alone does not generally have much of an impact on recycle gas purity. This means that it is not necessarily advantageous to try to

cool the cold separator to the lowest temperature possible with air or cooling water. A more moderate cold separator temperature will result in less exchanger surface and a more economical design. This means that cooling the reactor effluent to achieve a low separator temperature is not necessarily an optimum economical design.

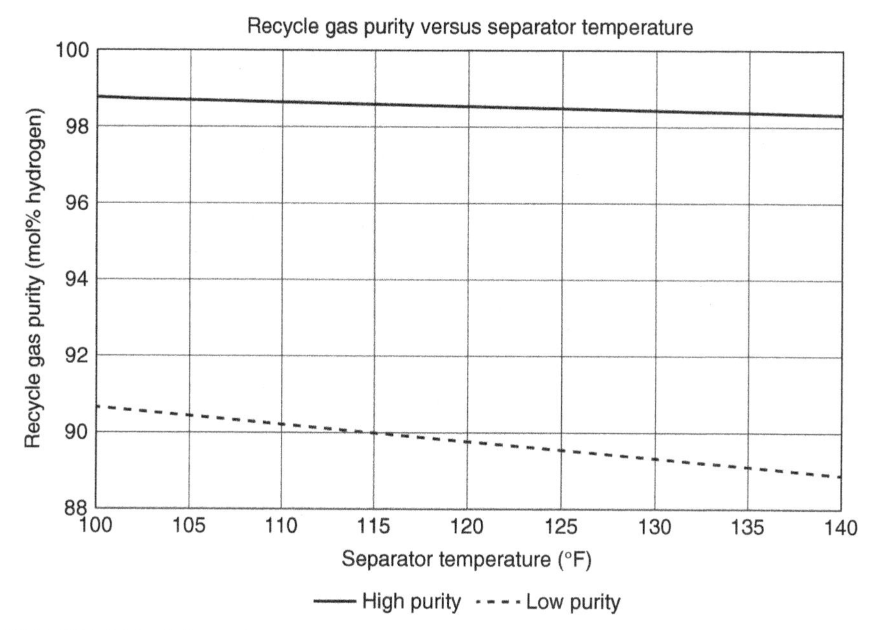

FIGURE 5.12. Effect of cold separator temperature on recycle gas hydrogen purity for a single separator hydroprocessing unit.

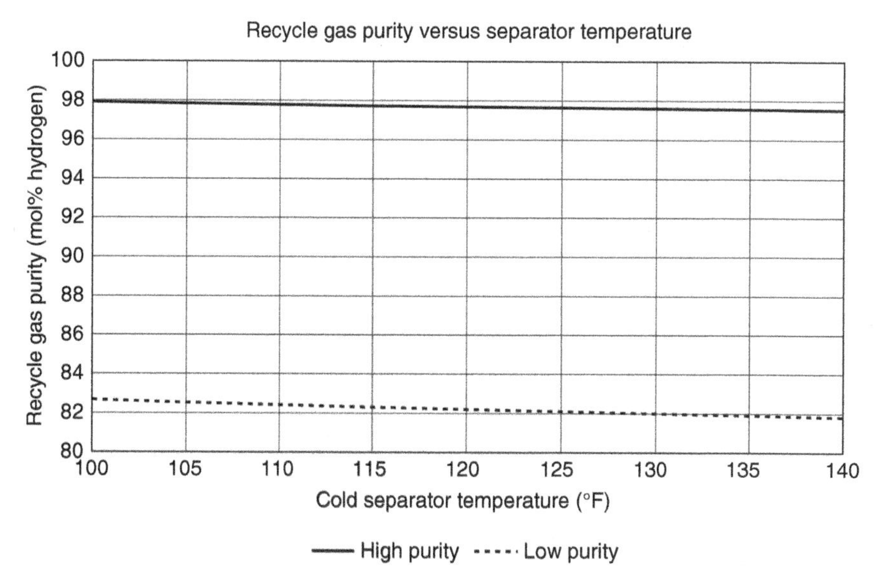

FIGURE 5.13. Effect of cold separator temperature on recycle gas hydrogen purity for a hydroprocessing unit with a hot separator.

When confronted with the issue of low recycle gas purity, several techniques are available to the designer. If the hydrogen purity is close but slightly below to the desired value, the operating pressure of the unit can be raised by a modest amount in order to achieve the required hydrogen partial pressure, rather than venting. If the increase is on the order of 3–4 bar (40–50 psig), this should have a minimal impact on the cost of the unit. Beyond that, other means must be employed. The next option is to provide a vent to maintain the required purity, but because most of the vent gas is valuable hydrogen, a means of recovering the hydrogen in the vent gas should be applied.

To maintain or increase recycle gas purity, the most difficult compounds to remove are methane and nitrogen because they have high equilibrium vaporization ratios, often called the *K* value. The parameter *K* is defined as the ratio of the concentration of a compound in the vapor phase to the concentration of the compound in liquid phase with which it is in equilibrium. The *K* value for a given compound is a function of temperature and pressure as can be seen from Figure 5.14, which shows the temperature dependency of the *K* value versus temperature for hydrogen and various hydrocarbon compounds.

Note that hydrogen exhibits a curious anomaly relative to other compounds. It is actually less volatile at higher temperatures than at lower temperatures. Methane and

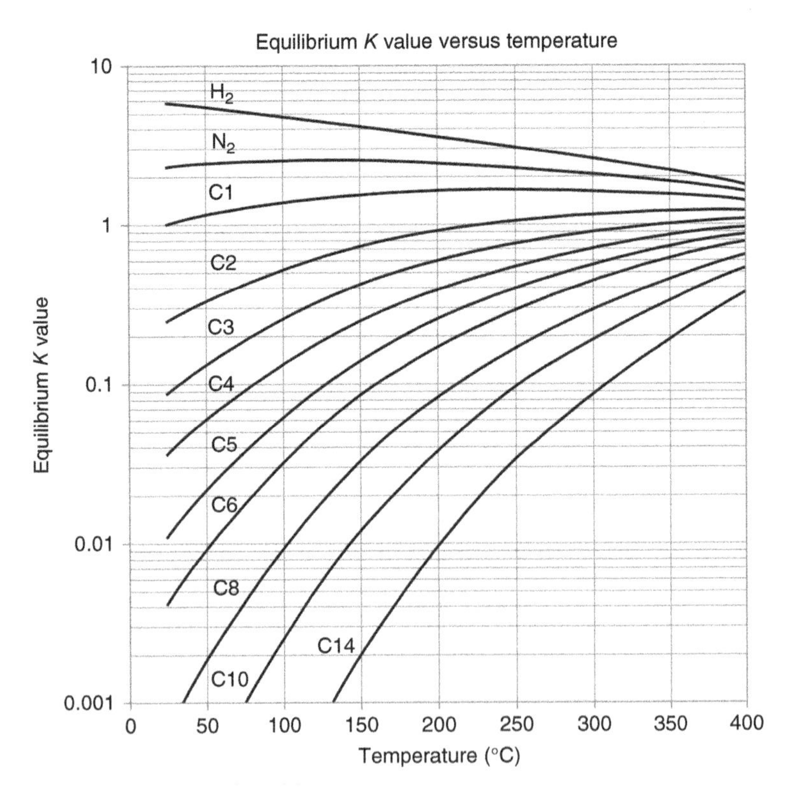

FIGURE 5.14. Equilibrium *K* versus temperature for hydrogen, nitrogen, and selected hydrocarbons.

nitrogen are almost constant and other compounds show a marked tendency to be more volatile at higher temperatures, with heavier compounds exhibiting the biggest change in volatility. Hydrogen's lower volatility at higher temperatures means that hydrogen is more likely to remain in the liquid phase at higher temperatures than at lower temperatures, and this translates into higher solution losses for a unit that has a hot separator versus a single separator. In addition, with a hot separator design, less condensable hydrocarbons will be present in the hot separator vapor and consequently, less hydrocarbon liquid will be present in the cold separator to adsorb methane and nitrogen. The consequence is that without venting to remove these compounds, they will build up to a level at which they will leave the system in the hot and cold separator liquids. Therefore, a hot separator flow scheme will tend to have a lower recycle gas purity for the same reactor yields and makeup gas composition, compared with a conventional flow scheme.

This effect is particularly evident for residuum hydrotreating units designed with a hot separator flow scheme, since only a small amount of condensable hydrocarbons are present in the hot separator vapor and, consequently, the cold separator liquid flow rate is small, even with the hot separator operating as high as 370 °C.

One technique has been used to address this involves circulating the hydrocarbon liquid from the cold flash drum to the inlet of the cold separator. This increases the amount of hydrocarbon available to adsorb nitrogen and methane and therefore provides a means of maintaining the recycle gas purity. This is illustrated in Figure 5.15. This technique is called enrichment.

The effect of enrichment on recycle gas hydrogen purity can be seen in Figure 5.16. This is a plot of recycle gas hydrogen purity versus enrichment ratio for a typical residuum hydrotreating unit operating at 2200 psig (150 bar). Enrichment ratio is defined as the volume of enrichment liquid divided by the volume of fresh feed. The source of the enrichment liquid is flash drum liquid at 225 psig (15 bar). As can be seen, a substantial amount of enrichment liquid is necessary to effect a significant

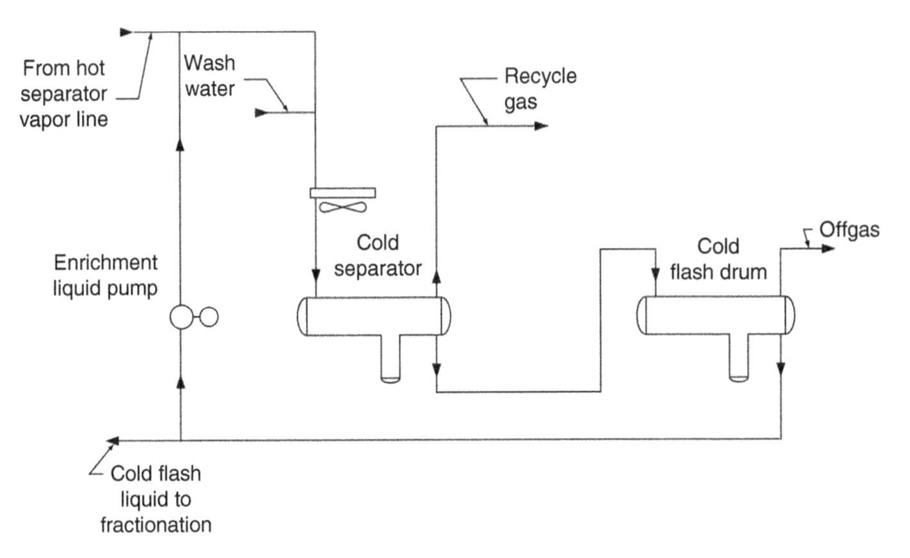

FIGURE 5.15. Enrichment flow scheme.

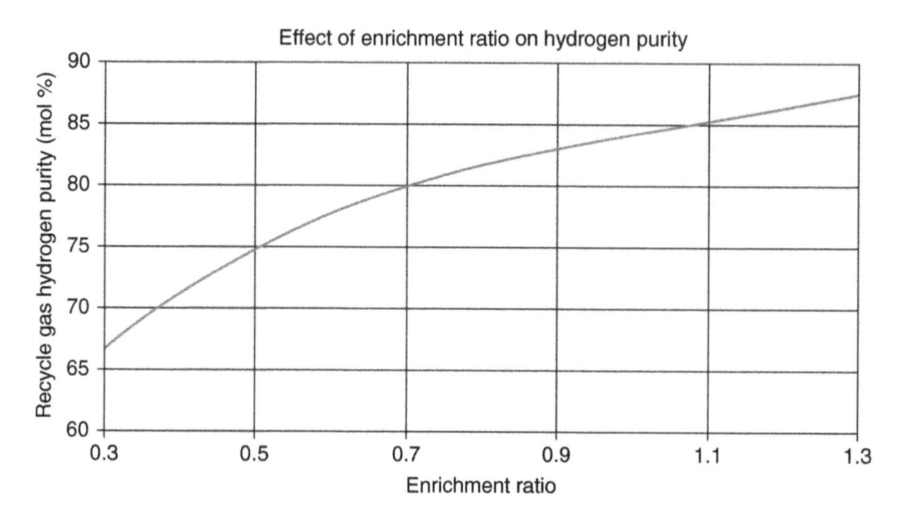

FIGURE 5.16. Effect of enrichment ratio on hydrogen purity.

increase in hydrogen purity. This extra liquid must be pumped by an additional high head pump, and the size of the cold separator and cold flash drum will be increased accordingly.

In addition, hydrogen solution losses increase substantially as the amount of enrichment liquid increases, thus further contributing to the hydrogen requirement. For these reasons, this has largely been abandoned.

With the advent of membrane and PSA technology, it is now more economical to vent recycle gas, recover hydrogen from the vent gas, and route it back to the makeup gas compressor. In addition, recovery of hydrogen in the cold flash drum offgas via PSA is now widely practiced and plays an important part in the economics of the unit, particularly for a hot separator configuration.

5.4 WASH WATER

If the feedstock contains sulfur and nitrogen compounds, they will be converted to hydrogen sulfide and ammonia in the reactor. As the reactor effluent material cools, there is a characteristic temperature at which these two compounds will sublimate as ammonium bisulfide, NH_4HS, and ammonium chloride, NH_4Cl, and deposit on piping and equipment. The point at which this occurs is a function of the product of the partial pressures of each compound in the reactor effluent material. Figures 5.17 (American Petroleum Institute, 2012a) and 5.18 (American Petroleum Institute, 2012b) show this relationship for ammonium bisulfide and ammonium chloride, respectively. In most cases, ammonium bisulfide will sublimate at temperatures most likely to occur in the air- or water-cooled effluent exchanger. If not removed, this material will foul the heat exchanger surface and result in a loss of heat transfer and a resultant increase in the separator operating temperature. Left uncorrected, plugging of exchanger tubes can eventually occur. Fortunately, ammonium bisulfide and ammonium chloride are both water soluble and are easily removed by injection

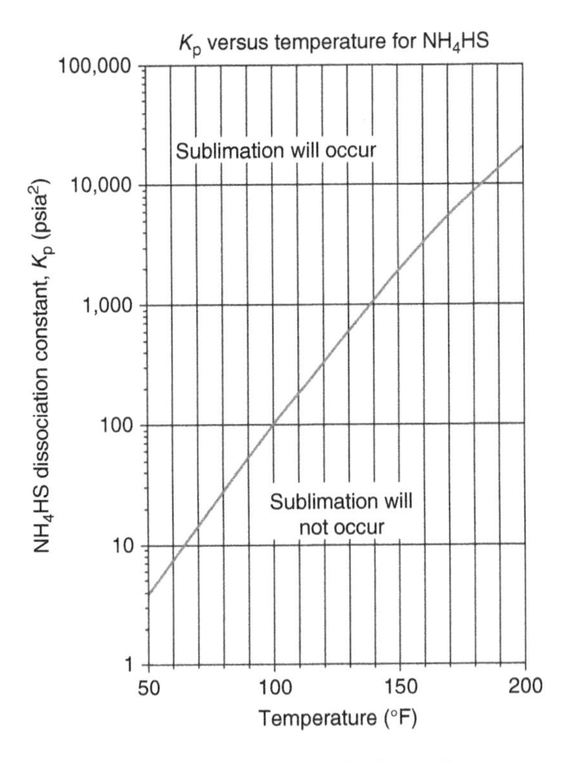

FIGURE 5.17. Ammonium bisulfide disassociation curve.

of a sufficient amount of water. Even water distribution through the tubes is another important consideration involved in the design. Because of the duty required, the effluent air cooler will normally be comprised of more than one bundle. It is critical to distribute the water evenly among the bundles, as insufficient wash water can cause fouling, deposit buildup, or in the worst case, localized high concentrations of ammonium bisulfide, which can cause erosion corrosion of the exchanger tubes or header boxes resulting in tube or header failure and loss of containment. The problem became apparent early in the history of operating hydrotreating and hydrocracking units. An early study on the effects of ammonium bisulfide corrosion was conducted by Roger Piehl (1976). This paper was a result of a survey of the problems with corrosion in reactor effluent air coolers encountered by a number of units operating different feeds with varying sulfur and nitrogen contaminant levels, air cooler configurations, wash water rates, and injection methods. Some of the key findings of the Piehl's work indicated the following:

(1) A factor that Piehl called K_p, which is the product of the mol% concentrations of ammonia and hydrogen sulfide, appeared to correlate with the corrosivity of the process fluid. Higher K_p values tended to lead to more corrosion.
(2) Another rough measure of corrosivity is the concentration of ammonium bisulfide in the separator sour water.
(3) Cyanides could have an effect on corrosion.

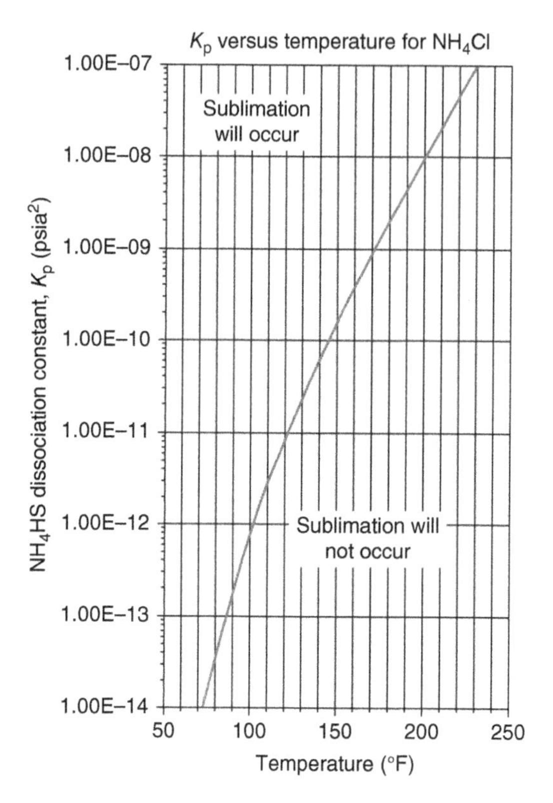

FIGURE 5.18. Ammonium chloride disassociation curve.

(4) The presence of oxygen in the wash water could cause accelerated corrosion.
(5) Corrosion decreases with increased wash water injection.
(6) Corrosion is influenced by fluid velocity, either too high or too low. A velocity range for the air cooler tubes of 10–20 ft/s (3–6 m/s) was proposed.
(7) Symmetrical piping to and from the air-cooled exchanger is the most desirable configuration. Nonsymmetrical piping can lead to unequal distribution of material to each bundle resulting in localized high ammonium bisulfide or ammonium chloride concentrations and subsequent corrosion of the air cooler bundles and associated piping. A symmetrical piping configuration is illustrated in Figure 5.19.

Building on this and other published work, the American Petroleum Institute put together a document known as API RP 932-B (2012a,b). This document includes some additional observations and recommendations:

(1) At least 20–25% of the injected water should remain in the liquid phase after injection.
(2) Ammonium bisulfide corrosion by erosion–corrosion can occur, and the flow regime can influence this effect.

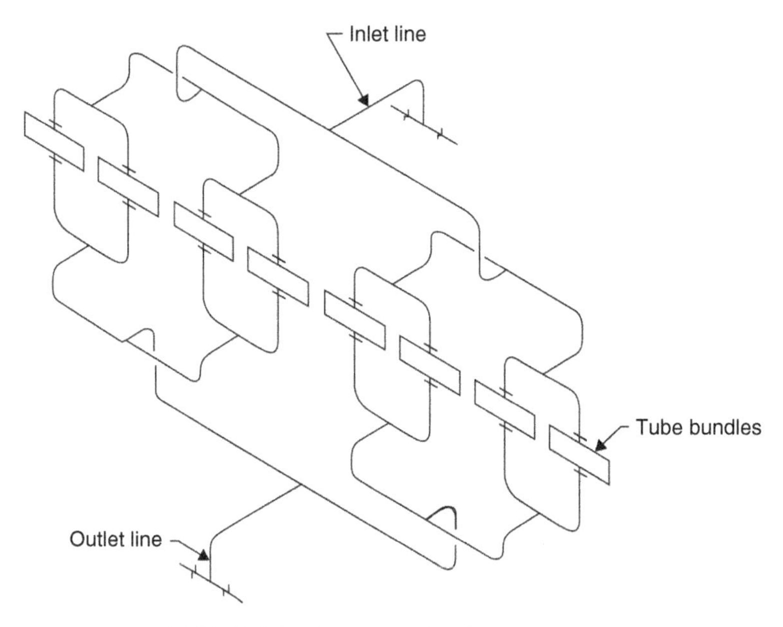

FIGURE 5.19. Symmetrical piping arrangement.

(3) Balanced, symmetrical header piping is better at distributing the material to the air cooler bundles.
(4) A single wash water injection point can be used with a balanced header design.
(5) Avoid nonflowing areas, such as pipe stubs, as these can be a source of corrosion.
(6) Oxygen content is a critical parameter of the wash water. A limit of 50 ppbw (parts per billion by weight) was proposed.

A Joint Industry Program (JIP) beginning in the late 1990s was undertaken by a number of companies to gain a better understanding of corrosion mechanisms of ammonium bisulfide on various materials. Some findings that arose from the program include the following (Kane et al., 2006):

(1) The partial pressure of hydrogen sulfide has a significant impact on the corrosion of carbon steel as well as other alloys.
(2) Wall shear stress is more effective in predicting corrosion rate than bulk fluid velocity.

Proprietary software, called PREDICT-SW®,[1] was developed to predict the ammonium bisulfide corrosion rates for various materials (Srinivasan et al., 2009). Required inputs include process data for all anticipated operating cases and the actual piping configuration at the inlet and outlet of the effluent air cooler.

[1]Predict-SW is a registered trademark of Honeywell International, Inc.

Another potentially serious problem can occur if the feedstock contains organic or inorganic chlorides. This is readily converted to chloride ions in the reactor. Chlorine ions will readily combine with ammonia to form ammonium chloride at a characteristic temperature that is also a function of the product of the respective partial pressures. From Figure 5.18, it can be seen that the sublimation temperature of ammonium chloride is much higher than ammonium bisulfide for the same product of partial pressures. This means that the sublimation, and, therefore, the point of wash water injection, must occur at a point farther upstream from that used to wash ammonium bisulfide salts. It is typical practice even for units not designed for the presence of chlorides in the feed to have at least one additional wash water injection point upstream of the last shell and tube exchanger before the air-cooled condenser. This provides the ability to add wash water in case a feed containing some chlorides is being processed.

If stripped sour water is used to supplement fresh water, other contaminants that would be of concern are cyanides, which can cause hydrogen blistering, phenols, which can cause foaming in the high-pressure separator, and chlorides. These compounds can originate from the crude and vacuum unit, fluid catalytic cracking unit, delayed coking unit, visbreaking unit, or other such units involving thermal cracking in a hydrogen-deficient environment. Stripped sour water from these units should not be used in a hydroprocessing unit. Typically, a so-called phenolic sour water stripper is provided to process sour water from these units and a nonphenolic sour water stripper processes sour water from the hydroprocessing units. To avoid buildup of dissolved and suspended solids, it is not recommended that stripped sour water alone be used as wash water.

For units processing feeds that have a low nitrogen content, typical when treating naphtha and kerosene, the production of ammonium bisulfide salt is very small. Applying the usual rules for minimum injection rate for wash water can result in a relatively large volume of sour water containing a rather dilute concentration of ammonium bisulfide. This has the effect of consuming excessive fresh water and expending extra utilities at the sour water stripping unit, as well as the utility associated with pumping the wash water into the unit. In this case, an alternate system, shown in Figure 5.20, could be considered. This system circulates the dilute sour water from the high pressure, combines it with an amount of fresh water and the mixture is routed to the wash water injection point. The principle behind this system is that a certain minimum wash water injection rate to the injection point is maintained so that reasonably good distribution is achieved in the exchangers to be washed and the use of fresh water is minimized. In designing such a system, it must be kept in mind that according to the conditions of the reactor effluent stream into which this combined water stream will be injected, some if not most of the injected water will be vaporized. The requirement that the amount of liquid remaining after the wash water injection still must be maintained. This means that the concentration of ammonium bisulfide will concentrate at the point of injection due to vaporization of the wash water. Therefore, the circulation rate and minimum fresh water makeup rate must be set to that the maximum ammonium bisulfide concentration in the sour water leaving the separator is kept below the desired value.

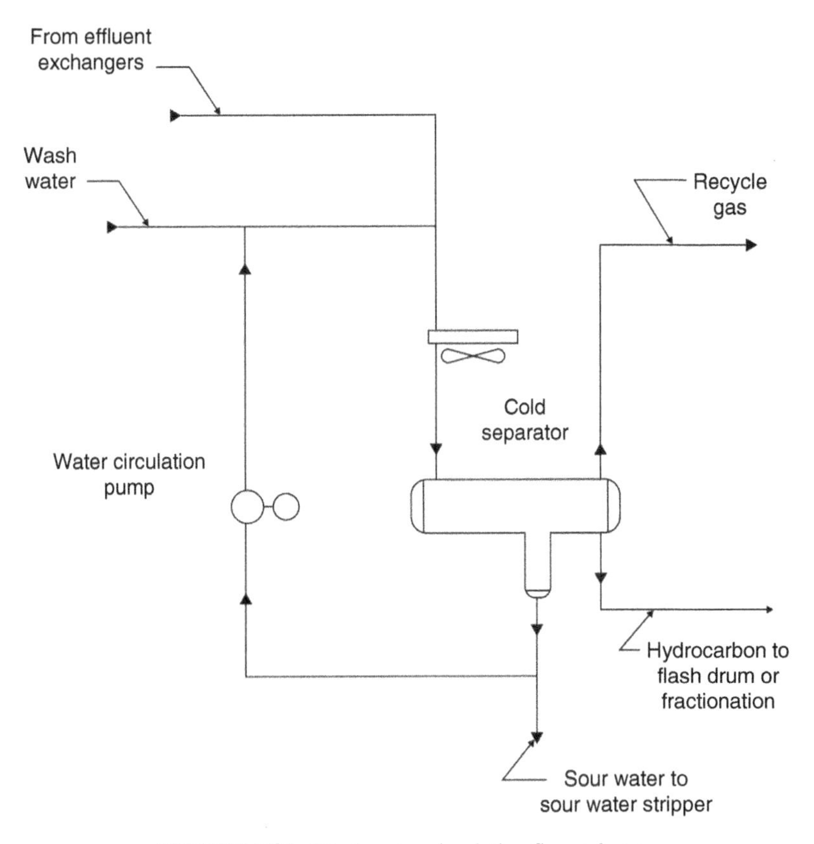

FIGURE 5.20. Wash water circulation flow scheme.

5.5 SEPARATOR DESIGN

Proper phase separation is necessary to the efficient function of the process. The two types of separation that take place in a hydroprocessing unit are the following:

(1) Separate either liquid hydrocarbon or water from vapor.
(2) Separate vapor, hydrocarbon liquid, and water.

Each requires the proper approach in order for the process to function properly. Failure to do so can result in overpressure of downstream equipment (failure to separate vapor from liquid), product loss or contamination (failure to separate hydrocarbon from water), or corrosion (failure to separate water from hydrocarbon). All three are undesirable; however, of the three, failure to separate water from hydrocarbon is the worst situation because it can lead to corrosion of downstream equipment that can result in a potential loss of containment.

Phase separation is typically governed by three equations: Stokes' law, Intermediate law, and Newton's law, each with a range of applicability. The selection of which equation to use is determined by a parameter called K (McCabe and Smith, 1967b), which has the following form:

$$K = D_{\mathrm{p}} \left[\frac{g_{\mathrm{c}} \rho (\rho_{\mathrm{p}} - \rho)}{\mu^2} \right]^{1/3}, \tag{5.18}$$

where

D_{p} = diameter of particle to be settled
g_{c} = acceleration of gravity
ρ = density of continuous fluid
ρ_{p} = density of particle to be separated from continuous fluid
μ = viscosity of fluid

In general, for values of K up to 3.3, the equation for Stokes' law is used. It is expressed as follows:

$$u_t = \frac{g_{\mathrm{c}}\, D_{\mathrm{p}}^2\, (\rho_{\mathrm{p}} - \rho)}{18\mu}. \tag{5.19}$$

For values of K between 3.3 and 43.6, the so-called Intermediate equation is used. It takes the following form:

$$u_{\mathrm{t}} = \frac{0.153 g_{\mathrm{c}}\, D_{\mathrm{p}}^{1.14}\, (\rho_{\mathrm{p}} - \rho)}{\rho^{0.29}\, \mu^{0.43}}. \tag{5.20}$$

For values of K above 43.6, the Newton's law equation is be used.

$$u_{\mathrm{t}} = 1.74 \sqrt{\frac{g_{\mathrm{c}}\, D_{\mathrm{p}}\, (\rho_{\mathrm{p}} - \rho)}{\rho}}, \tag{5.21}$$

where

u_{t} = droplet terminal velocity, ft/s
g_{c} = acceleration of gravity, 32.2 ft/s^2
D_{p} = diameter of droplet, ft
ρ_{p} = density of droplet to be separated from continuous fluid, lb/ft^3
ρ = density of continuous fluid, lb/ft^3
μ = viscosity of continuous fluid, lb/ft^2

This relation holds for K values up to about 2300, but it is expected that conditions encountered in the design of hydroprocessing separators will not approach that value.

The next decision to be made by the designer is the orientation of the separator. In general, gas/liquid separation can be effectively performed in a vertical separator. Liquid–liquid separation or three-phase separation (vapor, hydrocarbon liquid, and water) can be carried out in either a vertical or horizontal separator, although experience has shown that three-phase separation is carried out more effectively in a horizontal separator. This is because droplet terminal velocity in liquid–liquid separation situations tends to be lower requiring more residence time to effect the separation. A horizontal vessel accommodates this requirement more effectively than a vertical separator.

5.5.1 Vertical Separators

The designer will need to determine the vessel diameter and tangent length. In general, the diameter is set by the requirement to minimize entrainment of liquid droplets in the vapor phase. From equations (5.18)–(5.21), an acceptable liquid droplet size is determined and the resulting target upward vapor velocity is calculated by using one of the above formulae. The tangent length is determined by the liquid volume necessary for the required residence time, as well as mechanical constraints. A number of devices have been developed to enhance separation of entrained liquid droplets from vapor and thus minimize the diameter. This is particularly useful for a vapor–liquid separator where the expected amount of liquid is small relative to the vapor. Such a condition exists regularly in services such as compressor suction drums. The most common device to achieve this is a mesh blanket. It consists of small diameter wire woven into a blanket whose thickness can be from 6 in. (150 mm) to 1 ft (300 mm) in thickness. The blanket itself rests on a support grid that provides vertical stability for the blanket. If the vessel diameter is sufficiently large, one or more support beams are necessary to support the blanket. To prevent lifting of the mesh blanket due to the upward forces of the vapor flowing through the mesh blanket, a hold-down grid is usually provided above the blanket.

The object is to provide a surface on which the bubbles or droplets of liquid entrained in the vapor have a chance to coalesce, thus forming larger diameter droplets that will eventually drip down to the liquid in the bottom of the vessel. Such devices have a desirable range for the velocity of the upwardly flowing vapor over which they are most effective. The sizing equation (Mist Elimination, 2007) for a vertical separator vessel when using a mesh blanket is as follows:

$$u_{\text{opt}} = K\sqrt{\frac{(\rho_{\text{p}} - \rho)}{\rho}}. \tag{5.22}$$

Note that this equation is similar to the Newton's law equation (5.21). However, the value K is unrelated to the K value determined that equation. It takes into account the effect of the mesh blanket. Table 5.3 lists the K values for a typical mesh blanket. Specific vendors may have slightly different values, but they should not vary greatly from those listed.

Note that the K value is a function of the density of the vapor phase, so the K value varies with the operating pressure. Use of a mesh blanket generally means that a smaller vessel diameter can be used for the same liquid droplet separation. To illustrate this, two designs based on the same process data will be compared, one with and one without the use of a mesh blanket.

TABLE 5.3. Typical *K* Values for Mesh Blankets

Pressure (psig (bar))	Mesh Blanket, *K* Value
15 (1.0)	0.35
50 (3.4)	0.34
100 (6.9)	0.32
200 (13.7)	0.31
300 (20.7)	0.30
500 (34.5)	0.28
≥1000 (69)	0.27

Process Data:

Vapor flow: 85,300 lb/h
Vapor mole weight: 5.1
Flowing pressure: 265 psia
Flowing temperature: 122 °F = 582 °R
Vapor viscosity: $0.01\ \text{cP} = 6.72 \times 10^{-6}$ lb m/ft s
Liquid density: 48.03 lb/ft^3
Liquid droplet size: $150\ \mu\text{m} = 4.92 \times 10^{-4}$ ft

Determine vapor density:

$$\rho_v = \frac{P_{a\text{MW}}}{10.73\ T_a} = \frac{(265)\ (5.1)}{(10.73)\ (582)} = 0.216\,\text{lb/ft}^3. \tag{5.23}$$

From equation (4.4-1), determine K:

$$K = D_p \left[\frac{g_c \rho (\rho_p - \rho)}{\mu^2} \right]^{1/3}$$
$$= 4.92 \times 10^{-4} \left[\frac{(32.17)(0.216)(48.03 - 0.216)}{(6.72 \times 10^{-6})^2} \right]^{1/3} = 1.827. \tag{5.24}$$

Therefore, Stokes' law (equation (5.19)) will be used to find the terminal velocity for gravity settling of liquid droplets in vapor.

$$u_t = \frac{g_c\ D_p^2\ (\rho_p - \rho)}{18\mu} = \frac{(32.17)(4.92 \times 10^{-4})^2(48.03 - 0.216)}{(18)(6.72 \times 10^{-6})} = 3.08\,\text{ft/s}. \tag{5.25}$$

The diameter of the vessel has to be such that the upward vapor velocity will be at or below that value.

Determine the flowing volume of the vapor:

$$V_v = \frac{W_v}{3600\ \rho_v} = \frac{85{,}300}{(3600)(0.216)} = 109.7\,\text{ft}^3/\text{s}. \tag{5.26}$$

The required cross-sectional area of the separator is

$$A_{sep} = \frac{V_v}{u_t} = \frac{109.7}{3.08} = 35.55\,\text{ft}^2. \tag{5.27}$$

The required vessel diameter is

$$D_{sep} = \left(\frac{4}{\pi}A_{sep}\right)^{0.5} = [(1.273)(35.55)]^{0.5} = 6.72\,\text{ft (2048 mm)} \tag{5.28}$$

Generally, the diameter would be rounded up to the nearest 6 in. or 100 mm, so the vessel diameter would be 7 ft–0 in., or 2100 mm.

Now determine the vessel diameter when using a mesh blanket. From equation (5.22), find the optimum velocity for upward vapor flow through the mesh blanket:

At 265 psia (250 psig), $K = 0.31$, from Table 5.3. Therefore, the optimum velocity is

$$u_{opt} = 0.31\sqrt{\frac{(48.03 - 0.216)}{0.216}} = 4.61\,\text{ft/s}, \tag{5.29}$$

$$\text{A}_{sep} = \frac{V_v}{u_t} = \frac{109.7}{4.61} = 23.79\ \text{ft}^2. \tag{5.30}$$

The required mesh blanket diameter is

$$D_{sep} = \left(\frac{4}{\pi}A_{sep}\right)^{0.5} = [(1.273)(23.79)]^{0.5} = 5.50\,\text{ft (1677 mm)}. \tag{5.31}$$

The mesh blanket needs a support ring mounted around its perimeter, as well as support beams. Therefore, the vessel diameter must be increased to accommodate the ring, whose width is usually 2 in. (50 mm). Therefore, the vessel diameter is 5 ft–6 in. + 4 in. = 5 ft–10 in., rounded up to 6 ft–0 in., or in metric terms, 1677 + 50 = 1720 mm, rounded up to 1800 mm. Comparing the two sizing methods, it can be seen that a mesh blanket does have the potential to reduce the required size of a vertical vapor–liquid separator.

Diameter without mesh blanket	7 ft–0 in. (2100 mm)
Diameter with mesh blanket	6 ft–0 in. (1800 mm)

For applications where the liquid tends to foam, such as when separating heavy oils from hydrogen-rich vapor, the ability of the vapor bubbles to separate from the liquid phase also must be taken into account. Equations (5.18)–(5.21) would also be used.

As the viscosity of the hydrocarbon liquid increases, the value of K in equation (5.18) decreases and either Stokes' law (equation (5.19)) or the intermediate equation (equation (5.20)) would be used. This results in lower values for the terminal velocity for a given vapor bubble size. This would impose a maximum downward liquid velocity so as to be able to separate vapor bubbles of a certain size from the liquid and could even govern the diameter of the vessel.

The designer must also take into account the environment that exists in the separator. As mentioned, mesh blankets have the ability to reduce the required diameter due to their droplet coalescing effect. However, they are not effective if the environment is such that corrosion of the mesh blanket will take place, thus destroying their ability to perform the required coalescing function. Mesh blankets are available in a range of materials to be compatible with the processing conditions, but corrosion must be taken into account. As indicated earlier, mesh blankets are constructed of woven wires of small diameter. As such even a small amount of corrosion will cause breakage of the wires so that the function, as well as mechanical integrity, is impaired. One area in which this is particularly important is the hot separator, which separates hydrogen-rich vapor from the reactor effluent liquid. The operating temperature of this vessel tends to be fairly high, with temperatures in the range of 500–700 °F (260–370 °C). At these conditions, the presence of hydrogen sulfide, which is a common constituent in the reactor effluent from a hydroprocessing unit, will corrode most materials. The most common means of resisting corrosion in this kind of environment is via employing stainless steel, but even these materials will corrode to a certain extent in this environment. Most equipment can cope with this corrosion over time by utilizing a corrosion allowance, which is an increase in the required thickness of the piping or equipment to take into account that the environment will cause a certain amount of metal loss due to corrosion over the life of the vessel. In the case of mesh blankets, no such corrosion allowance is available due to the small diameter of the wire from which the mesh blanket is constructed. Therefore, it is a better practice not to use mesh blankets in such services and to base the sizing of the separator on gravity settling alone.

Having determined the required vapor or liquid velocities for separation of liquid droplets or vapor bubbles, the diameter of the vessel can then be determined. The next decision is to determine the liquid volume that the vessel will hold. There are two terms that are commonly used when sizing vertical separators: residence time or surge time. Residence time refers to the amount of time a given amount of liquid spends in the vessel. Surge time refers to the time that the liquid would rise or fall through a certain distance, usually the range of the instrument measuring the level in the vessel. For vertical vessels, residence time is important for liquids that are prone to degradation, such as materials that are likely to undergo thermal degradation at elevated temperatures. Surge time is important to assure that the unit can cope with short-term processing perturbations. In this case, a judgment must be made as to what constitutes a reasonable time such that the vessel size is minimized, but its ability to cope with upsets is not impaired. Typical surge times range from 2 to 5 min across the measured level range but can vary widely according to the service and required residence time, if applicable.

After determining the diameter and the level range, certain mechanical requirements enter into the sizing of the vessel. For example, for vertical vessels, the bottom level nozzle must be located a minimum distance above the vessel tangent line so that the nozzle weld can be made without interfering with the bottom head to shell weld seam. This distance is usually 6 in. (150 mm). The next consideration is the distance from the maximum indicated level to the bottom of the inlet nozzle. Often, this distance is determined by using a suitable surge time from the highest measured level to the bottom of the inlet nozzle. The placement of the mesh blanket near the vapor outlet, if provided, has to take into account that the vapor will exit the vessel through a nozzle usually much smaller than the mesh blanket itself. Therefore, it is not desirable to locate the mesh blanket too close to the top of the vessel, so that the flow pattern through the mesh blanket will be mostly uniform.

For three-phase separation, additional considerations enter into the design of the vessel, namely, the requirement to separate the aqueous phase from the hydrocarbon phase. The same equations that govern liquid droplet settling from vapor, or vapor bubble separation from liquid would be used for separation of water from the hydrocarbon phase and vice versa. Due to the relatively close proximity of the density of the hydrocarbon and water as well as the viscosities of the two fluids, the terminal velocity for settling of the two phases will usually be much slower than that of separating liquid droplets from water and vapor bubbles from liquid. To overcome this, a mesh blanket will often be used to facilitate the separation between the two liquid phases. The required velocity through the mesh blanket is a function of the type of coalescing medium used. The most common device being used for this purpose is a woven wire blanket. The vendor of the coalescing medium will advise the sizing criteria in this case, but it should be noted that velocities through the coalescing medium are usually much smaller than those employed for vapor–liquid separation. If this requirement is not taken into account, the separator will not perform as required. This is particularly the case for vertical three-phase separators. Figure 5.21 shows a typical vertical three-phase separator. The three-phase mixture enters usually in the side of the vessel, usually in the upper portion of the shell. The vapor exits the top of the vessel, usually through a mesh blanket. The hydrocarbon/water mixture flow downward on one side of a baffle and the hydrocarbon phase flows up through a coalescing medium for the purpose of separating water droplets from the hydrocarbon liquid phase. The coalesced water droplets formed as the material passes through the coalescing medium must be large enough to fall through the upward flowing liquid. This sets the required area for the coalescing medium. The water droplets fall to the bottom of the vessel where the water/hydrocarbon interface is maintained. Any hydrocarbon droplets that are dispersed in the water phase rise to the interface and enter the hydrocarbon phase. Since the entire diameter of the vessel is available for the separation of hydrocarbon droplets from water, it is not expected that this will govern the diameter of the vessel. However, the designer should confirm that this is the case by use of the appropriate gravity settling formula above.

Therefore, the most tightly constrained portion of the vessel is that which involves the separation of water from the hydrocarbon phase. If the vessel diameter is selected on the basis of vapor–liquid separation only and does not take into account the requirements for hydrocarbon/water separation, water entrainment in the hydrocarbon phase will occur. Separation of the water phase from a rising hydrocarbon liquid phase is an inherent disadvantage of the design shown in Figure 5.21.

While other configurations are possible, all suffer from the same inherent disadvantage of having insufficient area for hydrocarbon/water separation, thus limiting the size of the water droplets that can form.

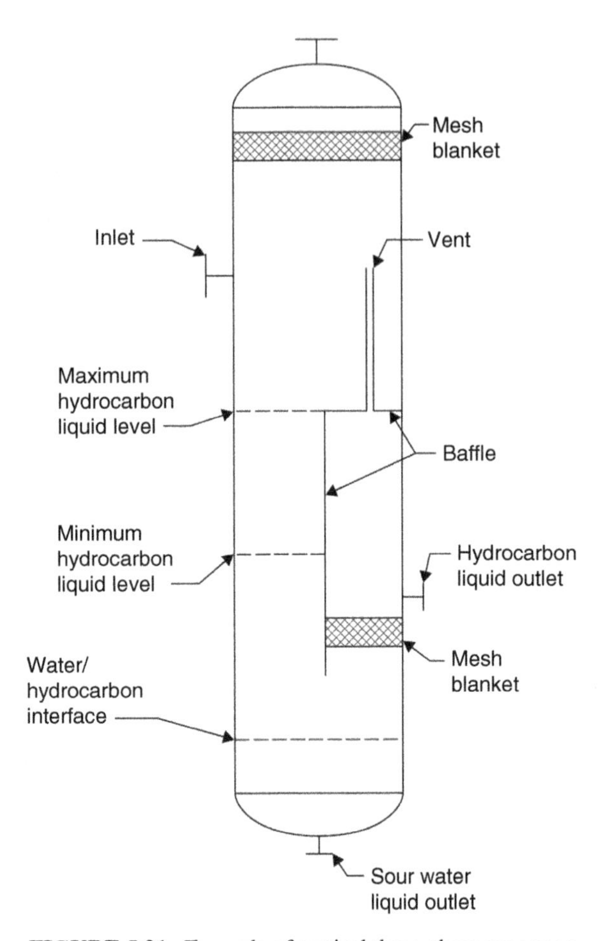

FIGURE 5.21. Example of vertical three-phase separator.

5.5.2 Horizontal Separators

Horizontal separators are often used when separating vapor, hydrocarbon liquid, and water. Figure 5.22 shows such an example.

In this design, the three-phase mixture enters the vessel near one end through some type of distributor device. In Figure 5.22, a slotted distributor is shown in which slots facing the vessel walls are provided. In some designs, a pipe elbow will be provided to direct the inlet material toward the inlet end. The aim is to distribute the material such that the vapor will be able to separate readily from the two liquid phases. The vapor then travels across the upper portion of the vessel and exits the vessel at an upper nozzle at the other end. Any vapor droplets entrained in the hydrocarbon liquid and water phases rise to the top interface and enter the vapor phase. The hydrocarbon/water mixture passes through a coalescing medium. Figure 5.22 shows a full diameter vertical mesh blanket, although some designs feature a partial mesh blanket whose height is set to the upper range of the hydrocarbon liquid level range. Water droplets contained in the hydrocarbon phase can coalesce and begin falling toward the bottom of the vessel where they join the majority of the water phase, which flows along the

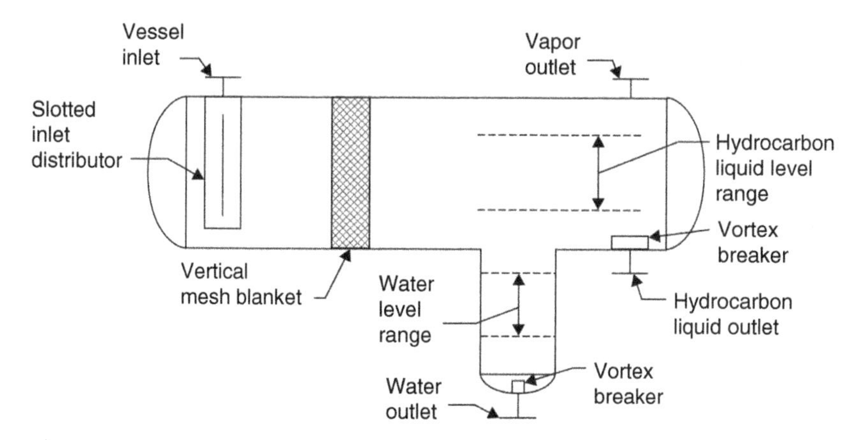

FIGURE 5.22. Example of horizontal separator.

bottom of the vessel. Hydrocarbon droplets contained in the water phase also will coalesce in traveling through the mesh blanket and rise to the hydrocarbon phase. It is necessary to provide enough distance from the coalescing medium to the inlet side of the water boot so that the droplets of the desired size will have enough time to fall from the top of the liquid phase to the bottom of the vessel. This distance is set by the velocity of the liquid through the coalescing medium and the terminal velocity of the liquid droplets of the desired size, as determined by equations (5.18)–(5.21). Normally, the water phase is small in relation to the hydrocarbon phase, so a water boot is located at an appropriate distance from the coalescing medium for collection of the water phase. The hydrocarbon/water phase is usually controlled within the water boot itself. The diameter of the boot is usually set so as to achieve separation of the required hydrocarbon droplet size from the water phase. The hydrocarbon phase flows past the water boot and exits the bottom of the vessel near the end opposite the inlet.

The sizing basis for a horizontal three-phase separator with a mesh blanket is different from that of a vertical separator. Instead of using equations (5.18)–(5.21), a horizontal velocity criterion is used. Typical values range from 3 to 6 ft/min (1–2 m/min) Residence times for the liquid phase range from 5 to 10 min, with the span across the hydrocarbon level range being equivalent to a liquid volume of up to 10 min. The level controller for the water phase usually is a standard 14 in. (180 mm) displacer type instrument.

5.6 MAKEUP GAS COMPRESSION

According to the operating pressure of a given unit, the hydrogen header pressure may or may not be sufficient to route the makeup gas directly to the process. Most often some amount of additional compression is required, usually by a means of a reciprocating compressor. There are two main constraints that govern the application of compression of hydrogen-rich gas. The first is maximum discharge temperature. As the gas is compressed, the temperature rises and the discharge temperature is a function of the suction temperature, C_p/C_v ratio of the gas, and the compression ratio.

The C_p/C_v ratio will be fairly constant, so that the discharge temperature is largely a function of the suction temperature and compression ratio. One industry standard that applies to compressor design is API RP 619. API RP 619 limits discharge temperature to 275 °F (135 °C). Lower temperatures are preferred, as it has been shown that compressor reliability improves with lower discharge temperatures. If the compression ratio is such that the maximum discharge temperature is exceeded before the desired discharge pressure is reached, it will be necessary to apply an additional stage of compression.

The preferred destination for the makeup gas is the recycle gas compressor discharge. This is because the adiabatic efficiency of a reciprocating compressor is higher than the polytropic efficiency of a centrifugal compressor, thus minimizing utilities. Also, the capacity and therefore the cost of the recycle gas compressor will be lower if it is only pumping the recycle gas, rather than the makeup and recycle gas together. However, if an additional stage of compression can be avoided by adding the makeup gas to the suction of the recycle gas compressor, instead of the discharge, it is advantageous to add it to the suction. The reduction in cost of the makeup gas compressor usually offsets the increased capacity and incremental utilities of the recycle gas compressor.

The design of the makeup gas compression system will usually include an allowance for spillback and an allowance for extra capacity for flexibility in processing alternate feeds. The typical values selected are 10% of the maximum process requirement for spillback and 10% of the process requirement for extra capacity. This means that a typical makeup gas compressor is designed for 120% of the maximum process requirement. The usual means to add the extra 10% capacity is via a clearance pocket. Figure 5.23 shows a typical design. It is mounted on the end of the cylinder and when open, the volumetric efficiency of the compressor is reduced such that the compressor pumps an amount of gas less than the rated capacity, typically 90% of rated. When the pocket is closed, the capacity of the compressor is equal to the full rated capacity of the compressor.

It can be easily recognized that pumping 10% extra gas for use as spillback can represent a major utility cost. To mitigate this, an enhancement is available to be able to match more closely the gas compressed by the makeup gas compressor with the process requirement. This system uses a special set of valve unloaders that act to activate the suction valve during part of the stroke cycle of the piston. By this means, the makeup gas compressor can be continuously varied from about 25–30% of rated capacity to full rated capacity, thus eliminating the need for spillback and saving additional utilities when the process requirement is less than the design of the machine. This has the potential to save substantial utilities cost. The system, called a stepless valve control system, is illustrated in Figure 5.24. In essence, this system is a very sophisticated compressor suction valve unloader and associated sensors and software to load and unload the compressor suction valves over a portion of the piston travel so that the amount of gas actually compressed by the machine is reduced from the rated capacity to an amount that matches the process requirement without having to utilize spillback or clearance pockets. The power consumption is also reduced roughly proportionally.

Figure 5.24 shows the pressure–volume envelope for the rated flow for a typical reciprocating compressor. When fully loaded, the entire shaded area represents the

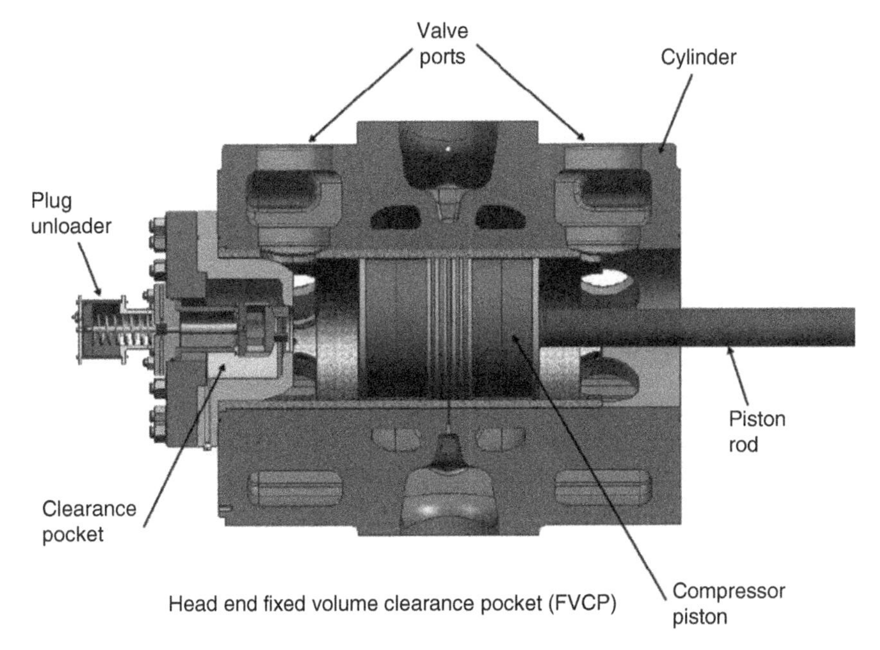

FIGURE 5.23. Clearance pocket for a reciprocating compressor. Source: Reproduced with permission of Dresser–Rand business, part of Siemens Power & Gas.

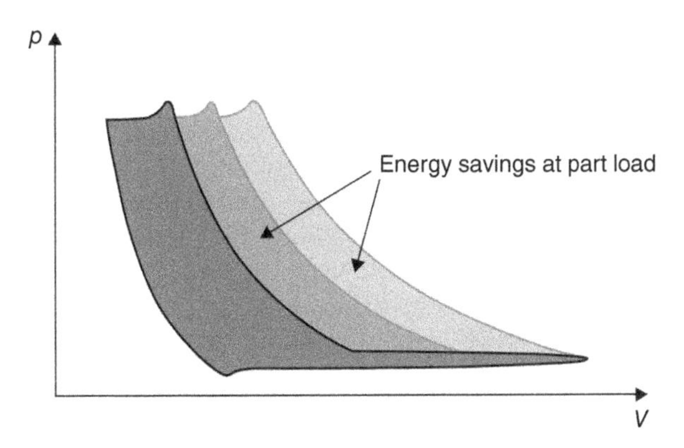

FIGURE 5.24. Operation of stepless valve unloading system. Source: Reproduced with permission of Hoebiger Ventilwerke GmbH & Co.

work done by the compressor at its rated capacity. If the required capacity is less than the rate value, the stepless valve control system acts to unload the suction valves so that the compressor delivers only the amount of gas required by the process. The compressor consumes less power, and the savings is represented by the green shaded areas in Figure 5.24.

Note that it is still necessary to design the makeup gas compression system with a full-sized spillback for capacity control, in case the stepless valve control system is off-line.

There are two main configurations for the spillback system itself. One is the single spillback system, shown in Figure 5.25.

The single spillback configuration features a single spillback valve and associated spillback coolers to route the compressor spillback from the last stage discharge to the first stage suction. The spillback control valve is controlled normally by a signal from the pressure controller located at the separator in the reactor section that passes to the control valve via a low signal selector. In case the makeup gas header pressure begins to fall, a signal from the pressure controller located at the first stage suction drum will become the lowest input signal to the low signal selector and, as a result, take over control of the spillback control valve. In this way, the compressor is protected in case of low makeup gas header pressure. The advantage of the single spillback system is its simplicity, but it has the disadvantage of not allowing for independent control of the interstage pressures. This can lead to problems if each stage of the compressor is not correctly designed. In addition, such a compressor has little or no flexibility for alternate operation, particularly in revamp situations.

The stagewise spillback configuration is shown in Figure 5.26.

Each stage of the compressor is provided with its own spillback valve. The system is arranged so that the compressor discharge coolers can serve also to cool the spillback flow. A pressure controller is provided at each suction drum. During normal operation, the pressure controller at the separator in the reactor section controls the last stage spillback control valve via low signal selector "C." The pressure controller is direct acting, so a low pressure will result in a low controller output and if sufficient hydrogen is available from the header, the separator pressure control signal will be the lower of the two inputs and, as a result, will control the third stage spillback control valve. The second stage spillback control valve is then controlled by a signal from the pressure controller located at the third stage suction drum via low signal selector "B." The first stage spillback control valve is controlled by a signal from the pressure controller located at the second stage suction drum via a signal from low signal selector "A." If the makeup gas header pressure begins to fall, the pressure at the first stage suction drum will begin falling. The pressure controller located on this drum will lower its output signal so that it becomes lower than the other input signal to low signal selector "A," thus taking over control of the first stage spillback valve to spillback gas to bring the pressure of the first stage suction drum back to the controller setpoint. This action will in turn cause the second stage suction drum pressure to decrease, which will cause the pressure controller signal to decrease such that it will be the lower of the two input signals to low signal selector "B," thus taking over control of the second stage spillback control valve to bring the pressure of the second stage suction drum back to the controller setpoint. This action will cause the pressure of the third stage suction drum to decrease, causing the pressure controller located on that drum to lower its output such that it becomes lower than the signal from the separator pressure controller, thus taking over control of the third stage spillback control valve to bring the pressure of the third stage suction drum back to the controller setpoint. The result of this action is that the operating pressure in the unit will begin to decrease, due to an insufficiency of available hydrogen. This situation would need to be addressed quickly, so as not to affect the operation of the unit. Either more hydrogen would need to be produced or the feed rate to the unit would need to be decreased to bring the hydrogen consumption in line with the hydrogen available.

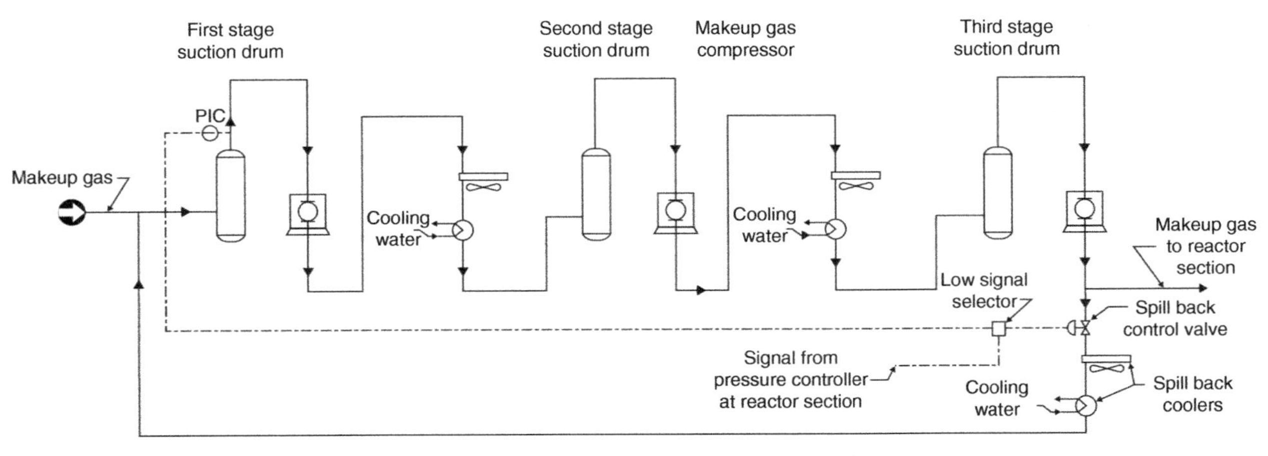

FIGURE 5.25. Single spillback compressor configuration.

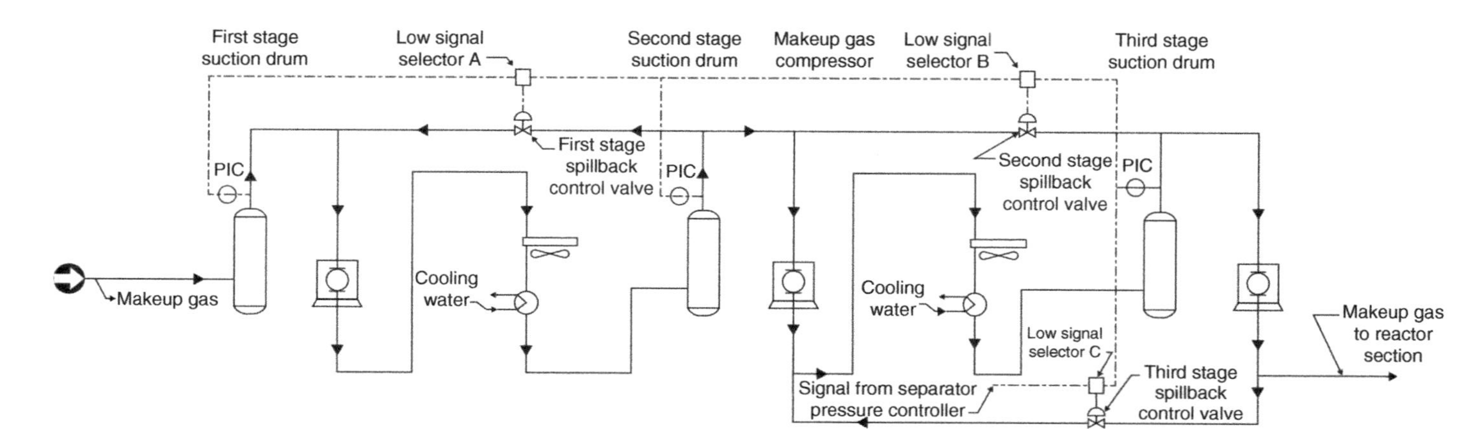

FIGURE 5.26. Stagewise compressor spillback configuration.

This system is more complicated than the single spillback system but provides the maximum flexibility in the operation and can easily accommodate different operations while assuring that the compression ratio of each stage is maintained at the desired value. This system is particularly suitable when the compressor is used to produce a separate makeup gas stream to another unit at an intermediate pressure.

REFERENCES

Aly F, Graven R., Lewis D. (1989) Distribution system for downflow reactors. US Patent 4,836,989.

American Petroleum Institute (2012a), *Design, Materials, Fabrication, Operation, and Inspection Guidelines for Corrosion Control in Hydroprocessing Reactor Effluent Air Cooler Systems*, API Recommended Practice 932-B, 2nd Edition, p. 40, American Petroleum Institute, Washington, DC.

American Petroleum Institute (2012b), *Design, Materials, Fabrication, Operation, and Inspection Guidelines for Corrosion Control in Hydroprocessing Reactor Effluent Air Cooler Systems*, API Recommended Practice 932-B, 2nd Edition, p. 41, American Petroleum Institute, Washington, DC.

Carson C (1972) Flow distributing apparatus. US Patent 3,685,971.

Charpentier JC (1978) Hydrodynamics of two-phase flow through porous media, Powder Europa Conference, Wiesbaden.

Charpentier JC, Favier M (1975) Some liquid holdup experimental data in trickle-bed reactors for foaming and nonfoaming hydrocarbons, *AIChE Journal* **21**(6), 1213–1218.

Ergun S (1952) Fluid flow through packed beds, *Chemical Engineering Progress* **48**(2) 89–94.

Gianetto A, Baldi G, Specchia V, Sicardi S (1978) Hydrodynamics and solid–liquid contacting effectiveness in trickle-bed reactors, *AIChE Journal*, **24**(6), 1087–1104.

Kane RD, Horvath RJ, Cayard MS (2006) Major improvements in reactor effluent air cooler reliability, *Hydrocarbon Processing*, 99–111.

Larkins R, White R (1961) Two-phase concurrent flow in packed beds, *AIChE Journal*, **7**(2), 231–239.

McCabe W, Smith J (1967a) *Unit Operations of Chemical Engineering*, McGraw-Hill, p. 160.

McCabe W, Smith J (1967b) *Unit Operations of Chemical Engineering*, McGraw-Hill, p. 168.

Midoux N, Favier M, Charpentier JC (1976) Flow pattern, pressure loss and liquid holdup data in gas–liquid downflow packed beds with faming and nonfoaming hydrocarbons, *Journal of Chemical Engineering of Japan*, **9**(5), 350–356.

Mist Elimination (2007) Bulletin MEPC-01. Koch-Glitsch LP, p. 6.

Piehl R (1976) *Materials Performance*, **15**(1), 15–20.

Srinivasan S, Ladag V, Kane R (2009) Evaluation of prediction tool for sour water corrosion quantification and management in refineries, corrosion 2009, Paper No. 09337, NACE International.

Tosun G (1984) A study of cocurrent downflow of nonfoaming gas–liquid systems in a packed bed. 2. Pressure drop: search for a correlation, *Industrial and Engineering Chemistry Process Design and Development*, **23**(1), 35–39.

6

DISTILLATE HYDROTREATING UNIT DESIGN

6.1 INTRODUCTION

Most ultra-low-sulfur diesel (ULSD) is produced by direct desulfurization of distillate range materials. This is because naturally occurring distillate range fractions of crude oil have sulfur levels much higher than required to meet ULSD standards.

Chapter 5 covered individual aspects of the design of a hydroprocessing unit. This chapter covers topics related to overall design and optimization of a distillate hydrotreating unit. As with the design of any processing unit, the selection of the optimum flow scheme must take into account factors such as feed type and properties, feed rate, and processing objectives. Once these are known, the designer can optimize the design so that the unit will meet the processing objectives over the required cycle length in an efficient and economical manner.

6.2 NUMBER OF SEPARATORS

One of the first decisions confronting the designer is whether to use a single high-pressure separator or a hot separator/cold separator configuration. Figure 6.1 shows a flow scheme for a single separator configuration. Feed is combined with recycle gas, exchanged against reactor effluent, and then heated to reaction temperature by a fired heater or hot oil, according to the required reactor inlet temperature and the available hot utility. The reactor effluent is exchanged with combined feed and is further cooled to the temperature at which the recycle gas is separated from the product liquid. This temperature is usually in the range of 100–130 °F (38–55 °C). The recycle gas is combined with makeup gas and recirculated to join with the

Hydroprocessing for Clean Energy: Design, Operation, and Optimization, First Edition.
Frank (Xin X.) Zhu, Richard Hoehn, Vasant Thakkar, and Edwin Yuh.

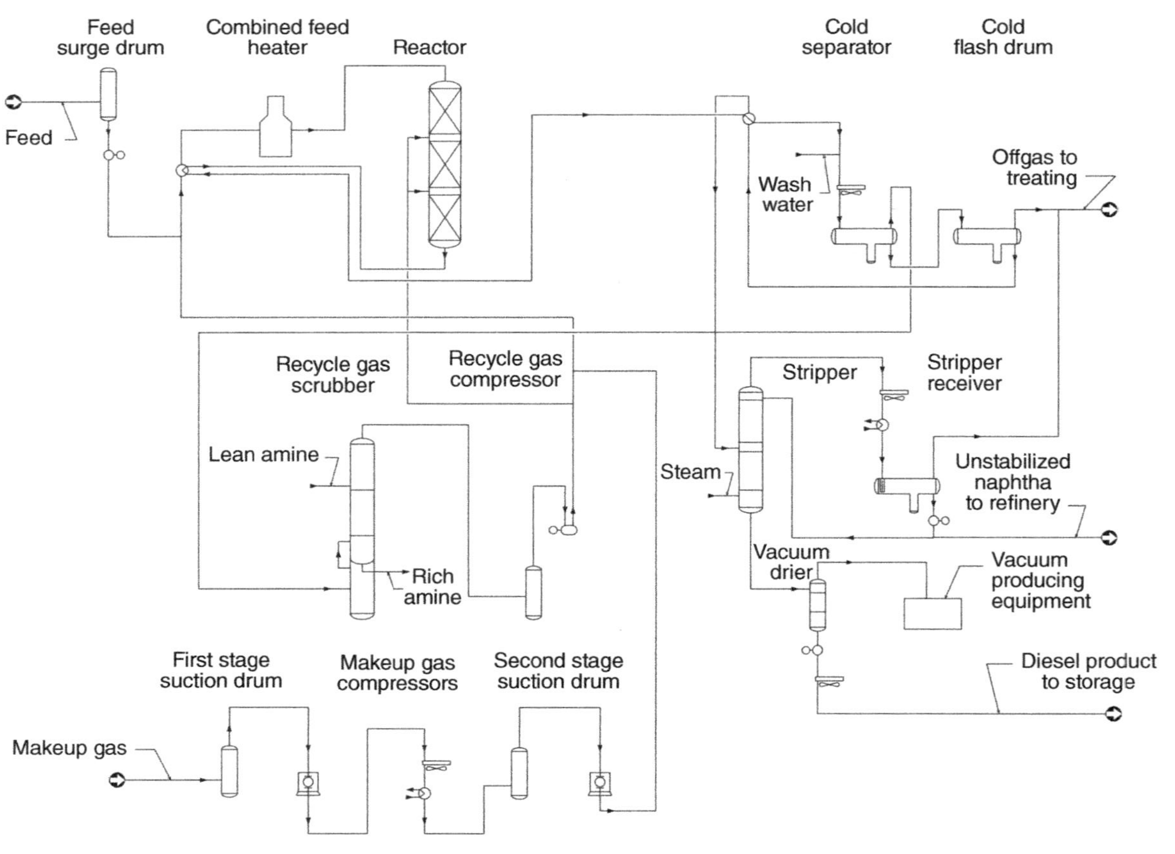

FIGURE 6.1. Single separator flow scheme.

fresh feed. The separator liquid is preheated and sent to the fractionation section for production of final products.

An alternative flow scheme is available and may be a better option in some cases. This involves the use of a hot separator in the reactor section flow scheme. Figure 6.2 shows the configuration of a hydroprocessing unit with a hot separator. The reactor effluent is partially cooled and the vapor and liquid are separated. The hot separator liquid either goes directly to the fractionation section or is let down in pressure and evolved gases are removed in a hot flash drum, and then the liquid is routed to the fractionation section.

The hot separator vapor is cooled to a suitable temperature, usually 130 °F (55 °C) or below and the condensed hydrocarbons are separated from the recycle gas in the cold separator.

The cold separator hydrocarbon is usually routed to the fractionation section directly or via a flash drum, where dissolved gases are evolved as a result of the pressure letdown and are separated. As a result of the partial cooling and separation of the reactor effluent material, it is not necessary to cool the entire reactor effluent stream to the temperature at which the final gas/liquid separation is made and then reheat the hydrocarbon liquid to the temperature required to feed the first fractionation column. This has the potential to make the unit more energy efficient. Table 6.1 was prepared to show a comparison of the estimated equipment cost

TABLE 6.1. Capital and Operating Cost Comparison for Hot Separator Flow Scheme versus Conventional Flow Scheme

Equipment Costs ($MM)	Conventional	Hot Separator
Heaters	1.83	1.74
Vessels		
Reactors	15.23	15.23
Other reactor section vessels	2.96	2.89
Fractionation section vessels	0.65	0.65
Exchangers	4.35	3.38
Pumps	0.81	0.79
Total equipment cost	25.82	24.66

Utility costs ($MM/year)	Conventional	Hot separator
Fuel	2.24	1.19
Fuel credit for PSA waste gas	−0.94	−2.20
Fuel credit for stripper offgas	−3.36	−3.55
Electricity	0.92	0.94
HP steam	1.96	2.07
MP steam	0.84	0.84
Cooling water	0.23	0.22
Makeup hydrogen	33.18	34.39
Credit for hydrogen recovered	−0.88	−1.61
Total annualized utilities cost	34.20	32.29

Utilities cost basis: Fuel: 470 US$/1000 kg; Electricity: 0.12 US$/kWh; Hydrogen: 1680 US$/MT (metric ton); Cooling water: 0.04 US$/$m^3$

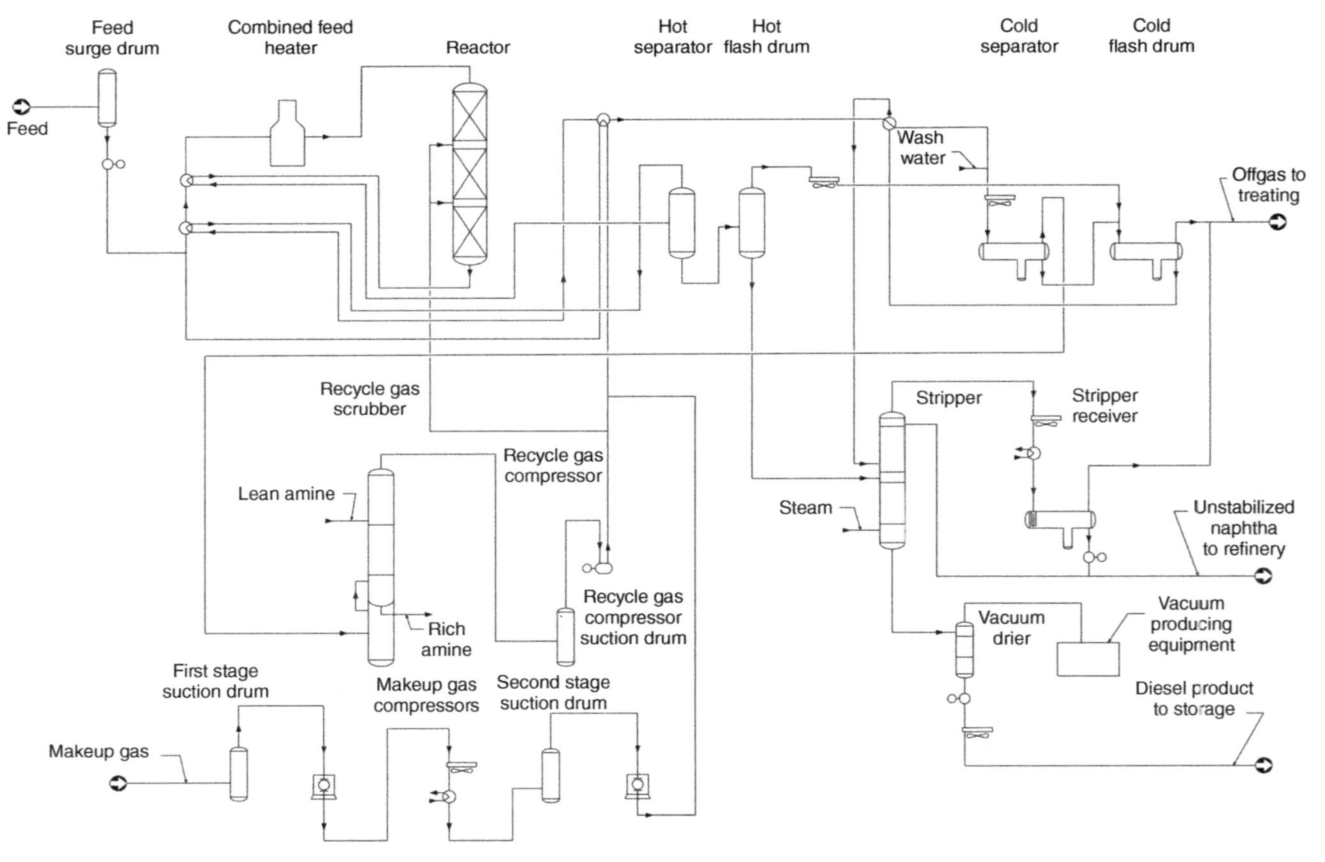

FIGURE 6.2. Hot separator flow scheme.

(US Gulf Coast basis) and utilities for a 35,000 BPSD distillate hydrotreating unit with either a conventional single high-pressure design versus a hot separator design.

It can be seen that a hot separator flow scheme is more energy efficient while also being less capital intensive than a conventional flow scheme for the 35,000 BPSD case studied. Although additional vessels are required for the hot separator flow scheme, the capital cost is still less because the heat exchange surface area requirement is less and the cold separator and cold flash drum can be of a smaller size. One observation can be made from the above data. The cost of hydrogen and its efficient use plays an important role in the economics of a hot separator design. This is related to the ability to maintain the required hydrogen partial pressure in the reactor. In the hot separator flow scheme, much of the liquid in the reactor effluent material is removed and routed directly to the fractionation facilities. Therefore, when the hot separator vapor is cooled to the final temperature of the cold separator, there is less hydrocarbon liquid present compared with a conventional single separator design. Less liquid present means a decreased ability to absorb nitrogen and methane; the former could be present in the makeup gas and the latter could be present in the makeup gas as well as a by-product of the conversion of the feed to products. This issue is treated in more detail in Section 5.3. Also, since the solubility of hydrogen increases with temperature, hydrogen solution losses are greater for a hot separator flow scheme than for a conventional flow scheme. Since most of the liquid in the reactor effluent in a hot separator flow scheme leaves as hot separator liquid, more hydrogen will be dissolved in this stream than would be the case for an equivalent amount of liquid at the lower temperature cold separator.

Hydrogen contained in the hot and cold separator liquid represents the loss of a valuable resource. In addition, as will be seen in the section on fractionation design, the presence of hydrogen can affect the recovery of light hydrocarbons. If the unit is designed with flash drums to recover some of the evolved hydrogen when the hydrocarbon liquid is let down in pressure as it is being routed to the fractionation facilities, then a hydrogen-rich gas can be produced at a pressure that is compatible with the makeup hydrogen supply pressure, in which case it can be recompressed and recycled to the unit. If the operating pressure of the unit is below a certain value, the amount of hydrogen contained in the separator liquids decreases to a point where the gas evolved is too small to justify the installation of flash drums. In that case, the hydrogen contained in the separator liquids is not generally recovered.

As can be seen from Table 6.1, the economics of a hot separator flow scheme depend on recovery of solution loss hydrogen to some extent. So while a hot separator flow scheme is potentially more economical and energy efficient, the economics are strongly influenced by the cost of hydrogen and how it is handled.

6.3 STRIPPER DESIGN

6.3.1 Introduction

Fractionation facilities for hydrotreating units are generally rather modest. Table 6.2 (Worldwide Fuel Specifications, 2013) lists the product specifications for diesel for the United States and Europe.

TABLE 6.2. Diesel Fuel Standards for the United States and Europe

Location	United States	Europe
Spec name	ASTM D 975-10c	EN 590:2009
Comment	No. 1-D S15	—
Cetane number	40[a]	47–51[b]
Sulfur, ppm, max	15	10
Total aromatics, vol%, max	35[a]	—
Polyaromatics, wt%, max	—	11
Density @ 15 °C, kg/m^3, min–max	—	800–845[b]
Distillation		
T95, °C, max	228	360
Flash point, °C, min	38	55
Cloud point, °C	—	−10 to −34[c]
Water, vol%, max	—	0.02

[a]Either cetane or total aromatics must be met.
[b]Depends on climate.
[c]For countries with severe winter conditions.

These standards can quite often be met by removing the hydrogen sulfide and sufficient light materials such that the flash point specification is met. The endpoint distillation temperature of the product is set by the upstream unit, which is usually the crude tower.

6.3.2 Stripper Column Design

Two major design options are available for the stripper. Figure 6.3 shows the configuration for a reboiled stripper. The operating pressure of the column is selected so as to be able to route the net gas from the receiver to the fuel gas system, thus avoiding an expensive overhead gas compressor. Since this gas stream will contain a significant amount of hydrogen sulfide, amine treating facilities are usually provided either within the hydrotreating unit itself or at a common location elsewhere in the refinery. This usually results in an operating pressure of the stripper receiver of 7 bar (100 psig) to 10 bar (150 psig). One important consideration when designing a reboiled distillate stripper is the outlet temperature of the reboiler. Because the focus is on sulfur reduction and not nitrogen reduction, and because nitrogen is generally more difficult to remove than most sulfur species, some nitrogen compounds will be present in the distillate product. These compounds are temperature sensitive and heating them above a certain temperature can cause the product to change color and thus make the product unfit for sale. This is particularly true for distillate products derived from thermal processes, such as delayed coking. For this reason, the reboiler outlet temperature is generally limited to about 350 °C (675 °F). (Note that diesel product from a hydrocracking unit does not normally exhibit this tendency, as the severity of the operation is such that nitrogen compounds are generally removed to a much greater extent and such color stability issues are not generally seen.)

If it is found that a reboiler outlet temperature in excess of the above-stated value would be required, then the pressure of the column should be reduced if possible until the reboiler outlet temperature falls within the limit stated above.

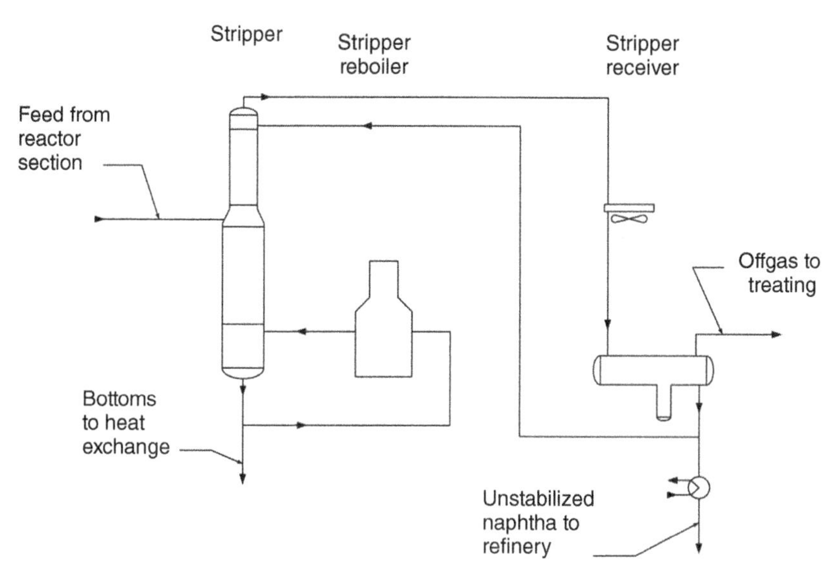

FIGURE 6.3. Reboiled stripper column.

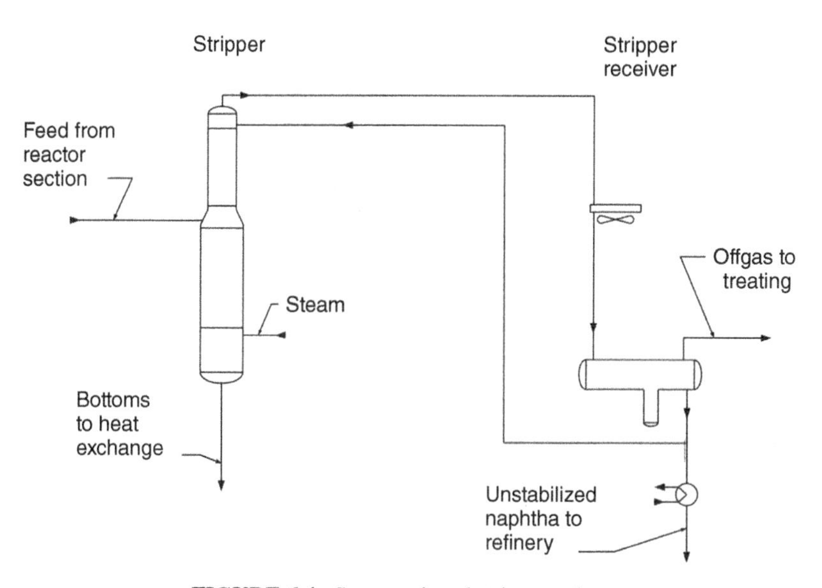

FIGURE 6.4. Steam stripped stripper column.

The other design option for the stripper is a steam stripped column. This is shown in Figure 6.4. The advantage of this design is that a fired heater is not required, thus reducing the number of fired heaters required for the unit. A source of steam at a pressure appropriate for the operating pressure of the column is routed to a point below the bottom tray. The amount of steam required is set to achieve the required product specifications. This can usually be accomplished with an amount of steam equal to 3–5 pounds of steam per barrel of bottoms liquid (8.6–14.2 kg/m^3). This amount represents a small mass flow relative to the bottoms liquid. As such, the enthalpy of the steam does not contribute significantly to the column heat input. The purpose of the steam is to reduce the partial pressure of the hydrocarbon at the bottom of the column, thus vaporizing a portion of it. This vaporized liquid, along with the stripping steam, constitutes the vapors rising up the column and thus provides the necessary stripping vapor to achieve the required separation.

Since there is no reboiler to supply heat to the column (the enthalpy supplied by the stripping steam is negligible), the heat necessary to operate the column has to originate from the feed stream(s).

Steam stripped columns have an important consideration that is not generally an issue for reboiled stripper columns. This has to do with the fact that most of the steam injected will continue up the column and exit the column via the overhead vapor (a portion of the injected steam will dissolve in the liquid phase and be carried out with the bottoms material). The column must be designed and operated so that the conditions are such that a water phase will not form on the trays.

This is usually addressed by requiring that the water dew point of the column vapor be lower than the actual column overhead vapor temperature by a certain amount. This is called the dew point margin and is calculated by the following process:

(1) Divide the moles of water present in the column overhead vapor by the total moles of overhead vapor.
(2) Multiply this value by the absolute pressure of the overhead vapor stream. This is the partial pressure of water.
(3) Look up the saturation temperature of water at the calculated partial pressure in a steam table.
(4) Compare the saturation temperature with the column overhead vapor temperature. The result is the dew point margin (or deficit).

The following is an example of this calculation:

(1) Total moles of column overhead vapor: 2864.01 lb mol/h
(2) Moles of water in column overhead vapor: 649.63 lb mol/h
(3) Mole fraction of water in overhead vapor = 649.63/2864.01 = 0.2268
(4) Pressure of overhead vapor: 166 psig = 180.7 psia
(5) Partial pressure of water: 180.7 × 0.2268 = 40.99 psia
(6) From steam table, saturation temperature of water at 40.99 psia = ~270 °F
(7) Column overhead temperature: 303 °F
(8) Dew point margin: 303 °F – 270 °F = 33 °F

A dew point margin of 25–30 °F is usually selected to provide some flexibility to account for normal operating variations. A closely related parameter is the column reflux ratio, expressed as the molar ratio of reflux to net liquid, sometimes expressed as molar ratio of reflux to feed. As the heat input to the column decreases, the reflux ratio decreases and less hydrocarbon exits the column in the overhead vapor. This causes a decrease in the dew point margin and if the column heat input is decreased further, at some point the water dew point is reached and water will begin to form on the trays at or near the top of column. At this point, an undesirable situation occurs. The water condensing on the trays has nowhere to go. It cannot vaporize and leave with the overhead vapor because the temperature is less than the dew point. The water cannot leave the bottom of the column because the temperature there is above the dew point and any water proceeding down the column will revaporize and travel back up the column. The result is a buildup of water at some point in the column and eventually the column will flood with a resulting upset in the column operation. Another undesirable situation is that the liquid water on the trays as well as hydrogen sulfide in the rising vapors can contribute to corrosion in the upper portion of the column. To mitigate this, trays near the top of the stripper are supplied with upgraded metallurgy. In addition, some portion of the top of the column, including the top head, will be lined with a corrosion-resistant material. The determination of which material to use will depend on the contaminants in the feed. For example, if the feed contains sulfur and nitrogen, the lining will consist of some type of stainless material. If chlorides are present in the feed, then the lining will usually be monel, as chlorides are known to attack stainless material.

As can be seen from the above discussion, maintenance of a dew point margin is important in the operation of a steam stripped column. Unfortunately, the operator does not always have a ready ability to determine the dew point margin of the water in the overhead vapor, leading to the problems described. Some work has been done to address this information deficit. A system has been devised to measure the partial pressure of water in the column overhead vapor and thereby the dew point margin. Figure 6.5 shows the basic configuration of this system (Hoehn, 2003).

An analyzer measures the density of the overhead vapor. The molecular weight can be determined by the following equation:

$$\mathrm{MW}_{\mathrm{ov}} = \frac{\rho R T_a}{P_a}, \tag{6.1}$$

where

MW_{ov} = molecular weight of the vapor
ρ = density of the vapor, lb/ft^3
R = ideal gas constant, 10.73 psi ft^3/°R
T_{a} = absolute temperature, °R
P_{a} = absolute pressure, psia.

The masses of sour water, net vapor, and total receiver hydrocarbon are also measured. The sum of these materials represents the total material exiting the column in the overhead vapor. The number of moles of overhead vapor is obtained by dividing the total mass by the molecular weight inferred by the density measurement.

$$N_{\mathrm{ov}} = \frac{W_{\mathrm{ov}}}{\mathrm{MW}_{\mathrm{ov}}}, \tag{6.2}$$

where

N_{ov} = moles of column overhead vapor
W_{ov} = mass of total overhead material.

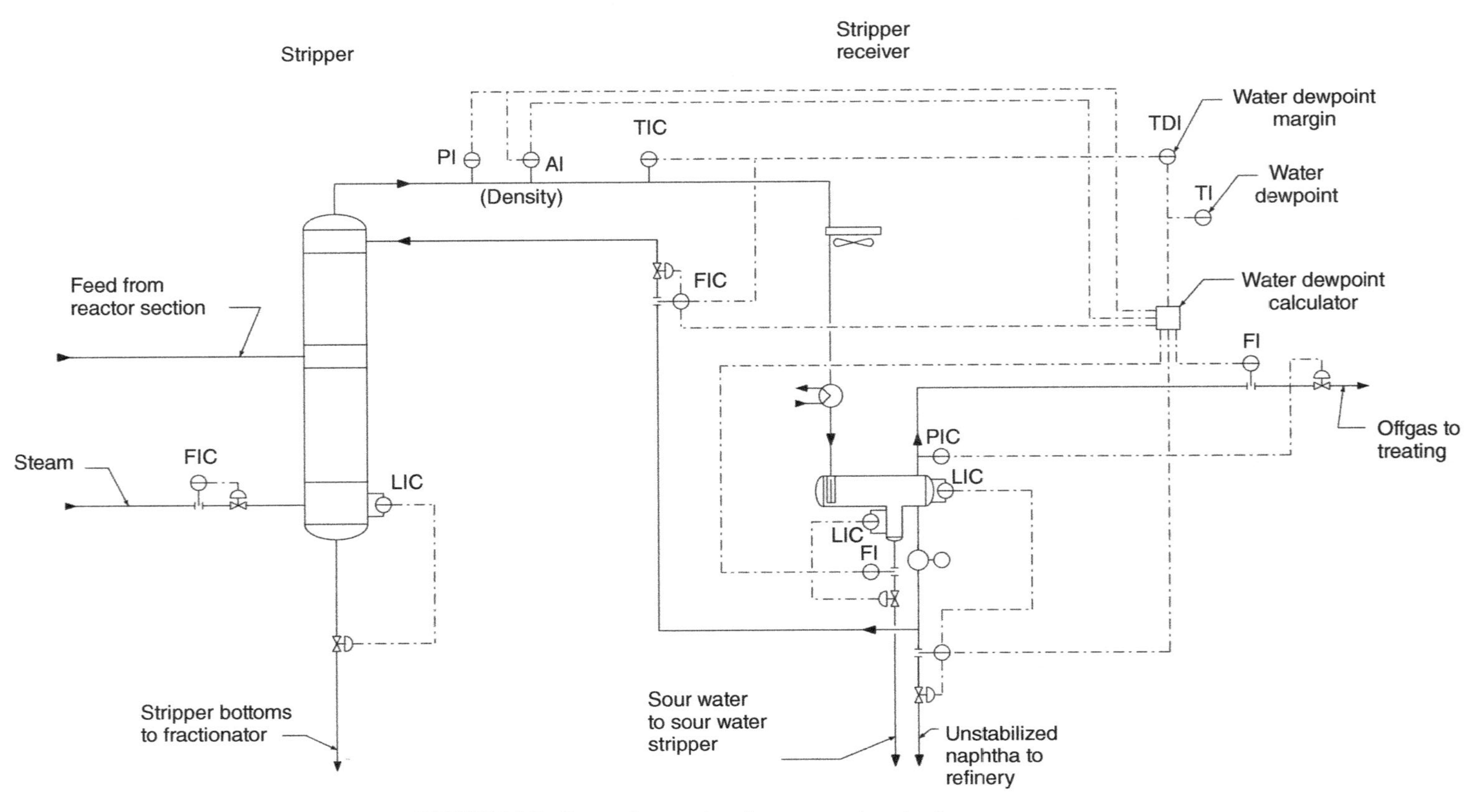

FIGURE 6.5. Dew point monitor for steam stripped columns.

This number is then divided into the number of moles of water leaving the receiver and when multiplied by the pressure of the overhead vapor, the partial pressure of water in the overhead vapor is obtained.

$$P_{H_2O} = \frac{(W_{H_2O}/18)}{N_{ov}} \times P_a, \tag{6.3}$$

where

P_{H_2O} = partial pressure of water in column overhead vapor
W_{H_2O} = mass of water leaving the overhead receiver.

Now that the partial pressure of water is known, the saturation temperature of water at the calculated partial pressure can be obtained using a correlation based on the steam tables.

This is easily calculated in the Distributed Control System and can be compared with the column actual overhead vapor temperature and the difference displayed as the operating margin above dew point. Such a system provides useful information to assist the refiner in operating a steam stripped column in a manner that avoids the problems mentioned above.

Steam stripped columns have another feature that must be taken into account in the design of the unit. The stripping steam will be in equilibrium with the column bottoms liquid at elevated temperature, and, therefore, the column bottoms liquid will be water saturated. The amount of water is a function of the temperature of the bottoms liquid. Figure 6.6 shows this relation for *n*-octane. Other hydrocarbons exhibit a similar tendency. Since the stripper bottoms temperature is typically in the range of 400–550 °F (477–530 °K), it can be readily seen that the water solubility can be quite substantial (Tsonopoulos, 1999).

As the bottoms liquid is cooled, the water solubility is decreased and a separate water phase will form. This is usually removed with the assistance of some type of coalescer device designed to separate the water phase from the hydrocarbon product. The hydrocarbon liquid is still water saturated and can have a tendency to form small water droplets when cooled further. This often happens when the product is routed to a storage tank and cools slightly from the temperature at which it leaves the unit. These small water droplets tend to remain suspended in the product and are called haze. Distillate material containing water haze cannot usually be sold. Therefore, some means must be applied to prevent this occurrence. There are basically two methods for mitigating this problem.

The first and simplest solution is to pass the cooled hydrocarbon product through a coalescer to remove any free water and then through a bed of salt. The salt is hygroscopic and will cause some of the soluble water to be removed. This has the effect of lowering the haze point, defined as the temperature at which small water droplets will begin to form in the hydrocarbon liquid, by about 10–15 °F (5–10 °C). This will be sufficient for most purposes. However, in cold climates, it is usually necessary to remove more of the water to prevent a haze formation. In this case, some form of vacuum drying is applied. This involves partially cooling the hydrocarbon product,

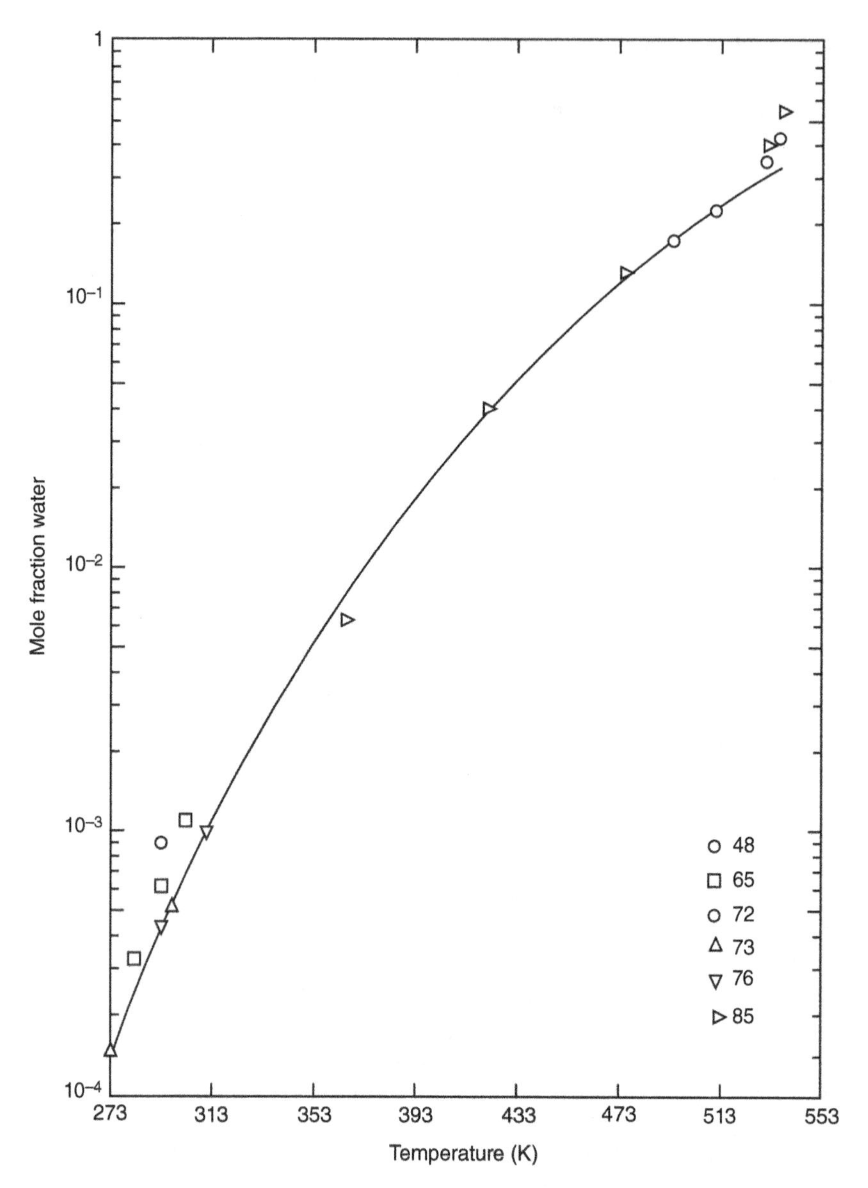

FIGURE 6.6. Solubility of water in *n*-octane versus temperature.

removing any water formed and then routing the hydrocarbon product to a vessel operated at subatmospheric conditions. The water will flash into the vapor phase and is withdrawn overhead, condensed, and produced as a separate water phase. The amount of vacuum that must be imposed is a function of the temperature at the inlet of the drier and the desired water content. Figure 6.7 shows the relationship between operating pressure, temperature, and product water content for a typical diesel range product.

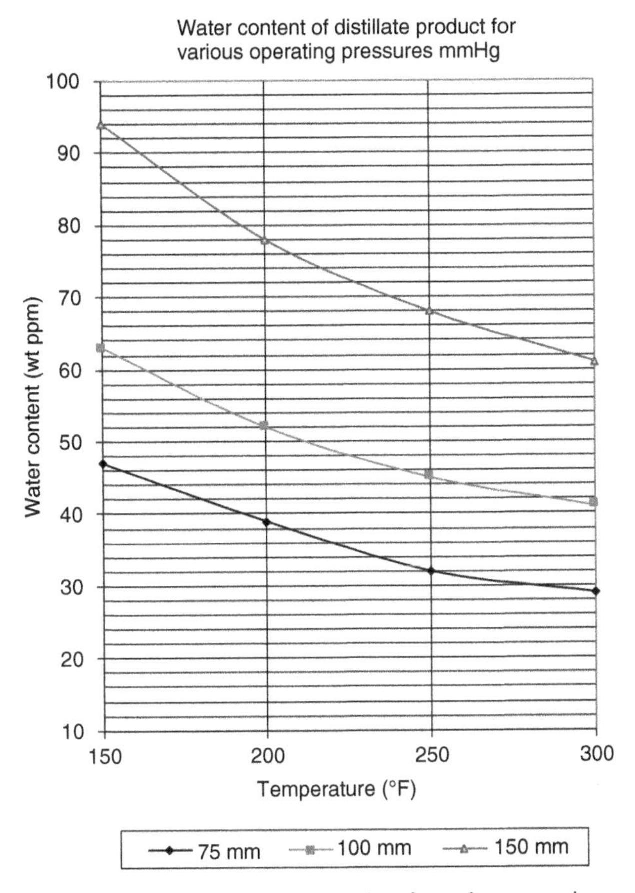

FIGURE 6.7. Water content of distillate product for various operating pressures.

6.4 DEBUTANIZER DESIGN

If sufficient light material is present in the feed or produced in the reactor, the stripper column will produce a net overhead liquid. This liquid will contain lighter components such as hydrogen, methane, ethane, hydrogen sulfide, in addition to LPG, and naphtha. The presence of these lighter components makes it unsuitable for sending directly to storage. Therefore, the overhead liquid must be "stabilized," which involves removing sufficient light material and hydrogen sulfide so that the material can be sent to storage. Often, the net overhead liquid can be processed in other facilities, such as the crude unit gas concentration plant. However, if other facilities do not have sufficient capacity to process this material, it will be necessary to add a debutanizer column to stabilize the net overhead liquid.

The first task is to set the operating conditions. Since light material will be produced overhead, it is necessary to select an operating pressure at which sufficient light material can be condensed to produce column reflux at the maximum expected dry bulb temperature for the site at which the unit will be located. This generally dictates an operating pressure of about 175–250 psig (12–17 bar).

When designing a column of this type, it is important to keep in mind that the column should not operate with a bottoms temperature that is too high. The reason for this is that the light components in the C_4 and C_5 range have critical temperatures that could be within the range of operating temperatures, particularly if the selected pressure is at the high end of the above-listed range. As the operating pressure approaches the critical temperature, it becomes more difficult to achieve the desired separation. To avoid this, the column operating pressure is usually set so that the reduced temperature, defined as the operating temperature divided by the critical temperature expressed as absolute temperatures, of the bottoms liquid is some fraction below 1. In other words, the bottoms temperature is sufficiently below the critical temperature so as not to prevent the column from operating as intended. The column is reboiled either with available heat, usually stripper bottoms, or with steam. The degree of separation will be dictated by the desired vapor pressure of the bottoms material, usually set by an RVP (Reid vapor pressure) specification, and the overhead material will usually have a specification on the maximum amount of pentane and heavier material. These specifications will determine the number of trays and reboiler duty.

6.5 INTEGRATED DESIGN

6.5.1 Introduction

Beyond the basic design concepts, there is an opportunity to optimize the overall design to minimize capital expenditure (CAPEX) and operating expense (OPEX). Given the life of most processing units – 20–30 years or more – this can have a significant and lasting effect on the long-term economics and profitability of the operation. This section covers several areas of hydrotreating unit design that merit consideration to achieve an optimized design.

6.5.2 Optimum Hot Separator Temperature

As mentioned in Section 6.2, most distillate hydrotreaters would be designed with a hot separator. The question then arises as to the proper operating temperature. This can be addressed by performing a rather quick and simple analysis. Basically, the operating temperature of the hot separator is selected to minimize the heat exchange surface requirement for a given heat recovery ratio, while still being able to achieve the proper stripper operation. If the stripper is reboiled, there is a trade-off between the reactor charge heater duty, stripper reboiler duty, and the reactor section exchange surface. If the stripper is steam stripped, nearly all of the heat required to operate the stripper column is present in the reactor section effluent streams. Therefore, the temperature of these streams must be such that the hydrogen sulfide specification in the stripper bottoms liquid and the dew point margin in the stripper overhead is maintained while minimizing the capital and operating costs.

While there are similarities in the optimization method, the following outlines the procedure for optimizing the hot separator temperature for a unit with a steam stripped stripper column. As noted above, almost all of the enthalpy needed to operate the stripper originates from the reactor effluent streams feeding the column. The enthalpy

TABLE 6.3. Stripper Operation versus Hot Separator Temperature

Hot Separator (°F)	Cold Flash Liquid (°F)	Stripping Steam (lb/bbl bottoms)	Bottoms H_2S Content (mol ppm)	Dew Point Margin (°F)
475	300	3.1	0.4	41
500	300	3.0	0.3	30
525	300	3.0	< 0.1	31
550	250	3.0	< 0.1	25
575	250	3.0	< 0.1	25

associated with the stripping steam is not significant. Its purpose is to lower the partial pressure of hydrocarbons and other compounds at the bottom of the column, thus facilitating vaporization, which together with the rising steam, provides the stripping vapor needed to effect the proper separation.

Due to the fact that there is relatively little light ends production in a typical distillate hydrotreating unit, the liquid originating from the hot separator will be the largest feed stream. The temperature to which the material from the cold separator has to be heated and the amount of stripping steam required is not a very strong function of the hot separator temperature, as can be seen from Table 6.3.

A few comments about the data in Table 6.3 are as follows:

(1) The flow scheme used for the study was designed with hot flash and cold flash drums, instead of feeding the hot and cold separator material directly to the stripper. The results would be expected to be essentially the same if the flash drums were not present.

(2) Less stripping steam is required as the hot separator temperature increases. However, at some point, it reaches a practical limit below which the tray design would be affected. This value is taken to be 3 lb steam/barrel of bottoms.

(3) Somewhat counterintuitively, the dew point margin actually decreased with an increase in hot separator temperature.

Figure 6.8 shows the effect of hot separator temperature on exchanger surface requirements. In this example, the reactor charge heater duty was kept constant and the exchanger requirements were then determined taking into account the cold flash liquid heating requirements (the hot separator liquid passes directly from the hot flash drum to the stripper).

Figure 6.9 is a graph of the total capital and operating costs versus hot separator temperature. The capital costs include the reactor section heat exchangers and stripper overhead condenser. Operating costs include fuel, steam, and electricity. In this case, the annual operating costs were multiplied by a factor of 3, which is a typical simple payback period for a trade-off between capital and operating costs.

Note that the shape of the curve in Figure 6.8 shows a minimum at about the same point as that in Figure 6.9. This means that the economics are driven mainly by the reactor section exchanger surface requirement and that this could be a quick optimization parameter on which to focus for a quick determination of the optimum hot separator operating temperature.

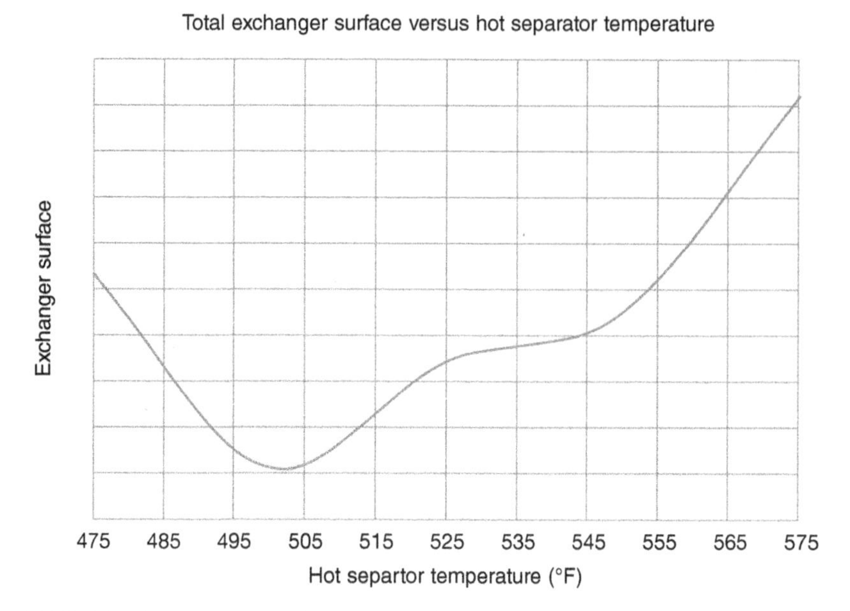

FIGURE 6.8. Exchanger surface area requirements versus hot separator temperature.

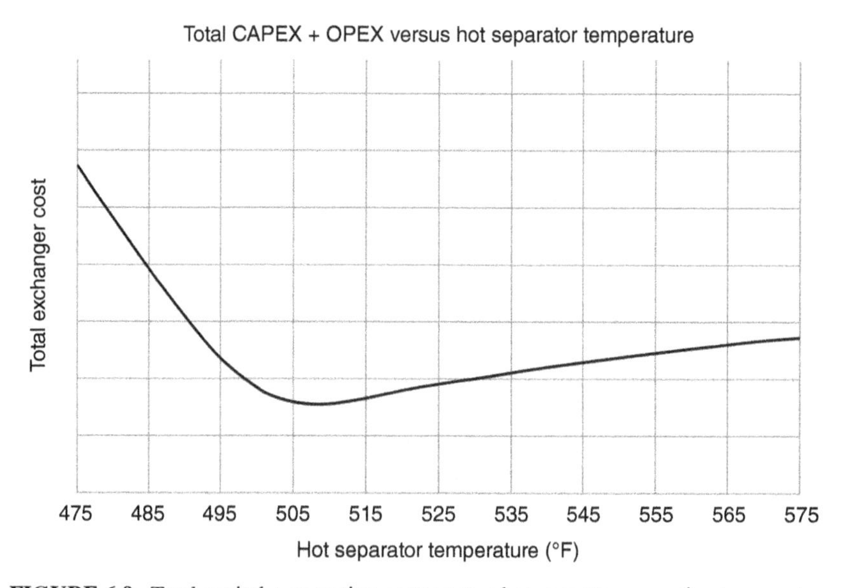

FIGURE 6.9. Total capital + operating costs versus hot separator operating temperature.

Another observation from Figure 6.9 is that after reaching a minimum, the total costs increase more slowly than for hot separator temperatures below the optimum. This is because exchange shifts from reactor effluent to stripper bottoms as the hot side material. Although stripper bottoms is at a lower pressure and such exchange would be expected to be cheaper, the results indicate that such is not necessarily the case.

6.5.3 Use of Flash Drums

It is often desirable to provide a flash drum for the hydrocarbon streams being let down in pressure. The flash drum has two purposes. The first is to provide a means of separation and removal of vapor evolved as the pressure of the liquid is reduced. This aspect is particularly useful in higher pressure units where hydrogen solubility, and therefore the amount of vapor evolved, would be rather large. Providing a flash drum operating at an appropriate pressure provides for the ability to route the flash gas, after treating to remove hydrogen sulfide, to facilities for recovering the hydrogen contained to lower the net hydrogen requirement for the unit. Table 6.4 shows the total hydrogen solution loss and percent of solution loss contained in the cold flash offgas for different operating pressures. The unit configuration has a hot separator with a low pressure hot flash drum and cold flash drum. The cold flash drum operates at 350 psig (24.1 bar).

As expected, solution losses increase with operating pressure. As the operating pressure of the unit decreases, the solution loss hydrogen recovered as flash gas, and the incentive to use flash drums, decreases. The pressure at which that trade-off occurs is a function of the size of the unit and the cost of hydrogen production.

In addition to providing a means for recovering solution loss hydrogen, the presence of flash drums reduces the amount of hydrogen contained in the feed to the fractionation section. The presence of hydrogen in the fractionation section affects the ability to recover light hydrocarbons, particularly propane and butane (LPG). Table 6.5 shows a comparison between the overall propane and butane (LPG) recovered as an overhead liquid product from the stripper column for a hydrotreating unit operating at three different pressures. The unit is configured with a hot separator flow scheme with and without flash drums.

Some conclusions can be drawn from the data. The recovery of LPG into the stripper overhead liquid is not very large. A significant amount of LPG range material

TABLE 6.4. Recoverable Hydrogen in Flash Gas for a Hot Separator Flow Scheme

Operating Pressure, psig (barg)	Total Solution Loss, SCFB (nm^3/m^3)	% Solution Loss in Flash Gas
1400 (94)	61 (10.3)	76.4
1200 (82.8)	52.3 (8.8)	72.4
1000 (69)	43.6 (7.3)	66.8
800 (55.2)	34.9 (5.9)	58.3
600 (41.4)	26.8 (4.51)	44.0

TABLE 6.5. Relative LPG Recovery with and without Flash Drums

Operating Pressure (bar)	LPG Recovery with Flash Drums (%)	LPG Recovery without Flash Drums (%)
94	17.4	6.9
70	18.5	8.8
50	18.5	11.0

leaves with the stripper net vapor. However, where LPG recovery is desired, the use of flash drums will significantly enhance the recovery. It can also be seen that as the pressure of the unit decreases, the LPG recovery actually increases, although for a unit with flash drums, the recovery seems to flatten out below 70 bar, but is still significantly higher than a comparable unit without flash drums. The reason for this is that at lower pressures, less hydrogen is dissolved in the separator liquid and therefore less hydrogen is present to interfere with LPG recovery.

For lower pressure hydrotreating units, the high-pressure separator operates at a low enough pressure such that providing a flash drum will be of minimal benefit in separating the relatively small amount of vapor produced when the hydrocarbon liquid is let down in pressure. In such cases, the hydrocarbon liquid is routed directly from the high-pressure separator to the fractionation section.

Another purpose of the flash drum is to provide a buffer between the high and low pressure systems. A flash drum provides a convenient location for the relief valve that would prevent an overpressure due to a loss of level in the upstream separator and subsequent large amount of vapor that would pass to the downstream equipment. This relief load is usually the controlling case. This is a very important consideration, as failure to do so can result in catastrophic consequences (UK Health and Safety Executive, 1987). In a similar manner, it may be more intrinsically safe to route the sour water from the cold separator to the flash drum as a means of limiting the pressure that would be imposed on the sour water rundown line to the sour water stripper. Routing the water directly from the cold separator to the sour water stripper requires that the piping either be provided with a relief valve to prevent overpressure in the event of a blockage in the rundown line, or that the design pressure of the piping up to the last block valve at the sour water stripper unit be rated for the same design conditions as the cold separator. Either design is less desirable than routing the sour water to the flash drum, where the pressure is limited to the flash drum relief valve set point. This pressure is usually low such that special protection of the sour water rundown line is not necessary. In any case, a thorough analysis of the safety and relieving requirements is essential to provide sufficient protection for downstream equipment and piping.

6.5.4 Hydrogen Recovery

Hydroprocessing units require extensive amounts of hydrogen. There are two main sources in a typical refinery: offgas from a catalytic reforming unit and hydrogen produced by steam–methane reforming (SMR). Often, a combination of the two is used to supply the hydrogen requirements of the refinery. Its cost is driven largely by the cost of the feedstock, which is primarily natural gas. Typical production cost can vary from about $1.75 per 1000 standard cubic feet (SCF) ($735 per metric ton) to as much as $7.00 per 1000 SCF ($3675 per metric ton), according to the local price of natural gas. To put this into perspective, for a 50,000 BPSD hydrotreating unit consuming an amount of hydrogen equal to 0.6 wt% of the feed, the annual hydrogen cost can vary from $9.7 to $39.6 MM for the hydrogen production costs mentioned above. For a hydrocracking unit of the same capacity consuming an amount of hydrogen equal to 2.3 wt% of the feed, the annual hydrogen cost would be between $37.2 and $151.8 MM for the same hydrogen production cost range. Therefore, efficient use of hydrogen is critical to the economic operation of any hydroprocessing unit and, indeed, the overall economics of the refinery.

Hydrogen consumption is a function of the level of impurities, such as sulfur and nitrogen, the amount of olefins and aromatics in the feed, and the required product quality. The sum total of the hydrogen required to accomplish these operations is called the chemical consumption. In addition, hydrogen has the ability to dissolve in hydrocarbon liquid and, to a lesser extent, in water. Thus, the liquid leaving a separator will contain some amount of dissolved hydrogen. The amount will be a function of the temperature and pressure of the separator, as well as the partial pressure of hydrogen in the vapor in equilibrium with the liquid. The hydrogen contained in the separator liquids is called solution loss. As was shown in Chapter 5, hydrogen exhibits a curious property in that unlike most compounds, its solubility in hydrocarbons increases with temperature. Therefore, a unit designed with a high-pressure hot separator and cold separator will have a higher solution loss than a unit designed with only a single high-pressure separator. Figure 6.10 compares the solution loss for a hydrotreating unit with a single separator and one with a hot separator. The solution loss is usually expressed in terms of standard volume of hydrogen loss per unit volume of feed to the unit.

As can be seen, the addition of a hot separator to the flow scheme has a major effect on the hydrogen solution loss. The amount is, of course, a function of the operating temperature of the hot separator.

Given the cost of producing hydrogen, there is incentive for recovering as much of the solution loss hydrogen as possible. As was shown in Section 6.5.3, use of flash drums can facilitate the recovery of solution hydrogen losses. If the cold flash drum is operated at a suitable pressure so that the gas can be fed directly to the hydrogen recovery facilities, the cost of the system is minimized. Table 6.4 shows that as the separator pressure decreases, less of the hydrogen solution can be recovered. Where is the remaining solution loss hydrogen? The answer is that almost all of the solution loss hydrogen (with the exception of the relatively small amount contained

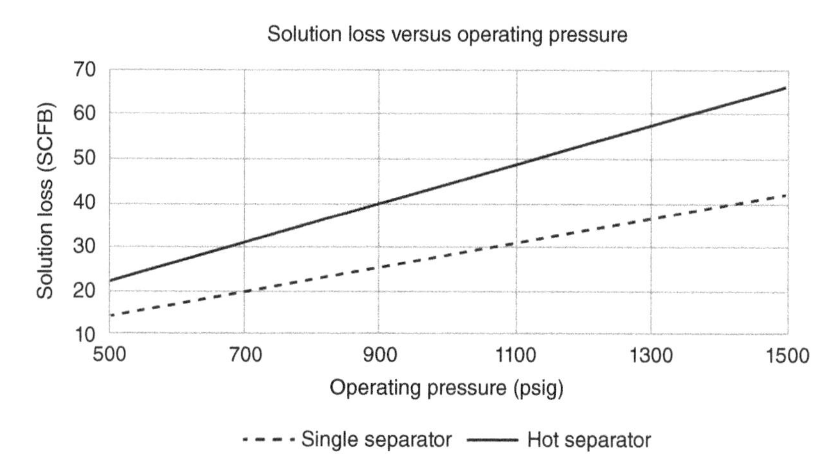

FIGURE 6.10. Solution loss for a hydrotreating unit with and without a hot separator.

TABLE 6.6. Recoverable Hydrogen Solution Loss versus Operating Pressure

Operating Pressure, psig (barg)	Solution Loss, % of Makeup Gas	% Solution Loss Recovered from Flash Gas	% Solution Loss Recovered from Flash + Stripper Offgas
1400 (94)	13.2	68.8	91.5
1200 (82.8)	11.6	65.2	91.7
1000 (69)	9.0	60.2	91.7
800 (55.2)	8.0	52.5	91.6
600 (41.4)	6.2	39.6	91.5

in the separator sour water and recycle gas scrubber amine streams) will be present in the stripper overhead vapor and liquid. Of these two streams, by far the larger amount of solution loss hydrogen is contained in the stripper offgas. Therefore, maximum utilization of hydrogen involves recovery of hydrogen in the stripper offgas as well as the flash gas (Hoehn, 2014). The stripper operating pressure is significantly below that of the flash drum and less than that necessary to be routed to the hydrogen recovery facility. Therefore, this stream must be compressed so that it can be joined with the flash gas. Table 6.6 shows the benefits of compressing the stripper offgas so that it can be routed to the hydrogen recovery facility along with the flash gas, if present.

Table 6.6 shows that although the recoverable solution loss decreases with operating pressure if a flash drum is the only means for recovery, compression of the stripper offgas results in an almost constant, high level of recovery. Furthermore, stripper offgas compression would be especially beneficial for units operating at relatively low pressures. Figure 6.11 shows the configuration for a unit configured without flash drums and Figure 6.12 shows the configuration for a unit that is designed with flash drums.

Two main methods for recovering hydrogen are available: pressure swing adsorption (PSA) and membrane separation. One of the primary selection criterion is the pressure at which the gas is available. PSA units produce the purified hydrogen at nearly the same pressure as the feed gas. This makes it more suitable for recovering hydrogen from flash drum gas. On the other hand, membrane units produce the purified hydrogen at a pressure much less than the feed gas and therefore is more suitable for purification of vent gas from the high-pressure recycle gas circuit of the reactor section. Table 6.7 provides a comparison of PSA and membrane capabilities (Whysall and Picioccio, 1999).

A third method for hydrogen purification, involving cryogenic cooling of the gas to be purified, is not usually applied in purifying vent gas from hydroprocessing units. The refrigeration required for the process is obtained by Joule–Thomson refrigeration, which is derived from throttling the condensed liquid hydrocarbons. Since the amount of impurities, particularly methane, is relatively low in hydroprocessing unit vent gas, cryogenic purification is less practical relative to PSA or membrane technology.

An optimized flow scheme takes into consideration minimizing hydrogen loss as operating conditions allow and maximizing its recovery.

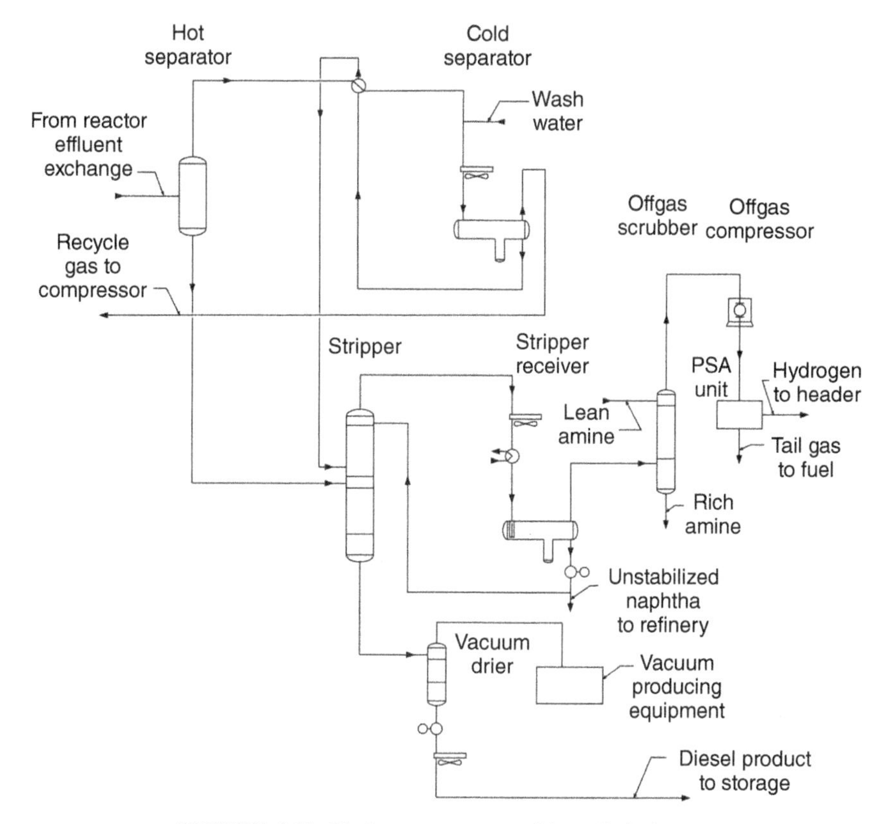

FIGURE 6.11. Hydrogen recovery without flash drums.

6.5.5 Design for No Heater Operation

The design of the heat exchange for the unit is a well-documented process using Pinch Technology developed a number of years ago and the reader is directed to Chapter 9 for more information on this subject. However, there is one issue not generally covered in a discussion of general Pinch Technology design philosophy. In the design of reactor section heat exchange network for units in which heat is evolved in the course of processing the feed, it is necessary to design for operational stability so as not to have an unrecoverable temperature excursion.

The heat generated in the reactor is a function of the amount of contaminants in the feed, such as sulfur or nitrogen, and the amount of thermally cracked material, such as coker gas oil, contained in the feed. Higher contaminant levels and cracked component content brings about the need to utilize multiple beds in the reactor. As discussed in Chapter 5, the design of the reactor system usually has a limit on the maximum temperature rise in a given reactor bed. This generally translates into a similar temperature difference between the reactor inlet and outlet temperatures. In addition, the reactor is usually designed to operate for equal peak temperature at the outlet of each bed. That means that unless the feed contains a very low level of contaminants, or no cracked material, the reactor outlet temperature will be a function of the inlet temperature and the allowable maximum bed temperature rise. This value is usually

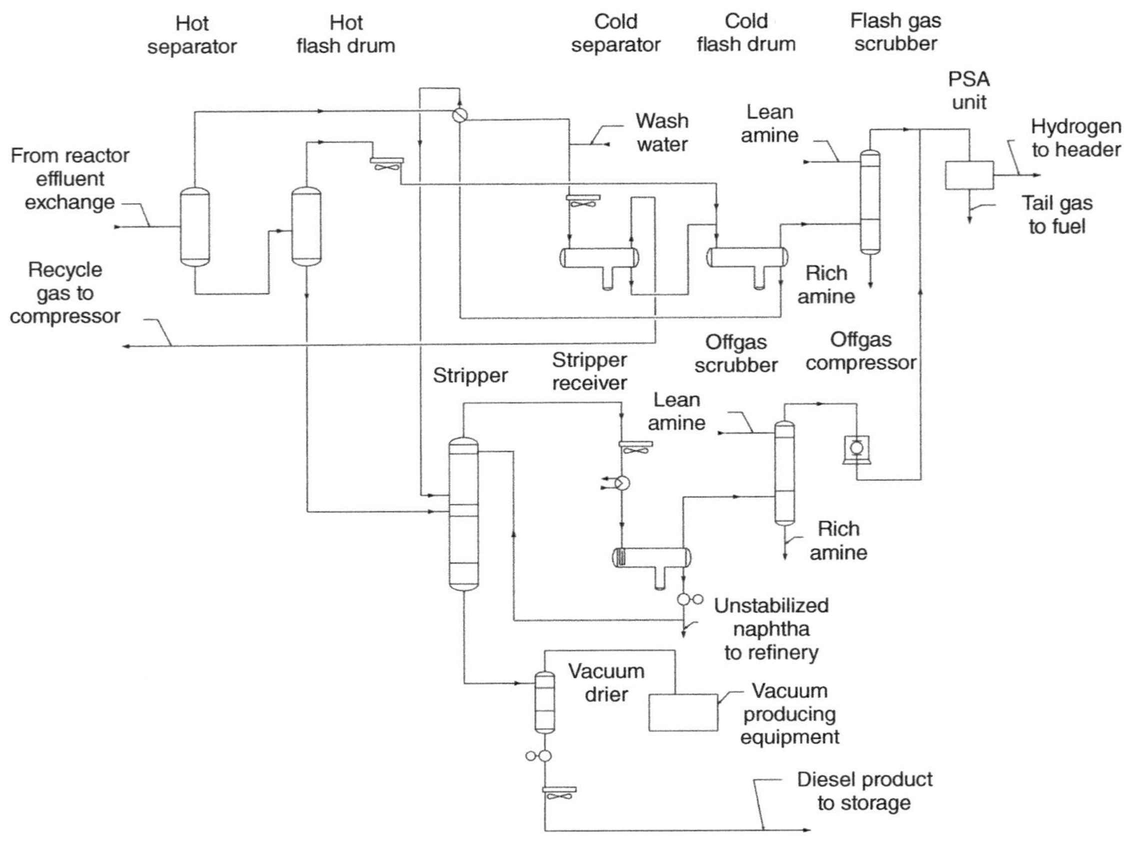

FIGURE 6.12. Hydrogen recovery with flash drums.

TABLE 6.7. Process Considerations for Hydrogen Purification Technology

Factors	PSA	Membrane
Minimum feed H_2 (%)	50	15
Feed pressure (psig)	150–1000	200–2000
H_2 purity (%)	99.9+	98 max.
H_2 recovery (%)	up to 90	up to 97
$CO + CO_2$ removal	Yes	No
H_2 product pressure	Approximately feed	Much less than feed

in the range of 50–75 °F (28–42 °C). Therefore, the reactor outlet temperature will be this range of values above the reactor inlet temperature. This temperature difference is sufficient to heat the reactants to the required reactor inlet temperature without the need to use fired heater duty, provided sufficient exchange surface is supplied. This is where Pinch Technology can assist in the design of the heat exchange network.

Before going further, it is necessary to introduce a concept that is important to the design of hydroprocessing units. Efficient use of heat in the reactor section involves using reactor effluent heat to recover and recycle heat to the reactants (recycle gas and feed) either by heating them separately or together. If the heat recovered from the reactor effluent is less than that required to achieve the required reactor inlet temperature, a fired heater is used to obtain the proper temperature. This is the configuration most commonly used. A measure of how much reactor effluent heat is "recycled" to the reactor inlet is the heat recovery ratio and is usually designated by the Greek letter β. It is defined as follows:

$$\beta = \frac{Q_{\mathrm{EX}}}{Q_{\mathrm{H}}}, \tag{6.4}$$

where

Q_{EX} = Exchanger duty for heating reactants
Q_{H} = Heater duty.

If there is no heat exchange and all of the heat is supplied by the heater, then $\beta = 0$. As the value of β increases, the fraction of the total heat requirement supplied by the heater follows the curve as shown in Figure 6.13.

It is readily seen that as the amount of heat recovered increases above a certain level, the heater duty decreases rapidly. Most traditional designs have maintained a certain minimum heater duty so that the heater can compensate for sudden changes in heat of reaction resulting in a higher outlet temperature.

For low fuel costs, this strategy works well and results in stable unit operation, while still being economical. However, as fuel costs increase, it becomes feasible to consider designing the unit for operation without the fired heater during normal operation. This represents an infinite heat recovery ratio. Such a design is feasible since, as stated above, there is usually a large enough temperature difference between the reactor inlet and outlet to provide sufficient driving force to design an exchanger network to heat the reactants to the required temperature. It does require more surface area than designs in which the heater would normally be operational and this is where

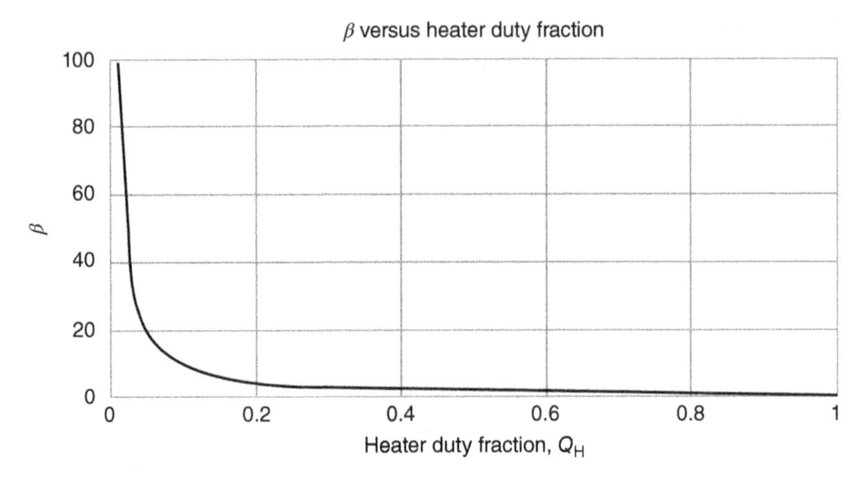

FIGURE 6.13. Heat recovery ratio (β) versus heater duty fraction (Q_H).

TABLE 6.8. Distillate Hydrotreater Design Comparison

Heat Recovery Ratio (β)	5	8	15	∞
Capital cost (MM$)	35.5	30.5	25.6	28.9
Operating cost (MM$/year)	84.9	77.1	76.6	75.4
CAPEX reduction (%)	Base	14	28	19
OPEX reduction (%)	Base	9	10	11

the utilities cost will largely determine the feasibility of such a design. Table 6.8 compares the capital and operating costs for several different design options for a distillate hydrotreating unit of approximately 54,000 BPSD capacity designed for different values of β. All designs incorporate a hot separator flow scheme. Note that although no fired heater duty is required during normal operation, a heater is still required to start up the unit.

Table 6.9 lists the utility data that was used to generate the table.

As can be seen from Table 6.8, the capital and operating costs decrease as the heat recovery ratio increases. Not surprisingly, the capital costs increase for a no-heater design (infinite β), but the utilities cost is less than the other options, because of the

TABLE 6.9. Utilities Cost Basis

Utility	Unit	Cost
Fuel	US$/MT	470
Electricity	US%/KWH	0.12
LP steam	US$/MT	20.87
MP steam	US$/MT	21.21
HP steam	US$/MT	24.3
Boiler feedwater	US$/MT	3
Cooling water	US$/m^3	0.04

lower fuel cost. The difference between the capital cost of a design at $\beta = 15$ and $\beta =$ is \$3.3 MM. The difference in utilities cost for these two cases is \$1.2 MM. This implies a simple payout of 2.75 years for the no-heater design and is therefore a viable design option.

Although economics would point to a no-heater design, there are at least two important issues to be taken into account when implementing such a design. The first of these is maintaining stability of operation. As stated above, traditional designs have more or less relied on varying the heater duty as a means of absorbing fluctuations in heat of reaction and consequent reactor outlet temperatures. If the heater is not running, the design of the exchanger bypass control scheme must be robust enough to provide this flexibility. Another consideration is how the unit will be operated. A heater must still be included in the design to provide the necessary heat input to obtain the proper reactor inlet temperature so that operations can be started. Therefore, there is a transition period during which the heater is taken out of service and the exchange network takes over maintenance of the required reactor inlet temperature. This has the potential to introduce a mild upset in the operation that can usually be accommodated, as hydrotreating catalysts are not usually very sensitive to small perturbations. However, hydrocracking units utilize catalysts systems that are more active and therefore more sensitive to small operating fluctuations that could lead to unwanted temperature excursions. Therefore, a no-heater design is not recommended for hydrocracking units.

REFERENCES

Hoehn (2003) Recovery of Hydrogen from Fractionation Zone Offgas, US Patent 6,640,161.

Hoehn (2014) Recovery of Hydrogen from Fractionation Zone Offgas, US Patent application 14/521,251.

Tsonopoulos C (1999) Thermodynamic analysis of the mutual solubilities of normal alkanes and water, *Fluid Phase Equilibria*, **156**, 21–33.

UK Health and Safety Executive (1987) *The Fires and Explosion at BP Oil (Grangemouth) Refinery Ltd*, pp. 15–43, HMSO Publications Centre.

Whysall M, Picioccio KW (1999) Selection and Revamp of Hydrogen Purification Processes, Comparison of PSA and membrane hydrogen recovery methods, paper 37e, presented at AIChE Spring Meeting.

Worldwide Fuel Specifications (2013) International Fuel Quality Center.

7

HYDROCRACKING UNIT DESIGN

7.1 INTRODUCTION

Hydrocracking is a more severe form of hydroprocessing in which the aim is not only product quality improvement but also conversion of high-molecular-weight hydrocarbons to more useful lower molecular weight hydrocarbons. Although some small amount of conversion to lighter materials is a consequence of hydrotreating reactions, hydrocracking units contain catalyst that is designed specifically for cracking high-molecular-weight molecules to lower molecular weight ones.

Conversion is defined by the following formula:

$$\text{Conversion, vol\%} = \frac{(\text{FF–UCO})}{(\text{FF})} \times 100\%, \tag{7.1}$$

where

FF = Fresh feed rate, volume units
UCO = Net unconverted oil, volume units

Conversion can range typically from about 30% to virtually 100%; however, as the conversion level approaches 100%, additional design considerations must be taken into account. The earliest applications of hydrocracking date from the early 1960s. The units designed at that time were mainly geared toward cracking distillate range material to naphtha for gasoline production. The process has since evolved to feed vacuum gas oil (VGO), coker gas oils, and deasphalted oils to produce

Hydroprocessing for Clean Energy: Design, Operation, and Optimization, First Edition.
Frank (Xin X.) Zhu, Richard Hoehn, Vasant Thakkar, and Edwin Yuh.

distillates, primarily ultra-low-sulfur diesel (ULSD), kerosene, and naphtha for use as transportation fuels. Some of these materials also often contain higher levels of sulfur, nitrogen, and metals, all of which place more stringent demands on the catalyst system. Cracking such heavy compounds requires more active catalyst and more severe conditions, such as higher reactor temperatures, compared with hydrotreating. In addition, higher molecular weight feeds charged to a modern hydrocracking unit have a higher tendency to form coke on the catalyst than lighter feeds. Consequently, it is generally necessary to operate at lower catalyst space velocities (larger catalyst volumes) to achieve the desired conversion and higher hydrogen partial pressures to suppress catalyst coking tendency. These aspects of hydrocracking units introduce unique and demanding requirements on the design of the unit. The design engineer must address a number of design issues to achieve a successful reliable design.

As originally practiced, a single bifunctional amorphous cracking catalyst was employed. This catalyst had the capability of removing sulfur, nitrogen, and other contaminants from the feed as well as achieving the required conversion. As currently practiced, hydrocracking units tend to utilize a suitable hydrotreating catalyst designed to remove sulfur, nitrogen, and other contaminants from the feedstock followed by a high-activity zeolitic cracking catalyst whose properties are selected to maximize the production of the desired product range material. Thus, the use of separate catalyst types can result in more optimal yields of the desired product range material because the treating and cracking catalysts can be more customized for optimal performance. Therefore, the selection of the optimum catalyst system to be used is one of the first steps in the optimization process and careful thought and attention is given to this detail in the early stages of the design.

There are two main flow scheme configurations utilized for the reactor section design of hydrocracking units: single stage and two stage.

Because the processing objectives of hydrocracking units differ from that of typical hydrotreating units, the flow scheme for both the reactor and fractionation sections tend to be more complex. Therefore, each section will be covered separately.

7.2 SINGLE-STAGE HYDROCRACKING REACTOR SECTION

Table 7.1 lists some representative feedstock properties and processing conditions for a typical VGO hydrocracking unit as currently designed. A simplified flow diagram for a typical single-stage hydrocracking unit is shown in Figure 7.1. The feedstock characteristics are similar to those of VGO hydrotreating, but the processing conditions are more severe and the overall space velocity is typically much lower. Similar to hydrotreating, feed is filtered and pumped into the reactor section where it is mixed with recycle gas, exchanged with one or more streams before being routed to a fired heater for heating the combined feed to the required reactor inlet temperature. The reactor system of a modern design hydrocracking unit consists of one or more beds of hydrotreating catalyst followed by one or more beds of cracking catalyst. One of the main functions of the hydrotreating catalyst is to remove sulfur and nitrogen, as well as other containments, such as metals. As a consequence of the hydrotreating reactions, some nominal conversion takes place, typically 15–20%.

TABLE 7.1. Typical Feedstock and Processing Conditions for VGO Hydrocracking

Property	Straight Run VGO
Specific gravity at 15 °C	0.937
Sulfur (wt%)	1.95
Nitrogen (wt ppm)	1580
Metals (Ni + V) (wt ppm)	1.9
Bromine number (gBr/100 g)	2.6
ASTM D-1160 distillation (°C)	
IBP	231
10%	400
30%	429
50%	457
70%	493
90%	541
EP	600
Separator pressure (barg)	121
Treating reactor average bed temperature (°C)	390–430
Cracking reactor average bed temperature (°C)	370–417
LHSV (h^{-1})	0.65
Treating reactor gas to oil ratio (nm^3/m^3)	600
Cracking reactor gas to oil ratio (nm^3/m^3)	1100
Min. H_2 in recycle gas (mol%)	92

After hydrotreating, the material is passed over the cracking catalyst, where the majority of the conversion takes place. The cracking catalyst is selected to meet the processing objectives, namely conversion level and preferred distillate range of the products. The heat of reaction for hydrocracking is most often much greater than that for hydrotreating, resulting in more reactor beds and more quench gas to maintain the catalyst at the required temperature level. The higher activity of the cracking catalyst requires careful monitoring of temperatures in the catalyst beds as well as systems designed to prevent or stop the occurrence of a temperature excursion. This is defined as an uncontrolled increase in reactor bed temperatures, which the normal heater and quench temperature control systems are not able to control. When this happens, the operator must take immediate action to prevent a serious situation. More recently, systems have been devised to initiate safety measures automatically in the event of a temperature excursion.

After passing through the reactor, the material is partially cooled by heat exchange with feed, recycle gas, or other suitable material and is then routed to a hot separator.

The hot separator vapor is cooled via exchange and wash water is injected upstream of the air-cooled hot separator vapor condenser. The wash water performs the same function as that in hydrotreating units.

A single-stage unit can be designed for a once-through operation, in which the feed passes only once through the hydrotreating and cracking catalyst beds. The reactor temperatures are adjusted to achieve the objective contaminant removal by the hydrotreating catalyst and conversion by the cracking catalyst. This configuration

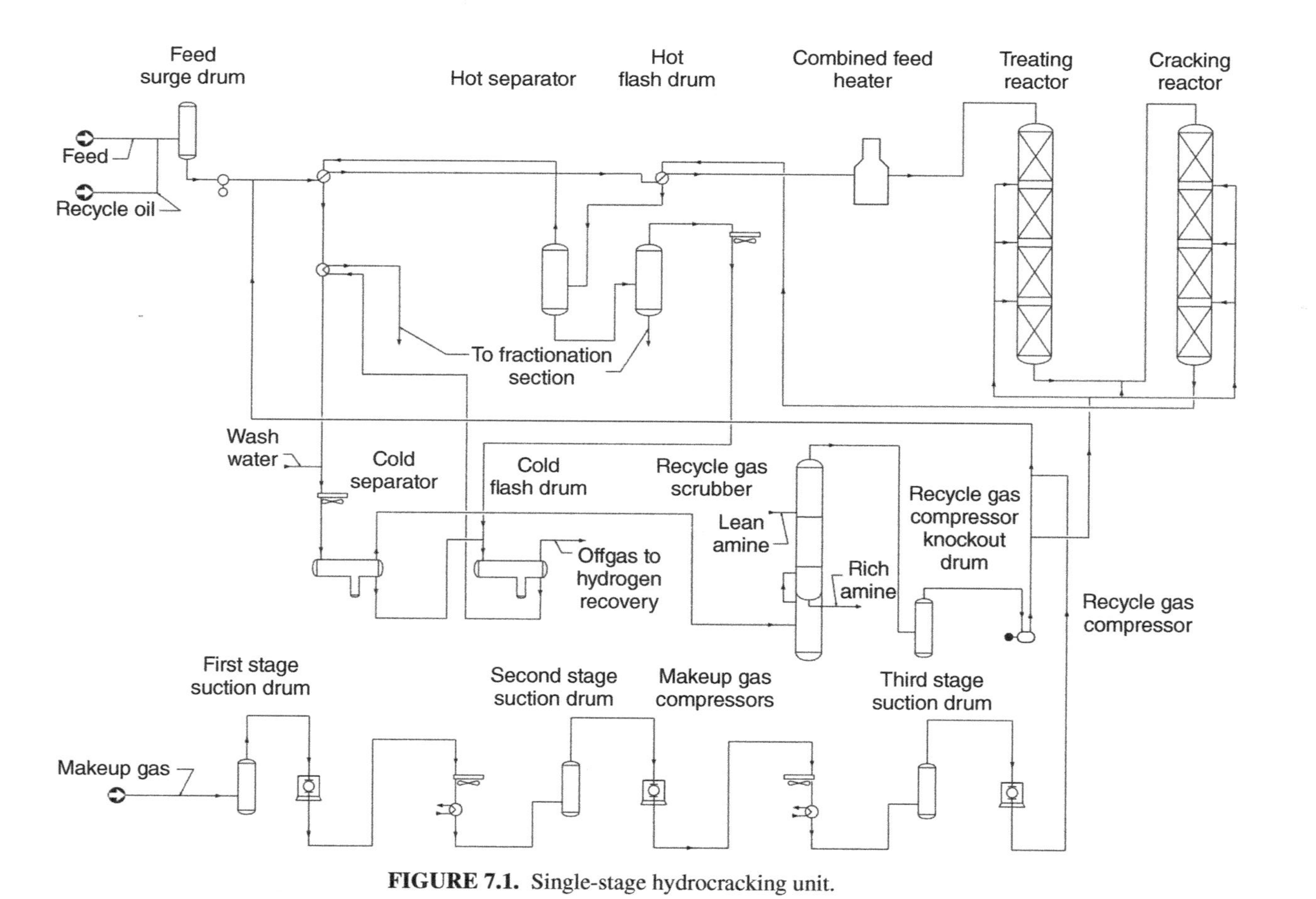

FIGURE 7.1. Single-stage hydrocracking unit.

can be utilized for conversions up to about 90% and tends to find application in cases where there is a specific use for the unconverted oil. For example, a once-through design would be used when the unconverted oil is intended to be used for lube oil production.

A single-stage unit can also be designed for recycle operation. Generally, a recycle flow scheme would be utilized for conversions above 60–70%, particularly when distillate range material is the desired product. Instead of performing all of the conversion of the feed in a single pass through the reactor, the system is operated such that the feed is partially converted and then sent to the fractionation section where the desired products range materials are recovered and at least a portion of the unconverted oil, defined as that material boiling above the heaviest desired product, is returned to the reactor section to mix with the feed to be further converted. Recycling of unconverted oil thus allows conversions up to almost 100%. The reason why this configuration is advantageous has to do with a relation between combined feed ratio (CFR) and distillate selectivity. This is shown in Figure 7.2.

CFR is defined as the ratio of fresh feed plus recycle liquid divided by the fresh feed. It is the reciprocal of the conversion per pass. For a once-through flow scheme, the conversion per pass is equal to one. As the CFR decreases, the conversion per pass increases and the resulting products tend to be lighter, meaning that the catalyst produces proportionally less distillate range material and much lighter range material, such as naphtha. If the objective of the unit is to maximize the production of distillate material, a higher CFR would be favored. However, there is a trade-off to be made, since increasing the CFR increases the volume of recycle oil that must be processed. This increases the hydraulic capacity of the unit, resulting in a higher capital cost. If the differential value of the desired product range material over other less desired products is great enough, then the expense can be justified.

A variation on the single-stage recycle flow scheme is shown in Figure 7.3. The distinguishing feature of this flow scheme is that the recycle oil from the fractionation section is routed directly to the cracking reactor, rather than combining with the fresh feed. This flow scheme has largely been abandoned because an additional high head pump for the recycle material and an extra fired heater are required.

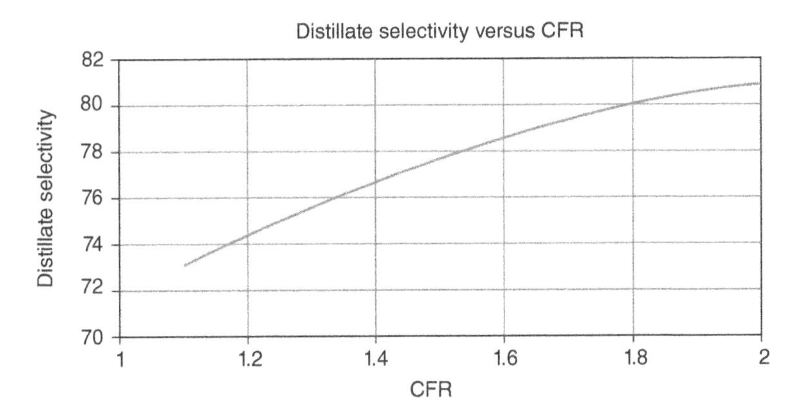

FIGURE 7.2. Relation between distillate selectivity and conversion per pass.

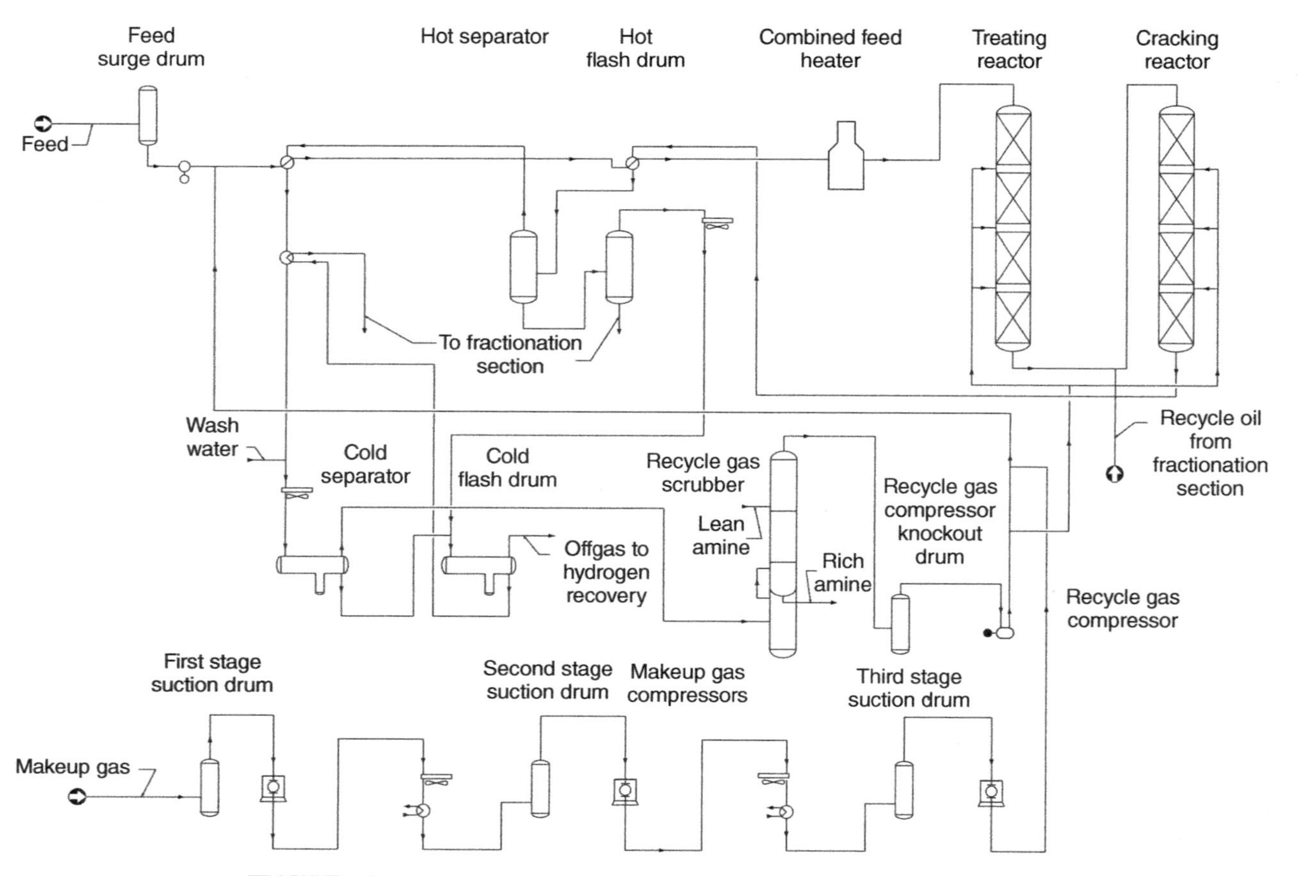

FIGURE 7.3. Alternative single-stage recycle hydrocracking flow scheme.

7.3 TWO-STAGE HYDROCRACKING REACTOR SECTION

Another flow scheme that is often used is shown in Figure 7.4, the so-called two-stage configuration. The first stage is, in essence, a once-through hydrocracking flow scheme. The conversion in the first stage is adjusted so that the product range material produced by the cracking reactions has properties that meet the required product requirements. The first stage reactor effluent is partially cooled and mixed with reactor effluent material from the second stage. The reactor products are routed to the fractionation facilities similar to that of a single-stage unit. Unlike a single-stage unit, the unconverted oil from the fractionation section is routed to the second stage, which consists of a separate reactor system. The recycle oil is heated via exchange with second stage reactor effluent, mixed with recycle gas, heated to the required reactor inlet temperature by a fired heater, and then processed in the second stage reactor. This operation occurs in parallel with the first stage with respect to the flow of recycle gas. The effluent from the second stage reactor effluent is partially cooled via heat exchange with recycle gas and the second stage feed and then joins the first stage reactor effluent material at the inlet of the hot separator. There are several advantages to this flow scheme. Impurities such as sulfur and nitrogen, which can affect the cracking catalyst activity, are largely absent in the feed to the second stage. Also, the recycle gas has usually been amine treated to remove hydrogen sulfide and any ammonia not removed by water washing. Since these contaminants tend to affect the activity of the cracking catalyst, less catalyst is required in the second stage for the same conversion level. A different catalyst type can also be utilized in each stage, thereby providing opportunities for further optimizing the production of the desired product range material. Another advantage of the two-stage design is that it can be utilized for units with higher nameplate capacities. This is because the recycle liquid is processed separately from the fresh feed and, therefore, the reactor sizes are a function of the feed to each stage, rather than the combined fresh feed and recycle oil rate. For larger units, the size limitations for reactors are important considerations that must be factored into the design.

One disadvantage of the two-stage flow scheme is that more equipment is required, such as an extra high head charge pump, fired heater, high-pressure exchangers, and the second stage reactor. In addition, the amount of recycle gas is increased significantly over a single-stage flow scheme, because recycle gas must be routed in parallel to each stage. As a result, capital cost and utilities for a two-stage unit tend to be higher than that for a single-stage unit of the same capacity. However, particularly for high capacities, processing the fresh feed and unconverted oil separately avoids the limitations on equipment size that would occur when designing for large throughputs. Also, the incremental higher yield of desired product range material, along with lower total catalyst volume requirement, can make the two-stage design more economically favorable, compared with a single-stage design.

A variation on the two-stage flow scheme is shown in Figure 7.5. This is called a separate hydrotreat flow scheme. It is essentially a two-stage flow scheme, but the first stage utilizes only hydrotreating catalyst. This flow scheme is used for feeds containing very high levels of sulfur and nitrogen. Sulfur in the fresh feed forms hydrogen sulfide and nitrogen forms ammonia in the hydrotreating reactor. The presence of

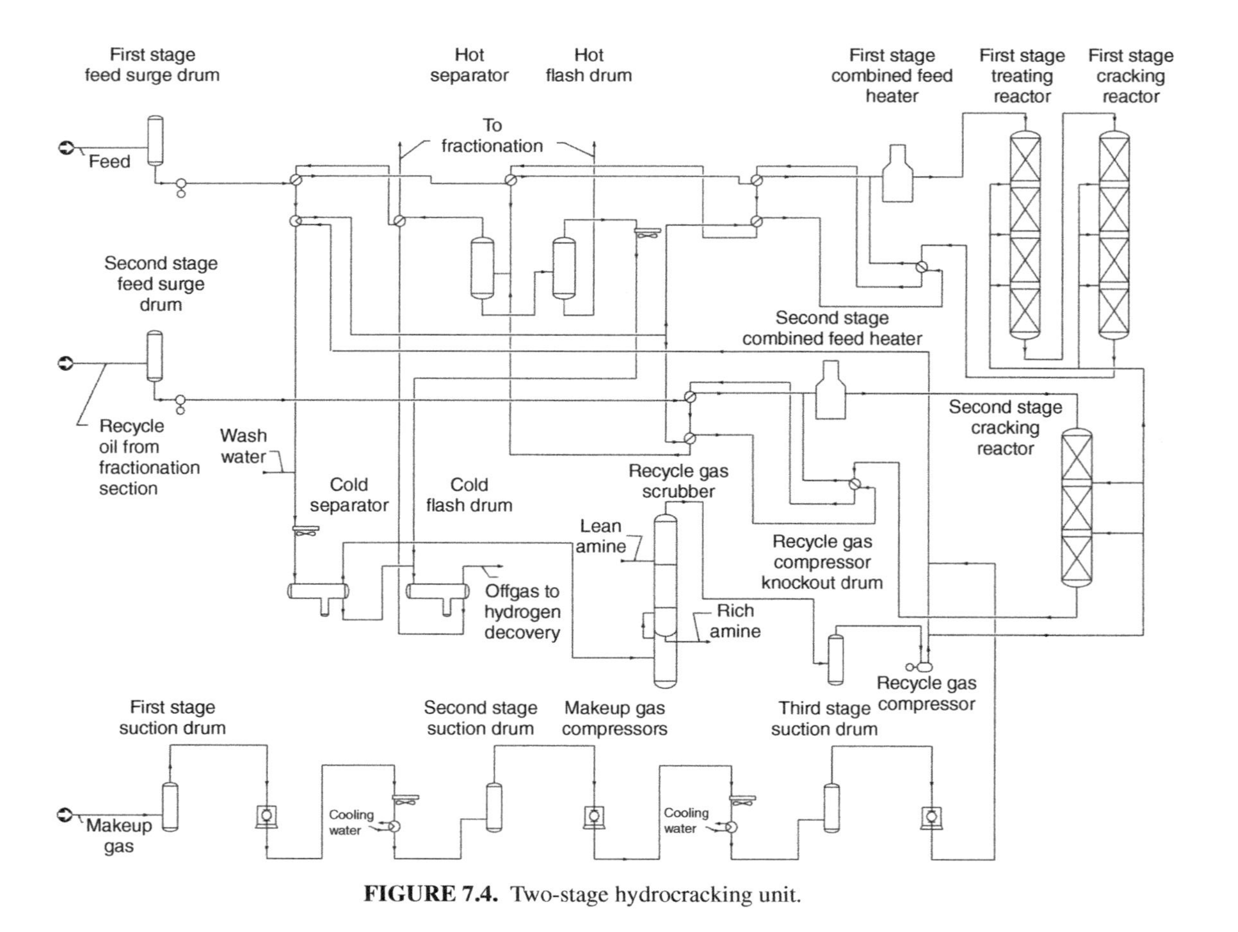

FIGURE 7.4. Two-stage hydrocracking unit.

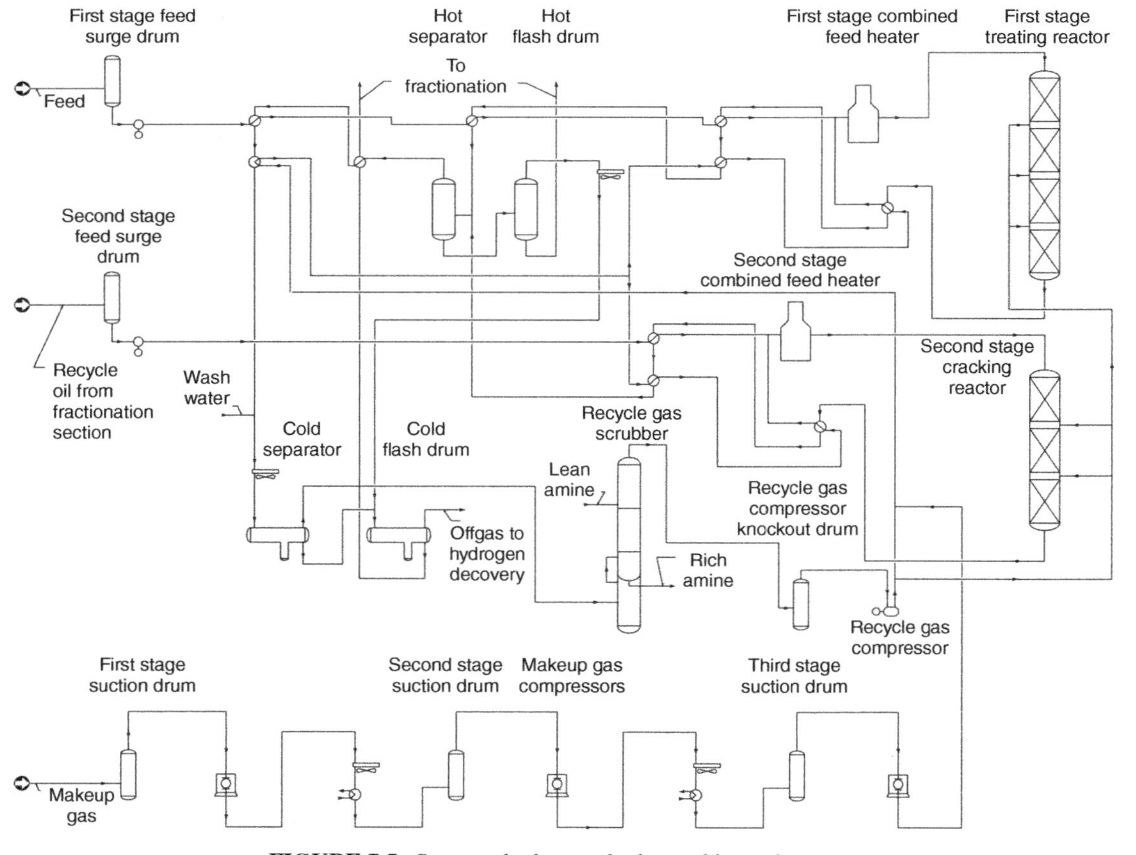

FIGURE 7.5. Separate hydrotreat hydrocracking unit.

ammonia has a strong effect on the activity of cracking catalyst, since it acts to neutralize the acid sites on the cracking catalyst. Up to a certain limit, the presence of ammonia can be compensated by increasing the amount of cracking catalyst and/or operating at higher temperatures. Above a certain feed nitrogen level, however, it becomes infeasible or uneconomical to do this. In this situation, the better option is to hydrotreat the feed to remove the sulfur and nitrogen before processing the feed further. The majority of the ammonia and hydrogen sulfide thus formed ends up in the recycle gas and are removed by water washing and amine scrubbing. The hydrotreated feed is sent to the fractionation section where any product range and lighter material is removed, including any remaining hydrogen sulfide and ammonia contained in the liquid from the reactor section. The combination of treated feed and scrubbed gas being charged to the cracking reactor creates a much more favorable environment in which the activity of the cracking catalyst is maximized. This, in turn, reduces the required amount of cracking catalyst for the same conversion. Note that even though the first stage contains only hydrotreating catalyst that has a lower activity than cracking catalyst, some nominal conversion does take place, usually on the order of 15–20%.

7.4 USE OF A HOT SEPARATOR IN HYDROCRACKING UNIT DESIGN

It is expected that most hydrocracking units, particularly those feeding VGO or heavier feeds, would be designed with a hot separator in addition to the cold separator. There are several reasons for this. As was explained in Chapter 6, use of a hot separator increases the energy efficiency of the unit by not requiring all of the reactor effluent to be cooled to the final temperature at which the recycle gas is separated from the liquid products, only to be reheated again for downstream separation of products. However, there are additional reasons for using a hot separator in the design of a hydrocracking unit that must be considered, particularly when processing heavy feed stocks. Heavy feeds, such as VGO and residuum can have densities approaching that of water. When using a single high-pressure separator, all of the reactor effluent is cooled to the separator temperature and wash water is injected, usually upstream of the air-cooled effluent condenser, to wash water soluble salts that are present. As the endpoint of the feed increases, so does its density and at some point the difference between the density of water and the hydrocarbon phase decreases to the point that it becomes difficult to separate the water phase from the hydrocarbon phase in the high-pressure separator. Another problem is due to the nature of feeds containing high-endpoint materials. These feeds can contain compounds called asphaltenes, which are high-molecular-weight compounds and the standard test for their presence is their insolubility in *n*-heptane. The term asphaltene does not describe a single compound, but rather a range of materials that have characteristics of high molecular weight, large number of condensed aromatic rings, combined with oxygen, nitrogen, and metals such as nickel and vanadium. One characteristic of these materials is that they can form an emulsion when mixed with water. Therefore, these compounds must be removed before the wash water is added so that the wash water can be separated from the hydrocarbon phase. Water foaming problems have been noted in some early hydrocracking and VGO hydrotreating units designed with a single high-pressure separator. A hot separator effectively removes these compounds in the liquid phase before wash water is injected so that problems with water emulsion formation are largely avoided, provided the hot separator is designed properly.

There is another important reason for using a hot separator for hydrocracking units, particularly those operating at high conversion. One of the inevitable by-products of hydroprocessing reactions is the formation of coke. The process is operated at elevated hydrogen partial pressure such that the amount of coke buildup on the catalyst, and subsequent catalyst deactivation, is reduced so that the required cycle length can be met. As it turns out, not all of the undesirable by-products end up as coke on the catalyst. High-molecular-weight compounds called heavy polynuclear aromatics (HPNAs) can be present in the reactor effluent. These compounds are similar to asphaltenes, in that they consist of multiple condensed ring aromatics. However, their molecular weight tends to be lower. They are composed primarily of carbon and hydrogen only. HPNAs are usually considered to be compounds with 11 or more condensed aromatic rings. These compounds can be either present in the feed or produced as a by-product of the cracking reactions and as such are considered coke precursors. Typically, the reactor system is operated so that a portion of the feed material is converted. If the unit is designed for high conversion, unconverted oil is recycled from the fractionation section back to the reactor section for further conversion to the desired products. As conversion increases, there comes a point where the generation of HPNA material in the reactor becomes greater than its removal via the net unconverted oil stream. The concentration of HPNA material will buildup in the recycle liquid and its presence can be seen as a characteristic reddish orange color of the recycle liquid.

If HPNA material is not removed by some means, it can cause problems in two areas. First, HPNA material present in the recycle liquid will eventually end up on the catalyst as coke, with the result that the end of run temperature is reached prematurely. The other problem is that since HPNA material has a low solubility in reactor effluent material, it will precipitate when a characteristic temperature and concentration is reached. If the unit is configured with a single high-pressure separator, the entire effluent material will be cooled to the separator temperature. If HPNA material is allowed to build up in the system, the HPNA solubility limit is reached first in the coldest part of the unit, which is the air-cooled reactor effluent air cooler. The result is a gradual but steady fouling of the exchanger surface with precipitated HPNA material. An indication of exchanger fouling is an increase in the outlet temperature of the effluent air cooler. The only solution is to shut down the unit and hydroblast the accumulated HPNA material from the inside of the tubes. This is, of course, very inconvenient and represents a large economic penalty due to lost production during the time the unit is shut down to perform this cleaning. Therefore, the solution that has been applied for high-conversion hydrocracking units is to utilize a hot separator operating at a temperature well above the HPNA solubility limit. Since HPNA compounds are essentially nonvolatile, they exit with the hot separator liquid and are not subjected to cooling that would bring about precipitation. As they are nonvolatile, they will end up in the unconverted oil from the fractionator, from which they will be routed back to the reactors. Unfortunately, this will increase catalyst deactivation due to coke deposition, unless some means of HPNA removal is applied. This is discussed further in the section on fractionation section design.

As can be seen, there are a number of factors that determine the basic flow scheme and each must be considered in the process of setting the final configuration.

7.5 USE OF FLASH DRUMS

In Chapter 6, the role of flash drums in aiding hydrogen and liquefied petroleum gas (LPG) recovery from reactor and fractionation section offgas streams was explained. The same holds for hydrocracking unit design. Indeed, the incentive is much greater to use them in hydrocracking units because solution losses are much greater and the difference between the cold separator and cold flash drum operating pressures means that a greater portion of the solution loss hydrogen will be present in the flash gas. The flash drum operates at an appropriate pressure that allows the gas to be treated and then routed directly to the hydrogen recovery facilities, which operate essentially at the refinery hydrogen header pressure. Table 7.2 shows the amount of hydrogen solution loss and hydrogen purity in the flash drum offgas obtained for recycle and once-through hydrocracking units at varying pressures. All units utilize a hot separator operating at 525 °F (274 °C) and a low-pressure hot flash drum and cold flash drum. The cold flash drum operates at 350 psig (24.1 bar).

As can be seen, solution loss is a function of operating pressure and CFR. Even for moderate-pressure hydrotreating units, the hydrogen contained in the flash gas contains a significant percentage of the makeup gas added to the unit.

As is the case for hydrotreating units, flash drums operating at an appropriate pressure provide a convenient means for recovering solution loss hydrogen and thus reduces the amount of hydrogen contained in the feed to the fractionation section, which would affect recovery of light hydrocarbons. The presence of hydrogen in the fractionation section affects the ability to recover light hydrocarbons, particularly propane and butane (LPG).

TABLE 7.2. Solution and Recoverable Hydrogen via Flash Drums

Operating Pressure (psig (barg))	CFR	Total H_2 Solution Loss, SCFB (nm^3/m^3)	H_2 Solution Loss (% of Makeup Gas)	H_2 in Flash Gas (% of Solution Loss)
1700 (117)	1	87.1 (14.7)	4.7	82
	1.5	124.7 (21.0)	6.6	82
	2	157.9 (26.6)	8.2	82
2100 (145)	1	111.6 (18.8)	5.9	86
	1.5	158.3 (26.7)	8.2	86
	2	200.0 (33.7)	10.1	86
2500 (172)	1	135.9 (22.9)	7.1	88
	1.5	191.4 (32.3)	9.7	88
	2	240.2 (40.5)	11.9	88

7.6 HYDROCRACKING UNIT FRACTIONATION SECTION DESIGN

Now that the feed has been converted in the reactor section, it is necessary to separate the effluent material as efficiently as possible into the various desired products, all meeting the required product qualities. Fractionation design practices continue to evolve, and the following discussion analyzes the advantages and disadvantages of using different types of fractionation flow scheme configurations. Also described below are the recently developed dual zone stripper and dual fractionator designs that represent a significant utility and capital savings.

7.7 FRACTIONATOR FIRST FLOW SCHEME

Figure 7.6 shows the basic configuration of a fractionator first flow scheme. This configuration is derived from the fact that the effluent material from the reactor section has a wide range of products, from methane to the heaviest portions of the unconverted oil. As such, the feed can be considered to be similar to that of a crude unit.

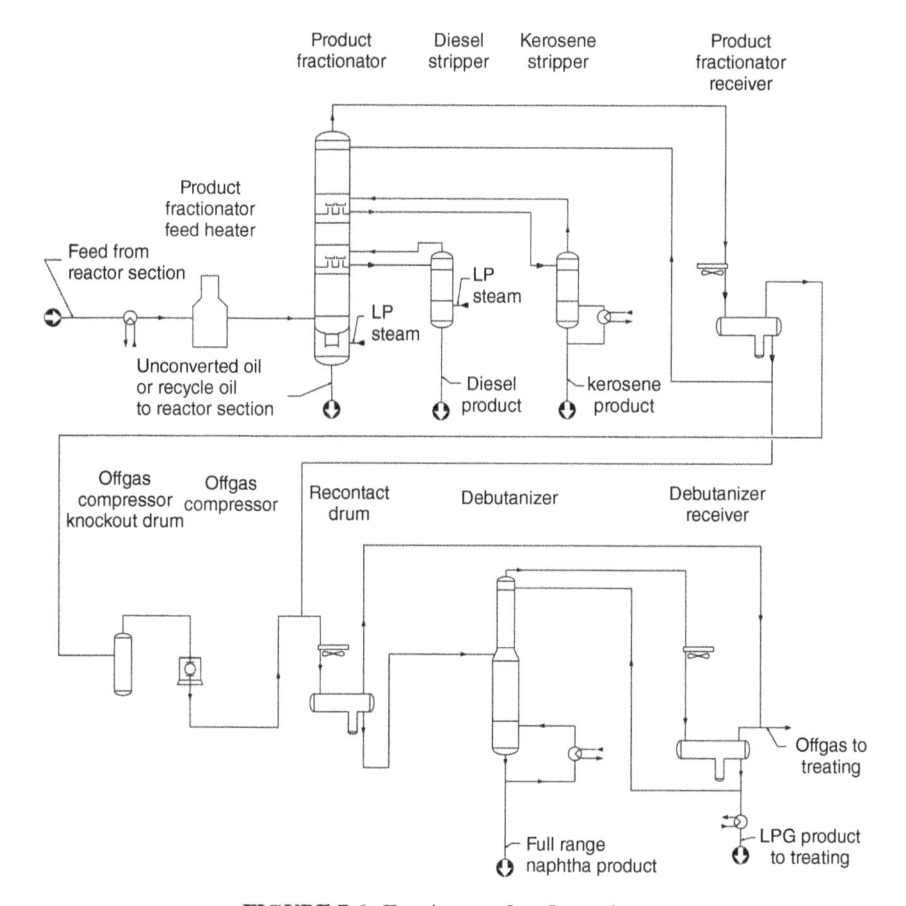

FIGURE 7.6. Fractionator first flow scheme.

The main difference between this column and a crude unit fractionator is that in the case of a hydrocracking unit there will be significant amounts of hydrogen sulfide, light hydrocarbons, and hydrogen present in the feed to the column. This will have a definite effect on the design of this flow scheme.

Similar to a crude column, the feed is preheated by exchange with products, heated to the final temperature via a fired heater, and then routed to the product fractionator. The feed is heated to a temperature where all of the desired products and some of the column bottoms material are vaporized. Vaporized product travels up the column from which kerosene and diesel are generally withdrawn at an appropriate intermediate location, stripped in a sidecut stripper to remove the desired amount of light material, cooled, and then produced as a product.

The naphtha and lighter products are recovered as an overhead product. Because the feed to the fractionator contains hydrocarbons and other components that would not condense at the operating conditions of the fractionator receiver, a net vapor will be produced. This vapor will need to be compressed, since the operating pressure of the fractionator is generally lower than the pressure of the normal destination of the offgas, which is usually the fuel gas system. The overhead vapor contains a substantial portion of valuable hydrocarbons, such as C_3/C_4 (LPG), as well as naphtha range material, generally making it desirable to recover these hydrocarbons, if possible. To facilitate recovery, the net vapor is boosted to a suitable pressure and recontacted with the fractionator net overhead liquid. This enables readsorption of these desirable compounds to be reabsorbed into the liquid phase where they can be recovered by subsequent fractionation. Hydrogen sulfide and some ammonia will also be present, so it will be necessary to amine-treat the net vapor, as well as the LPG range material so that these products meet the required specifications. This treating can be performed either in the hydrocracking unit or at suitable downstream facilities, such as the crude unit.

One of the disadvantages of this flow scheme is that the feed to the fractionator must be heated to a level at which all of the products and some of the fractionator bottoms material will be vaporized. This involves temperatures well above 650 °F (343 °C) and often above 700 °F (370 °C). At these temperatures, the hydrogen sulfide present in the feed will be very corrosive. Therefore, it is necessary to upgrade the tube metallurgy of the feed heater as well as the line from the heater to the column to stainless steel. It is also necessary to provide a stainless steel lining of at least the bottom portion of the fractionator in order to keep corrosion at a minimum. In addition, the net gas compressor, which must be spared, represents a substantial capital and operating expense and the net vapor, which typically contains hydrogen sulfide and water, is a source of maintenance problems due to corrosion that can occur in the piping to the compressor. These two aspects of the design make this flow scheme option less favorable compared to others and is not generally used for modern designs. Table 7.3 provides a comparison of equipment costs (Gulf Coast basis) for a fractionator first versus stripper first flow scheme for a hydrocracking unit of approximately 55,000 BPSD capacity.

As can be seen from the table, the fractionator feed heater is significantly more expensive for the fractionator first flow scheme than that for the stripper first flow scheme. There are two reasons for this. The first is that the heater must be constructed of more expensive type 321 or 347 austenitic stainless steel versus 9% chrome for the

TABLE 7.3. Comparison of Capital Cost for Fractionator First and Stripper First Flow scheme

Equipment	Fractionator First Flow scheme (US$MM)	Stripper First Flow scheme (US$MM)
Product fractionator feed heater	4.59	0.90
Stripper and receiver	—	0.74
Product fractionator, receiver and sidecut stripper	1.01	0.76
Shell and tube exchangers	0.49	1.12
Air-cooled exchangers	1.11	1.09
Pumps	0.47	0.54
Offgas compressor and associated equipment	1.74	—
Total	9.41	5.15

TABLE 7.4. Utilities Comparison for Fractionator First and Stripper First Flow scheme

	Fractionator First	Stripper First
Fuel fired (MM kcal/h)	22.15	4.15
Electricity (KW)	560	240
Steam (kg/h)	7,182	11,050

stripper first flow scheme. The second is that in order to limit the offgas compressor design to two stages, the product fractionator must be operated at a slightly higher pressure (1 bar receiver pressure), than in the stripper first flow scheme (0.35 bar), where there is no net overhead vapor to route to a destination. The higher operating pressure requires a higher heater outlet temperature for the same separation in the product fractionator and thus a higher design heater duty.

A second factor that increases the cost of the fractionator first flow scheme is the cost of the product fractionator offgas compressor. Its cost, including the associated equipment, is more than that of the stripper and associated equipment.

Utilities are also higher for the fractionator first versus stripper first flow schemes. Table 7.4 compares the utilities for the two flow schemes.

As noted in Table 7.4, fuel and electricity requirements for the fractionator first flow scheme are higher than for a stripper first flow scheme, even taking into account that steam usage is higher for the stripper first flow scheme.

The conclusion is that for most applications, a stripper first flow scheme is generally more economical in both capital and operating costs, compared with a fractionator first flow scheme.

7.8 DEBUTANIZER FIRST FLOW SCHEME

Figure 7.7 shows the basic flow scheme for a debutanizer first configuration. Instead of heating the feed and fractionating it at low pressure, the feed is preheated and routed to a debutanizer column. The column operates at a pressure such that the main net overhead liquid is butane and lighter hydrocarbons. This usually involves

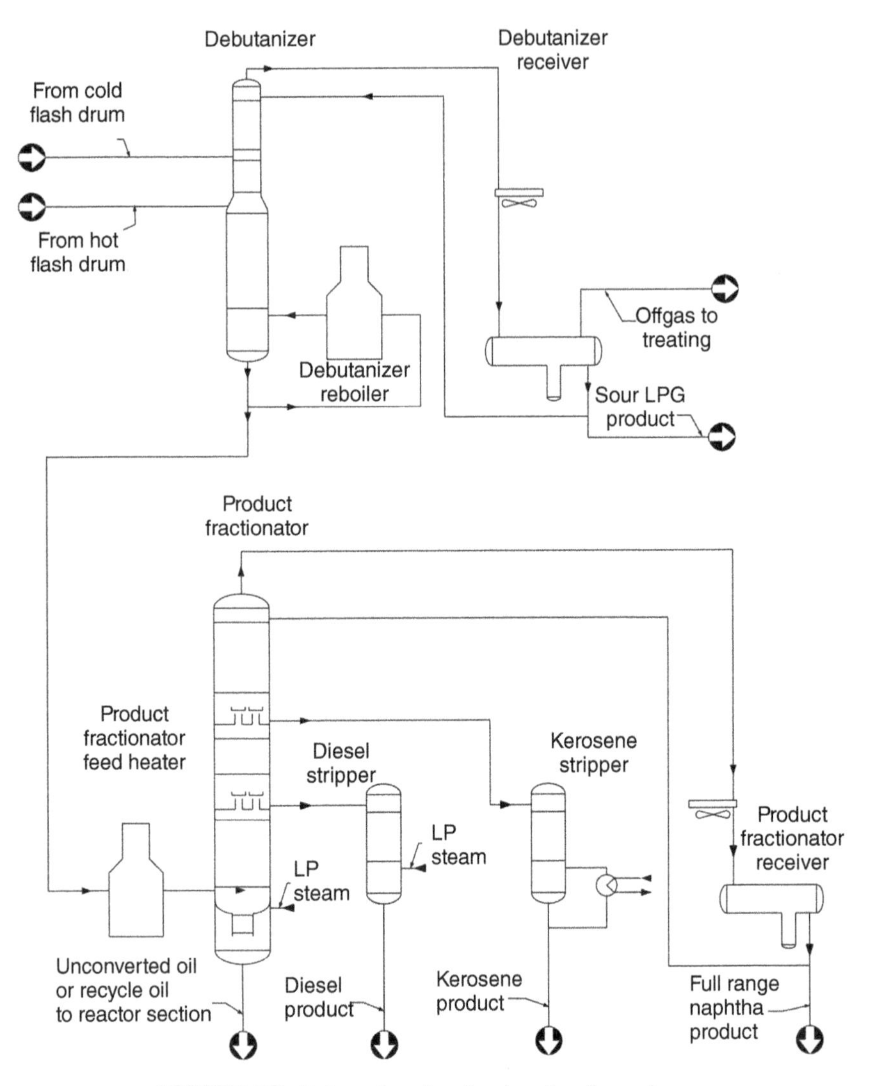

FIGURE 7.7. Debutanizer first fractionation flow scheme.

operating at a pressure of about 14 bar (200 psig) or higher. This configuration has the advantage of removing the hydrogen sulfide from the reactor effluent liquid before routing the reactor products are heated and routed to the fractionator, where the presence of hydrogen sulfide at the higher temperatures present in the fractionator would cause severe corrosion without expensive metallurgy upgrades.

The feed to the debutanizer column is usually preheated and routed to the column at an appropriate location. The column is reboiled by a fired heater to generate the necessary stripping vapor. The net overhead vapor from the column consists of water, hydrogen, hydrogen sulfide, and light hydrocarbons. LPG range product is produced as a net overhead liquid. The allowable C_5 and heavier specification of the LPG product sets the design of trays and reflux required between the feed point and overhead receiver. According to the LPG product vapor pressure specification, it may be necessary to route the product LPG to a deethanizer column (not shown) to remove ethane and lighter compounds. In addition, the LPG range material must be amine-treated and caustic-washed to remove hydrogen sulfide. These functions are often performed in the crude unit gas concentration section.

The debutanizer column bottoms material is then routed to the product fractionator, via a feed heater. As with a fractionator first flow scheme, all of the distillate products and some of the net bottoms liquid are vaporized where it travels up the column and is removed at a point appropriate with the desired boiling range. Usually, sidecut strippers are employed to strip the kerosene and diesel sidecut products of lighter hydrocarbons. The fractionator net overhead liquid contains both light and heavy naphtha liquid and is usually routed to a naphtha splitter. The naphtha splitter column is utilized to provide a separate light and heavy naphtha product. Light naphtha is either blended directly into the gasoline pool, while the heavy naphtha cut is generally sent to the catalytic reforming unit for additional upgrading. Occasionally, the fractionator will utilize a heavy naphtha sidecut stripper and draw the light naphtha as the overhead product.

While a debutanizer column, in principle, is a straightforward design that is capable of producing an LPG product directly from the column, practical issues make this configuration undesirable. One of the reasons is that the column must operate at elevated pressure, usually between 150 and 200 psig (10.3–13.8 barg), in order to be able to condense the LPG range material that comprises the net overhead liquid and reflux. For a maximum distillate hydrocracking unit, the feed to the debutanizer contains relatively little LPG and naphtha. However, in order to generate reboiler vapor, the composition at the bottom of the column must be such that sufficient reboiler vapor can be generated at the required operating pressure. Therefore, the system will adjust the amount of material in the bottom of the column to achieve that. Figure 7.8 shows the composition profile for a debutanizer column in a debutanizer first fractionation

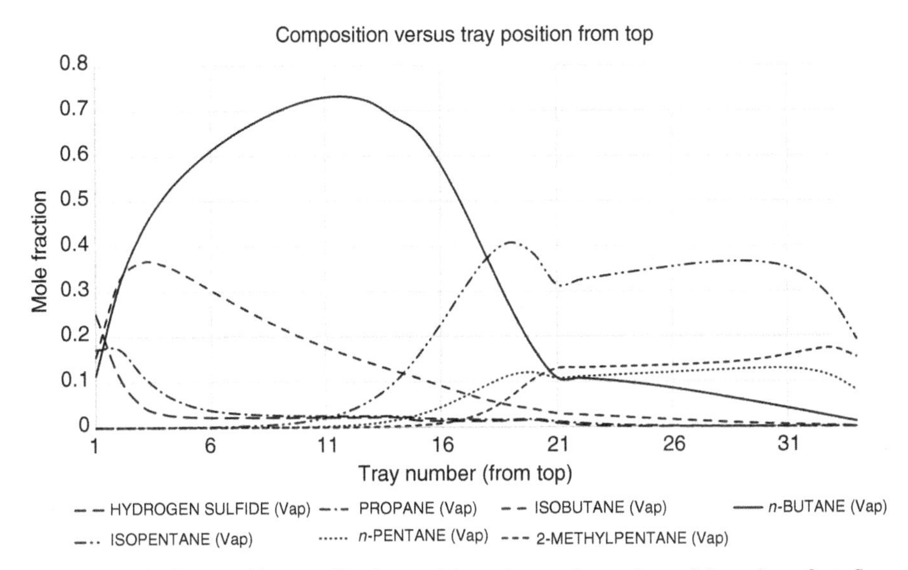

FIGURE 7.8. Composition profile for a debutanizer column in a debutanizer first flow scheme.

flow scheme for a maximum distillate hydrocracking unit. This particular unit has a preflash drum upstream that produces a liquid, which feeds the column on tray 20 (trays are numbered from top to bottom in this case), and a vapor stream that is fed to tray 14. Note how much of the light hydrocarbons (C_3 and C_4 material) are present in the bottom section, even though the endpoint of the bottoms material can be over 1000 °F (537 °C). The biggest problem with a debutanizer first design is that it is very hard for the operator to know that sufficient stripping vapor exists to remove the hydrogen sulfide and light hydrocarbons. In addition, a sizable amount of the reboiler duty is taken with heating the bottoms material from the feed temperature to the net bottoms temperature. That means that a lot of the potential stripping vapor produced by the reboiler is condensed as the liquid proceeds from the feed tray to the bottom of the column. The remaining vapor is available to strip the hydrogen sulfide and lighter hydrocarbons. The result is that the temperature difference across the reboiler can approach as much as 100 °F (55 °C). In addition, operators are accustomed to operating fired heater reboilers according to outlet temperature, being mindful of the need to limit the maximum heater tube wall temperature so as not to reduce the life of the heater. However, the reboiler outlet temperature does not necessarily correlate with vaporization, which is a function of the composition.

In addition, since the yield of LPG range hydrocarbons is small relative to the yield of distillate range material, small variations in actual yields or particularly startup operations can result in much less LPG material being present, thus affecting the effective operation of the column, particularly during startup and the initial operation on fresh catalyst. This means that the desired operation is not always achieved with the result that significant amounts of hydrogen sulfide can get into the column bottoms liquid, especially during off-normal operations, resulting in poor column performance and corrosion of the downstream product fractionator.

The alternative to reboiling is steam stripping. However, the stripping steam rate is usually determined by its ratio relative to the bottoms liquid, usually mass of steam per unit volume of bottoms liquid. For a maximum distillate hydrocracking unit, the amount of bottoms relative to overhead material is quite large, and the result would be that there would not be sufficient hydrocarbon above the column feed point to avoid a water dew point at or near the top of the column.

7.9 STRIPPER FIRST FRACTIONATION FLOW SCHEME

Another configuration for the fractionation section is shown in Figure 7.9. This features a steam stripped column followed by a product fractionator. The function of the stripper is to remove hydrogen sulfide and light hydrocarbons overhead so that the feed to the fractionator contains only naphtha and heavier distillate material, as well as the unconverted oil. Unlike a reboiled column that relies on vaporizing the column bottoms material, stripping vapors are readily established in the bottom of the column by the injection of steam.

The stripper column is typically designed and operated such that the hydrogen sulfide content of the bottoms liquid will be virtually nil, so that hydrogen sulfide

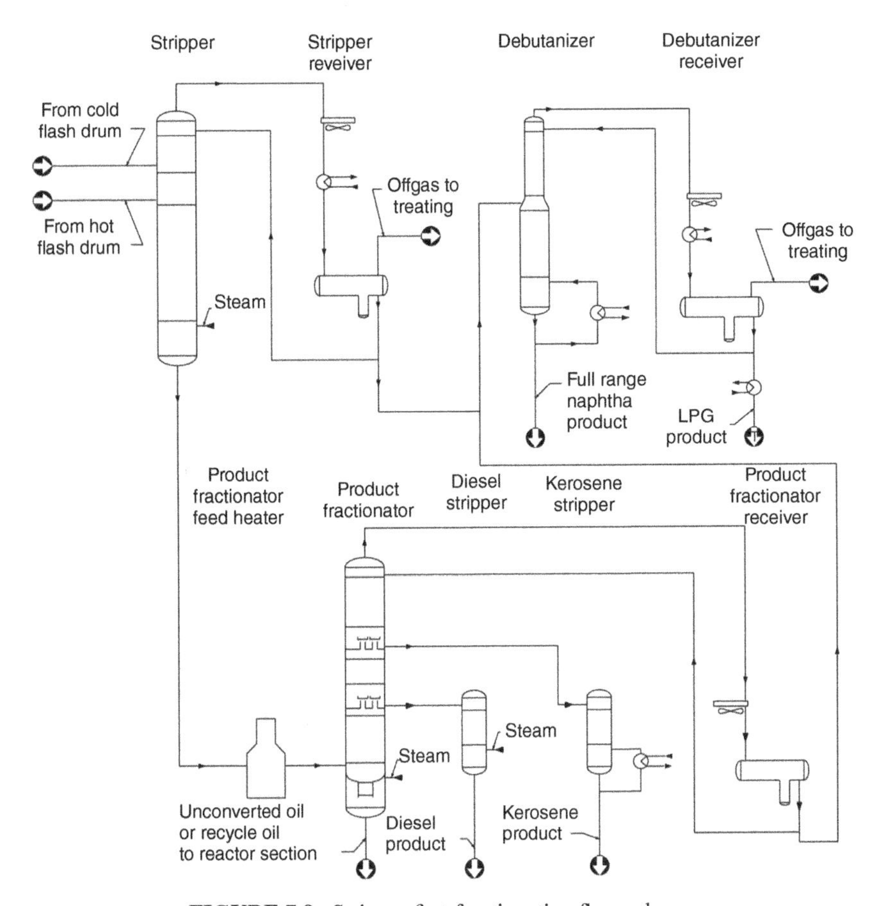

FIGURE 7.9. Stripper first fractionation flow scheme.

corrosion would not be an issue in the design of the fractionator column, thus allowing the product fractionator to be constructed of carbon steel, without the need for stainless lining or other measures to prevent corrosion.

The stripper column also usually includes a specification on the endpoint of the net overhead liquid. This stripper net overhead liquid is usually routed to a debutanizer to remove the LPG and lighter materials in the overhead. Therefore, the net overhead liquid will have an endpoint specification, which is set by the heavy naphtha product requirements. The debutanizer bottoms liquid is routed to a naphtha splitter, if desired, to produce separate light and heavy naphtha products.

If the reactor section utilizes a hot separator flow scheme, the hot flash material and cold flash material are fed separately to the stripper, since the composition of each is quite different. Figure 7.10 shows a composition profile for a typical stripper design for a maximum distillate hydrocracking unit. In this plot, the cold flash liquid is fed to tray 5 and hot flash liquid is fed on tray 8.

A comparison of Figure 7.8 with Figure 7.10 shows that there is much less light material in the column bottoms in the case of the stripper design. This is to be

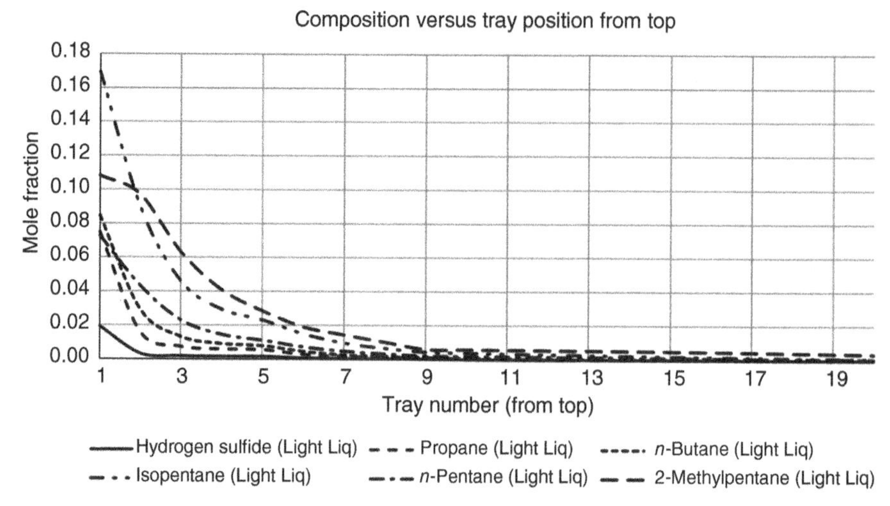

FIGURE 7.10. Composition profile for stripper column in a stripper first configuration.

expected, in that the function of the stripper column is to remove hydrogen sulfide and hydrocarbons boiling in the heavy naphtha range and lighter, versus butane and lighter in the case of the debutanizer design. The absence of lighter hydrocarbons in the bottoms liquid provides some margin to ensure that hydrogen sulfide is removed. In addition, the use of stripping steam provides a more reliable source of stripping vapor, rather than relying on vaporizing the bottoms liquid itself.

7.10 DUAL ZONE STRIPPER FRACTIONATION FLOW SCHEME

A recent variation to the stripper first design has been developed (Hoehn et al., 2015). This concept takes advantage of the initial separation that takes place in the hot separator. Figure 7.11 shows typical distillation data for the hot flash and cold flash liquids from a hydrocracking unit processing a vacuum gasoil feed. As can be seen, the composition of the two liquids is quite different. Routing both streams to the same column

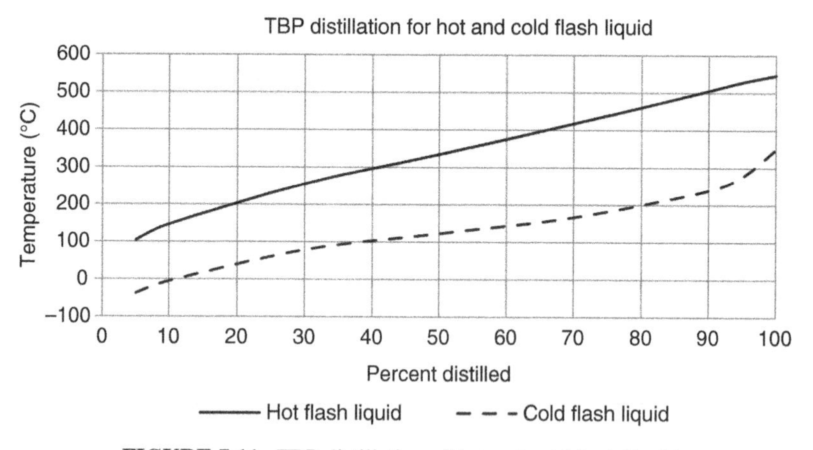

FIGURE 7.11. TBP distillation of hot and cold flash liquid.

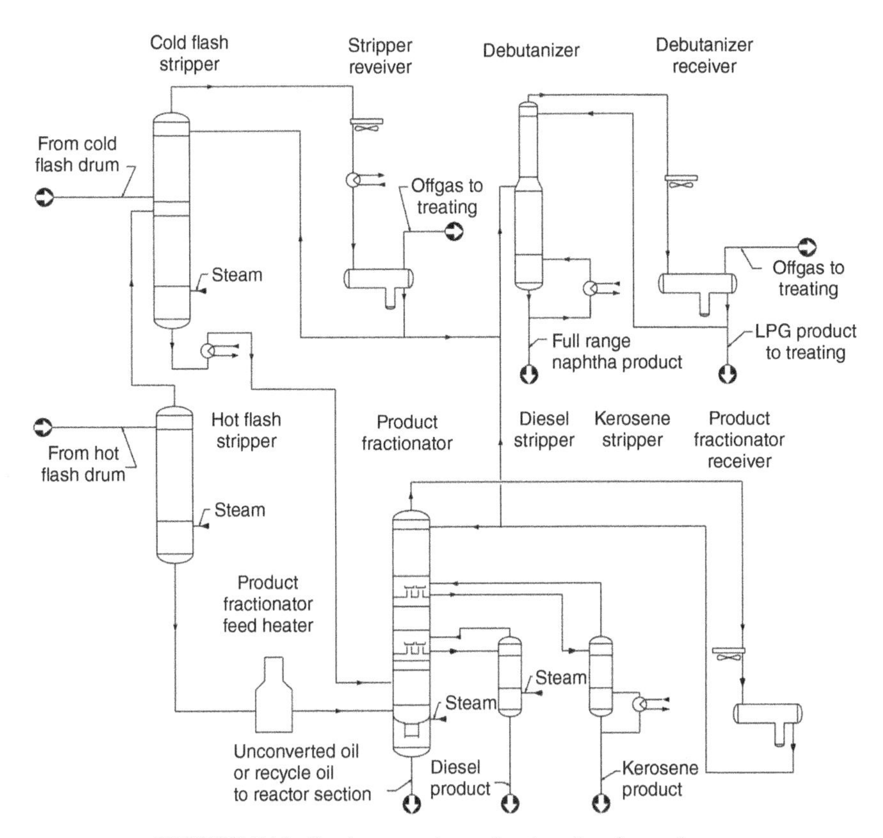

FIGURE 7.12. Dual-zone stripper fractionation flow scheme.

would negate the effects of that initial separation. With this in mind, the flow scheme of Figure 7.12 was developed. This concept takes advantage of the fact that an initial separation, albeit only that of a single stage, has taken place and it is desired to maintain and build on that initial separation. The stripper column is split into two zones. The upper zone processes the cold flash liquid and the lower zone processes the hot flash liquid. Both sections are steam stripped separately. The vapors from the lower hot flash stripping section are routed to the upper cold flash stripping section. The composition of this vapor contains negligible amounts of material heavier than diesel and thus does not increase the endpoint of the cold flash stripper bottoms liquid. The hydrogen sulfide and light hydrocarbons from the both zones pass to the top of the upper zone and exit with the cold flash stripper overhead vapor. The upper cold flash stripper is designed to maintain a sufficient dew point margin to prevent condensation of water near the top of the column. This, in effect, sets the required feed temperature of the cold flash liquid. In addition, the column is designed and operated to limit the endpoint of the net overhead liquid to a value not greater than that of the heavy naphtha product. The bottoms liquid from each stripping zone also must reduce the hydrogen sulfide content to a value that will minimize corrosion in the downstream product fractionator.

One of the main benefits of the dual-zone stripper configuration is that the amount of material that is sent to the product fractionator feed heater, and, therefore, the fired

TABLE 7.5. Utilities Comparison for Single Stripper versus Dual-Zone Stripper Designs

	Single Stripper (MM Btu/h)	Dual-Zone Stripper (MM Btu/h)
Fuel	163.0	76.3
Electricity	14.2	14.47
Cooling water	0.71	0.71
MP steam	−12.13	−12.13
LP steam	62.34	62.34
BFW	14.11	14.11
Total FOE	**242.23**	**155.76**

TABLE 7.6. Capital Cost Comparison for Single Stripper Versus Dual-Zone Stripper Designs

	Single Stripper ($MM)	Dual-Zone Stripper ($MM)
Fractionator feed heater	7.73	5.66
Stripper column (Hot)	2.50	1.90
Stripper column (cold)	—	1.06
Total capital cost	**10.23**	**8.62**

fuel requirement is significantly reduced. The cold flash stripper bottoms liquid is preheated by available process heat before being routed to the product fractionator and does not require complete vaporization. The vaporization of the distillate products in this stream comes partly from exchange with available process streams, such as diesel product or pumparound. The rest comes from excess heat available in the fractionator. Table 7.5 shows a comparison of utilities for the single stripper versus dual-zone stripper designs for a hydrocracking unit of about 50,000 BPSD capacity. Values are in fuel oil equivalent (FOE). Table 7.6 compares the capital cost for each flow scheme.

From Tables 7.5 and 7.6, it can be seen that the dual-zone stripper configuration offers a significant operating and capital cost advantage.

7.11 DUAL ZONE STRIPPER – DUAL FRACTIONATOR FLOW SCHEME

The dual-zone stripper concept has been developed further (Ladkat et al., 2016), as shown in Figure 7.13, again capitalizing on the principle of maintaining the initial separation that took place at the hot separator in the reactor section. This flow scheme involves not only a dual-zone stripper, but also two product fractionators. The first, called a light fractionator, processes mainly the bottoms material from the upper stripping zone. A second fractionator, called a heavy fractionator, processes the bottoms material from the lower stripping zone.

The light fractionator produces a kerosene product as a sidedraw and naphtha range material produced as an overhead product. The bottoms material from the hot flash stripper is routed to a heater and then to the heavy fractionator. The heavy

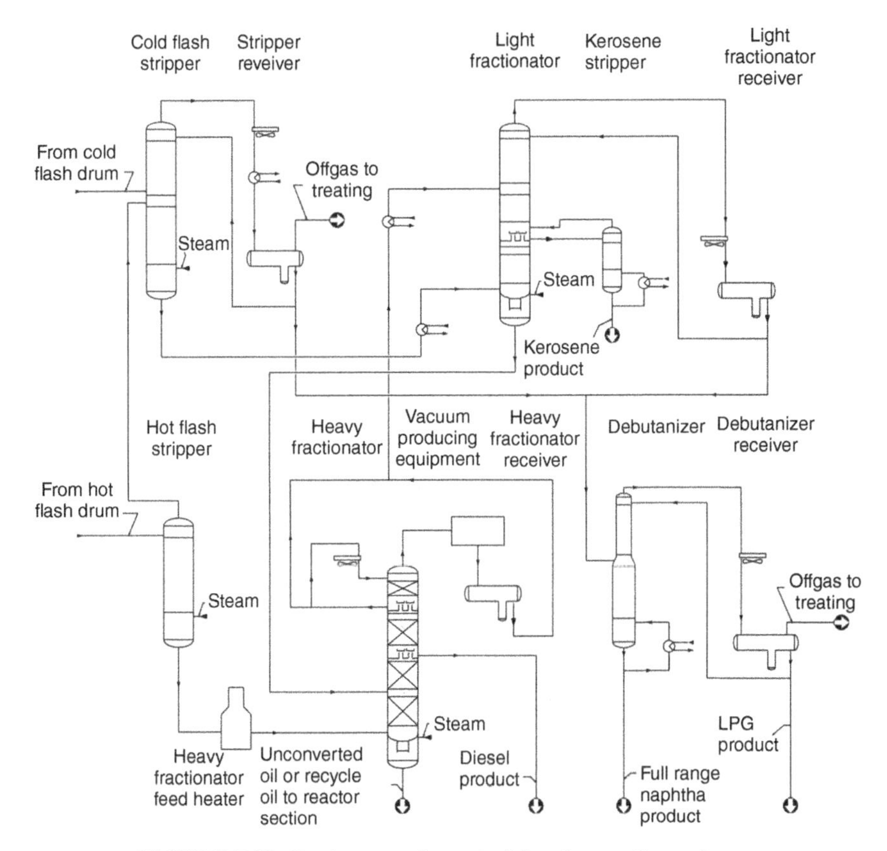

FIGURE 7.13. Dual-zone stripper dual-fractionator flow scheme.

fractionator is operated at a partial vacuum to facilitate the separation between the diesel sidecut product and the unconverted oil bottoms product. The bottoms of the light fractionator contains principally diesel and unconverted oil. This material is fed to the heavy fractionator below the diesel draw. The top of the heavy fractionator contains a partial contact condenser. Heavy fractionator overhead vapors are further condensed to produce an overhead liquid, which is mainly the kerosene and lighter material that was present in the lower stripping zone bottoms material. This liquid is routed, along with net liquid from the contact condenser zone at the top of the heavy fractionator, to the light fractionator. Table 7.7 compares the utilities for single stripper, dual-zone stripper, and dual-zone stripper/dual-fractionator design for a hydrocracking unit of about 55,000 BPSD capacity. Values are in FOE.

As can be seen from Tables 7.5–7.7, the dual-fractionation configuration represents a significant capital and operating cost saving over other alternatives.

7.12 HOT SEPARATOR OPERATING TEMPERATURE

As was seen in Chapter 6, the operating temperature of the hot separator also has an effect on the economics of the unit. The hot separator temperature is selected to minimize the heating requirement of the cold separator liquid feed to the downstream

TABLE 7.7. Utilities Comparison for Single Stripper, Dual-Zone Stripper, and Dual-Zone Stripper/Dual-Fractionator Designs

	Single Stripper (MMBtu/h)	Dual-Zone Stripper (MMBtu/h)	Dual-Zone Stripper/Dual Fractionator (MMBtu/h)
Fuel	163.0	76.3	35.0
Electricity	14.20	14.47	12.0
Cooling water	0.71	0.71	2.26
MP steam	−12.13	−12.13	11.53
LP steam	62.34	62.34	7.77
BFW	14.11	14.11	4.64
Total FOE	**242.23**	**155.76**	**73.19**

fractionation facilities, which can be either a steam stripped or reboiled stripper column, while also minimizing the fired utility in the reactor section. If the stripper is reboiled, there is a trade-off between the reactor charge heater duty, the stripper reboiler duty and the reactor section exchange surface. If the stripper is steam stripped, nearly all of the heat required to operate the stripper column is present in the reactor section effluent streams. Therefore, the temperature of these streams must be such that the hydrogen sulfide specification in the stripper bottoms liquid and the dew point margin in the stripper overhead is maintained while minimizing the capital and operating costs. In hydrocracking units, there is usually much more naphtha and lighter material produced, so that the amount of cold flash liquid is greater and its effect on the design of the stripper is greater. In this case, there is an inverse relationship between the operating temperature of the hot separator and the temperature to which the cold flash liquid must be heated. This is illustrated in Figure 7.14, which was generated for a full conversion hydrocracking unit with a CFR of about 2.

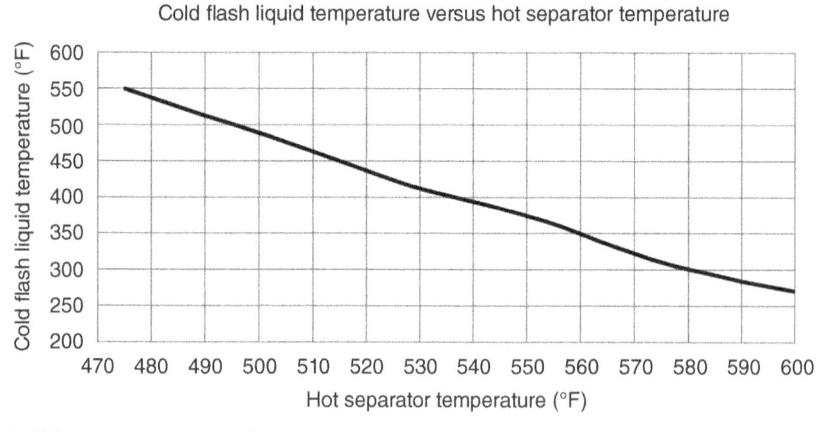

FIGURE 7.14. Cold flash liquid temperature requirement versus hot separator operating temperature.

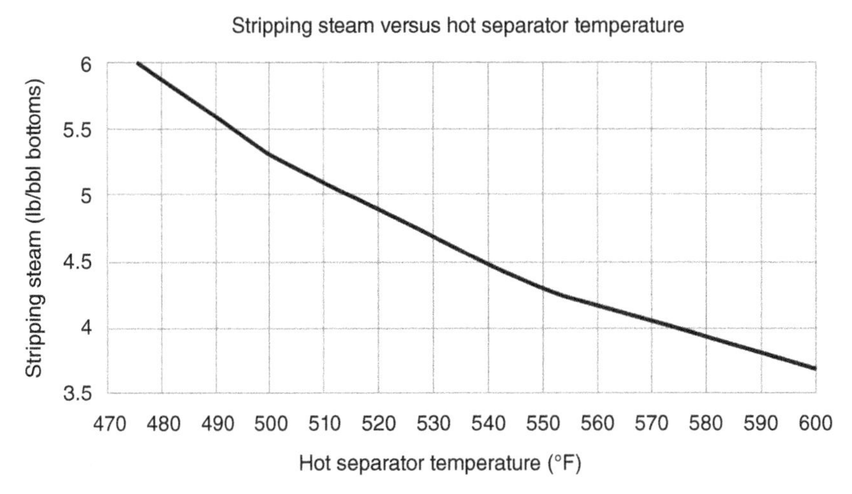

FIGURE 7.15. Stripping steam requirement versus hot separator operating temperature.

Note that for hot separator operating temperatures below about 500 °F (260 °C), it is necessary to heat the cold flash liquid above the hot separator operating temperature. The stripper design is also affected by the hot separator operating temperature. For example, Figure 7.15 shows the effect of hot separator operating temperature on the stripping steam requirement.

Figure 7.16 shows the effect of hot separator temperature on reactor section exchange surface requirements. In this example, the reactor section heater duty was kept constant and the exchanger requirements were then determined taking into account for the cold flash liquid heating requirements (the hot separator liquid passes directly from the hot flash drum to the stripper).

Figure 7.17 is a graph of the total capital and operating costs are considered versus hot separator temperature. The capital costs include the reactor section heat

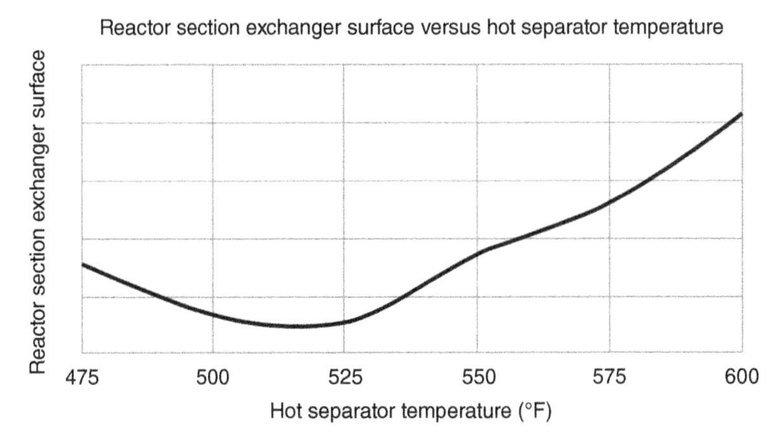

FIGURE 7.16. Reactor section exchanger surface versus hot separator operating temperature.

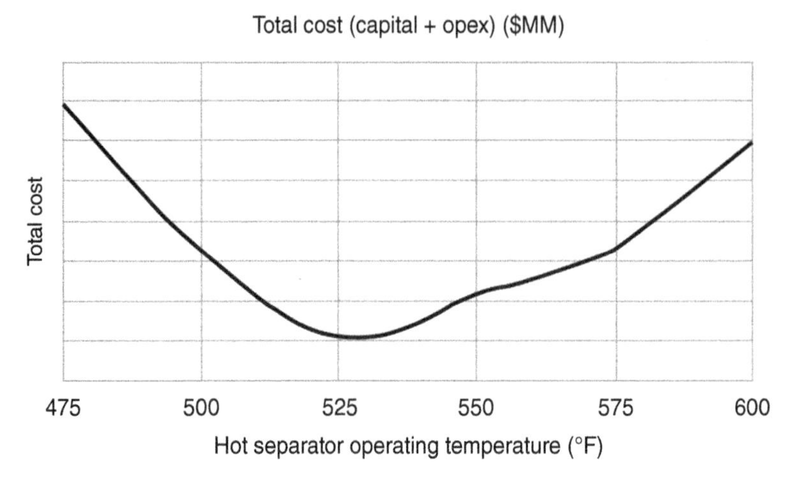

FIGURE 7.17. Total capital + operating costs versus hot separator operating temperature.

exchangers and stripper overhead condenser. Regarding the operating costs, the major variant is the stripping steam cost since the reactor section heater duty was held constant over the operating temperature range of the hot separator. In this case, the annual operating costs were multiplied by a factor of 3, which is a typical simple payback period for a trade-off between capital and operating costs.

Note that the shape of the curve in Figure 7.16 shows a minimum at about the same point as that in Figure 7.17. This means that the economics are driven mainly by the reactor section exchanger surface requirement and that this could be a quick optimization parameter on which to focus for a quick determination of the optimum hot separator operating temperature.

7.13 HYDROGEN RECOVERY

Hydrotreating and hydrocracking reactions consume hydrogen. Hydrocracking units tend to require much more hydrogen per barrel of feed than do hydrotreating units, because both hydrotreating and hydrocracking reactions take place. The additional hydrogen associated with hydrocracking is a function of overall conversion and target cutpoint. For example, a hydrocracking unit designed for maximum conversion of VGO into naphtha will use much more hydrogen than one designed to produce maximum diesel.

As was seen in Chapter 6, hydrogen can be an expensive material to produce and its efficient use is essential to the overall economics of the unit. Assuming hydrogen at $2.00/1000 SCF, or a 50,000 high-conversion hydrocracking unit capacity consumes an amount of hydrogen equal to about 2.3 wt% of the feed, the annual hydrogen cost could be between $37.2 and $151.8 MM, according to the cost of hydrogen production. Therefore, efficient use of hydrogen is critical to the economic operation of any hydroprocessing unit, and indeed the overall economics of the refinery.

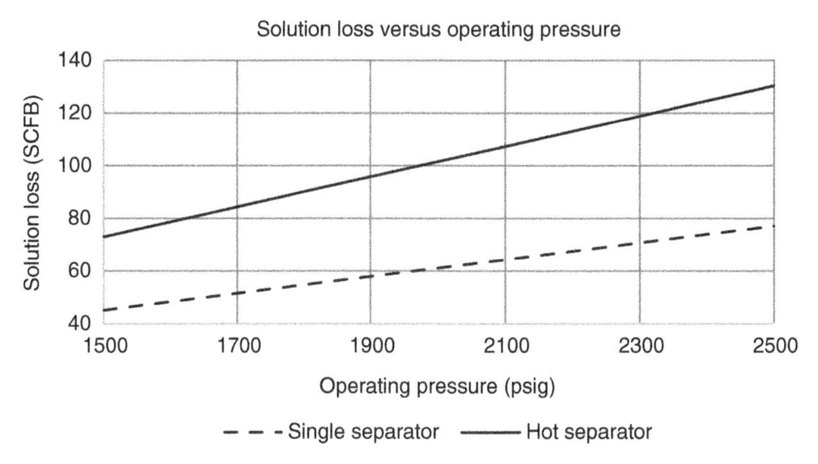

FIGURE 7.18. Solution loss for a hydrocracking unit.

The sum total of the hydrogen required to accomplish the desired reactions is called the chemical consumption. As was demonstrated in Chapter 5, hydrogen has the ability to dissolve in hydrocarbon liquid and, to a lesser extent, in water. Thus, the liquid leaving a separator will contain some amount of dissolved hydrogen. The amount of hydrogen dissolved in the separator liquid will be a function of the temperature and pressure of the separator, as well as of the partial pressure of hydrogen in the vapor in equilibrium with the liquid. The hydrogen contained in the separator liquids is referred to as solution loss. Hydrocracking units will generally be designed with a hot separator flow scheme. Since hydrogen solubility in hydrocarbon increases with temperature, a unit designed with a hot separator will have a higher solution loss than a unit designed with only a single separator. However, the solution loss will also be affected by the CFR. A once-through design (CFR = 1) will have a lower solution loss than one designed for a recycle operation and a CFR of 2. Figure 7.18 compares the solution loss for these two unit designs as a function of operating pressure. The solution loss is usually expressed in terms of standard volume of hydrogen loss per unit volume of feed to the unit.

As discussed , the CFR has a major effect on the hydrogen solution loss.

The hydrogen recovery methods detailed in Chapter 6 also pertain to hydrocracking units. Generally, hydrocracking units, because of their tendency to operate at higher pressures, will be provided with flash drums. Therefore, a configuration similar to Figure 6.12 would be used for units that are not designed to maximized LPG recovery.

7.14 LPG RECOVERY

LPG recovery presents another area of optimization during the design of the process. Most hydrocracking units are designed for maximum distillate production. Although not necessarily a main product of a hydrocracking unit, in certain regions of the world,

TABLE 7.8. Typical Yields for a Maximum Distillate Hydrocracking Unit

Product	Wtl% Yield
C_1	0.30
C_2	0.35
C_3	0.92
C_4	1.92
C_5	2.51
C_6	3.53
Heavy naphtha	9.43
Kerosene	22.30
Diesel	58.86

recovery of propane and butanes (LPG) from the hydrocracking reactor products is economical and desirable.

Typical yields for a maximum distillate hydrocracking unit fed with a heavy VGO with a 1100 °F (594 °C) endpoint are as follows (Table 7.8):

Over 80% of the products are in the distillate range (kerosene and diesel). Even so, 2.8% of the products are LPG range material. For a hydrocracking unit of about 55,000 BPSD capacity, this can represent almost 1500 BPSD of LPG.

Some losses of LPG range material are inevitable from the flash drum offgas, but for areas where LPG has a high market value or where existing facilities elsewhere in the refinery cannot accommodate the light offgases for LPG recovery, there is a definite benefit to designing the unit to include it. The following table 7.9 shows the amount of light material present in the flash drum offgas, stripper net vapor, and stripper net overhead liquid for a maximum distillate hydrocracking unit of about 50,000 BPSD capacity with a stripper first flow scheme.

From Table 7.9, the benefits of the presence of the flash drum can clearly be seen. Almost 78% of the hydrogen and 62% of the methane is present in the flash gas. These compounds, particularly hydrogen, would interfere with the efficient recovery of LPG in the fractionation section. However, the flash gas does contain about 13% of

TABLE 7.9. Composition of Flash Gas and Stripper Overhead Streams

	Flash Drum Offgas (lb mol/h)	Stripper Net Vapor (lb mol/h)	Stripper Net Liquid (lb mol/h)
H_2	1032.6	285.1	3.54
H_2S	19.70	36.46	16.90
C_1	92.35	45.59	4.22
C_2	31.43	37.89	14.64
C_3	28.70	62.17	65.44
i-C_4	14.02	38.81	85.55
n-C_4	8.50	24.53	73.58

the total LPG components, so some losses are inevitable, leaving about 87% available for recovery.

Efficient recovery of the LPG from the reactor effluent products requires some additional fractionation equipment. In general, the flow scheme of choice for a maximum distillate unit would incorporate a stripper first fractionation configuration. In such a flow scheme, it is noted that only about 64% of the LPG range material in the stripper overhead is present in the overhead liquid, with the remaining 36% present in the stripper net vapor. From this, it follows that maximizing LPG recovery involves extracting as much LPG range material from the stripper vapor as is practical. The usual method for achieving this is by the use of a suitable material, called sponge oil, to absorb the desired components, while minimizing loss of the absorbent liquid in the offgas. Economics dictates that the flow scheme selected should be as simple as possible and use an effective sponge oil. To minimize utilities, the chosen sponge oil should not require vaporization in order to recover the absorbed LPG. In other words, it should be possible to recover the absorbed LPG by stripping the absorbent liquid to remove the desired components. The only heat required would be that required to heat the sponge oil from the operating temperature of the absorber to the temperature at which the LPG components are stripped from the sponge oil. Since this incurs only sensible heat, rather than latent heat, the energy investment is much less. Light naphtha (mainly C_5 and C_6) hydrocarbons would be the most effective, but would incur the largest losses of these compounds to the offgas. Full range naphtha would also be a good absorbent, but the light naphtha components present would also have a tendency to be lost to the net offgas. An additional disadvantage to the use of light naphtha is that light and heavy naphtha are usually produced as separate products. This would mean that the light naphtha used for sponge oil would have to be vaporized in a naphtha splitter column, and this would adversely affect the economics of the use of that material. Therefore, the material that appears to fit this criterion most closely is heavy naphtha.

One flow scheme that uses heavy naphtha as the sponge liquid is shown in Figure 7.19.

With the selection of the flow scheme, the main variable is the amount of lean oil used to recover the LPG. Circulating more lean oil will tend to recover more LPG but at the cost of higher capital and utilities. Figure 7.20 shows the relation between lean oil circulation rate and overall LPG recovery for a hydrocracking unit of about 55,000 BPSD capacity whose yields are defined in Table 7.8.

As can be seen, an increase in lean oil rate does improve the recovery of LPG, although for very high lean oil rates, the recovery flattens out and actually decreases. This is due primarily to recovery losses in the Debutanizer. According to the value of LPG, the optimum LPG recovery would be expected to be somewhere in the range of 10,000–20,000 BPSD lean oil rate.

7.15 HPNA REJECTION

Hydrocracking units designed before about 1985 tended to have a single high-pressure separator design. As design conversions increased toward 100%, problems with fouling of equipment began to be noted. The problems became evident as

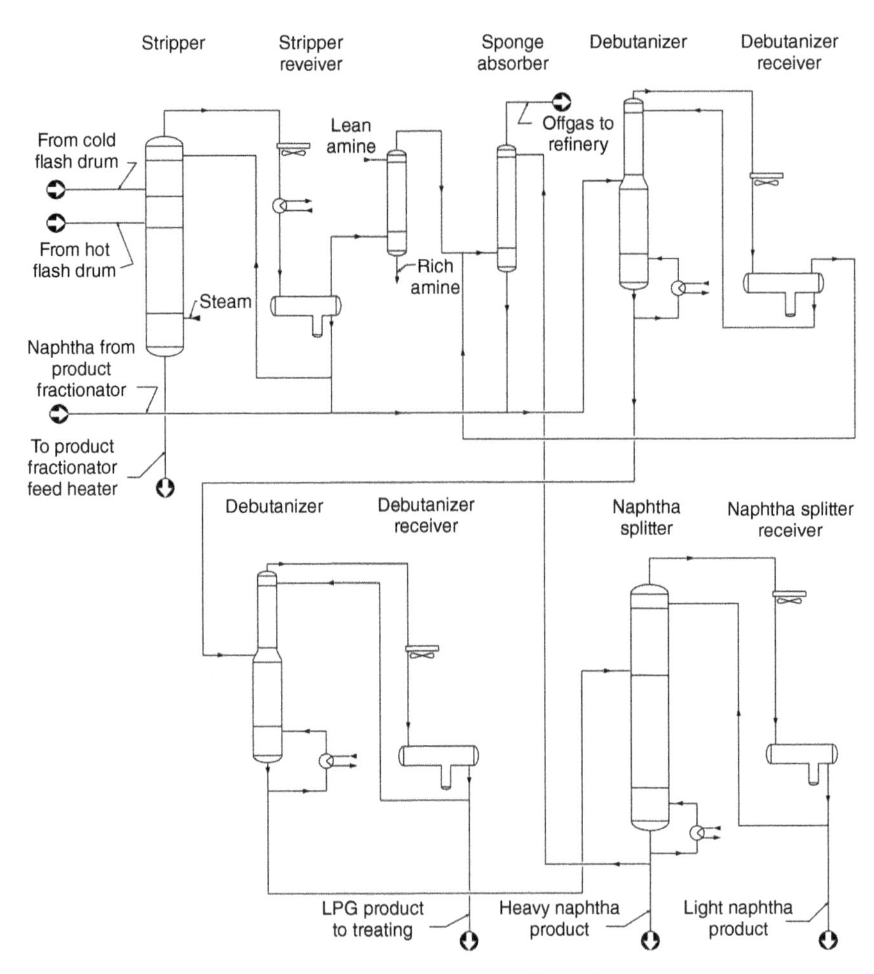

FIGURE 7.19. LPG recovery flow scheme.

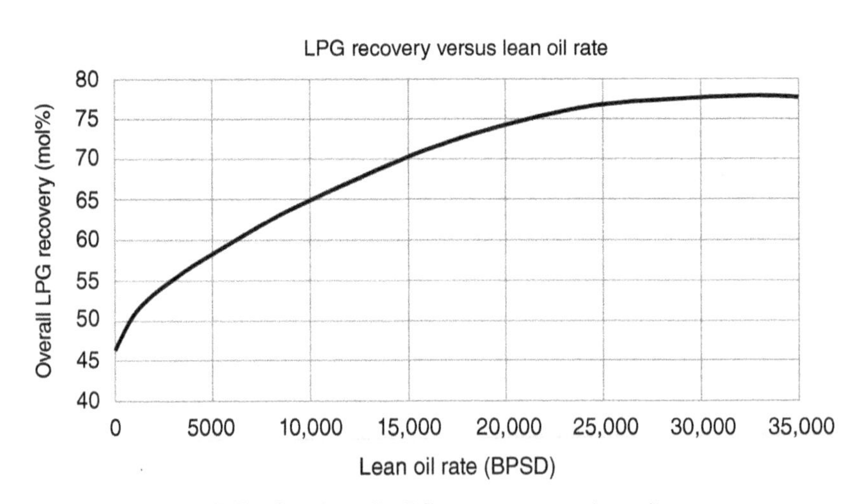

FIGURE 7.20. Overall LPG recovery versus lean oil rate.

fouling of the reactor effluent air cooler with a reddish, waxy material and as an increased deactivation of the catalyst. The problem was particularly acute when processing feeds with high endpoints, such as deasphalted oils. Some units experienced air cooling that was rapid enough to cause a shutdown for cleaning in a matter of weeks. Analysis of the material indicated that it consists of high-molecular-weight polycyclic compounds. Compounds with seven or more rings were termed HPNAs. Subsequent work established that these materials have very low solubility in both aromatic and aliphatic hydrocarbons and the only effective way to clean fouled equipment was by mechanical means. The presence of HPNA material in the recycle liquid causes its appearance to change. As the amount of HPNA material increases, the recycle liquid will begin to appear somewhat reddish (Figure 7.21-left bottle). With increasing HPNA concentrations, a definite reddish-brown color is seen. The middle and right bottle in Figure 7.21 are essentially free of HPNA material. It was also established that because of its relatively low solubility, HPNA compounds were precipitating at a temperature corresponding to values typically seen in the reactor effluent air cooler, namely, 300–400 °F (150–200 °C). The solution to this problem therefore was to design high-conversion units with a hot separator, thus removing the issue of fouling of equipment.

Although a hot separator design resolved the issue of premature shutdown due to fouling of equipment, the issue of catalyst deactivation due to deposition of HPNA material in the recycle material on the catalyst as coke remained unsolved. One solution to this is to remove or purge a portion of the recycle oil to limit the concentration of HPNA material in the recycle oil. For heavy feeds, the amount of purge required approaches 7–8% of the feed, thus leading to obvious economic penalties due to lost yields of desirable distillate products. There was, therefore, incentive to develop a means to reject HPNA compounds so that conversion could be maximized

FIGURE 7.21. Recycle oil sample from a high conversion hydrocracking unit.

while minimizing operational problems. To this end, two different methods have been commercialized.

7.15.1 HPNA Rejection via Carbon Adsorption

In the late 1980s, work was done to determine if the HPNA material could be removed from the recycle liquid. One method that showed promise was adsorption by activated carbon. Certain types showed an affinity, particularly for the heavier HPNA material. Working with a major carbon supplier, the necessary operating parameters and design requirements were determined. Carbon bed adsorption was first implemented in 1990 for a high-conversion hydrocracking unit in Thailand. Subsequent operation revealed that it not only prevented the fouling problem but also significantly reduced catalyst deactivation, thus providing two tangible economic benefits: longer run lengths and higher conversion. Figure 7.22 shows an installation of an HPNA adsorption system for a hydrocracking unit in New Zealand.

Although a number of installations are operating successfully, acceptance of this method of HPNA rejection has been low. Two main reasons for this are carbon handling and disposal. As seen in Figure 7.22, the size of the carbon chambers can be

FIGURE 7.22. HPNA adsorption chamber installation. Source: Reproduced with permission of New Zealand Refining, Ltd.

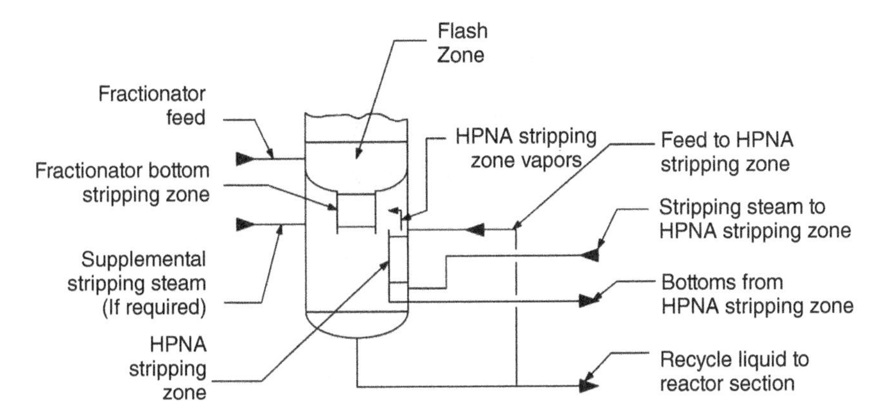

FIGURE 7.23. Schematic of HPNA stripping zone.

rather large and carbon change-out frequency is on the order of 3–6 months. Thus, the refiner is faced with a regular, periodic handling of large amounts of carbon and its disposal. Acceptance has been greater, for refiners that are located near an installation that provides a convenient destination for the spent carbon, such as a cement plant.

7.15.2 HPNA Rejection by Fractionation

As the foregoing discussion indicated, the logistics of carbon handling and disposal have prevented widespread implementation. Due to that, an alternative method of HPNA removal was sought. The outcome of that work was the development of a fractionation technique (Hoehn et al., 2005, 2013) by which the HPNA material in the recycle liquid is concentrated and rejected as a small purge stream amounting as little as 0.5 vol% of the feed. A slipstream of recycle material is routed to a small HPNA stripping zone and the lighter hydrocarbons are stripped and returned to the product fractionator and the concentrated HPNA-rich bottoms is purged from the unit. The overhead vapor from the HPNA stripping zone contains steam and hydrocarbon, and the steam can be reused as stripping steam for the main fractionation section. Figure 7.23 shows a schematic of the system. Because of its relatively small size, the HPNA stripping zone can be placed in the bottom of the product fractionator in the annular area between the outer shell of the product fractionator and the bottom stripping zone. The overhead vapors can be easily routed to the bottom of the bottoms stripping zone. Figure 7.24 shows a sample of the feed to the HPNA stripping zone (the left-hand tray) and bottoms from the HPNA stripping zone (right-hand tray).

7.16 HYDROCRACKING UNIT INTEGRATED DESIGN

From a design standpoint, various issues arise from the properties of the feed, the catalyst system chosen, and the required processing conditions. These include the following:

(1) Distillate Product Maximization – The goal of maximizing distillate products means that feed not converted during the first pass through the reactors will

FIGURE 7.24. Feed and product from an HPNA stripping zone.

need to be reprocessed until the desired level of conversion is achieved, in this case, 99.5 vol%. Due to the high feed rate, a two-stage flow scheme would be the best choice to achieve this, since a single-stage flow scheme in which the fresh feed and recycle material are combined and fed to a single reactor system would result in reactors and other equipment being larger than current vendor capabilities.

(2) Reactor Size – Reactor bed temperature rise limitations and catalyst volume requirements mean that there will be five beds of hydrotreating catalyst and two beds of cracking catalyst in the first stage and three beds of cracking catalyst in the second stage. Generally, the first stage would consist of separate hydrotreating and hydrocracking vessels. However, given the number of beds and required amount of hydrotreating catalyst, it is necessary to confirm that the reactor weight would not be exceeded, or in the absence of that, whether reactor vendor limitations would not be exceeded. Current limitations range from 1500 metric tons (The Japan Steel Works Ltd., 2004) to 2000 metric tons (Harima Works, Kobe Steel Ltd., 1995; Heavy-Wall Reactors and Pressure Vessels, GE Oil & Gas, 2011) or more, according to the vendor. The reactors must also take into account vendor limitations regarding diameter. A practical limitation for reactor size is 6 m based on the outside diameter. The inside diameter will, of course, be smaller by the wall thickness required for the material selected and the design temperature and pressure. This, in effect, sets the limitation to the maximum charge rate that can be handled by a single reactor train. In the case of this unit, the estimated reactor diameter and length is just able to meet the diameter and weight limitations with a single reactor train.

(3) Heat Exchange Configuration – The heat exchange configuration for the reactor section also must take into account the size limitations for high-pressure shell and tube exchangers. This is generally taken to be about 48 in. (1200 mm). Given the size and expected number of heater passes for this unit, parallel sets of exchangers heating feed and recycle gas separately are used. The feed and recycle gas combine at the heater inlet. To determine the duty each service, a suitable tool for pinch analysis would be used. There are several commercially available tools available for this analysis. The design will be based on an economical minimum approach temperature. As it is expected that there will be a significant amount of heat evolved due to the reactions involved, additional heat can be available for recovery into streams in the fractionation section and the heat exchange design should take this into account.

(4) Separator Design Considerations – Pinch analysis has proven to be very effective in the design of heat exchange networks. However, there is another important concept that pinch analysis cannot address: the optimum operating temperature for the separators. While the operating temperature of the reactors and fractionation equipment will be dictated by the processing objectives, the operating temperature of the hot separator is less constrained. As seen in Section 7.2, there exists an operating temperature for the hot separator that minimizes the high pressure exchange requirements, while also meeting the feed temperature requirements to the fractionation section. Also, as was seen in Section 5.3, the operating temperature of the cold separator can be varied over a fairly wide temperature without having a major effect on the recycle gas hydrogen purity. This means that the separator temperature can be set such that the hot separator vapor condenser can be an air-cooled exchanger with the outlet temperature set taking into account the maximum dry bulb temperature of the site where the unit will be located.

(5) Coker-Derived Feed Components – The feed contains a significant amount of coker-derived material. Coker gas oil contains coke fines that must be removed as much as possible to avoid premature shutdown for catalyst bed skimming due to pressure drop buildup. Close attention to feed filtration and use of specific graded bed material will also be needed to address this. Specifically, backwash type feed filters would be indicated, as cartridge filters would quickly foul and require very frequent replacement. Even so, the particle size distribution is such that additional fines retention capacity be provided in the top bed of the first stage hydrotreating reactor. As explained in Chapter 5, there are a number of suitable materials available to provide this capability.

(6) Feed Endpoint – The feed endpoint is typical for a deep-cut VGO. A hot separator configuration for the reactor section will be required to avoid water separation problems in the cold separator. A hot separator configuration is also necessary to avoid deposition of HPNA material, which would occur if the reactor effluent, particularly for the second stage, were to be cooled below a certain point.

(7) Feed Sulfur and Nitrogen Content – The feed sulfur indicates that a recycle gas amine scrubber would be necessary to minimize the concentration of hydrogen sulfide in the system in order to maximize catalyst activity. The feed nitrogen will have an effect on the required space velocity of the first stage cracking catalyst, as ammonia tends to neutralize the acid sites. In addition, the concentration of hydrogen sulfide and ammonia in the reactor effluent material may have implications on the metallurgy in certain parts of the unit. In the case of this unit, the parameter *Kp*, which is a product of the hydrogen sulfide and ammonia partial pressures, is such that upgraded metallurgy is not required, due to the presence of the recycle gas scrubber and its action in removal of hydrogen sulfide, as well as the removal of most of the ammonia and some hydrogen sulfide by the wash water.

(8) Feed Metals Content – The metals in the feed are moderately high and will require specific metals removal catalysts. One type of catalyst will be required to adsorb the nickel and vanadium and a different material may be required to adsorb the silicon originating from the coker gas oil.

(9) Hydrogen Management – Hydrogen is an expensive material to produce and its economical use is essential to an efficient design. Accordingly, the design would be expected to recover the hydrogen present in the flash drum offgas. To achieve this, the gas is first amine-scrubbed to remove hydrogen and then sent to a pressure swing adsorption (PSA) unit for recovery of the hydrogen, as explained in Chapter 6. The operating pressure of the flash drums will be such that the hydrogen recovered by the PSA unit can be routed directly to the hydrogen header.

(10) Product Separation – As was seen in Section 7.10, the dual-zone stripper/dual-fractionation flow scheme represents the most economical method for separating the products in a maximum distillate hydrocracking unit and such a flow scheme would be expected in this case.

(11) In addition to maximum distillate production, the design basis calls for maximum LPG recovery. Therefore, flow scheme shown in Figure 7.20 would be employed. The sponge oil rate would be varied to maximize recovery while minimizing utilities and capital.

(12) Energy Efficiency/Recovery – Efficient recovery of heat dictates that exchange between reactor section and fractionation section streams be considered. This may involve a high-pressure hot stream exchanging heat with a low-pressure cold stream. With proper analysis and safeguards, such exchange combinations are possible and should be used to maximize the recovery of heat in the unit.

(13) HPNA Management – Due to the high endpoint of the feed and desire to maximize conversion to achieve the target distillate yields, HPNA rejection will be necessary. The method selected is an HPNA stripping zone, as shown in Figure 7.23.

From the above-listed conditions, the basic reactor section flow scheme can be derived. It is shown in Figure 7.25. The fractionation section flow scheme is shown in Figures 7.26 and 7.27.

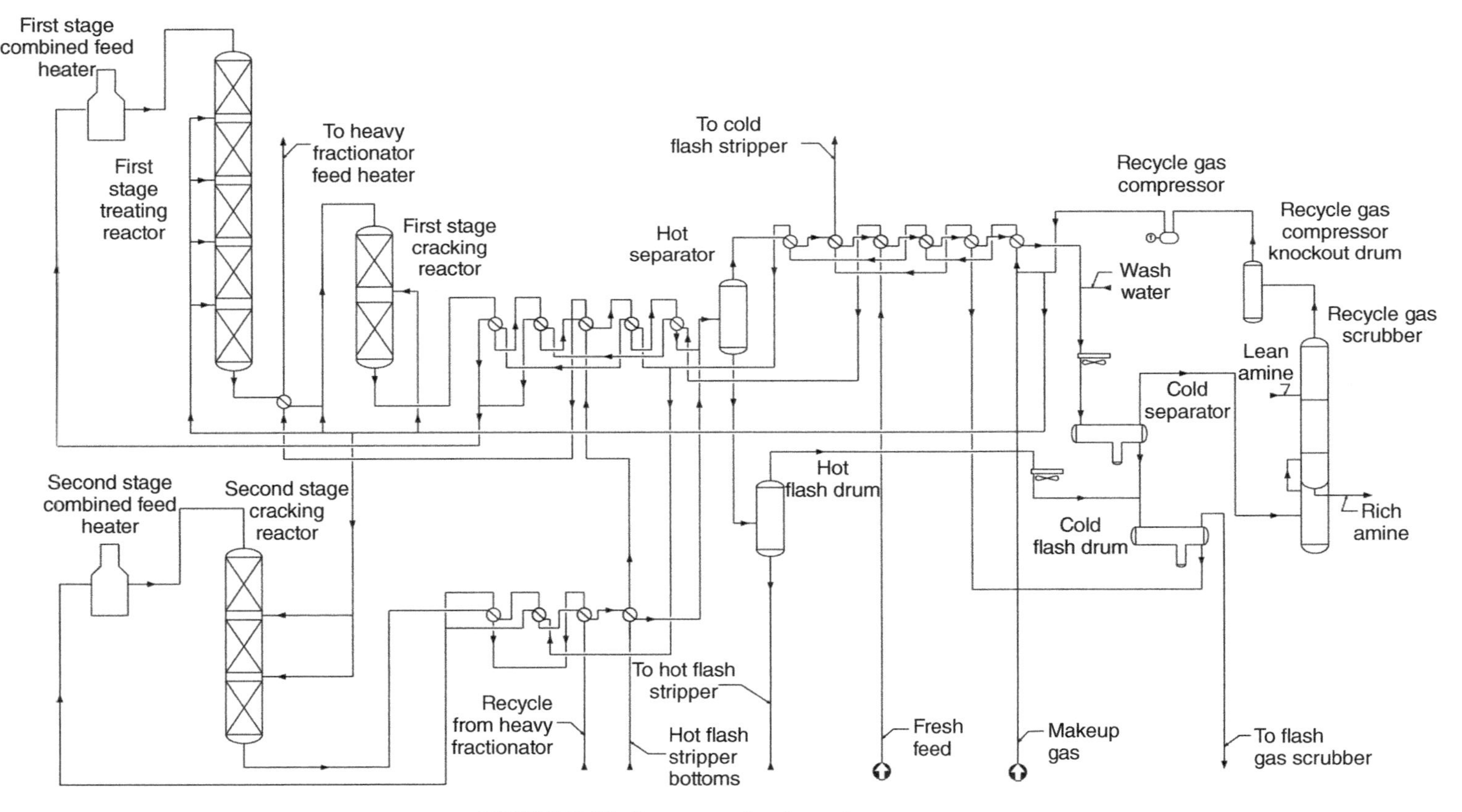

FIGURE 7.25. Reactor section flow scheme.

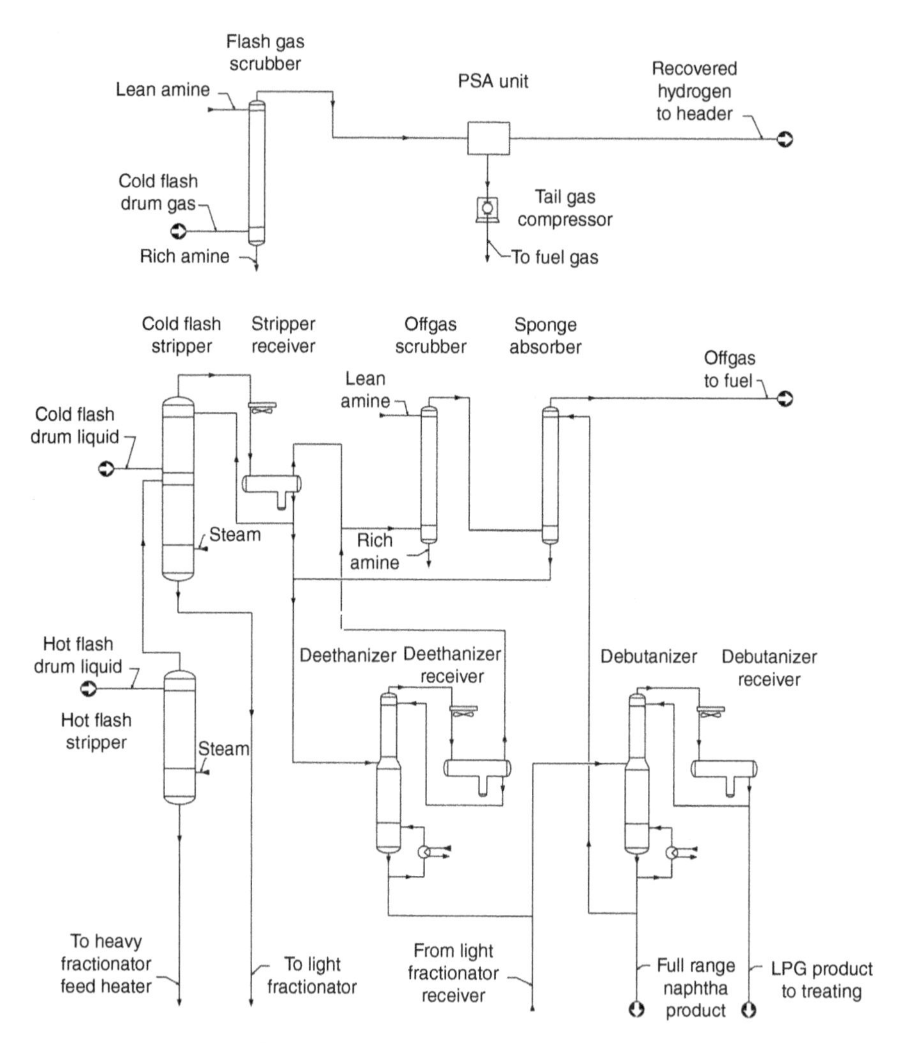

FIGURE 7.26. Light fractionation section.

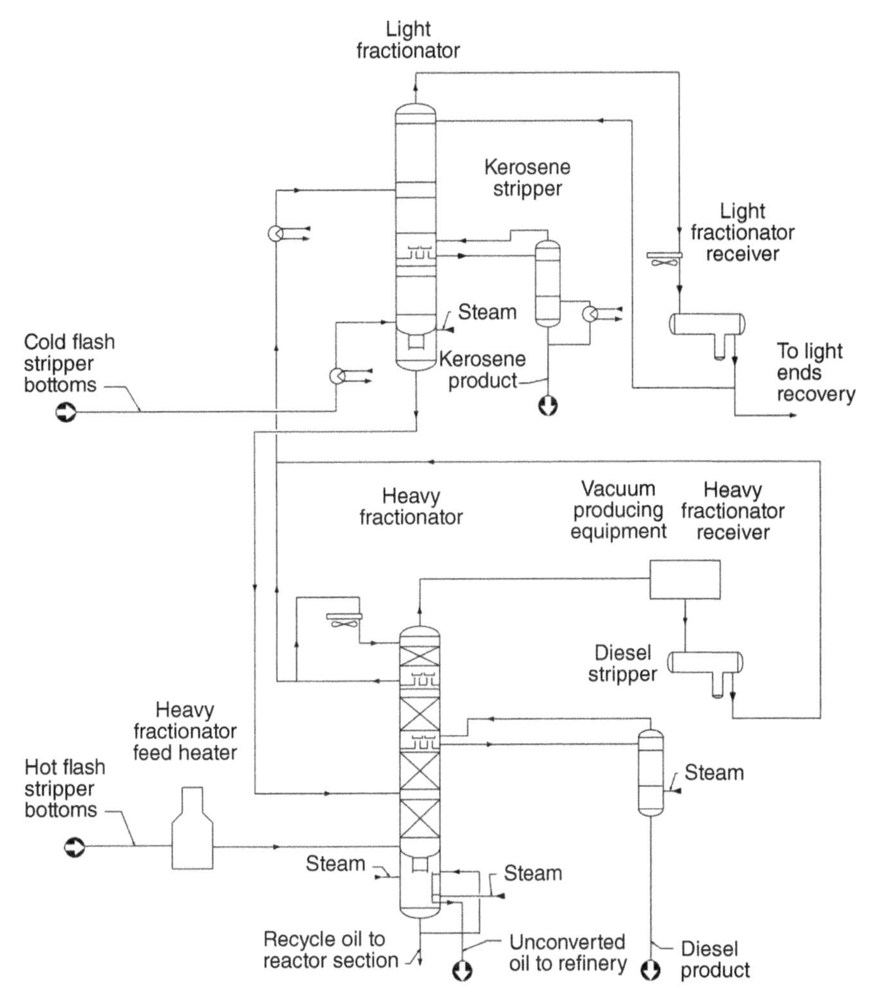

FIGURE 7.27. Heavy fractionation section.

REFERENCES

Harima Works, Kobe Steel, Ltd. (1995) 2000 ton Reactor Manufacturing Shop in Kobe Steel.

Heavy-Wall Reactors and Pressure Vessels, GE Oil & Gas (2011).

Hoehn R et al. (2005) Hydrocracking process. US Patent 6,858,128.

Hoehn R et al. (2013) Process for removing heavy polynuclear aromatic compounds from a hydroprocessed stream. US Patent 8,574,425.

Hoehn R et al. (2015) Dual stripper column apparatus and methods of operation. US Patent 9,074,145.

Ladkat K et al. (2016) Process and apparatus for hydroprocessing with two product fractionators. US Patent 9,234,142.

The Japan Steel Works, Ltd. (2004) JSW's Current Activity and Manufacturing Experience of Pressure Vessel.

PART 3

ENERGY AND PROCESS INTEGRATION

8

HEAT INTEGRATION FOR BETTER ENERGY EFFICIENCY

8.1 INTRODUCTION

After the process configuration is determined, process heat integration must be conducted to determine process heat recovery in order to minimize energy loss. To do this, two questions must be answered. First, what is the maximal heat recovery target for a process unit? Second, what improvements in heat recovery are to achieve the target? The first question can be answered by the energy targeting method, which is a very powerful tool for identifying energy saving opportunities. If the opportunity is large enough, it will warrant a more detailed assessment to determine what improvements are required.

It requires two methods to answer the second question. To obtain the most energy-efficient design for a grassroots process, the heat recovery design approach by Zhu et al. (1995) will be adopted, while for retrofit of an existing process, the heat recovery retrofit method by Zhu and Asante (1999) will be applied. These two methods together with energy targeting method will be the focus of this chapter. These methods have been implemented into several commercial software including Aspen HxNet and Honeywell UniSim ExchangerNet.

8.2 ENERGY TARGETING

Identifying saving opportunity for process heat recovery should be the first step for process energy optimization, and this can be accomplished by energy targeting based on composite curves.

Hydroprocessing for Clean Energy: Design, Operation, and Optimization, First Edition.
Frank (Xin X.) Zhu, Richard Hoehn, Vasant Thakkar, and Edwin Yuh.

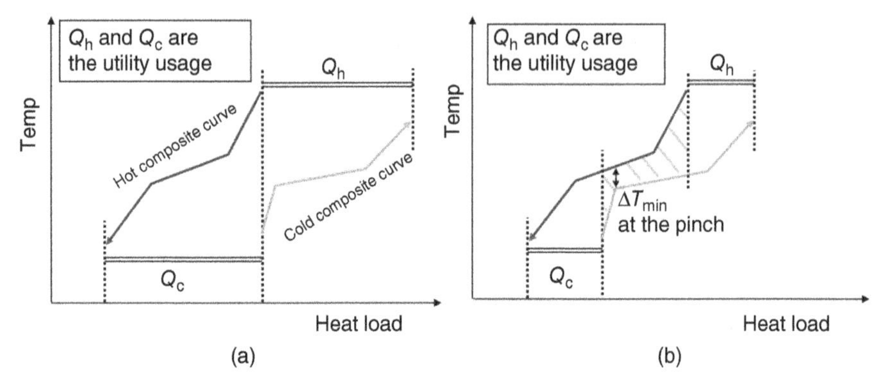

FIGURE 8.1. Composite curves: heat demand (grey) versus heat availability (dark) profiles. (a) No heat recovery case; (b) heat recovery (hatched area).

8.2.1 Composite Curves

Composite curves were developed for heat recovery targeting (Linnhoff et al., 1982). The word "composite" reveals the basic concept behind the composite curves method: system view of the overall heat recovery system. The problem of assessing a complex heat recovery system involving multiple hot and cold streams is simplified as a problem of two composite streams (see Figure 8.1). One hot composite stream (shown in dark) represents all the hot process streams, while one cold composite stream (shown in grey) represents all the cold process streams. In essence, the hot composite stream represents a single process heat source, while the cold composite stream a single process heat sink. The composite curves can indicate the case of no process heat recovery (Figure 8.1a) as well as the targeted heat recovery (Figure 8.1b). The difference between these two is a measure of the potential for heat recovery.

Generation of composite curves starts with identification of a representative base case for the process. The preliminary heat and mass balances for the base cases are then generated, to determine the critical process conditions such as reaction temperature and pressure, separation temperatures and pressures, conditions of feeds, and products and recycle streams. The stream data from this base case is then used to build the composite curves.

8.2.2 Building the Composite Curves

To illustrate applications of the composite curves in reducing the heating duty of the process, let us consider the example in Figure 8.2. The design involves reactors and separation columns and features heat recovery between hot and cold process streams.

A hot composite curve is built for the process, which includes three hot streams and two cold streams. The composite stream should have two features: (1) it should go through the exact same temperature changes as the three streams it represents and (2) it should have the same total heat load as the summation of the heat loads of three streams.

To satisfy these two conditions, these three streams are plotted on the *T*/*H* axis with their starting and terminal temperatures corresponding to the enthalpy changes (heat

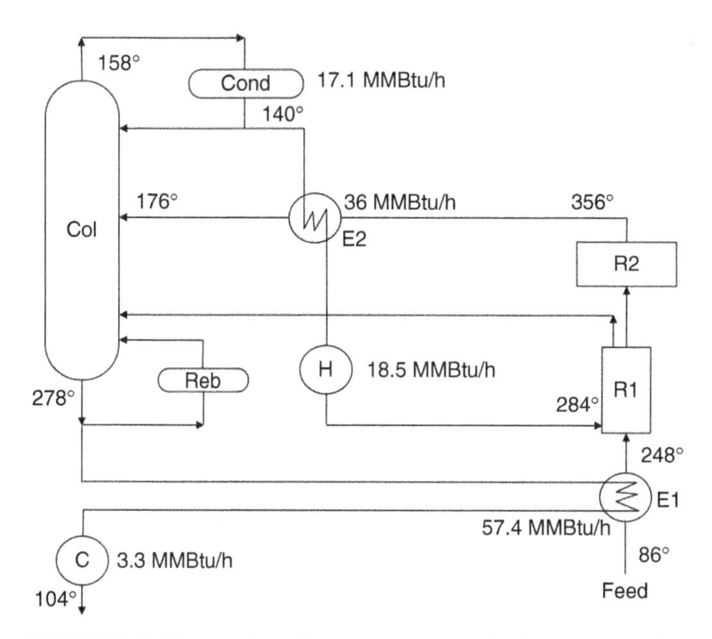

FIGURE 8.2. Process flow diagram as an example of energy targeting.

loads), respectively, as shown in Figure 8.3a. As a result, five temperature intervals are created from the starting and terminal temperatures of each stream. In interval 1, only stream C exists, and, thus, the enthalpy change $\Delta H_1 = (T_1 - T_2) \times \mathrm{CP}_C$. However, in interval 2, streams A and C contribute to it, and, thus, the enthalpy change is the summation of two individual enthalpy changes, that is, $\Delta H_2 = (T_2 - T_3) \times (\mathrm{CP}_A + \mathrm{CP}_C)$. In interval 3, only stream A exists, $\Delta H_3 = (T_3 - T_4) \times \mathrm{CP}_A$. For interval 4, streams A and B coexist and then $\Delta H_4 = (T_4 - T_5) \times (\mathrm{CP}_A + \mathrm{CP}_B)$. Finally, in interval 5, only stream A exists, so $\Delta H_5 = (T_5 - T_6) \times \mathrm{CP}_A$.

By plotting the five temperature changes and corresponding enthalpy changes in sequence on the T/H axis, a hot composite stream is obtained and the resulting T/H diagram (Figure 8.3b) is the hot composite curve, which satisfies the aforementioned two conditions. Clearly, the composite hot curve represents the three hot streams in terms of temperature and enthalpy changes.

Similarly, the T/H representation for the two cold streams can be shown in Figure 8.4a and the composite curve for the two cold streams can be constructed in Figure 8.4b. By plotting both hot and cold composite curves on the T/H axis, we obtain the T/H diagram, which is the so-called composite curve (Figure 8.5).

8.2.3 Basic Pinch Concepts

Composite curves can reveal very important insights for a heat recovery problem regarding process heat recovery, pinch point, hot and cold utility targets, which can be visualized in Figure 8.6. The basic concepts are explained as follows.

(1) *Minimum temperature approach* $\Delta T_{\min}$: For a feasible heat transfer between the hot and cold composite streams, a minimum temperature approach must be

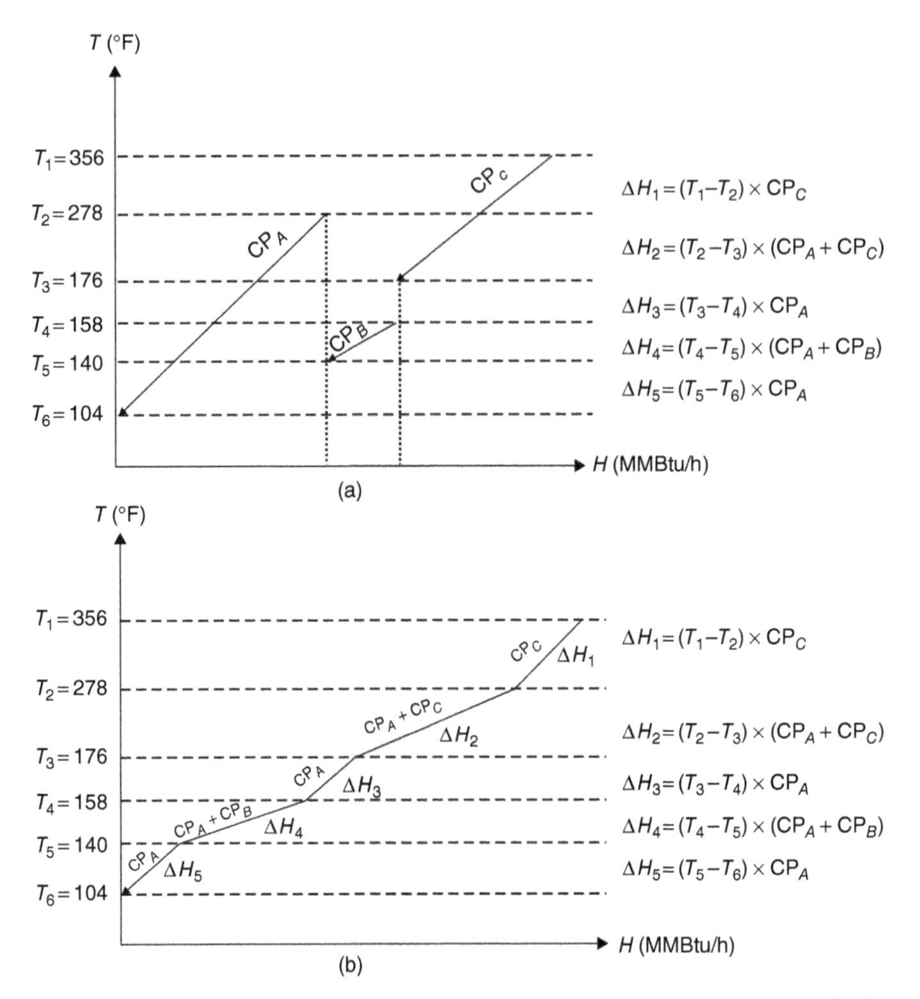

FIGURE 8.3. (a) *T/H* representation of three hot streams; (b) *T/H* representation of a hot composite stream.

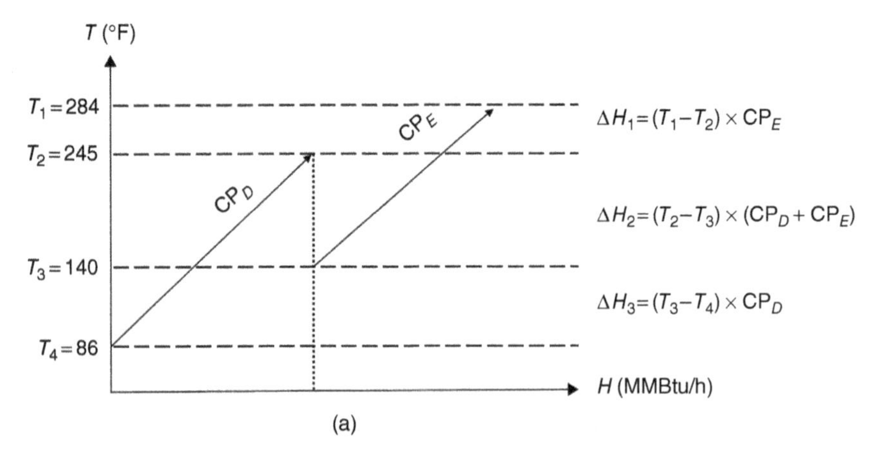

FIGURE 8.4. (a) *T/H* representation of two cold streams; (b) *T/H* representation of a cold composite stream.

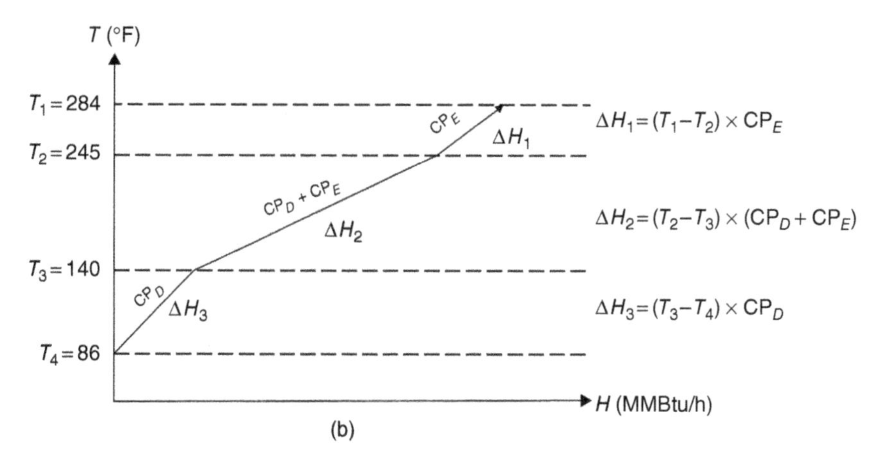

FIGURE 8.4. *(Continued)*

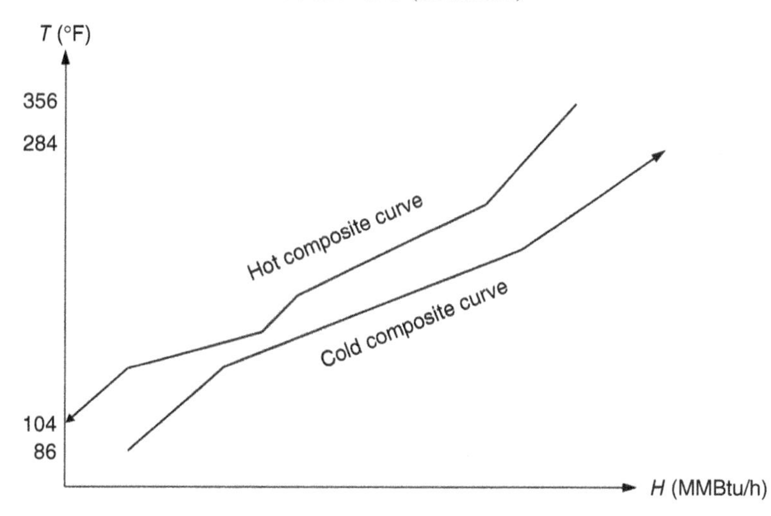

FIGURE 8.5. Composite curves representing the three hot and two cold streams.

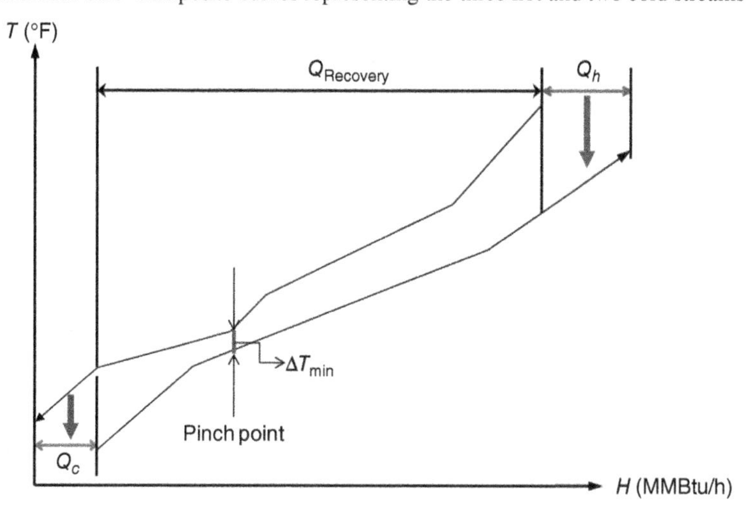

FIGURE 8.6. Basic concepts of composite curves.

specified, which corresponds to the closest temperature difference between the two composite curves on the *T*/*H* axis. This minimum temperature approach is termed as the network temperature approach and defined as ΔT_{min}.

(2) *Process heat recovery*: The overlap between the hot and cold composite curves represents the maximal amount of heat recovery for a given ΔT_{min}. In other words, the heat available from the hot streams in the hot composite curve must be heat-exchanged with the cold streams in the cold composite curve in the overlap region for achieving maximum heat recovery.

(3) *Hot and cold utility requirement*: The overshoot at the top of the cold composite represents the minimum amount of external heating (Q_h), while the overshoot at the bottom of the hot composite represents the minimum amount of external cooling (Q_c).

(4) *Pinch point*: The location of ΔT_{min} is called the process pinch. In other words, the pinch point occurs at the minimum temperature difference indicated by ΔT_{min}. When the hot and cold composite curves move closer to reach ΔT_{min}, the heat recovery reaches the maximum and the hot and cold utilities come to the minimum. Thus, the pinch point becomes the bottleneck for further reduction of hot and cold utilities. Process changes must be made to change the shape of the composite curves if further utility reduction is pursued.

8.2.4 Energy Targeting

By assuming a practical ΔT_{min}, the composite curves can indicate targets for both hot and cold utility duties. This task is called energy targeting.

The procedure of obtaining the composite curves and energy targets can be summarized. First, the base case is determined from which stream data are collected based on the heating and cooling requirements of the process streams in terms of temperatures and enthalpies. Next, the hot streams are plotted on temperature–enthalpy axes and then individual stream profiles are combined to give a hot composite curve (Figure 8.3). This step is repeated for the cold streams to generate the cold composite curve (Figures 8.4). Finally, the two composite curves are plotted together to obtain the composite curves (Figure 8.7a) for a given ΔT_{min} and thus the minimum hot and cold utility targets for the process can be determined.

The general trend is that a large ΔT_{min} corresponds to higher energy utility but lower heat transfer area and thus lower capital cost, and vice versa (Figure 8.7b). The calculations for capital cost for a heat recovery system can be seen latter.

8.2.5 Pinch Design Rules

Once the pinch is identified, the overall heat recovery system can be divided into two separate systems: one above and one below the pinch, as shown in Figure 8.8a. The system above the pinch requires a heat input and is therefore a net heat sink. Below the pinch, the system rejects heat and so is a net heat source. When a heat recovery system design does not have cross-pinch heat transfer, that is, from above to below the pinch, the design achieves the minimum hot and cold utility requirement under a given ΔT_{min}.

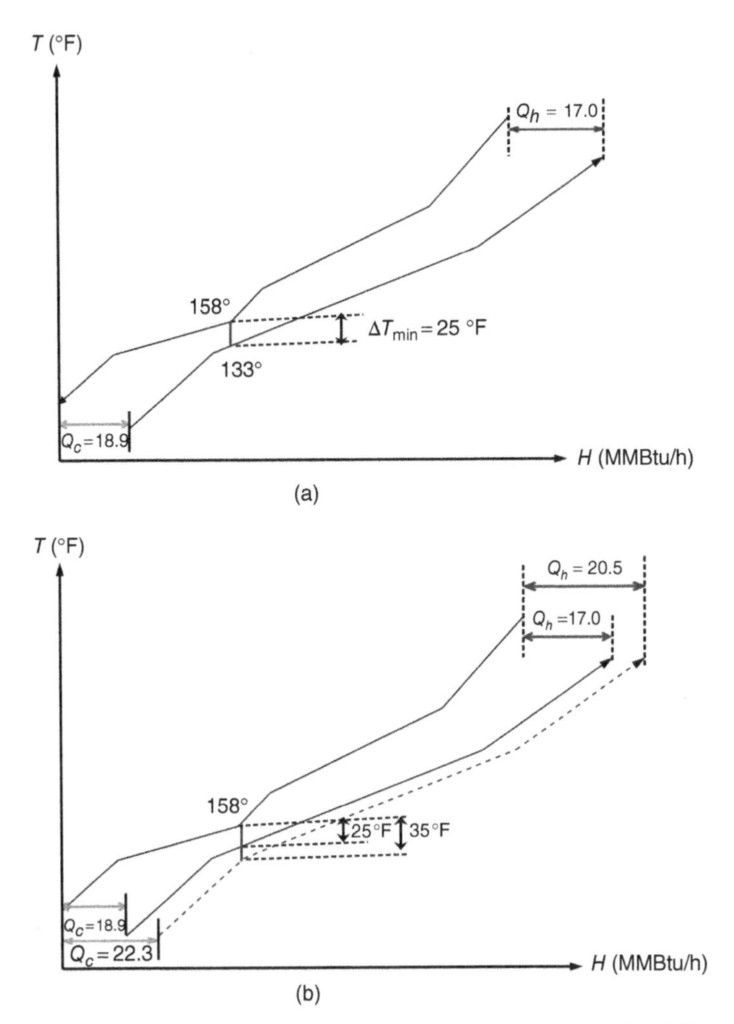

FIGURE 8.7. (a) Energy targets for a specified ΔT_{min}; (b) energy targets for different ΔT_{min}.

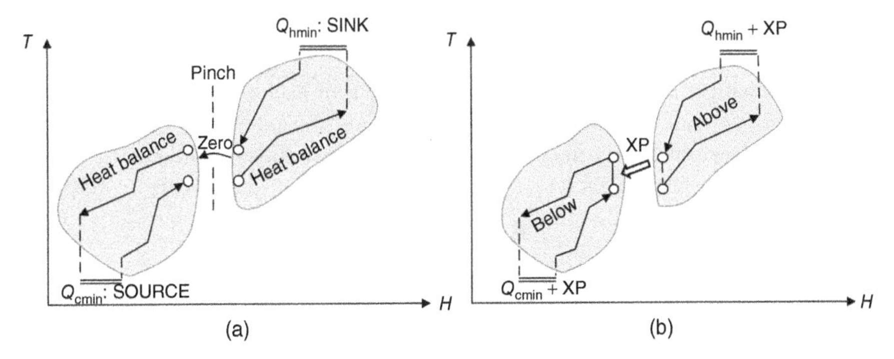

FIGURE 8.8. Pinch principle: penalty of cross-pinch heat transfer.

On the other hand, if cross-pinch heat transfer is allowed (Figure 8.8b), XP amount of heat is transferred from above to below the pinch. The system above the pinch, which was before, in heat balance with Q_{hmin}, now loses XP units of heat to the system below the pinch. To restore the heat balance, the hot utility must be increased by the same amount, that is, XP units. Below the pinch, XP units of heat are added to the system that had an excess of heat; therefore, the cold utility requirement also increases by XP units. The consequence of a cross-pinch heat transfer (XP) is that both the hot and cold utilities will increase by the cross-pinch duty (XP).

Based on the same principle, if external cooling is used for hot streams above the pinch, it increases the hot utility demand for the cold streams by the same amount. Similarly, external heating below the pinch increases the cold utility requirement by the same amount.

To summarize, there are three basic pinch golden rules that must be followed in order to achieve the minimum energy targets for a process:

(1) Heat must not be transferred across the pinch.
(2) There must be no external cooling above the pinch.
(3) There must be no external heating below the pinch.

Breaking any of these rules will lead to cross-pinch heat transfer resulting in an increase in the energy requirement beyond the target.

8.2.6 Cost Targeting: Determine Optimal ΔT_{min}

The optimal ΔT_{min} is determined based on the trade-off between energy and capital such that the total cost for the heat recovery system design is at minimum. The total annual cost for a heat recovery system consists of two parts, namely the capital cost and the energy operating cost:

- The energy operating cost includes energy expenses for both hot and cold utilities, which is billed regularly in $/year.
- The capital cost of the network includes surface area costs for all individual heat exchangers, water coolers, air coolers and refrigeration, fired heaters and steam heaters, as well as related costs including foundation, piping, instrumentation, and control. Thus, it is a total investment ($) required to build the entire heat transfer system.

Heat exchanger capital cost is estimated based on exchanger surface area. The overall exchanger area can be directly calculated from the composite curves using the area model (Townsend and Linnhoff, 1983). To do this, utilities are added to composite curves to make heat balance between hot and cold composites. Then, the balanced composite curves are divided into several enthalpy intervals and each enthalpy interval must feature straight temperature profiles (Figure 8.9).

There could be several hot and cold streams within an enthalpy interval. For each heat exchange involving hot stream i and cold stream j in the kth interval, the surface area and cost can be calculated by equations (8.1) and (8.2), respectively,

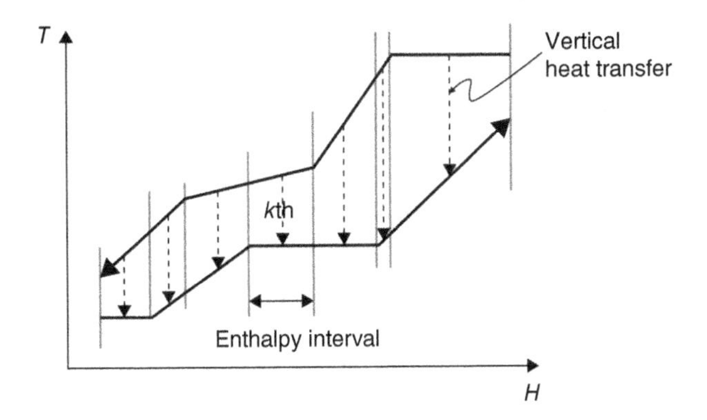

FIGURE 8.9. Calculation of surface area from the composite curves.

as follows:

$$A_{i,j,k} = \frac{Q_{i,j,k}}{U_{i,j,k}\mathrm{LMTD}_{i,j,k}}, \tag{8.1}$$

$$C_{i,j,k} = a_{i,j,k} + b_{i,j,k}(A_{i,j,k})^{c_{i,j,k}} = a_{i,j,k} + b_{i,j,k}\left(\frac{Q_{i,j,k}}{U_{i,j,k}\mathrm{LMTD}_{i,j,k}}\right)^{c_{i,j,k}}, \tag{8.2}$$

where

A = surface area, ft^2
Q = heat load, MMBtu/h
LMTD = logarithmic mean temperature difference, °F
U = overall heat transfer coefficient, MMBtu/h/(ft^2 °F)
C = exchanger cost, \$
a = fixed cost for exchanger, \$
b = surface area cost, \$/ft^2
c = economic scale factor, fraction

Thus, total surface area and purchase cost for all exchangers in the kth interval can be calculated through equations (8.3) and (8.4)

$$A_k = \sum_{i,j} A_{i,j,k} = \sum_{i,j} \frac{Q_{i,j,k}}{U_{i,j,k}\mathrm{LMTD}_{i,j,k}}, \tag{8.3}$$

$$C_k = \sum_{i,j}\left[a_{i,j,k} + b_{i,j,k}\left(\frac{Q_{i,j,k}}{U_{i,j,k}\mathrm{LMTD}_{i,j,k}}\right)^{c_{i,j,k}}\right], \tag{8.4}$$

where

A_k = total surface area for kth interval, ft^2
C_k = total exchanger cost for kth interval, \$

And the overall surface area and capital cost for the network can be calculated by equations (8.5) and (8.6). It is important to point out that the exchanger equipment costs must be converted into installed costs, which include exchanger purchase cost, foundation, piping, instrumentation, control, and erection.

$$A_{\text{Network}} = \sum_{k}^{\text{intervals}} A_k = \sum_{i,j,k} \frac{Q_{i,j,k}}{U_{i,j,k}\text{LMTD}_{i,j,k}}, \tag{8.5}$$

$$\begin{aligned} C_{\text{Network}}^{\text{Cap}} &= \sum_k C_k + \sum_p C_p + \sum_q C_q \\ &= \sum_k \sum_{i,j} I_{i,j,k} \left[a_{i,j,k} + b_{i,j,k} \left(\frac{Q_{i,j,k}}{U_{i,j,k}\text{LMTD}_{i,j,k}} \right)^{c_{i,j,k}} \right] \\ &\quad + \sum_p (a_p + b_p Q_p{}^{c_p}) + \sum_q (a_q + b_q Q_q{}^{c_q}) \end{aligned} \tag{8.6}$$

where

A_{Network} = total surface area, ft^2
C_p = capital cost for heater p, \$
$C_{\text{Network}}^{\text{Cap}}$ = total installed cost, \$
$I_{i,j,k}$ = installation factor for exchanger between streams i and j in the kth interval
Q_p = heat load for heater p, MMBtu/h
a_p = fixed cost for heater, \$
b_p = heater cost based on duty, \$/MMBtu
c_p = economic scale factor, fraction
Q_q = heat load for cooler q, MMBtu/h
a_q = fixed cost for cooler q, \$
b_q = heater cost based on duty, \$/MMBtu
c_q = economic scale factor, fraction

On the other hand, utility consumption and costs can be calculated for both hot and cold utilities, respectively, as

$$Q_{\text{h,Network}} = \sum_m Q_{\text{h},m}, \tag{8.7}$$

$$Q_{\text{c,Network}} = \sum_n Q_{\text{c},n}, \tag{8.8}$$

$$C_{\text{Network}}^{Q} = \sum_m c_{\text{h},m} Q_{\text{h},m} + \sum_n c_{\text{c},n} Q_{\text{c},n}, \tag{8.9}$$

where

$Q_{\text{h},m}$ = heat load for hot utility m, MMBtu/h
$Q_{\text{c},n}$ = heat load for cold utility n, MMBtu/h
$c_{\text{h},m}$ = cost for hot utility m, \$/MMBtu
$c_{\text{c},n}$ = cost for cold utility n, \$/MMBtu
C_{Network}^{Q} = total utility cost, \$/h

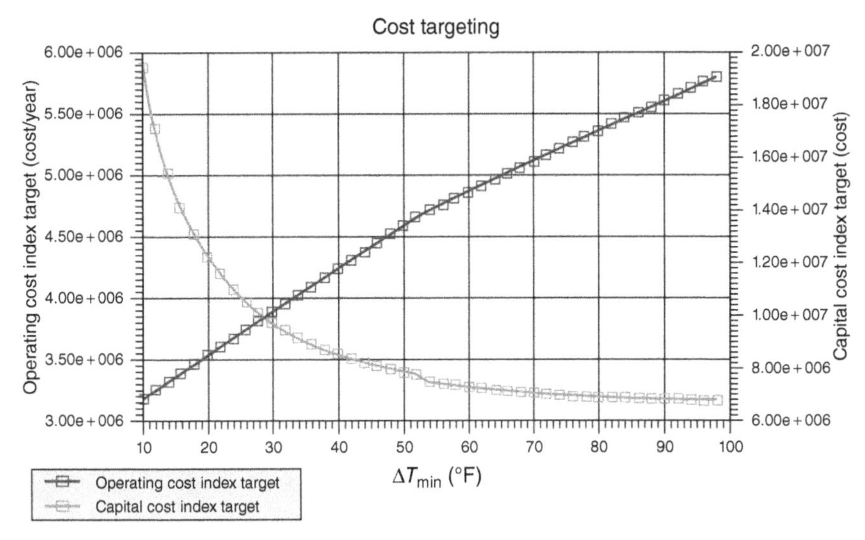

FIGURE 8.10. Capital and energy trade-off.

The capital cost for each $\Delta T_{\min}$ can be calculated based on equation (8.6) while the energy cost by equation (8.9). By calculating both energy and capital costs for different $\Delta T_{\min}$, two cost curves, namely the energy and capital costs curves for a range of $\Delta T_{\min}$, can be obtained as shown in Figure 8.10.

The total annualized cost for the entire heat recovery system can then be defined as

$$C_{\text{Network}}^{\text{Total}} = F \times C_{\text{Network}}^{\text{Cap}} + KC_{\text{Network}}^{Q}, \tag{8.10}$$

where

F = capital annualized factor, 1/year
K = time annualized factor = 24 h/day* operating days/year
$C_{\text{Network}}^{\text{Total}}$ = total annualized cost, \$/year

For different $\Delta T_{\min}$, it will be expected to have different capital cost and utility cost. When $\Delta T_{\min}$ increases, capital costs drop as exchanger LMTD increases and thus surface area reduces. At the same time, the utility consumption goes up. The impact of reduced $\Delta T_{\min}$ is opposite: capital costs go up as LMTD reduces while utility consumption goes down. Thus, there is a trade-off between utility cost and capital cost as shown in Figure 8.10. This trade-off can be better visualized by plotting the total annualized cost by equation 8.10 versus $\Delta T_{\min}$ on a graph in Figure 8.11. The optimal network approach, denoted as $\Delta T_{\min,\text{opt}}$, is the one corresponding to the lowest total annualized cost. In many cases, there is a range of $\Delta T_{\min}$ values in which total costs are similar and thus selection of $\Delta T_{\min,\text{opt}}$ in this range should be made toward design simplicity.

The significance of $\Delta T_{\min,\text{opt}}$ is in setting the design basis for where the heat recovery design should start and what to expect for the utility and capital costs to result from design. If the design comes up with much high costs than the targeted costs, a design evaluation must be conducted to find out why and figure out design changes which can bring the total cost down correction.

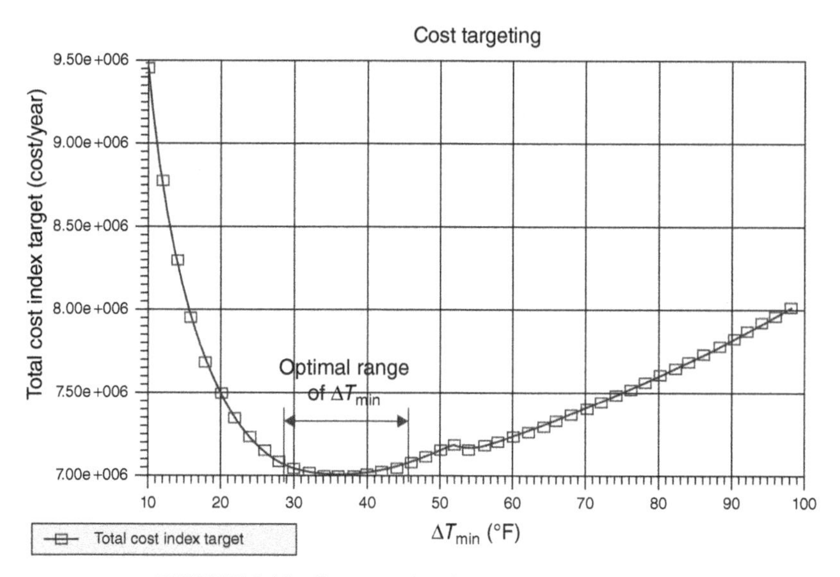

FIGURE 8.11. Cost targeting for determining $\Delta T_{\text{min,opt}}$.

8.3 GRASSROOTS HEAT EXCHANGER NETWORK (HEN) DESIGN

After the optimal ΔT_{min} is determined, this $\Delta T_{\text{min,opt}}$ becomes the design basis for process heat recovery system. At this stage, process configuration (reaction and separation sequences) and conditions (feeds intermediate products and products as well as temperature and pressure for reaction and separation) are already determined. Thus, the process streams such as reaction effluent, feeds, products, as well as the process streams representing column overhead condensers and reboilers are available. The task of process heat recovery system design is to determine the heat exchange matching schemes or heat exchanger network (HEN) design. It is a difficult task to obtain the "optimal" HEN design with the lowest total cost as it could involve many design choices. The best design must contain two features. The first one is to achieve the maximal heat recovery corresponding to $\Delta T_{\text{min,opt}}$. The second feature is to obtain design simplicity for minimal equipment count. To accomplish these design features, the block decomposition method was developed by Zhu et al. (1995) and is explained as follows.

8.3.1 The Concept of Block Decomposition

As discussed previously, the composite curves provide guidelines for achieving minimum surface area design based on enthalpy intervals, which is defined by the "kink" points (Figure 8.12a). Equation 8.5 can be rewritten as

$$A_{\text{Network}} = \sum_{k}^{\text{intervals}} \left[\sum_{i,j}^{\text{streams}} \frac{Q_{i,j,k}}{U_{i,j,k}\text{LMTD}_{i,j,k}} \right], \tag{8.11}$$

This minimum surface area is achieved based on vertical heat transfer. If a HEN is synthesized using the enthalpy intervals defined by the "kinks" and the temperature

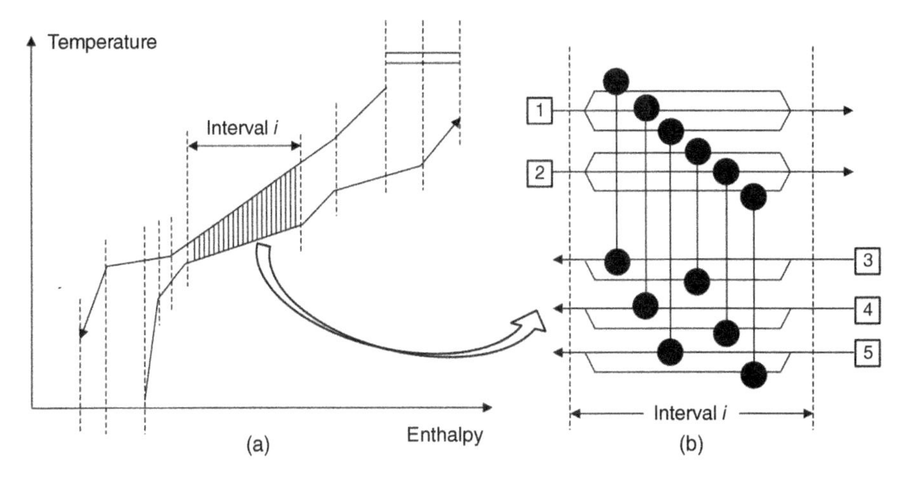

FIGURE 8.12. General stream splitting and matching based on the composite curves.

profiles of the matches exactly follow the composite curves, then the resulting HEN will achieve both the area and utility targets. However, such networks possess an excessive number of heat exchangers, which is known as a "spaghetti structure" (Figure 8.12b; Ahmad, 1985).

To approach both exchanger area and utility targets with near minimum units, a block that contains several small intervals featuring similar temperature–enthalpy profiles is used. Using the block decomposition, the composite curves are partitioned into a smaller number of such blocks with larger enthalpy intervals than that defined by "kink" point-based enthalpy intervals. Typically, each block spans a number of "kink point" enthalpy intervals and the hot (cold) composite segments within a "block" are represented by straight lines as shown in Figure 8.13. These straight

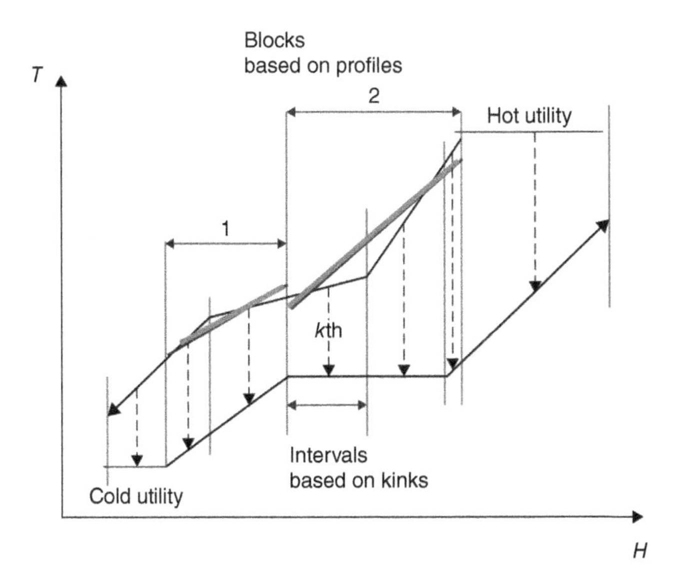

FIGURE 8.13. Block decomposition.

TABLE 8.1. Stream Data for the Illustrative Example

Stream	Supply Temperature (°C)	Target Temperature (°C)	Heat Capacity Flowrate (kW/K)	Film Coefficient (W/(m^2 K))
1	159	77	2.285	100
2	267	80	0.204	40
3	343	90	0.538	500
4	26	127	0.933	10
5	118	265	1.961	500
Steam	300	300	—	50
Water	20	60	—	200

Source: Data from Zhu (1994).

Cost data:

Installed heat exchanger cost (\$) = $3800 + 750A^{0.83}$ where A = exchanger area (m^2);

Cost of hot utility = \$110/(kW year); Cost of cold utility = \$10/(kW year);

Rate of interest = 10% per annum; equipment lifetime = 6 years.

lines approximating the composite curves in each block are called quasi-composite curves. Next, for each block, a "within-block" subnetwork is then synthesized. Obviously, the number of enthalpy intervals used to synthesize a HEN is greatly reduced compared to the "spaghetti structure" as shown in figure 8.12b.

Block decomposition can be determined automatically by software based on profiles of the composite curves. Enthalpy intervals with similar profiles may be combined into a single block. For given a set of blocks, the synthesis of a HEN proceeds block by block. Network synthesis commences at the most constrained block (either at a pinch point or the block containing the pinch point). The design based on block decomposition becomes the initial design for further nonlinear optimization in order to minimize the total cost by simplifying the network design further.

8.3.2 Illustrative Example

The stream data to this problem is shown in Table 8.1 based on the study of Rev and Fonyo (1991). This example demonstrates how differing block structure influences the initial design and determines the final design. Using this data, an optimal $\Delta T_{\min}$ equal to 30 K may be deduced. This $\Delta T_{\min}$ and its associated targets determine the design conditions, which are summarized in Table 8.2. The composite curves are provided as in Figure 8.14. Three distinct regions may be identified, and these are used to define a block. The initial design based on these parameters is shown in Figure 8.15. In summary, the energy consumption is less than the target by 4.7%, the total area required exceeds the area target by 7.0%, and there are nine exchangers with five process–process exchangers and four utility exchangers.

The network is deduced by applying the Import/Export rule. Consider block 2; this block includes sections of the three hot streams (1–3) and the two cold streams (4–5). The enthalpy of hot stream 1 (22.62 kW) dominates the remaining hot streams' contribution (stream 2 – 2.02 kW and stream 3 – 5.33 kW). Likewise, the content of cold stream 5 (21.57 kW) is much greater than that of stream 4 (8.4 kW). To remove all

TABLE 8.2. Design Performance for the Illustrative Example

$\Delta T_{min,opt} = 30\,K$	Hot Utility (kW)	Cold Utility (kW)	Area (m^2)	Exchanger (unit)	Total Cost ($/annum)
Target	145.7	124.8	299	6	48,975
Initial design (Figure 8.15)	139.4	118.5	320	9	51,188
Final design (Figure 8.16)	144.3	123.4	289	7	46,786

Source: Data from Zhu (1994).

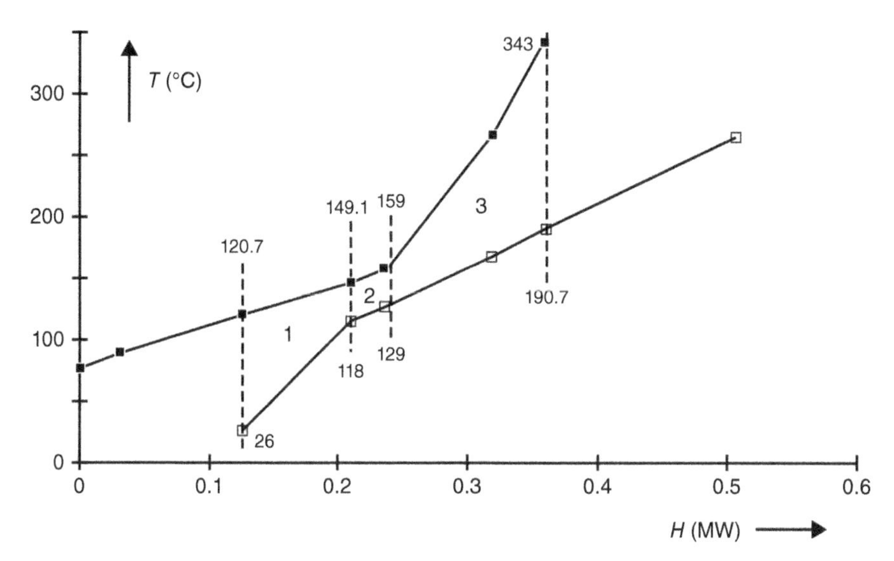

FIGURE 8.14. Composite curves for the illustrative example. Source: From Zhu (1994).

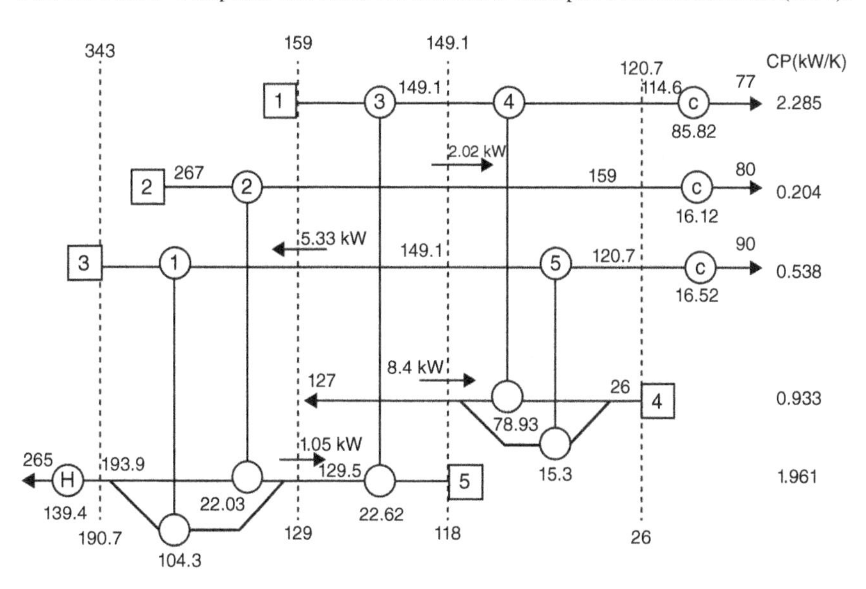

FIGURE 8.15. Initial design for the example: $Q_H = 139.4\,kW$, area $= 320\,m^2$, total cost $=$ $51,188/year. Source: From Zhu (1994).

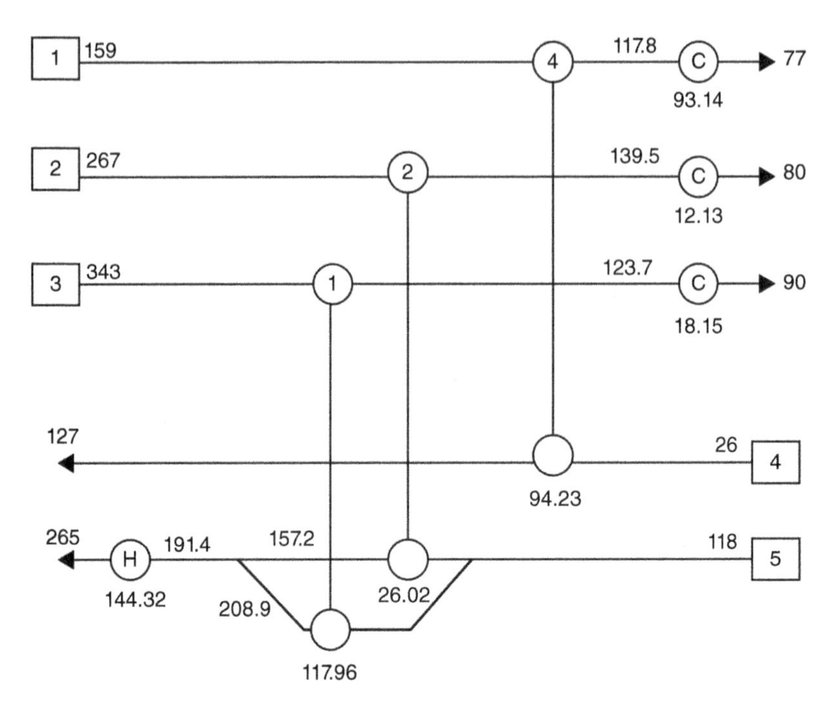

FIGURE 8.16. Optimized design for the example: $Q_H = 144.3$ kW, area = 289 m^2, total cost = \$46,786/year. Source: From Zhu (1994).

the energy from stream 1 in this block necessitates an import of 1.05 kW for stream 5. To deal with the residual energy within the block requires the addition of two small exchangers [i.e., transfer of 7.35 kW from hot streams 2 and 3 to heat cold stream 4 (8.4 kW available)]. To avoid introducing these small units, energy may be exported to neighboring blocks. The enthalpy associated with stream 2 is exported to block 1 because this block has closer temperature approaches than block 3. However, as stream 3 has relatively large enthalpy, it may be exported to block 3; these transfers are illustrated on Figure 8.15. The next step is to apply a nonlinear optimizer to the network to seek an optimum design. Matches 3 and 5 in the initial network are removed producing a simple but low-cost network (Figure 8.16). This network has seven units (one more than the unit target) and its total cost is 4.5% below the cost target. The design performance for both designs in Figures 8.15 and 8.16 are given in Table 8.2.

8.4 NETWORK PINCH FOR ENERGY RETROFIT

We have discussed grassroots design of HEN. For an existing process in operation, is it possible to increase the heat recovery of a system by adding exchanger surface area to some of the existing exchangers? The answer is that for a fixed network structure, there is a limit for heat recovery increase regardless of the surface area to be added. The exchanger imposing the limit is called the pinching exchanger, and the temperature point at which this limit occurs is termed as the network pinch, which is the bottleneck of an existing network (Zhu and Asante, 1999). Briefly, heat recovery for a process can be increased to a certain limit when surface area is added to existing exchangers in a HEN without making changes to the HEN structure. This limit is called the network pinch. Beyond this limit, any additional surface area will not increase the heat recovery.

8.4.1 Definition of the Network Pinch

The network pinch identifies the heat recovery limit inherent in an existing HEN and therefore indicates the requirement of structural changes to the HEN. In other words, when a network reaches its network pinch limit, the only way to overcome it for increased heat recovery is to make structural changes to the HEN; another option is process modifications.

However, not all network structural changes are able to overcome the network pinch; but only those that can move heat from below to above the network pinch can. This principle provides a valuable guideline for screening out nonbeneficial modification options, which is discussed as follows.

It must be noted that the process pinch is different from the network pinch. This is because the process pinch is defined by the process conditions such as temperatures and heat capacities, and the process pinch is mainly used for energy targeting in grassroots design. However, for an existing heat recovery network, heat exchangers are already in place and the network configuration will impose a limit for increased heat recovery. These limits (pinching matches) can only be identified by the network pinch method (see Figure 8.17 for the network pinch vs the process pinch illustrated in Figure 8.7, for the same heat recovery scheme in Figure 8.2).

A practical exchanger minimum approach temperature (EMAT) should be used for identification of pinching matches. The definition of pinching matches excludes "pinched" exchangers whose limiting driving forces can be relaxed by shifting heat

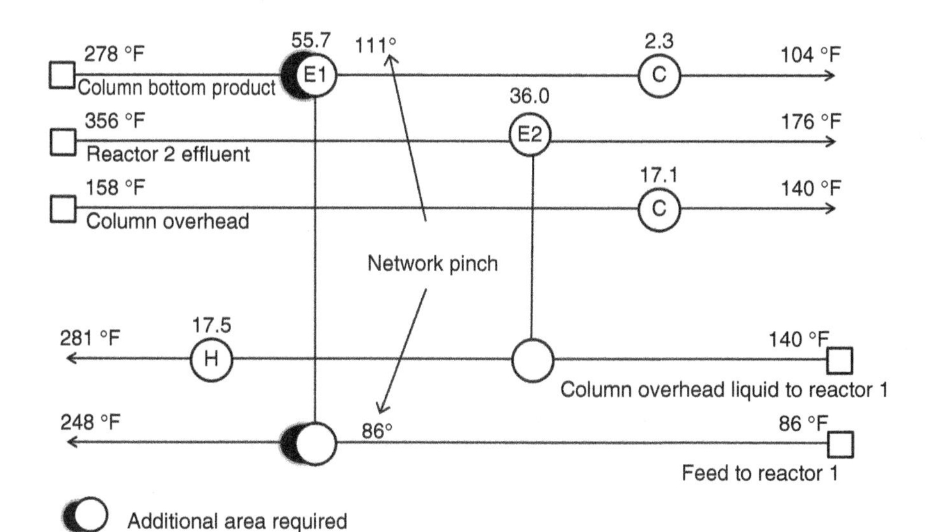

FIGURE 8.17. An illustration of a network pinch.

around a loop. Therefore, it is important to understand that pinch matches for a given network structure are unique and they can be determined mathematically by minimizing the number of pinch matches.

8.4.2 Identifying Possible Modifications

Only changes to the network structure can overcome the network pinch bottleneck. Four types of modifications can be considered for HEN. These are as follows:

- *Resequencing*: The order of two exchangers can be reversed, and this sometimes allows better heat recovery. Usually, this option involves relative minor piping changes and hence the least cost required.
- *Repiping*: This is similar to resequencing, but one or both of the matched streams can be different in pairing to the current matches.
- *Adding a new match*: This can be used to change the load on one of the streams in the pinching match. This option could be relatively expensive as it requires new foundation, piping, and control system.
- *Splitting*: Split a stream, again reducing the load on a stream involved in the pinching match. In practice, the stream split will be very asymmetric in both flow and temperature and a special piping configuration would be needed.

In most cases, many structural changes can help overcome the network pinch and usually, they produce different energy saving and capital costs. To select the best modification(s), among many alternatives, it is necessary to define the criterion for optimization – ideally, cost. However, it would be a very daunting task to estimate costs of piping, labor, foundation, and installation for all potential modifications prior to design.

As an alternative to a cost-based optimization, maximum heat recovery is used in the identification stage. The modification options identified is then further evaluated for capital costs, and impact on operation and safety, to make sure the modifications selected meets costs as well as operation and safety criteria.

8.4.3 An Illustrative Example

The retrofit objective in this example was to debottleneck the HEN to accommodate with a 10% increase in crude throughput for the crude distillation unit as shown in Figure 8.18. The main retrofit constraint is the maximum furnace duty of 100 MW. Figure 8.19 shows the grid diagram for the existing network design. The minimum temperature considered for retrofit was at 10 °C.

In generating the base case in Figure 8.19, the crude feed rate (stream C1) is increased by 10% and the product streams H1–H7 are increased accordingly. Then, the heat recovery limits for the increased feed rate case with the original network topology are established, which are shown in Figure 8.20. The furnace duty at the maximum heat recovery for the given topology is a 102.5 MW and exchangers 5 and 6 are pinched. As the furnace duty at the maximum heat recovery (R_{max}) for the given topology is at 102.5 MW, which is above the maximum allowable design duty

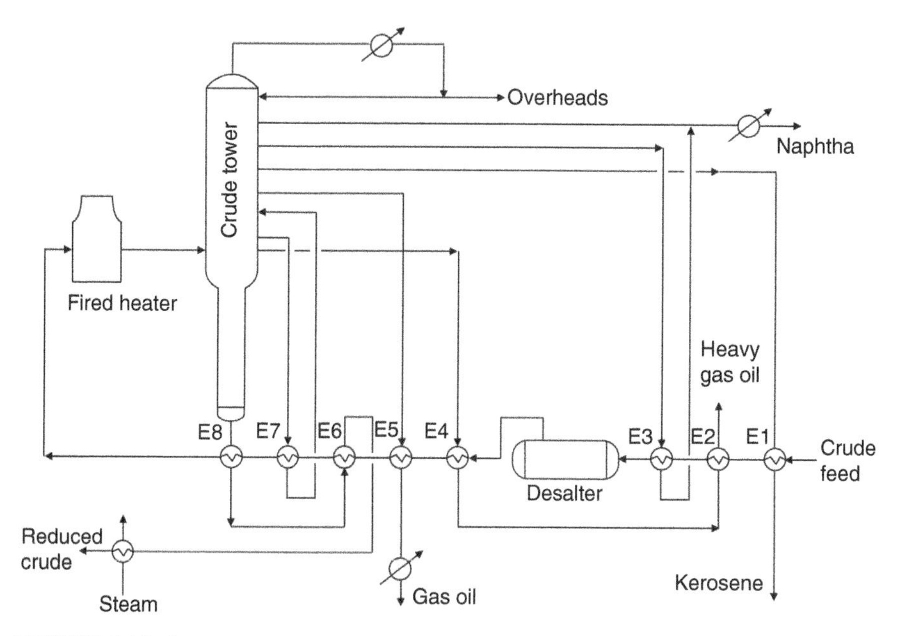

FIGURE 8.18. Process flow diagram for the crude distillation unit. Source: Zhu (1999). Reproduced with permission of AIChE.

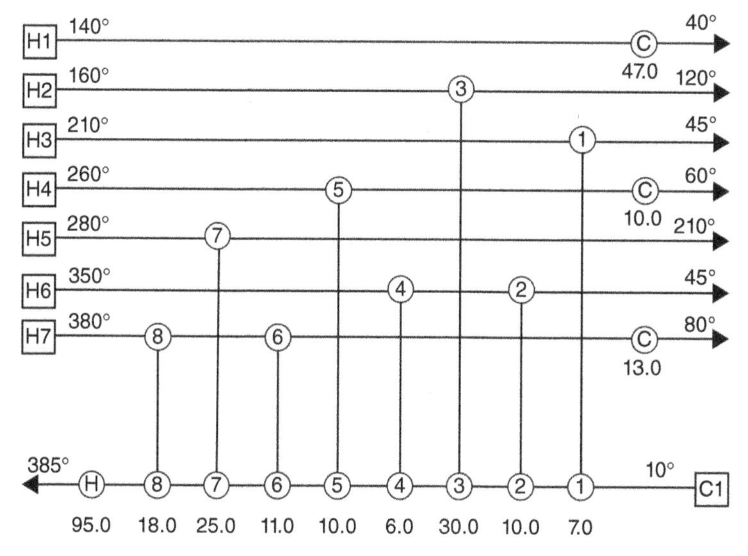

FIGURE 8.19. Base case heat exchanger network. Source: Zhu (1999). Reproduced with permission of AIChE.

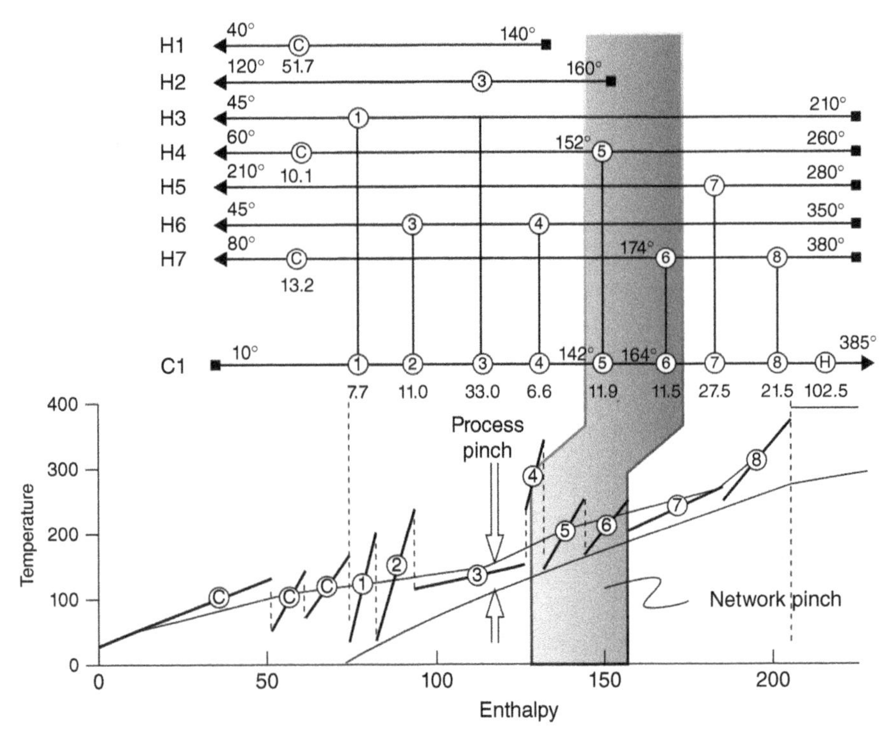

FIGURE 8.20. Heat recovery limits in the base case network. Source: Zhu (1999). Reproduced with permission of AIChE.

of 100 MW, changes to the topology are needed in order to reduce the furnace duty below the design limit.

The first modification option considered is exchanger resequence, and the option selected by the method is the resequence of exchanger 4 as illustrated in Figure 8.21. This modification produces a 4.4 MW increase in heat recovery and reduces the minimum furnace duty to 98.1 MW. Although the minimum furnace duty after this resequence modification is below the design limit of 100 MW, it is quite close to it. It can be expected that a retrofit design close to the topology R_{max} would require excessive exchanger surface area. Thus, another topology change is sought to further increase heat recovery, and consequently, reduce the exchanger areas required for the retrofit.

The topology produced after the resequence of exchanger 4 features three adjacent pinch matches (exchangers 4/5/6) as shown in Figure 8.22. As the first of these pinching matches occurs at the process pinch, the process and network pinches become coincidental, and this fulfills the conditions of stream split heuristic. Thus, the stream split is implemented. Exchangers 4/5/6 being the three adjacent pinching matches are placed in parallel with each other to produce a 1.8 MW increase in heat recovery. Although the three exchangers are initially placed in parallel with each other (Figure 8.23), the optimal slit configuration will be determined during the optimization stage.

The search for modifications could be stopped at this point, and the resulting topology is submitted to the optimization stage. If, however, the search is continued

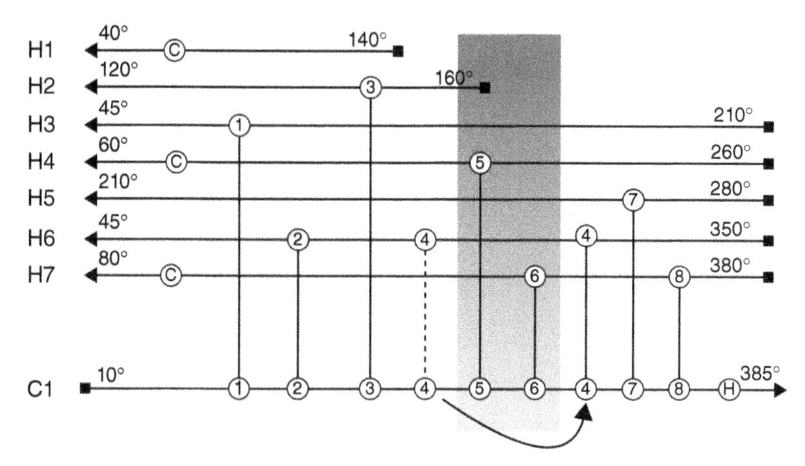

FIGURE 8.21. Resequence of exchanger 4: min $Q_H = 98.08$; $\Delta Q_{Rec} = 4.4$. Source: Zhu (1999). Reproduced with permission of AIChE.

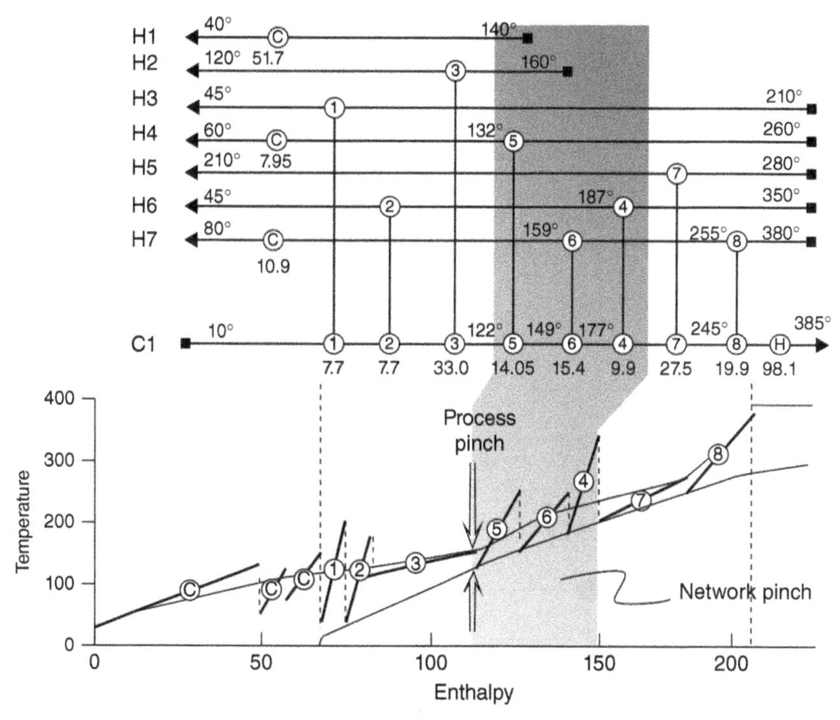

FIGURE 8.22. Heat recovery limits for the network after resequence of exchanger 4. Source: Zhu (1999). Reproduced with permission of AIChE.

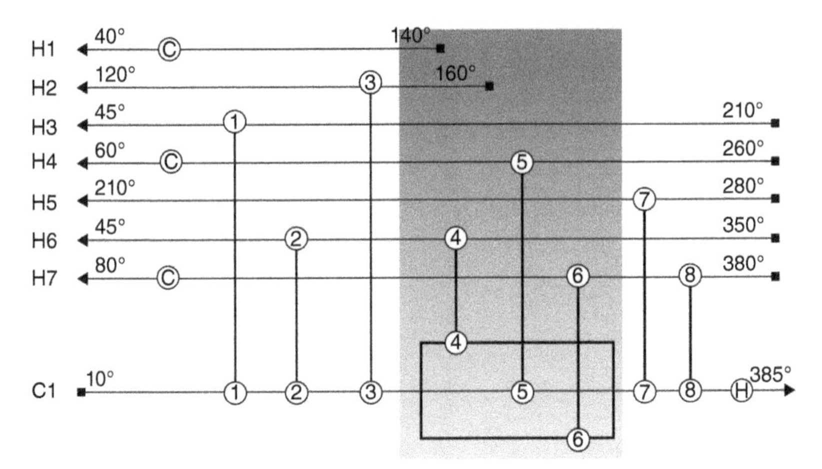

FIGURE 8.23. Stream split: min $Q_H = 96.3$; $\Delta Q_{Rec} = 1.8$. Source: Zhu (1999). Reproduced with permission of AIChE.

for another modification and it will be the addition of a new exchanger, which further increases the heat recovery by 3.9 MW to give a minimum furnace duty of 92.4 MW. It is not possible at this stage to determine whether the addition of the new exchanger can be economically justified. The optimization stage will provide the necessary clarification.

After the optimization at a fixed furnace duty of 99.6 MW the same as the existing design, the retrofit design without the new exchanger requires 1974 m^2 of additional surface area in total, while the design with the new exchanger requires only 1265 m^2 (Figure 8.24). With this information, the retrofit designs can be effectively compared

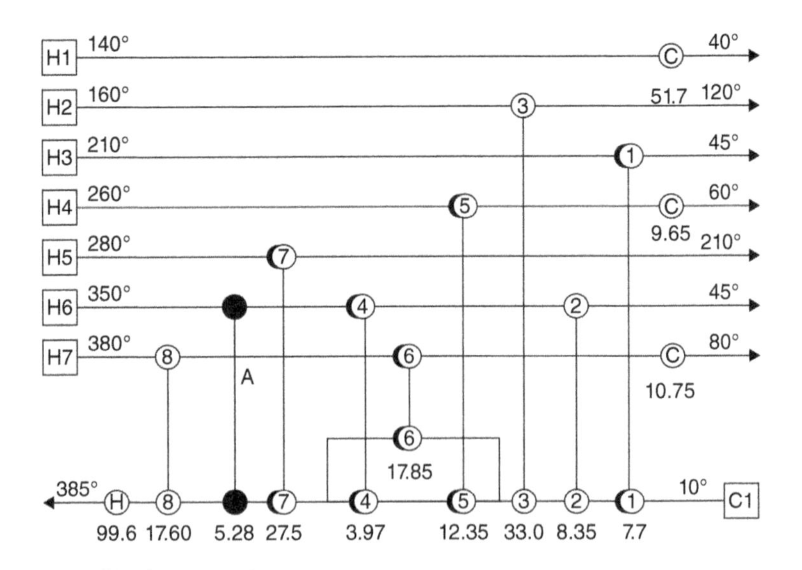

FIGURE 8.24. Retrofit design developed by the network pinch method. Source: Zhu (1999). Reproduced with permission of AIChE.

to identify the best retrofit design. The installation cost of the new exchanger can be fairly accurately estimated at this stage to assist in the decision process, which is much simpler than initiating the design process by making such cost estimation for all new exchanger options possible.

NOMENCLATURE

a	fixed cost for exchanger ($)
A	surface area (ft^2)
b	surface area cost ($/ft^2)
$c_{\mathrm{h},m}$	cost for hot utility m ($/MMBtu)
$c_{\mathrm{c},n}$	cost for cold utility n ($/MMBtu)
C	cost ($)
CP	specific heat capacity (Btu/lb-°F)
F	capital annualized factor (1/year)
H	enthalpy (MMBtu/h)
$I_{i,j,k}$	installation factor for exchanger between streams i and j in kth interval
K	time annualized factor = 24 h/day* operating days/year
LMTD	logarithmic mean temperature difference (°F)
Q	heat load (MMBtu/h)
T	temperature (°F)
U	overall heat transfer coefficient (MMBtu/h/(ft^2 °F))

REFERENCES

Ahmad S (1985) Heat exchanger networks: Cost trade-offs in energy and Capital, PhD thesis, University of Manchester, UK.

Linnhoff B, Townsend DW, Boland D, Hewitt GF, Thomas BEA, Guy AR, Marsland RH (1982) *A User Guide to Process Integration for the Efficient Use of Energy*, IChemE, UK.

Rev E, Fonyo Z (1991) Diverse pinch concept for heat exchanger network synthesis: the case of different heat transfer conditions, *Chemical Engineering Science*, **46**, 1623–1634.

Townsend, DW, Linnhoff, B (1983) Heat and power networks in process design Part II: Design procedure for equipment selection and process matching, *AIChE Journal*, **29**(5), 748–771.

Zhu XX (1994) Strategies for optimization in heat exchanger network design, PhD thesis, The University of Adelaide, Australia.

Zhu XX, Asante NDK (1999) Diagnosis and optimization approach for heat exchanger network retrofit, *AIChE Journal*, **45**(7), 1488–1503.

Zhu XX, O'Neill BK, Roach JR, Wood RM (1995) A method for automated heat-exchanger network synthesis using block decomposition and nonlinear optimization, *Chemical Engineering Research and Design*, **73**(A8), 919–930.

9

PROCESS INTEGRATION FOR LOW-COST DESIGN

9.1 INTRODUCTION

In the nineteenth century, oil refineries processed crude oil primarily to recover kerosene for lanterns, and the refinery design was very simple. The mass use of the automobile, since early twentieth century shifted the demand to gasoline and diesel, which remain the primary refined products today. Fuel production for automobiles required more conversion and hence increased plant complexity. In the 1990s, stringent environmental standards for low sulfur drove refineries to include more hydroprocessing, leading to greater refinery complexity. In recent times, higher energy efficiency and lower greenhouse emission requirements have resulted in designs incorporating more heat recovery with additional equipment and thus have further elevated the process complexity and capital cost. This trend is illustrated in Figure 9.1.

The major challenge is how to achieve first-class energy-efficiency process design with simplified process design and low capital costs. Recent work (Zhu et al., 2011) has pointed to the areas of process and equipment innovations, which play important roles to overcome this challenge. Implementing process and equipment innovations often result in combined benefits in process yields, throughput, energy efficiency, and reduced capital costs at the same time. Many of these areas include optimizing process flowsheet and condition as well as use of advanced equipment. The powerful methodology for connecting these innovations, to achieve high energy efficiency and simpler designs with low capital is the process integration.

Hydroprocessing for Clean Energy: Design, Operation, and Optimization, First Edition.
Frank (Xin X.) Zhu, Richard Hoehn, Vasant Thakkar, and Edwin Yuh.

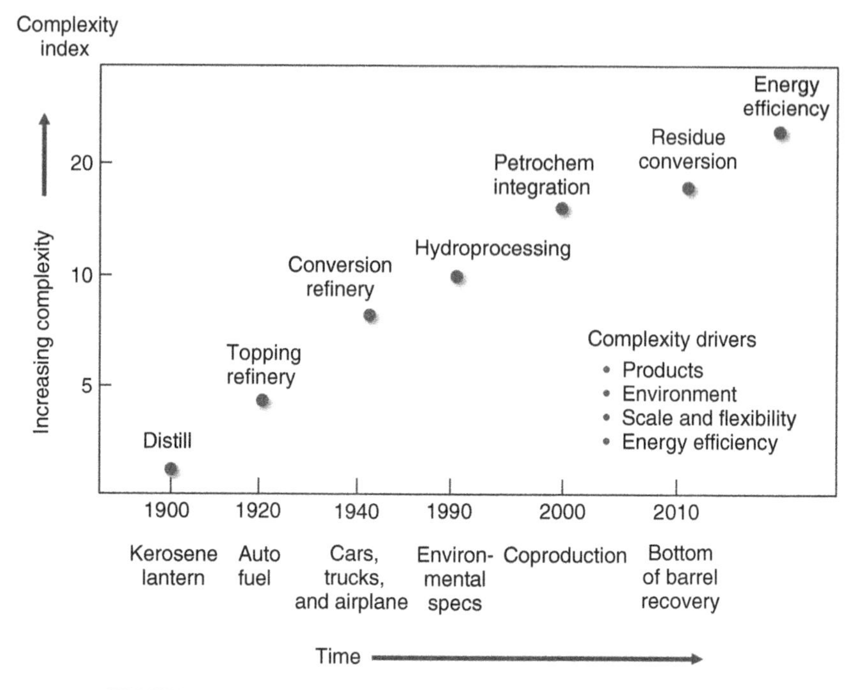

FIGURE 9.1. The trend of increased refinery complexity over time.

9.2 DEFINITION OF PROCESS INTEGRATION

To define the concept of process integration, we need to briefly review the traditional design approach, which can be described by the so-called "onion diagram" as shown in Figure 9.2 that provides an overall view of the traditional design procedure.

The design of a process complex starts from defining a design basis. This step consists of defining physical and chemical conditions for feeds, products, and utilities. The design then concentrates on the chemical reaction system. The reaction system is the core of a process complex where the conversion of feeds to products takes place. The goal of a reaction system design is to achieve a desirable product yield structure via selection of catalyst and design of reactors. As reaction effluents contain a large amount of heat at high temperature, the heat recovery of reaction effluent is a major consideration for process energy efficiency.

After the reaction, the reaction effluent goes through a separation system in order to separate desirable products from by-products and wastes. For separating multiple products, a separation system is required. The majority of energy use came from separation. Thus the objective of design is to select the most appropriate separation method for low energy use with a reasonable capital cost. Heat recovery from products makes significant contributions to process energy efficiency. At the same time, there could be large amount of excess heat available in fractionation columns where multiple products are produced. It is essential that this excess heat is removed from pumparound and used for process heating purpose.

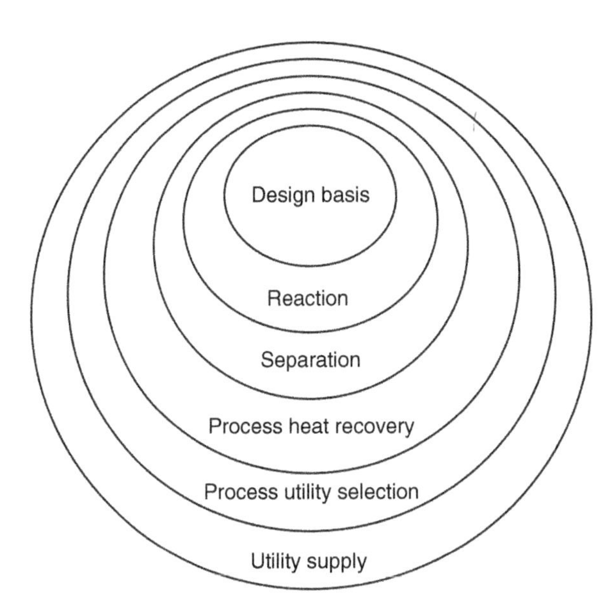

FIGURE 9.2. Sequential process design: traditional design approach.

Process heat recovery design comes after design of reaction and separation systems. The design of reaction and separation defines the basis for process energy demand. In the heat recovery design step, the goal is to minimize overall process energy use (fuel, steam, and power) for a given process energy demand. This is achieved by heat recovery between those process streams with heat available (such as reaction effluent, separation products, and column overhead vapor), and process streams with the need of heat (such as reaction feed, separation feed, and reboiling).

After the process heat recovery is done, the next design step is to determine the utility supply in terms of heating and cooling, and power based on the needs and characteristics of process energy demand. In this step, the means of energy supply for reaction and separation system will be addressed. For example, a choice for the reboiling mechanism must be made for a separation column between a fired heater and steam heater. Similarly, a choice of process driver between steam turbine and motor will be determined. Selection is made based on operation considerations, reliability and safety limits, and capital cost. Selection of process utility supply defines the basis for design of a steam and power system.

The above steps complete the process design in the process battery limit. The last design step is to design utility system, which is mainly the steam and power system. The main design consideration of the steam and power system is power generation technology selection in terms of combined cycle (gas turbine plus steam turbine) or steam ranking cycle (steam turbine). At the same time, fuel selection, system configuration, and load optimization of steam and power system need to be determined. Furthermore, offsite utility demand should be addressed, and this involves feed and product tank farm design with proper insulation and heating.

In short, the traditional design approach adopts a sequential design approach. In contrast, process integration methodology for process design takes a different approach in that process design aspects in the inner part of the onion diagram are allowed to change, which may enhance the possibility of heat recovery and enable more energy savings in the utility system in the outer part of the onion.

Therefore, the discussions in this chapter focus on effects of process changes on energy usage. It can be found that the pinch analysis method (Linnhoff et al., 1982) is powerful in evaluating process changes in the early stage of design without waiting for completion of process design.

9.3 GRAND COMPOSITE CURVES (GCC)

The pinch analysis tool that can be used for assessing process changes is called the grand composite curve (GCC), which can be constructed based on the composite curve that was discussed in Chapter 8. The first step of building GCC is to make adjustments in the temperatures of the composite curves in Figure 9.3a to derive the shifted composite curves of Figure 9.3b. This involves increasing the cold composite temperature by $\frac{1}{2}\Delta T_{min}$ and decreasing the hot composite temperature by $\frac{1}{2}\Delta T_{min}$.

Due to this temperature shift, the hot composite curve moves down vertically by $\frac{1}{2}\Delta T_{min}$, while the cold composite curve moves up by $\frac{1}{2}\Delta T_{min}$. As a result, shifted hot and cold composite curves touch each other at the pinch (see Figure 9.3b). In doing so, the minimum approach (ΔT_{min}) condition is built in for the shifted composite curves, which makes the task easier for utility selection on GCC (this will become self-evident later).

The GCC is then constructed from the enthalpy (horizontal) differences between the shifted composite curves at different temperatures (shown by the distance α in Figure 9.3b and c). The GCC provides the same overall energy target as the composite curves, that is, targets are identical in Figure 9.3a and c. Furthermore, GCC represents the difference between the heat available from the hot streams and the heat required by the cold streams, relative to the pinch, at a given shifted temperature. Thus, the GCC is a plot of the net heat flow for any given shifted temperature, which can be used as the basis for assessing process changes and intermediate utility placement.

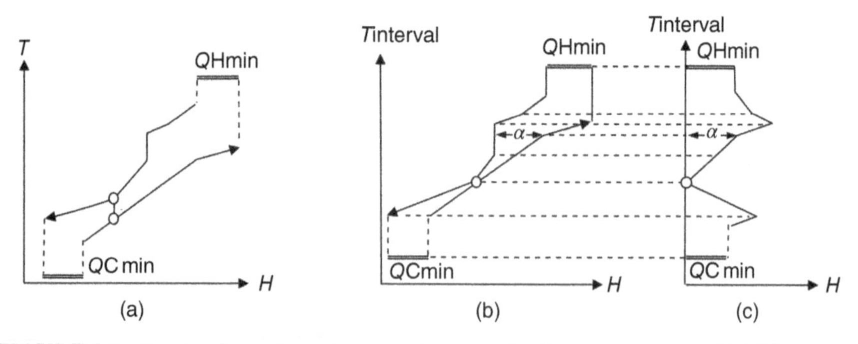

FIGURE 9.3. Construction of grand composite curve. (a) Composite curves. (b) Shifted composite curves. (c) Grand composite curve.

9.4 APPROPRIATE PLACEMENT PRINCIPLE FOR PROCESS CHANGES

Let us see how GCCs are applied for process evaluation.

9.4.1 General Principle for Appropriate Placement

Assume there is a hot utility that can be used for process heating at any temperature levels. Where should we place it for process heating? Of course, we don't want to use it below the pinch according to the pinch golden rule (Chapter 8): do not use hot utility below the pinch. To be smart, we should consider minimizing its use since the hottest utility is the most expensive. If intermediate utilities are available, we should consider maximizing the use of the utility at lowest temperature first (e.g., low-pressure steam) and then the second lowest temperature (e.g., medium-pressure (MP) steam) and so on (e.g., high-pressure (HP) steam) above the pinch prior to the hottest utility (e.g., furnace heating).

Similarly, the cooling utility at the highest temperature should be used first (e.g., air cooling) and then second highest temperature (e.g., cooling water (CW)) and so on (e.g., chilled water) below the pinch prior to the coldest utility (e.g., refrigeration).

The above discussions point to the general principle for *Appropriate Placement for process changes*. The penalty of violating this principle is that both hot and cold utility requirements go up and the process no longer achieves its energy targets.

This general principle was developed for utility selection in terms of the correct levels and loads. But it is much less obvious that the principle also applies to process changes. For better illustration, application of this principle for utility selection will be discussed first and then discussions will cover unit operations such as reactors, separation columns, and feed preheating.

9.4.2 Appropriate Placement for Utility

When multiple utility options are available, the question is which utility is to be selected to reduce overall utility costs. This involves setting appropriate loads for the various utility levels by maximizing cheaper utility loads prior to use of more expensive utilities. The GCC is an elegant tool for accomplishing this purpose.

Consider a specific process that requires heating and cooling. HP steam is sufficient for heating at any temperature levels and likewise, refrigeration is sufficient for cooling at any temperature levels. The simplest way of utility selection is to use HP steam everywhere for heating and refrigeration everywhere for cooling as explained in Figure 9.4a. However, this could be a very costly option as HP steam and refrigeration are expensive. However, there exist intermediate utilities for use. If MP steam and CW can be used, a GCC can be constructed as shown in Figure 9.4b. The target for MP steam is set by simply drawing a horizontal line at the MP steam temperature level starting from the vertical (shifted temperature) axis until it touches the GCC. Remember that the minimum approach temperature is built in when constructing the GCC via shifting hot composite curves as explained previously. The remaining heating duty is then satisfied by the HP steam. This maximizes the use of MP steam before HP steam and therefore minimizes the total hot utility cost as MP steam is

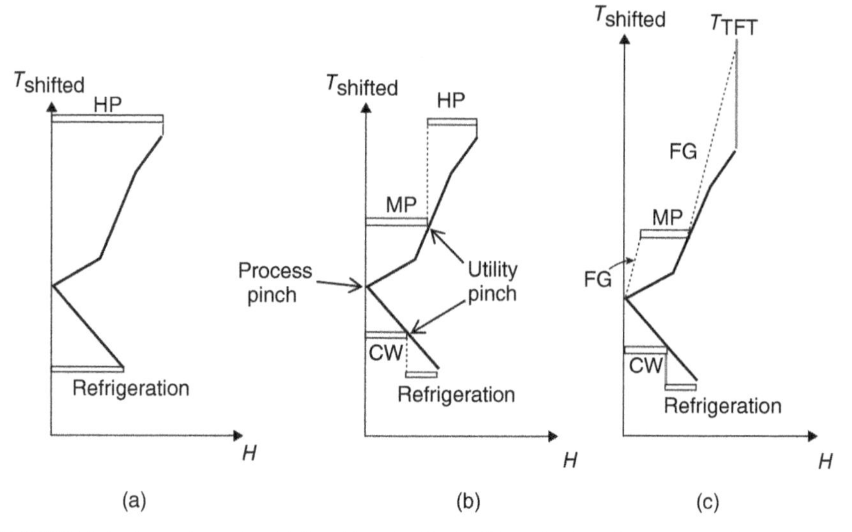

FIGURE 9.4. Selection of multiple utility. (a) Bad utility selection, (b) proper utility selection, (c) utility involving furnace.

cheaper than HP steam. The additional benefit from using MP steam versus HP steam is that higher latent heat is available in MP steam, which reduces the MP steam rate to meet the same duty requirement. Similarly, maximal use of CW before refrigeration reduces the total cold utility costs. The points where the MP and CW levels touch the GCC are called the "Utility Pinches."

If the process requires furnace heating at high temperature, how can the furnace duty be reduced in design because furnace heating is more expensive? Figure 9.4c shows the possible design solution where the use of MP steam is maximized. In the temperature range above the MP steam level, the heating duty has to be supplied by the furnace flue gas. The flue gas flow rate is set as shown in Figure 9.4c by drawing a flue gas (FG) sloping line starting from the MP steam temperature to theoretical flame temperature (T_{TFT}).

The above discussions lead to the design guidelines for minimizing utility costs as follows:

- Minimize furnace heating or HP steam via maximizing the use of lower quality hot utility first.
- Minimize refrigeration or chilled water by maximizing the use of air and water cooling first.
- Maximizing generation of higher quality utility first.

9.4.3 Appropriate Placement for Reaction Process

Reaction integration implies appropriate heat integration of reaction effluent. The reactor integration can be evaluated by the process GCC that is constructed without the reaction effluent stream and then the reaction effluent stream is placed on top of

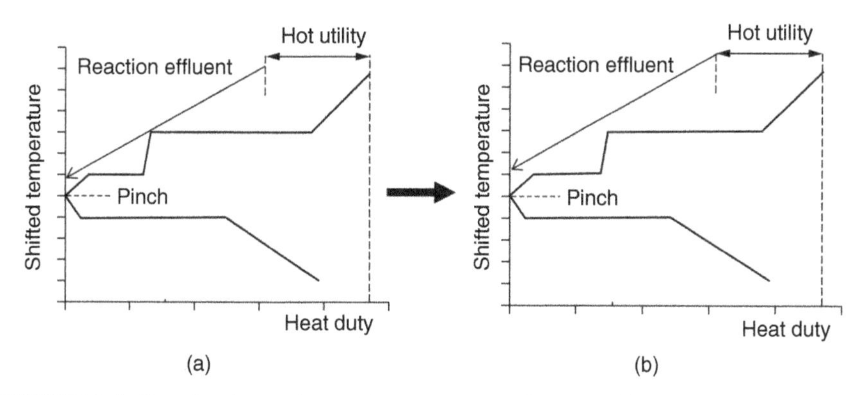

FIGURE 9.5. Reaction integration against process. (a) Poor reaction integration. (b) Better reaction integration.

the GCC. The general Appropriate Placement principle states: The heat of reaction effluent should be released above the process pinch.

With guidance provided from the GCC, the reactor integration for new and existing process designs can be assessed. For existing processes, the process GCC is fixed, but reaction temperature might be adjusted to a small degree for better integration (from figure 9.5a to 9.5b). The general guideline is to maximize heat recovery of reaction effluent heat above the pinch.

For new design, if reaction effluent stream does not fit well with the background process (Figure 9.5a), the reaction conditions such as temperature may be required to vary. However, only small changes in reaction conditions may be tolerated because any significant change would impact on conversion and product yields, which usually outweighs energy costs. Thus, in grassroots design, there is little opportunity to change the desired reaction temperature that is determined based on yield. However, instead of changing the reaction temperature, can we modify the background process in order to have better reaction integration (Figure 9.5b)? This topic is discussed in more details by Glavic et al. (1988).

9.4.4 Appropriate Placement for Distillation Column

There are several key opportunities for column integration that include reflux ratio improvement, pressure changes, feed preheating, side reboiling/condensing, and feed stage location. A pinch tool called column grand composite curve (CGCC) (Dhole and Linnhoff, 1993) was developed to aid in evaluation of these improvements.

9.4.4.1 The Column Grand Composite Curve (CGCC) The CGCC can be constructed based on a converged column simulation as shown in Figure 9.6a. From the simulation, the column stage-wise data is extracted and this data is then organized to generate the CGCC in Figure 9.6b. The stage-wise data relates to "ideal column" design. For ideal column design, the column requires infinite number of stages and infinite number of side reboilers and condensers as shown in Figure 9.6c, which represents minimum thermodynamic loss in the column. In this limiting condition, the

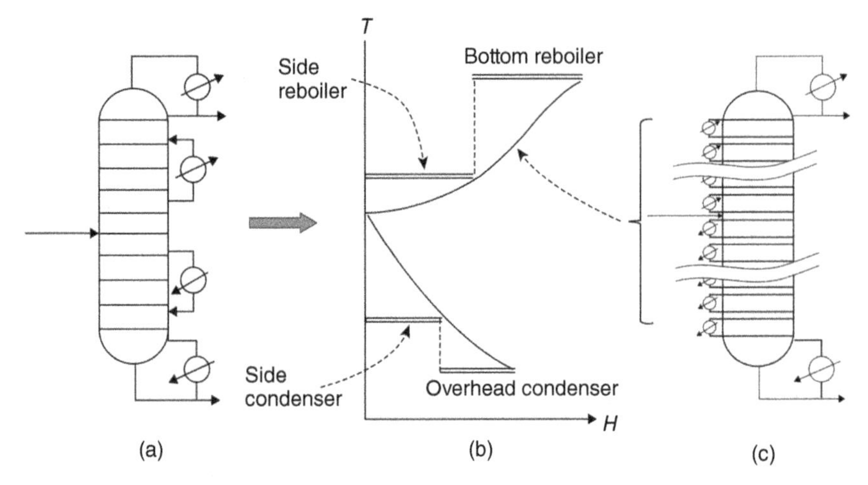

FIGURE 9.6. Construction of column grand composite curve. (a) Converged simulation. (b) Column grand composite. (c) Ideal column.

energy can be supplied to the column along the temperature profile of the CGCC instead of supplying it at extreme reboiling and condensing temperatures. The CGCC is plotted in either *T–H* (*T* = temperature; *H* = enthalpy) or stage-H diagrams. The pinch point on the CGCC is usually caused by the feed.

Similar to the GCC for utility selection for a process, the CGCC provides a thermal profile for evaluating heat integration ideas for a column such as side condensing and reboiling (Figure 9.6b). In a practical column, energy is supplied to the column at feasible reboiling and condensing temperatures.

9.4.4.2 Column Integration Against Background Process Column integration implies heat exchange of the column heating/cooling duties against background process or the external utility available. The principles of Appropriate Placement of columns against a background process can be explained as follows.

Let us look at Figure 9.7a where the reboiler receives heat above the pinch of the background process while the condenser rejects heat below the pinch. The background process is represented by its GCC. Therefore, this distillation column is working across the pinch. In this case, Figure 9.7a represents a case of no integration of the column against the background process. The column is therefore inappropriately placed as regards its integration with the background process.

Assume the pressure of the distillation column is raised and the condenser and reboiler temperatures can increase accordingly. As a result, the column can fit entirely above the pinch. This case represents a complete integration between the column and the background process via the column condenser as shown in Figure 9.7c. The column is now on one side of the pinch (not across the pinch). The overall energy consumption (column plus background process) equals the energy consumption of the background process. In energy-wise, the column is running effectively for free. The column is therefore appropriately placed as regards its integration with the background process. Alternatively, lowering the column pressure so that its temperature drops will make the column fit below the pinch. Placing the column above or below the pinch is another application of the Appropriate Placement principle.

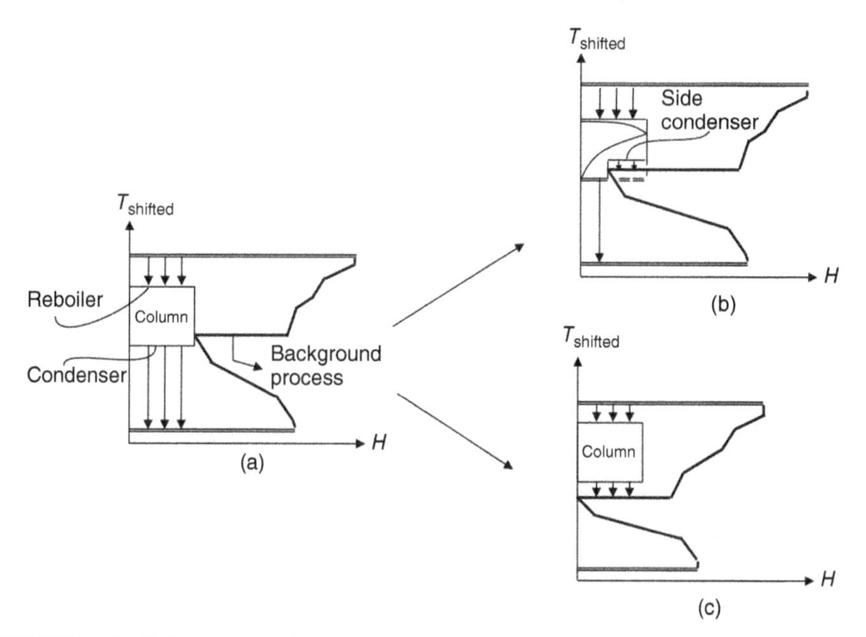

FIGURE 9.7. Column integration with process. (a) Inappropriate placement. (b) Using column modification for integration. (c) Appropriate placement.

In reality, big change to the operating pressure of the distillation column is rarely possible due to the process limits such as product specifications, capital costs, safety, or other considerations. However, there are other ways of reducing heat transferred across the pinch. One option is to install intermediate condenser so that it works at a higher temperature than the main condenser at the top of the column. Figure 9.7b shows the CGCC of the column. The CGCC indicates a potential for side condensing. The side condenser opens up an opportunity for integration between the column and the background process. Compared to Figure 9.7a, the overall energy consumption (column plus background process) has been reduced due to the integration of the side condenser. Alternatively, use of intermediate reboiler or pumparound can be considered.

In summary, the column is inappropriately placed if it is located across the pinch because the column has no heat integration with the background process. On the other hand, the column is appropriately placed if it is placed on one side of the pinch and can be integrated against the background process. Although appropriate column integration can provide substantial energy benefits, these benefits must be compared against associated capital investment and difficulties in operation. In some cases, it is possible to integrate the columns indirectly via the utility system that may reduce operational difficulties.

9.4.4.3 Design Procedure for Column Integration The design procedure for column integration is shown in Figure 9.8, which can be applied for new and revamp projects. Let us walk through the procedure as follows.

(a) *Feed stage optimization*: The feed stage location of the column is optimized first in the simulation prior to the start of the column thermal analysis since the feed

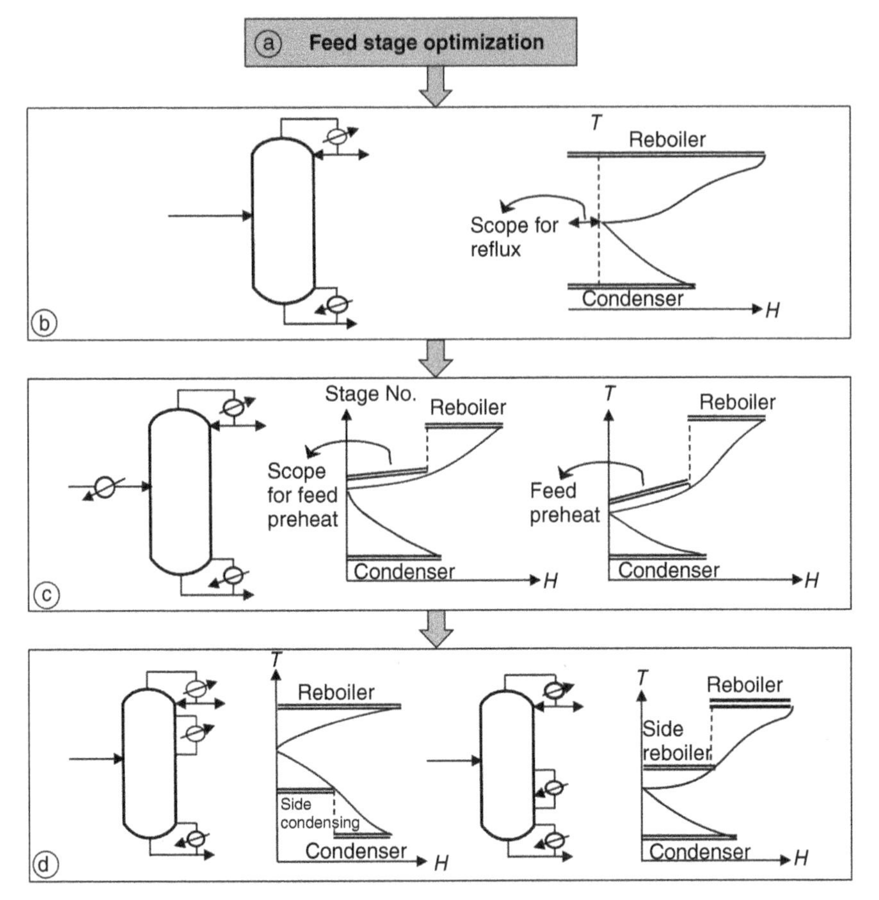

FIGURE 9.8. Procedure for column integration with process. (a) Feed stage optimization. (b) Reflux modification. (c) Feed conditioning. (d) Side condensing/reboiling.

stage may strongly interact with the other options for column improvements. This can be carried out by trying alternate feed stage locations in simulation and evaluating its impact on reboiling duty.

In principle, there could be several stages that can be used as feed stages. When the feed stage is too low, there is a big jump in temperature in the region below the feed stage since too much change in composition is happening than necessary. To get the composition change needed to meet bottoms spec, more reboiler duty is required leading to higher boilup and liquid and vapor traffic in the bottom section. Because of the higher flow rates, the bottom section will have a larger diameter. Having the feed too high does not have the dramatic change as having the feed too low.

Since the objective for feed stage optimization is to minimize energy use or reboiling duty for the separation without the need of additional trays, a plot of reboiler duty versus stage number can be obtained as Figure 9.9. The optimal feed stage should be in the flat region away from the steep change.

After the feed stage optimization is accomplished, the CGCC for the column is then obtained, which is used as the basis for the next step optimization.

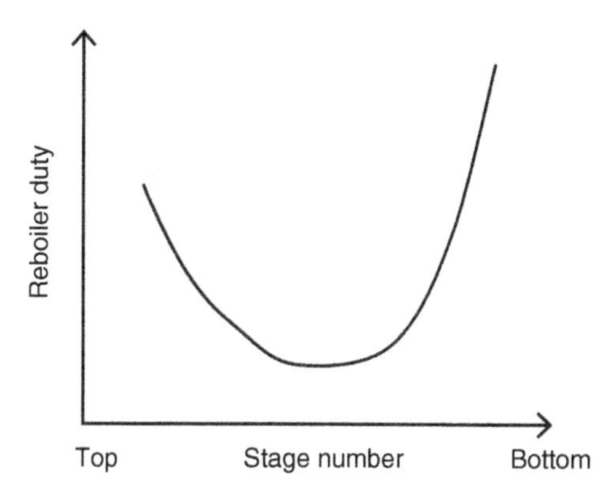

FIGURE 9.9. Feed stage optimization.

(b) *Reflux rate optimization*: The next step is to optimize reflux rate for the column as shown in Figure 9.8b. The horizontal gap between the vertical axis and CGCC pinch point is the scope for reflux improvement. The CGCC will move closer to the vertical axis when the reflux ratio is reduced. The reflux rate optimization must be considered first prior to other thermal modifications since it results in direct heat load savings from the reboiler and the condenser. In an existing column, the reflux can be improved by adding stages or by improving the efficiency of the existing stages.

(c) *Feed conditioning optimization*: After reflux improvement, the next step is to address feed preheating or cooling. In general, feed conditioning offers a more moderate temperature level than side condensing/reboiling. Also feed conditioning is external to the column and is therefore easier to implement than side condensing and reboiling. Feed conditioning opportunity is identified by a "sharp change" in the stage-H (H: enthalpy) CGCC close to the feed point as shown in Figure 9.8c. The extent of the sharp change approximately indicates the scope for feed preheating. Successful feed preheating allows heat load to be shifted from reboiler temperature to the feed preheating temperature. An analogous procedure applies for feed precooling.

(d) *Side condensing/reboiling optimization*: Following the feed conditioning, side condensing/reboiling should be considered. Figure 9.8d describes CGCCs that show potential for side condensing and reboiling. An appropriate side reboiler allows heat load to be shifted from the bottom reboiling to a side reboiling without significant reflux penalty.

9.5 DIVIDING WALL DISTILLATION COLUMN

A dividing wall column (DWC) is a special type of distillation columns containing a vertical partition wall inside the column, allowing three or more products to be produced using a single vessel. This type of distillation column can not only reduce both the capital and energy costs of fractionation systems comparing with simple columns but also produce higher purity products than the products from a simple sidedraw column. The process integration theory via the GCC can be applied to DWCs.

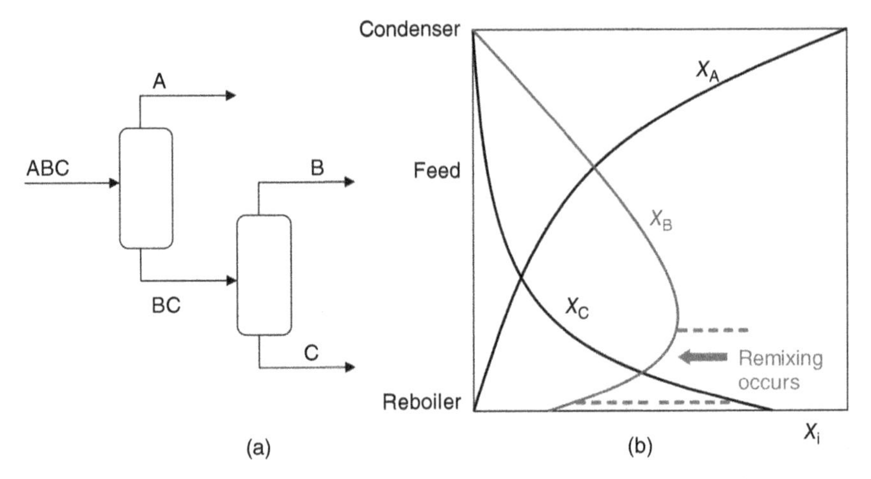

FIGURE 9.10. Thermal inefficiency in direct sequence. (a) Direct distillation sequence. (b) Component profiles for the columns.

9.5.1 DWC Fundamentals

Consider a mixture consisting of three components, A/B/C, where A is the lightest and C is the heaviest. Figure 9.10a shows how a direct sequence of two distillation columns can be used for this separation. For some mixtures, for instance when B is the major component and the split between A and B is roughly as easy as the split between B and C, this configuration has an inherent thermal inefficiency, as illustrated in Figure 9.10b for a generic example. In the first column, the concentration of B builds to a maximum value at a tray near the bottom of the column. On trays below this point, the amount of the heaviest component, C, will continue to increase, diluting B so that its concentration profile will now decrease on each additional tray toward the bottom of the column. Energy has been used to separate B to a maximum purity on an intermediate tray in the first column, but because the B has not been removed at this point, it is remixed and diluted to the concentration at which it is removed in the bottoms. This remixing effect creates a thermal inefficiency.

Figure 9.11 shows a configuration that eliminates this remixing problem. This prefractionator arrangement performs a sharp split between A and C in the first column, while allowing B to distribute between the overhead and bottoms streams. All of the A and some of the B are removed in the overhead of the smaller prefractionation column, while all of the C and the remaining B are removed in the bottoms of the prefractionation column. The upper portion of the second column then separates A from B, while the lower portion separates B and C. This design scheme leads significant energy saving. This saving can be about 30% for a typical design but can reach 50% or 60% for unconventional designs.

The prefractionator arrangement (Figure 9.11a or Figure 9.12a) can be thermally integrated, creating what is known as the Petlyuk arrangement (Figure 9.12b). Vapor and liquid streams from the second (main) column are used to provide vapor and liquid traffic in the prefractionator. This system has only one condenser and one

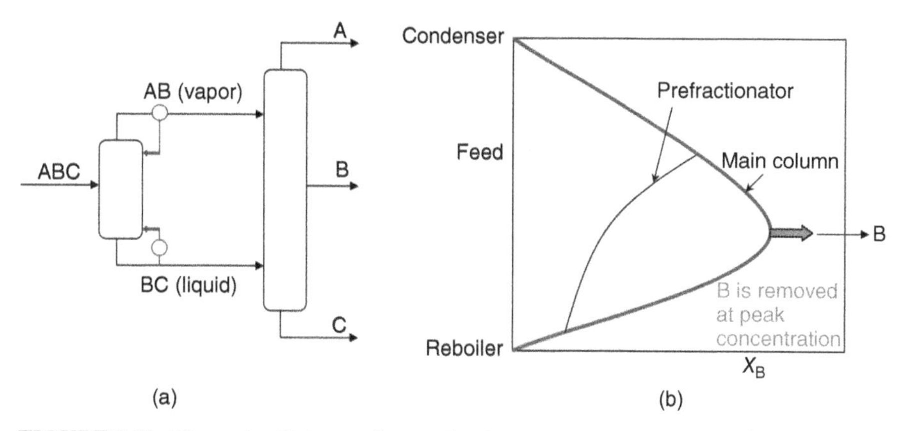

FIGURE 9.11. Thermal efficiency for prefractionator arrangement. (a) Prefractionator arrangement. (b) Component profiles for the columns.

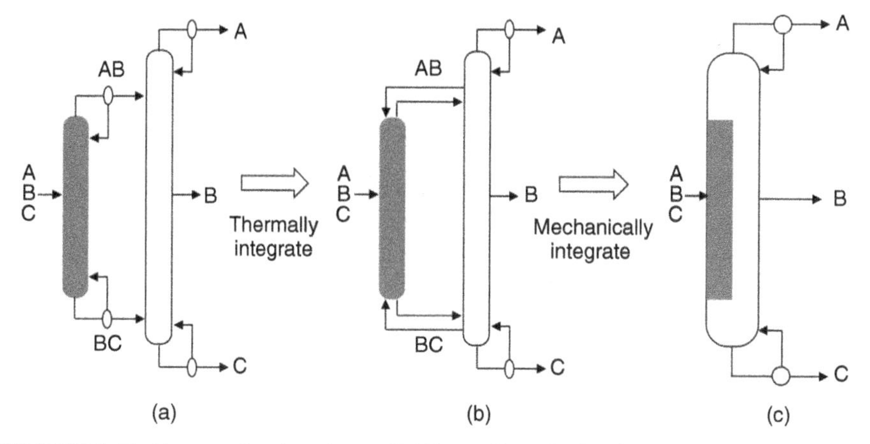

FIGURE 9.12. From prefractionation to dividing wall. (a) Prefractionator. (b) Thermally coupled columns (Petlyuk column). (c) Dividing wall column.

reboiler, and both are attached to the second column. Because the Petlyuk arrangement has fewer major equipment items than the conventional two-column sequence, the total capital costs may be reduced. The further improvement is that the prefractionation column can be integrated into the same shell as the main column, as shown in Figure 9.12c, forming the DWC. Assuming that heat transfer across the dividing wall is negligible, a DWC is thermodynamically equivalent to a Petlyuk arrangement. When compared to a conventional two-column system, reboiling duty saving and capital cost saving of up to 30% is typical.

9.5.2 Guidelines for Using DWC Technology

It seems DWC has significant advantages over simple distillation columns. Can it be applied to replace simple columns when three products are produced? The answer depends as there are certain conditions for applying DWC. When evaluating the possibility of using a DWC, it is important to consider the properties and composition of the separating material as well as the product specifications. The following guidelines (G) (Tsai et al., 2016) describe situations when a DWC may be beneficial.

Guideline for feed composition (G1): Components A, B and C should be comparable in flow rate.

Guideline for relative volatility (G2): The relative volatility ratios of A/B and B/C should be comparable.

Guideline for pressure (G3): The pressure for the corresponding two-column system should not be too different.

Guideline for material construction (G4): The corresponding two-column system should not require distinct metallurgies with vastly different costs.

As with any heretic rules, there are exceptions, and, in certain situations, the rules may contradict each other, but these guidelines can be useful when identifying applications in which a DWC may be better than other distillation options. Two case studies are discussed below to show how these guidelines are applied and what benefits could be achieved by DWC.

9.6 SYSTEMATIC APPROACH FOR PROCESS INTEGRATION

Traditional energy efficiency improvements have a narrow focus on energy recovery alone with little consideration of interactions with process flowsheeting, equipment design, and process conditions. As a result, energy-saving projects usually have limited economic benefits and thus have difficulty in competing with capacity and yield-related projects. In contrast, the process integration methodology as explained in Figure 9.13 takes a different approach in that energy optimization is closely integrated with changes to both process flowsheeting and conditions, as well as equipment design. The goal of this methodology is to enhance throughput and yields via energy optimization with lowest capital cost possible.

This methodology consists of four core components, namely process simulation development, equipment rating analysis, process integration analysis, and synergy optimization. This methodology has been applied to numerous design projects (Zhu, 2014) with common features such as the following:

- Enhanced throughput and yields
- Reduced capital investment and operating cost to achieve the objectives
- Simplified process design and enhanced equipment performance.

The purpose of process simulation development is to represent current plant design for the base case, defined in terms of key operating parameters and their interactions. Thus, the simulation can provide the specifications for equipment rating assessment.

The key role of equipment rating analysis is to assess equipment performance and identify equipment spare capacity and limitations. Utilization of spare capacity can enable expansion up to 10–20% in general and accommodate improvement projects with low capital cost investment. When equipment reaches hard limitations – for

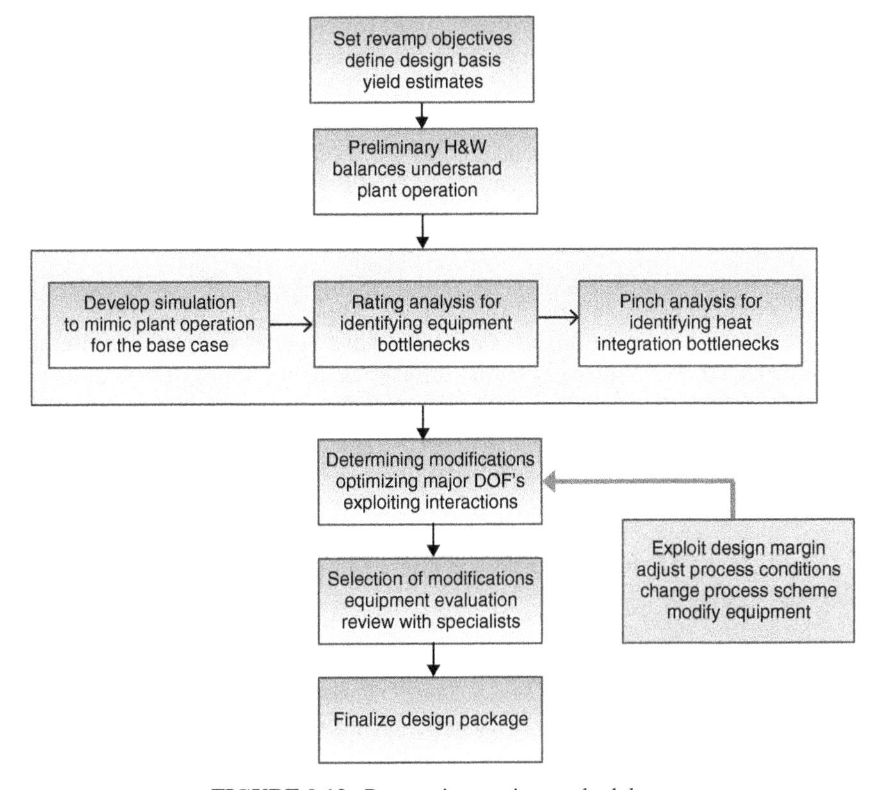

FIGURE 9.13. Process integration methodology.

example, a fractionation tower reaches its jet flood limit, or a compressor reaches its flow rate limit, or a furnace reaches its heat flux limit – it could be expensive to replace or install new equipment. The important part of a feasibility study is to find ways to overcome these constraints, which is accomplished in the next two steps.

The third step is to apply the process integration methods to exploit interactions and identify changes to process conditions, equipment, process redesign, and utility systems, with the purpose of shifting plant bottlenecks from more expensive to less expensive equipment. By capitalizing on interactions, it is possible to utilize equipment spare capacity, push equipment to true limits in order to avoid the need to replace existing equipment or install new equipment. This is a major feature of this process integration methodology.

A simple and effective example is fractionation tower feed preheat. A tower reboiler could reach a duty limit. With a tower feed preheater, the required reboiling duty is reduced. A column assessment may show the effects on separation with increased feed preheat and reduced reboiling at the bottom. If the effects are acceptable, this modification by adding a feed preheater could eliminate the need of installing a new reboiler, which is expensive.

The fourth step is synergy optimization, and the driver is to exploit interactions between equipment, process redesign, and heat integration. Making changes to process conditions provides a major degree of freedom to achieve this. One direct benefit of optimizing process conditions in this context is that spare capacity available in existing assets can be utilized. Process redesign provides another major degree of freedom as it can increase heat recovery and relax equipment limitations.

In summary, major changes to infrastructure and installation of key equipment such as a new reactor, main fractionation tower, and/or gas compressor could form a major capital cost component. In many cases, it is possible that the level of modification to major equipment could be reduced or even avoided by exploiting design margins for existing processes and optimizing degrees of freedom available in the existing design and equipment. It is the goal of the process integration methodology to achieve minimum operating and capital costs. Applying this integration approach can give results in three categories. First, alternative options for each improvement will be provided. Second, any potential limitations, either in process conditions or equipment, will be flagged. Third, solutions to overcome or relax these limits will be obtained by exploiting interactions between process conditions, equipment performance, process redesign, and heat integration.

9.7 APPLICATIONS OF THE PROCESS INTEGRATION METHODOLOGY

The process integration principles outlined provide the guidelines for process changes in general. By applying the process integration methodology, the design is no longer confined in a subsystem, from reaction system to separation system, heat exchanger network, and site heat and power systems. In many studies, the energy savings from process change analysis far outweigh those from heat recovery projects. The following examples are used to demonstrate the effectiveness of the process integration methodology. The first example involves improving energy efficiency for a single process design, while the second one deals with optimizing an overall complex consisting of multiple process units.

9.7.1 Catalyst Improvement

Catalyst with high distillate selectivity in the hydrocracking process can also improve energy efficiency while they improve yield. These kinds of catalysts reduce chemical H_2 consumption, lower natural gas consumed, lower makeup gas power, and reduce need for quench. In this example, as shown in Figure 9.14 (Zhu et al., 2011), natural gas is reduced by 80,000 Btu/bbl feed due to lower consumption in the H_2 Plant. Makeup gas compression power is reduced by 4000 Btu/bbl on fuel equivalent basis. Recycle gas compression power is reduced by 500 Btu/bbl on fuel equivalent basis. The total energy savings are approximately 85,000 Btu/bbl on fuel equivalent basis translating to an energy saving of approximately $10 million per year for a 50,000 BPD hydrocracking unit.

9.7.2 Process Design Improvement and Equipment Simplification

Traditional hydrocracking process design features one common stripper that receives two feeds containing very different compositions, one from the cold flash drum and

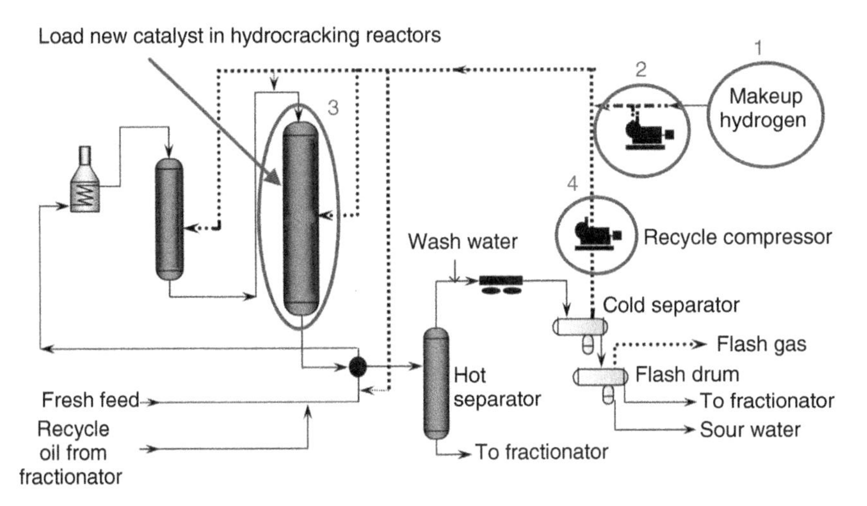

Benefits of catalyst change:

1. Lower chemical H_2 consumption reduces natural gas consumed at H_2 plant
2. Lower chemical H_2 consumption reduces makeup gas compression energy
3. Higher distillate selectivity reduces heat release in the new distillate catalyst; less quench is required
4. Lower quench requirement reduces the recycle gas compressor utilities

FIGURE 9.14. Effects of catalyst improvements on process energy efficiency.

the other from the hot flash drum; and a typical design is shown in Figure 9.15. The processing objective of the stripper is to remove H_2S from the feeds. These two feeds originate from the reaction effluent that first comes to the hot separator. The overhead vapor containing light products of the hot separator goes to the cold flash drum, while the bottom containing relative heavy products of the hot separator goes to the hot flash drum. Eventually, the liquids of both hot and cold drums are fed to the same stripper. Then the stripper bottoms become the feed for the main fractionator. The shortcoming of this process sequence can be summarized as separation and then mixing and separation again.

Thus, the inefficiency of this single-stripper design is rooted in mixing of the hot flash drum and cold flash drum liquids, which undo the separations upstream. To avoid this inefficient mixing, it is proposed to use two strippers (Hoehn et al., 2014), namely a hot stripper that receives the hot flash drum liquid as the feed and a cold stripper that is used for the cold flash drum liquid. Furthermore, the cold stripper bottoms does not pass through the main fractionator feed heater but goes directly to the main fractionators. Only the hot stripper bottoms goes to the main fractionator feed heater. This proposed design is shown in Figure 9.16. The idea sounds promising; but how to quickly evaluate the effect of this idea?

Let us turn to composite curves for getting the answer. The stream data for both single-stripper and two-stripper designs are obtained separately and the data includes both the reaction and fractionation sections. For the relative comparison purpose, the composite curves are plotted based on zero degree $\Delta T_{\min}$, absolute pinch point. The single-stripper design is described by the CC in Figure 9.17, which indicates

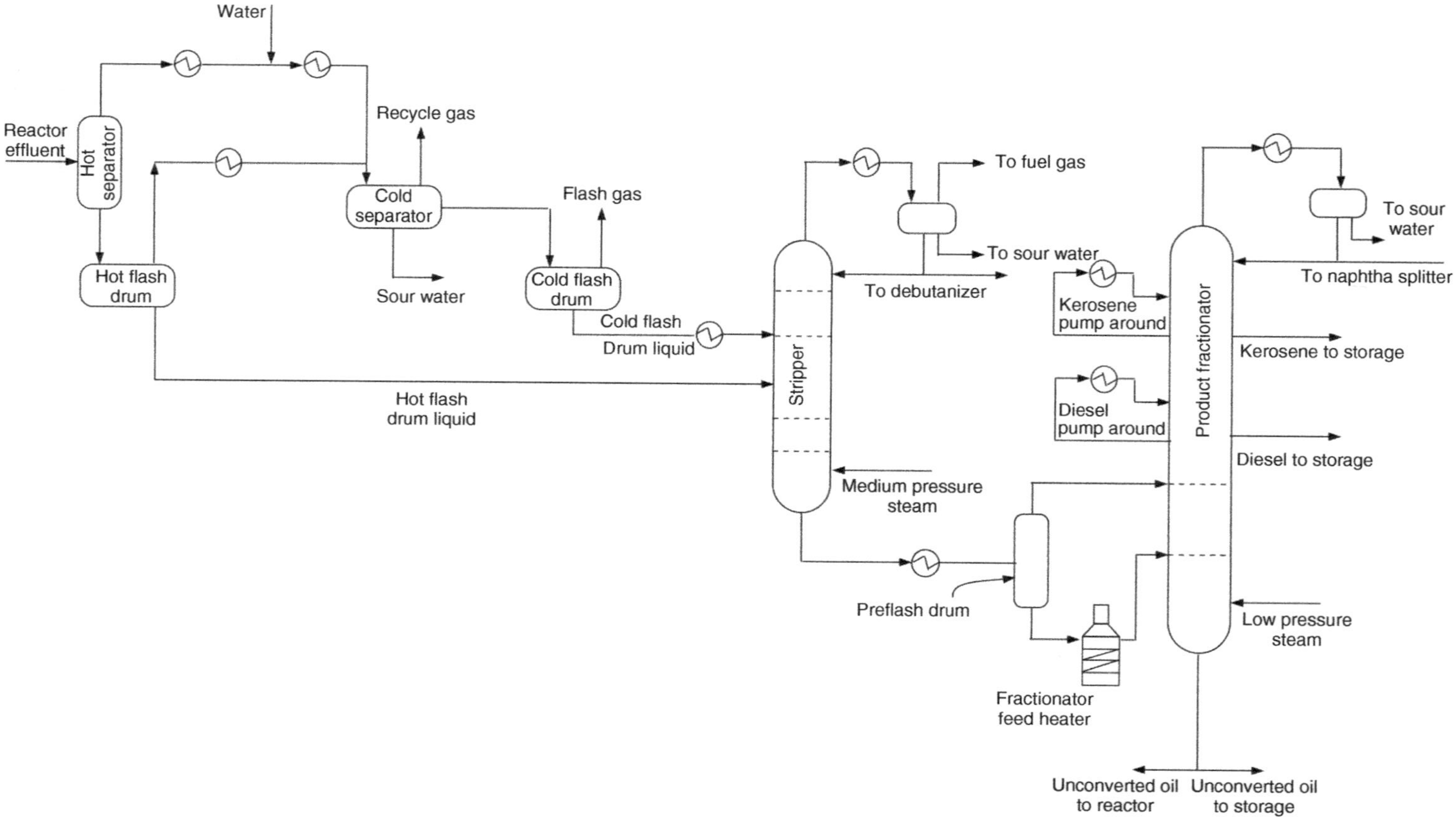

FIGURE 9.15. Single-stripper fractionation scheme.

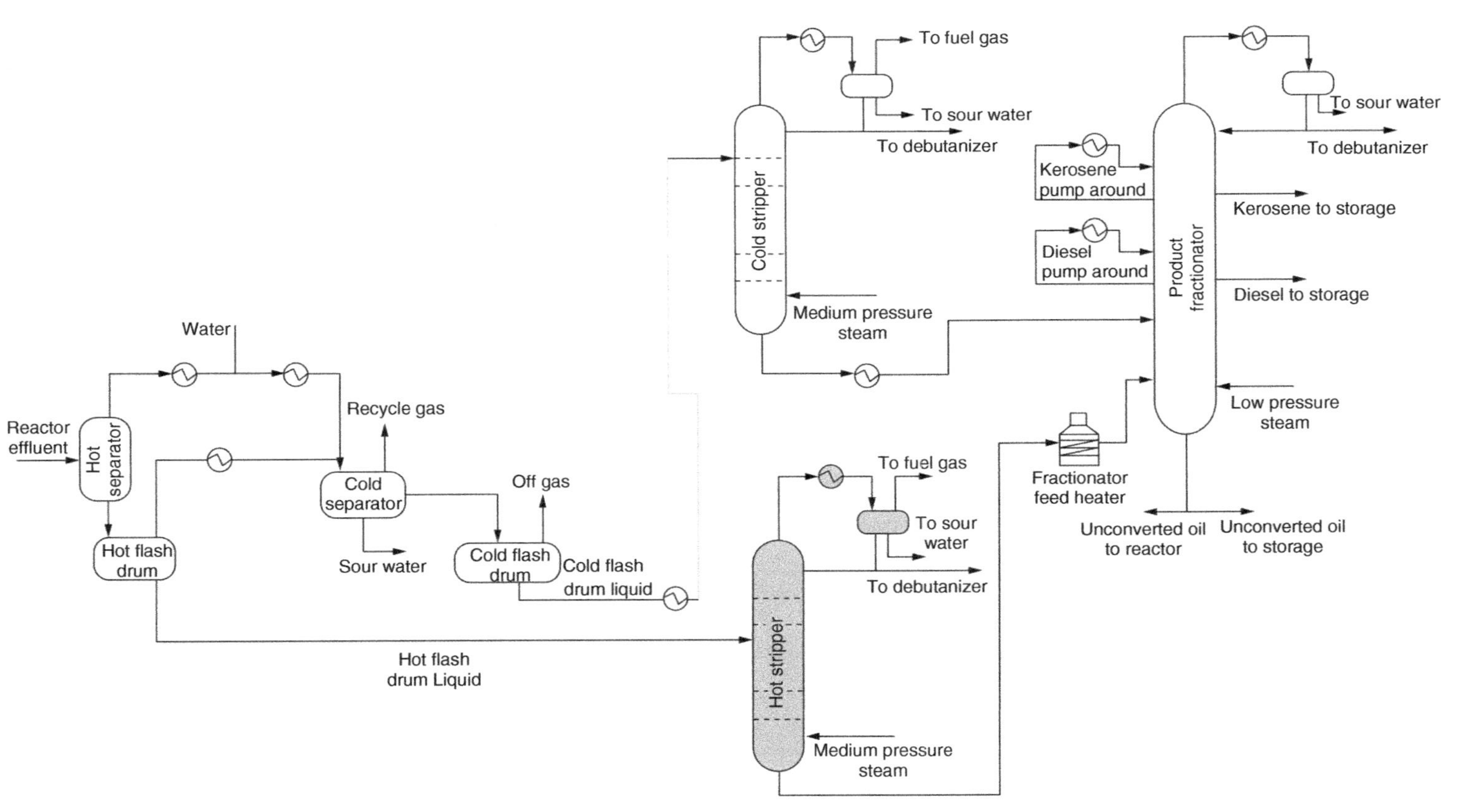

FIGURE 9.16. Proposed two-stripper fractionation scheme.

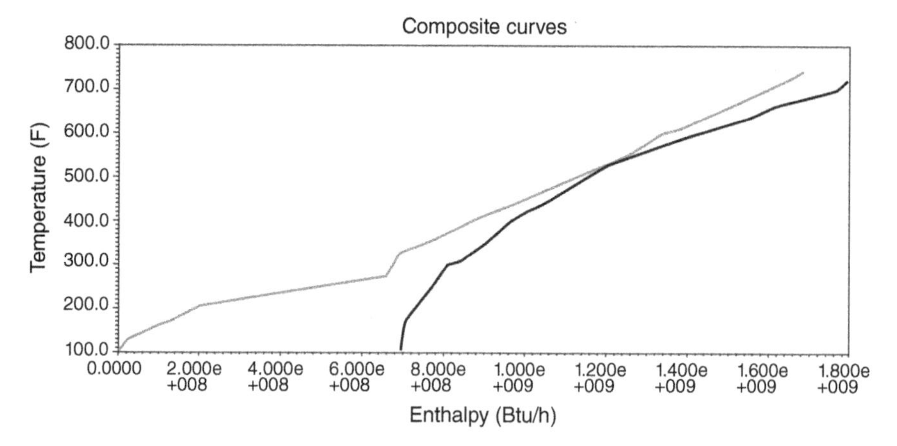

FIGURE 9.17. Composite curves representing the single-stripper fractionation scheme.

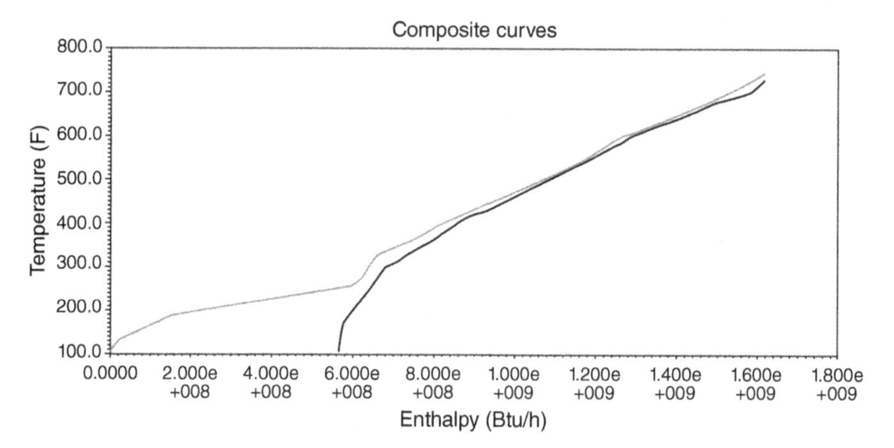

FIGURE 9.18. Composite curves representing the two-stripper fractionation scheme.

the net total hot utility (fuel in this example) will be 110 MMBtu/h. In contrast, the two-stripper design depicted by the CC in Figure 9.18 shows zero net hot utility required at zero ΔT_{min}. The net difference between two designs is 110 MMBtu/h by the proposed two-stripper design, which is 42% reduction of the fractionator heater duty or 23% of total energy reduction for the hydrocracking unit. This improvement is significant.

Furthermore, the process GCC curves are generated, one for the single-stripper design (Figure 9.19) and the other for the two-stripper design (Figure 9.20). The pinch point is the zero enthalpy point on the GCC. As can be observed from the single-stripper GCC (Figure 9.19), the process heat curve (above the pinch point) moves away from zero enthalpy, which indicates the need of large amount of external heating. In contrast, the GCC for the two-stripper design indicates moves closer to zero enthalpy, implying the reduction of external heating and cooling demand.

In practical terms, the main reason for reduced fractionator heater duty is that the flow rate going through the heater is reduced in the two-stripper design (Figure 9.16)

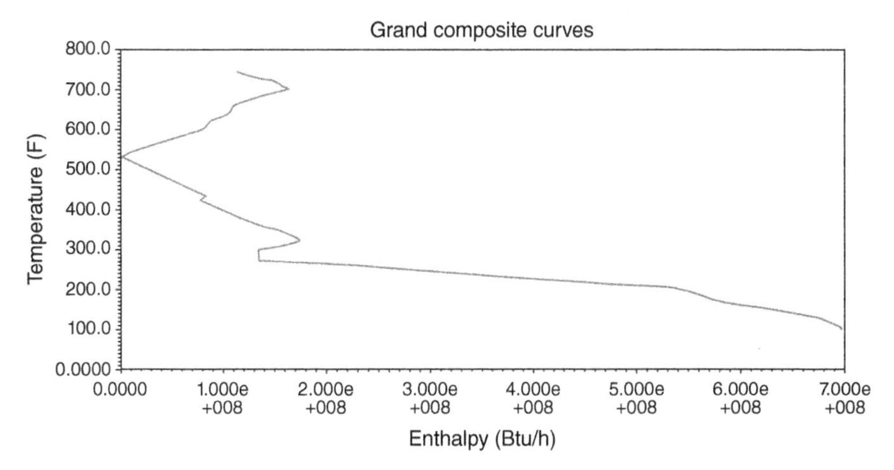

FIGURE 9.19. Grand composite curve for the single-stripper fractionation scheme.

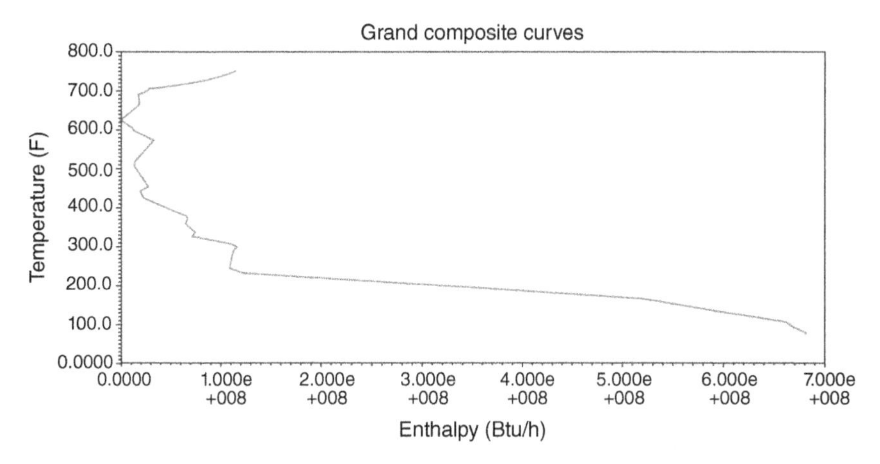

FIGURE 9.20. Grand composite curve for the two-stripper fractionation scheme.

comparing with that in the single-stripper design (Figure 9.15). The cold stripper bottom is preheated by a process stream before entering the product fractionator.

Before jumping into a detailed design for the two-stripper design, the composite curve method is able to show the significance of the two-stripper design at the early stage. Since the benefit is revealed to be significant, further investigation of details is warranted. Although it is a very significant improvement, the process design becomes more complex as a new stripper column must be added together with associated equipment. The next question is: can we simplify the design and reduce the capital cost while improving energy efficiency?

To simplify the overall process design, a stacked column design is adopted as shown in Figure 9.21 in which the cold stripper is stacked over the top of the hot stripper to form a single column with a common overhead system (Hoehn et al., 2015). The vapor from the hot flash stripper, which is below the cold flash stripper, leaves the side of the column. The trays for the cold flash stripper are full diameter and if necessary,

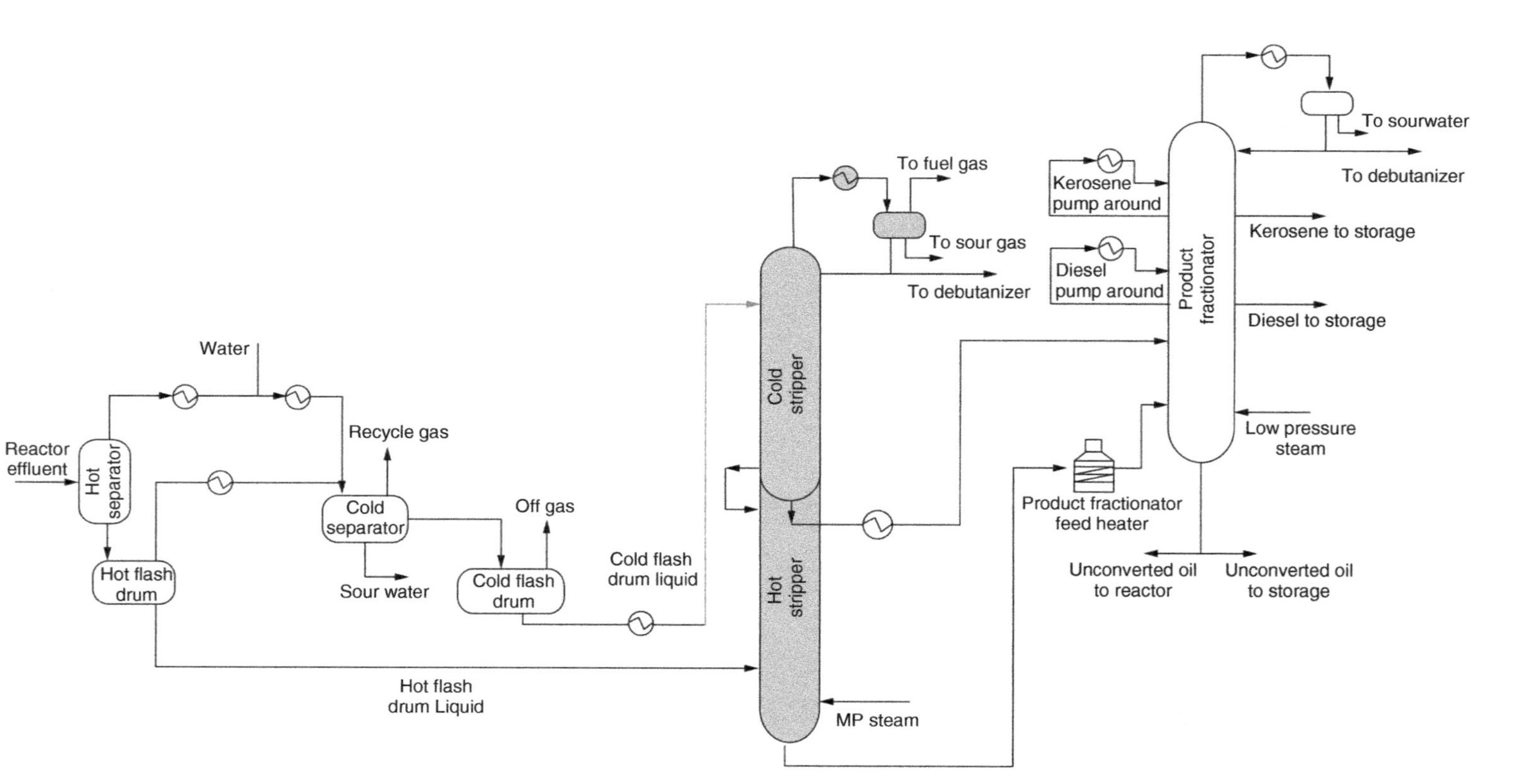

FIGURE 9.21. Stacked two-stripper fractionation scheme.

the column can be swaged down for this stripper. The stripped cold flash liquid leaves the side of the column just above the hot flash stripper top tray. An internal head separates the two columns. This simplified design (Figure 9.21) reduces equipment count, simplifies column overhead operation, and reduces capital cost from the initially generated two-stripper design in Figure 9.16. A more detailed cost/benefit analysis indicates operating cost saving of $2.5 MM/year as well as equipment cost saving of $3.0 MM compared with the base design of Figure 9.15. The capital cost saving is obtained from much reduced size of the feed heater for the main fractionators. Although there is capital cost increase from the stacked stripper column versus the simple stripper column, it is insignificant in comparison with the capital cost saving from the feed heater.

So far, we introduced composite and GCCs as the means for energy targeting. The composite and GCCs will remain the same for fixed process flowsheet and conditions. The energy targeting, in this case, will indicate the pinch point and minimum hot and cold utilities required for a given ΔT_{min}. However, process design engineer may feel constrained with fixed stream flow rates and temperatures that restrict his/her ability to design an integrated heat recovery system. This desire motivates him/her to explore process changes in order to improve the current process design since large benefits could be gained if good process changes can be found.

Obviously, process modifications will have a profound impact on stream data such as flow rates and temperatures. Consequently, composite and GCCs will be very different from the base case. The above example has shown that these targeting curves could become powerful tools to demonstrate the benefit of process changes.

9.7.3 Application of Dividing Wall Column

Naphtha isomerization processes typically use distillation columns for both prefractionation of the feed and postfractionation of the reactor products. Depending on a refiner's gasoline blending requirements, C5–C6 isomerization unit prefractionation schemes can include depentanizers (DPs) and deisopentanizers (DIP). The DP and DIP columns are designed to recover *n*-pentane and *i*-penta from the feed. These tower columns are large in dimension and consume a significant amount of utilities.

Currently, the DP and DIP towers are combined into a single 3-cut tower (Figure 9.22) with sidedraw. The sidedraw product is designed to recover 94% of the *n*-pentane in the feed. The remaining bottoms stream contains C6+ material. There are savings in plot space and capital when using a single tower, total utility requirements are similar to the two-column arrangement. Product purities are also reduced when going to a single tower with sidedraw. There is consideration to find a more economical solution for cases where we want to produce separate iC5, nC5, and C6 rich streams for isomerization complexes.

The sidecut column in Figure 9.22 is designed with 72 theoretical stages or 80 real stages with a thermosiphon reboiler per the column specifications above. To achieve the purity specification for the *n*-pentane, the resulting reflux to feed (R/F) must be high at 2.96 with a required reboiler duty of 248 MMBtu/h. To reduce the reboiling duty, the number of real stages is increased to 107 with reboiling duty reduced to 209 MMBtu/h. Beyond this, any further addition of stages generates very small reduction in reboiling duty.

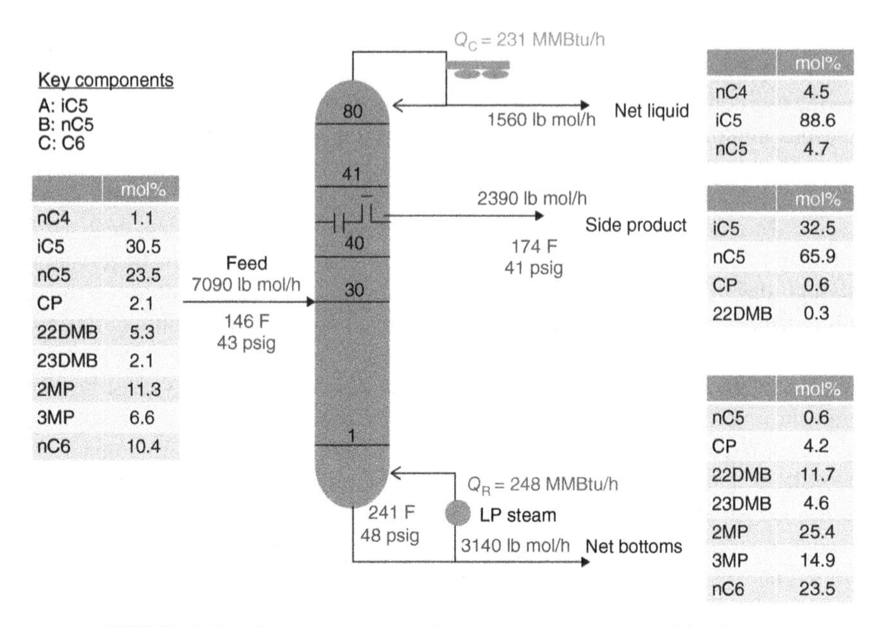

FIGURE 9.22. Current design: single column depentanizer with sidedraw.

Potential application of DWC for this case is verified based on the guidelines provided above. Although the sidecut B is not in excess, it has 23.5 mol%, which is not small quantity. Thus, G1 is loosely satisfied. Furthermore, the most important guideline is G2, which is satisfied with $\alpha(\text{iC5/nC5}) = 1.15 - 1.21$ while $\alpha(\text{nC5/C6}) = 1.4 - 1.5$ across the pressure range of the distillation. This implies the separation of components B and C in the bottom section is easier than A and B, which is the fundamental reason why a DWC can achieve significant reduction in reboiling duty.

The DWC concept is considered to maximize the recovery of iC5 in the overhead product while minimizing reboiler duty. The advantage of utilizing a DWC for this application is that the light key components (iC5) would need to boil up over the wall and condense back down to the sidedraw location. Thus, this allows more stages of separation to ensure minimal amount of iC5 is in the sidedraw while keeping the fractionation utility minimal.

Therefore, a DWC (Figure 9.23) to combine the DIP and DP columns instead of combining them into a simple column with sidedraw (Figure 9.22). The DWC was designed to the specifications outlined above and optimized toward reducing reboiler duty via optimizing feed tray location, location of the top of the dividing wall, location of the bottom of the dividing wall, liquid reflux to each side of the wall, and vapor split to each side of the wall.

For a recent C5–C6 isomerization unit design, we found that using a DWC instead of a simple column with sidedraw reduced capital cost by \$4.6 MM USD (32%) and, more importantly, the required reboiler duty was reduced by 30%, resulting in a savings of US\$2.6 million in operating costs per annum for a 2000 KMTA capacity unit.

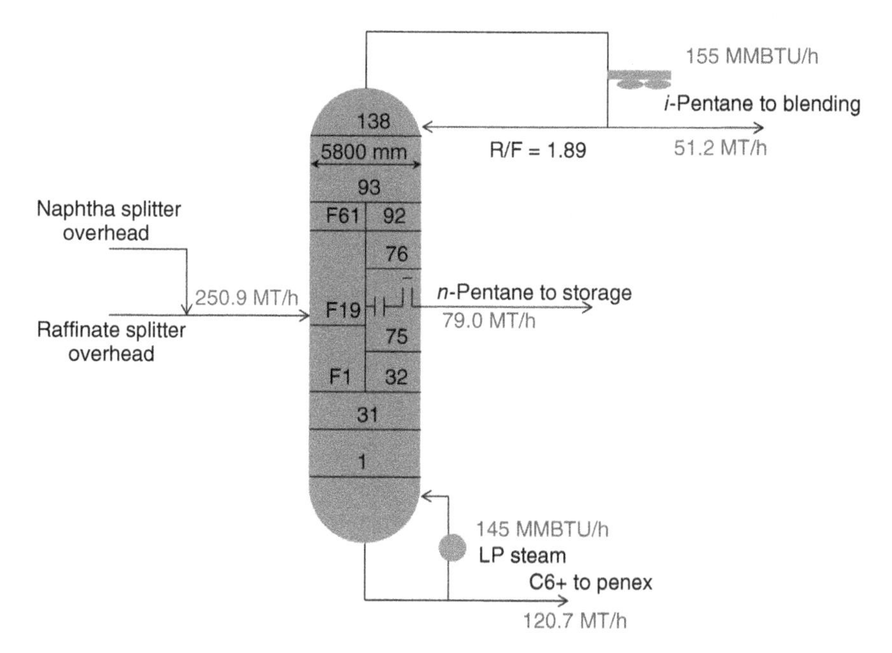

FIGURE 9.23. Dividing wall column for the depentanizer.

9.7.4 Dividing Wall Column Application in the Process System Context

This example is about use of DWC for saving both energy and capital required in product fractionation. Due to market opportunity, an existing refinery wants to make four naphtha products from two products currently it makes. If conventional design and technology are used, two new columns are required. The showstopper for the revamp is the plot space, and the capital required for new equipment and related supporting infrastructure is considered as prohibitive. The innovations discussed in this case not only resolve the plot space issue but also reduce capital and energy use in a very significant way. Let us see how these improvements could happen.

9.7.4.1 Description of the Base Case Fluid catalytic cracking (FCC) is widely employed to convert straight-run atmospheric gas oils, vacuum gas oils, certain atmospheric residues, and heavy stocks into high-octane gasoline, light fuel oils, and olefin-rich light gases. The reactor effluent from the reaction section of the FCC unit is sent to the main fractionator. The overhead of the main fractionator that includes unstabilized gasoline and lighter material is processed in the FCC Gas Concentration section or the unsaturated gas plant. A detailed description of the FCC process is given by Meyers (2004).

The existing unsaturated gas plant is shown in Figure 9.24, which consists of absorbers and fractionators to separate main fractionator overhead into gasoline and other desired light products. According to this figure, overhead gas from the FCC main fractionator receiver is compressed and mixed with primary absorber bottom and stripper overhead gas and directed to the HP separator. Gas from this separator is sent to the primary absorber, where it is contacted by unstabilized gasoline from

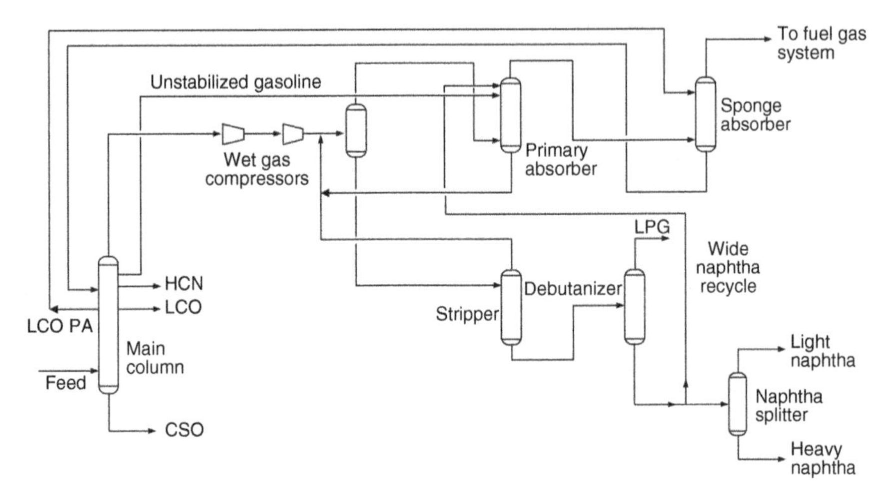

FIGURE 9.24. Existing naphtha separation for 2-naphtha products.

the main fractionator overhead receiver to recover C3 and C4 fractions. The primary absorber offgas is directed to a secondary or "sponge" absorber, where a circulating stream of light cycle oil (LCO) from the main column is used to absorb C5+ material and the remaining C3 and C4 fractions in the feed. The sponge absorbed rich oil is returned to the FCC main fractionator. The sponge absorber overhead is sent to fuel gas or other processing.

Liquid from the HP separator is sent to a stripper column where C2-material is removed overhead and recycled back to the HP separator. The bottoms liquid from the stripper is sent to the debutanizer, where an olefinic C3–C4 product is separated. Typically, 25–50% of the debutanized gasoline (full range naphtha) is recycled back to the primary absorber to control the recovery of light hydrocarbons. Currently for an existing refinery, two gasoline cuts are demanded; the debutanizer bottom, which is stabilized gasoline, is sent to naphtha splitters to make light and heavy naphtha products.

With increased market demand, the refinery wants to make four naphtha products from two products. If using the conventional design, two new columns are required with a direct sequence of three columns in order to make four naphtha products as shown in Figure 9.25. However, the showstopper for this design is the plot space for installing two new columns. The question then becomes thus: Are there any advanced separation technologies available that can resolve the plot space issue?

9.7.4.2 ***Dividing Wall Technology*** Dividing wall technology received the attention of the refinery because multiple products can be produced in one column with dividing wall inside the column. Applications of DWCs are discussed in the preceding section.

To verify the benefit by using DWC, a simulation model was conducted for two columns for naphtha separation as the base case (Figure 9.26a). On the other hand, a DWC where a vertical partition separates the column sections is simulated for making three naphtha products (Figure 9.26b). With the partition of column into two sections, separation of light naphtha from the rest occurs in the partition on the left-hand side

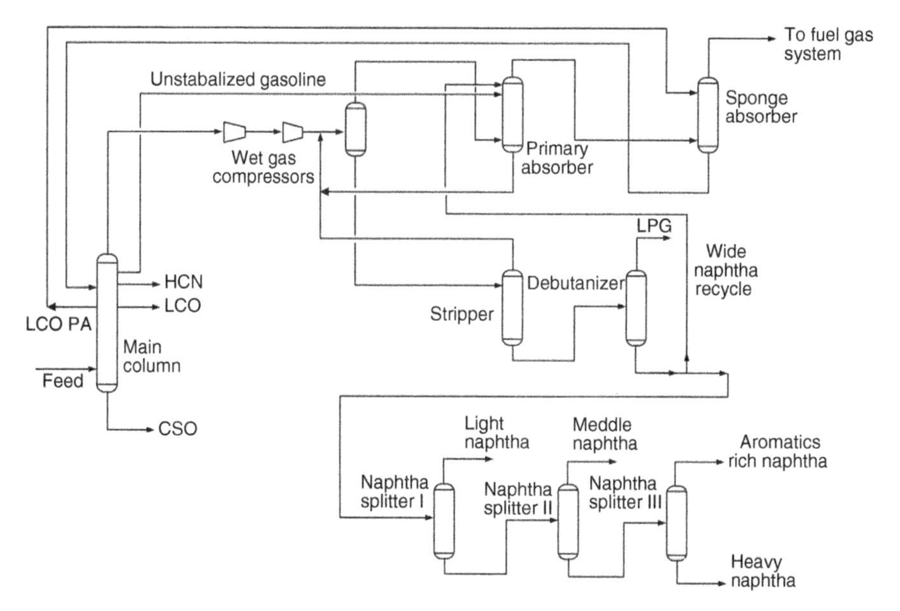

FIGURE 9.25. Typical design scheme of naphtha separation for 4-naphtha products.

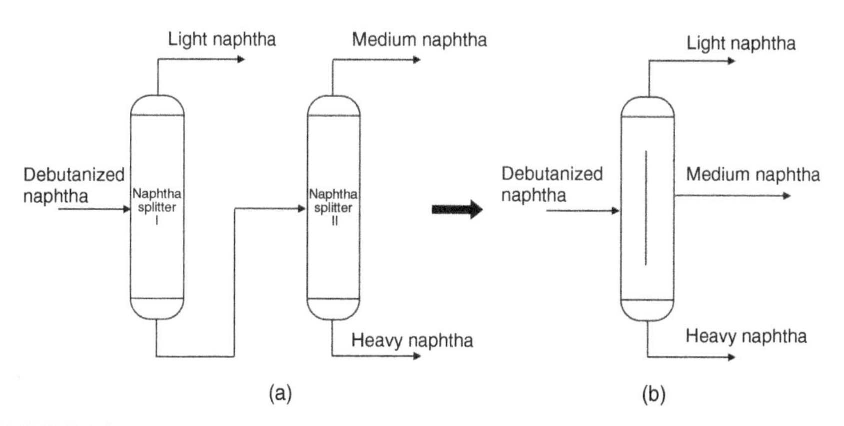

FIGURE 9.26. Separation of three naphtha products. (a) Typical sequence of two splitter columns. (b) Dividing wall column.

of the column similarly to what happens in Splitter 1, while the partition on the right-hand side acting similarly to Splitter 2 mainly deals with separation between medium and heavy naphtha. With medium naphtha withdrawn in the middle of the column where its concentration peaks, there is no remixing occurring. The remaining light naphtha separated travels up to the top and joins that on the left side to go to the overhead condenser. In this case, reboiling duty is reduced by 25% with a single DWC versus two simple columns. This proof of principle based on a simulation is demonstrated by the benefit of applying dividing wall technology for naphtha separation in general. However, it was not certain if it was a clear-cut case for applying DWC for this existing naphtha system.

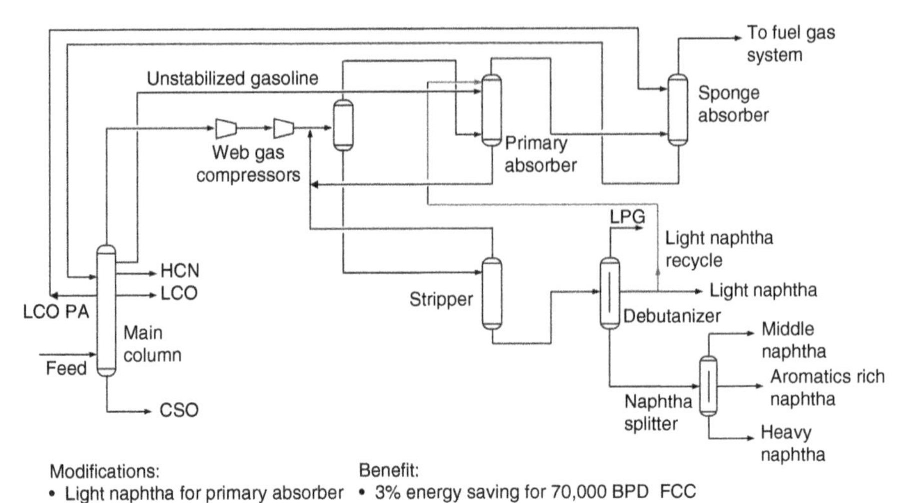

FIGURE 9.27. Applying dividing wall to naphtha separation.

By assessing the naphtha separation from the FCC process, it was found that the guidelines given above could be satisfied readily because there is no strict specification for the two naphtha products from middle sidedraws and the quantity of these two middle draws are relatively large. Since dividing wall technology has been successfully applied to many processes, this boosted the confidence for the refinery to apply the technology. Therefore, it was concluded from the search of advanced separation technology that DWC was determined the most promising for the naphtha separation.

9.7.4.3 Application of Dividing Wall Technology Now the question becomes more specific: how to apply the technology to the existing gas plant?

Due to the restriction of plot space, the revamp team was looking into revamp of the existing debutanizer column for a DWC (Figure 9.27). By retrofitting the debutanizer column with dividing wall technology, liquefied petroleum gas (LPG) can be recovered from the overhead and light naphtha withdrawn as an intermediate product. Part of the light naphtha is recycled and used as sponge liquid in the primary absorber instead of full range naphtha as is traditionally used. Light naphtha has much better absorbing ability than the full range naphtha; consequently, the recycle can be significantly reduced. The bottom is directed to the downstream separation. Naturally, the existing naphtha splitter was also revamped for dividing column by inserting a metal wall to make partition of the column. With the naphtha splitter using dividing wall technology, three more naphtha products are produced, that is, medium naphtha, aromatics grade naphtha, and heavy naphtha.

The DWC strategy reduces the overall reboiler duty by 30%. At the same time, considerable cost saving is realized as an entire separation column is eliminated. With

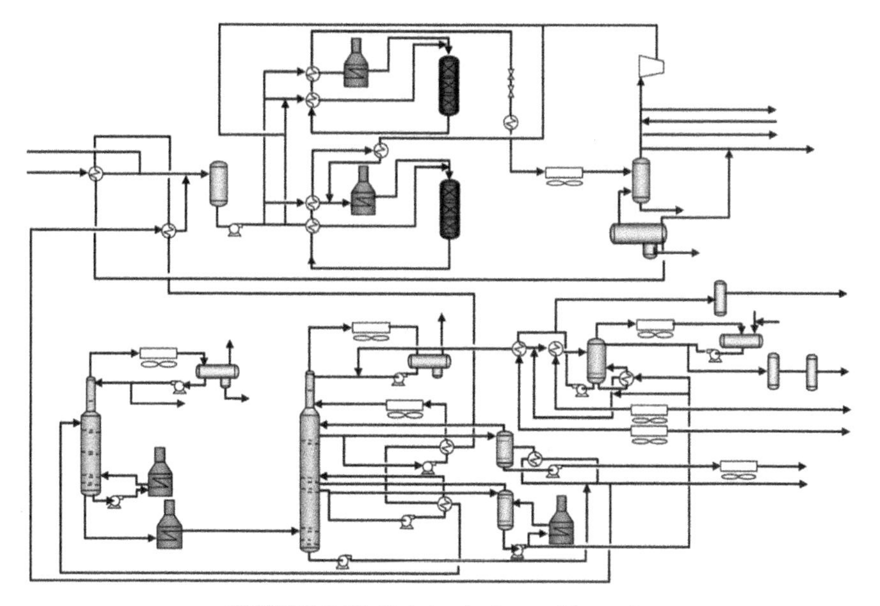

FIGURE 9.28. Existing hydrocracking unit.

the revamped naphtha separation scheme by applying the dividing wall technology, four naphtha products can be made from two products currently made without the need of installing two new columns. This resolves the plot space issue. Consequently, capital cost was reduced significantly around 25% although major revamp for the existing two columns were required.

9.7.5 Integrated Energy and Process Optimization

This example is about how energy retrofit projects can support capacity expansion and yield improvement (Zhu et al., 2011). A refinery wishes to increase hydrocracking capacity by 15% in order to meet new diesel demand in the region; the existing hydrocracking unit is shown in Figure 9.28. A screening study based on simulation and equipment rating indicated several major bottlenecks, which could require significant capital investment and make the expansion too expensive. In the screening study, it was found that the hydrocracker was unable to handle a 15% expansion in feed rate because of the following:

- The heaters for the reaction and fractionation sections would be too small.
- The space velocity would be too high for the existing reactors.
- The fractionation tower and debutanizer tower would have severe flooding.

By applying process and energy integration methodology, the bottlenecks are overcome with acceptable capital costs. The results are summarized as follows.

9.7.5.1 Removing the Heater Bottlenecks To do this, the network pinch method (Zhu and Asante, 1999) was applied to the existing heat exchanger network. The modifications for reducing the reactor charge heaters and the fractionator heater were identified. As a result, installation of four heat exchangers (A–D) was required that use reactor effluent heat for preheating the reactor feed and the fractionation feed (Figure 9.29). These new exchangers can reduce the total heater duty by 20% or 100 MM Btu/h with a payback of less than 2 years based on energy saving alone. Thus, the need of revamping existing heaters was avoided. At the same time, an opportunity of using a liquid expander for power recovery was identified. In the existing design, HP liquid at 2200 psig was throttled to around 450 psig through a valve.

9.7.5.2 Removing the Reactor Bottlenecks The hydrocracking unit was designed in the 1970s. It was not surprising to find that the existing reactor internals experienced poor distribution. With installation of better mixing and distribution devices, gas and liquid distribution was improved and the existing reactors could handle high space velocity from feed rate increase (Figure 9.30). A new catalyst was also used to deal with diesel cold flow property issues.

9.7.5.3 Removing the Fractionation Bottlenecks In evaluating both the fractionation tower and debutanizer tower, the simulation model predicted too high liquid loading and thus a downcomer backup flooding. To avoid this, UOP ECMD trays were considered for replacement because the ECMD trays feature multiple downcomers and equalized loadings for downcomers. This feature could mitigate downcomer backup flooding and thus allow towers to accommodate more than 15% feed rate increase.

During evaluation, a new opportunity was identified. In the current operation, a 58 °C overlap exists between the recycle oil and the gas oil product, which corresponds to a 340 °C true boiling point (TBP) cut point. Newly designed units will typically have a TBP cut point of 380–410 °C and a gap of 10–30 °C. The poor separation in the fractionation column results in approximately 13 wt% of the recycle oil being in the gas oil boiling range in order to meet the specifications for the gas oil product. Due to the large amount of gas oil range material that is being recycled, the distillate yield is reduced from overconversion of the gas oil and excess hydrogen is consumed. In addition, a higher reactor severity is needed, which leads to a higher activity catalyst (or shorter catalyst life) and higher temperatures compared to a unit with better separation. Also, more heavy polynuclear aromatic (HPNA) components are made at the higher reactor severity. Recovery of gas oil from the recycle oil (fractionation column bottom) could worth $20 MM per year for the refinery margin. A possible solution was to add several trays in the bottom. The good news is that there was enough space available to accommodate a few trays in the column bottom section (Figure 9.30).

9.7.5.4 Modification Summary In summary, the modifications for the process unit include the following:

- Four new heat exchangers to reduce both the reaction fractionation heaters' duty more than 20%. Thus, the heater bottlenecks were resolved.

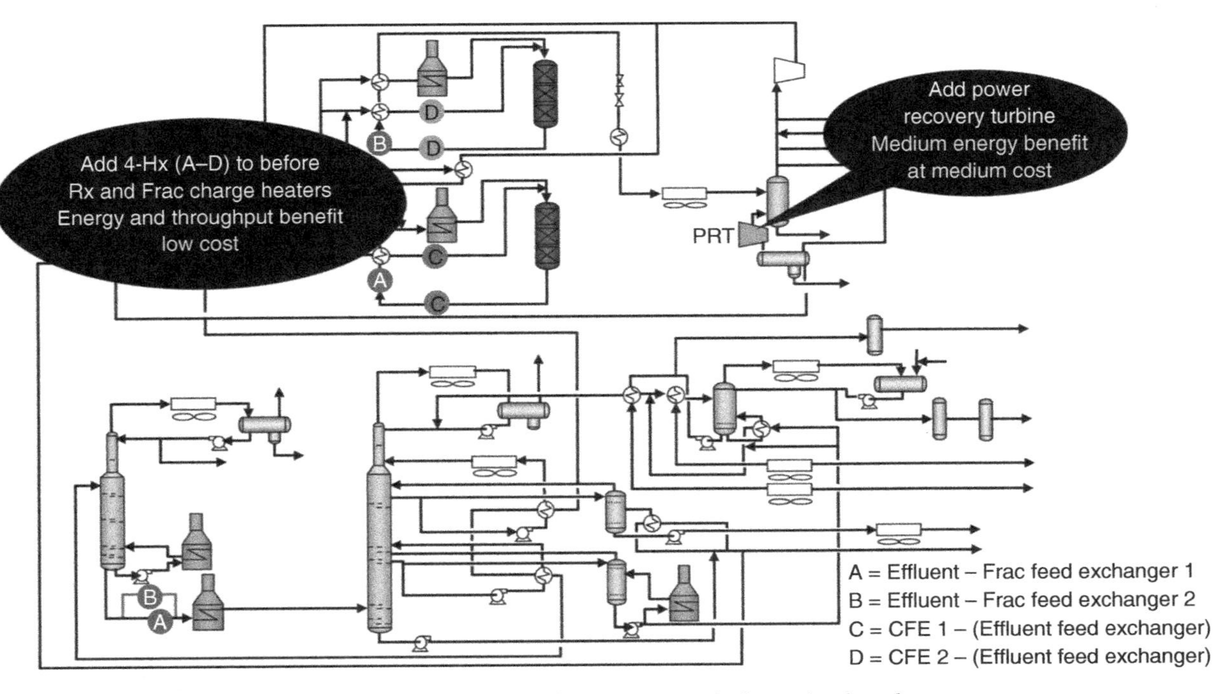

FIGURE 9.29. Energy-saving projects to remove the heater bottlenecks.

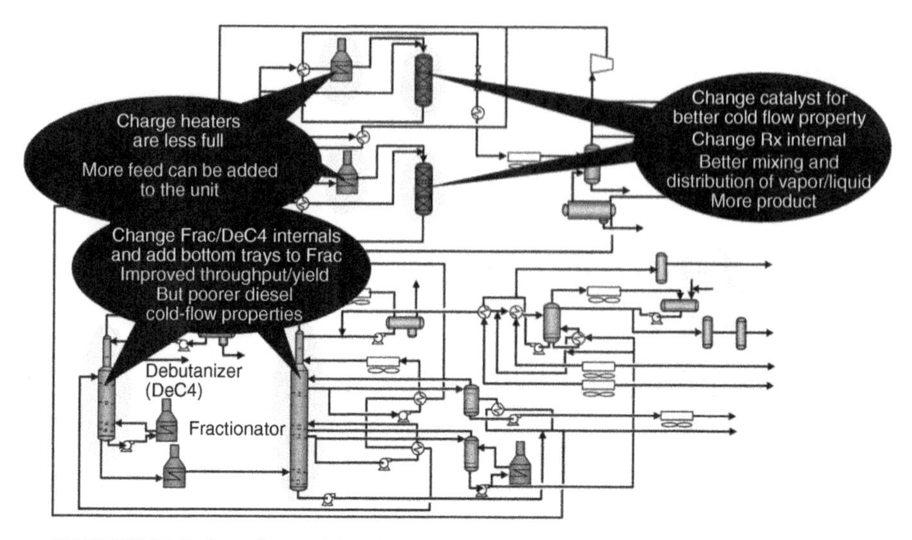

FIGURE 9.30. Reaction and fractionation projects to remove the process bottlenecks.

- Reactor internals with mixing and distribution devices to avoid addition of new reactor or major revamp of existing two reactors.
- New catalyst to deal with diesel cold flow property issues.
- Replacement of part of tower internals for fractionators and debutanizer to mitigate liquid loading to avoid severe downcomer flooding.
- Adding several trays in the bottom section of the column to reduce the gas oil in the recycle oil from 13 wt% to less than 4 wt%.
- The power recovery turbine was not selected in this revamp package as it was not required for feed expansion. But it could be considered as a major energy-saving project in future.

The overall study identified energy savings and allowed the desired throughput increase to be attained with minimal capital costs because expensive modifications were avoided. The synergy between energy savings and process technology know-how was critical to achieving these results.

9.8 SUMMARY OF POTENTIAL ENERGY EFFICIENCY IMPROVEMENTS

Table 9.1 combines all of the potential saving opportunities to provide a perspective on the level of benefits that could be achieved for a typical refinery of 100,000 barrels per day (BPSD) by adopting the comprehensive process integration methodology. Recent studies (Sheehan and Zhu, 2009) indicate that typical benefits of 15% energy reduction are expected and may rise to 30% for refinery plants operating in the fourth quartile in energy efficiency. The resulting energy cost-saving opportunity is in the range of \$11–23 MM per year. The carbon credits associated with the reduction in

TABLE 9.1. Typical Energy Saving from Different Categories of Opportunities

Saving Opportunities	Energy Improvement (%)	Energy Saving (MM$/year)	CO_2 Reduction (kMT/year)
Group 1: Get basics right	3–5	2.2–3.7	36–60
Group 2: Improved operation and control	2–5	1.5–3.7	24–60
Group 3: Improved heat recovery	5–10	3.7–7.5	60–120
Group 4: Advanced process technology	3–7	2.2–5.2	36–84
Group 5: Utilities optimization	2–3	1.5–2.2	24–36
Total	15–30	11–23	180–360

Basis: 100,000 BPSD refinery; energy price: $6/MMBtu; 330 operating days/year

greenhouse gas emissions range from 180,000 to over 360,000 metric tons per year. For US oil refining capacity at 9 million barrels per day, the energy cost-saving opportunity could be in the range of $1–2 billion per year, and greenhouse gas emissions reductions could be in the range of 16–32 million metric tons per year.

REFERENCES

Dhole VR, Linnhoff B (1993) Distillation column targets, *Computers and Chemical Engineering*, **17**(5/6), 549–560.

Glavic P, Kravanja Z, Homsak M (1988) Heat integration of reactors; 1 Criteria for the placement of reactors into process flowsheet, *Chemical Engineering and Science*, **43**, 593.

Hoehn RK, Bowman DM, Zhu XX (2013) Process for recovering hydroprocessed hydrocarbons with two strippers, U.S. Patent 8,940,254, 2015.

Hoehn RK, Bowman DM, Zhu XX (2014) Apparatus for recovering hydroprocessed hydrocarbons with two strippers in one vessel, U.S. Patent 8,715,596.

Linnhoff B, Townsend DW, Boland D, Hewitt GF, Thomas BEA, Guy AR, Marsland RH (1982) *User Guide on Process Integration for the Efficient Use of Energy*, IChemE, Rugby, UK.

Meyers, RA (2004) Chapter 3, *Handbook of Petroleum Refining Processes*, 3rd edition, McGraw-Hill.

Sheehan BP, Zhu XX (2009) Improving energy efficiency, *PTQ*, **Q2**, 29–37.

Tsai R, Zhu XX; Steacy P (2016) Dividing-wall column screening guidelines and applications, AICHE Annual Meeting, November, San Francisco.

Zhu XX (2014) *Energy and Process Optimization for the Process Industries*, Wiley/AIChE.

Zhu XX, Asante NDK (1999) Diagnosis and optimization approach for heat exchanger network retrofit, *AIChE Journal*, **45**, 7, 1488–1503.

Zhu XX, Maher G, Werba G (2011) Spend money to make money. *Hydrocarbon Engineering*, September Issue.

10

DISTILLATION COLUMN OPERATING WINDOW

10.1 INTRODUCTION

Distillation column tray design determines the geometry of the column such as column diameter, tray type and spacing, and downcomer layout, which defines the column operating capability or operating window. Understanding of the capability diagram helps engineers to know the operating limits for their columns and identify opportunities to improve column performance.

What criteria should be used for assessing a distillation tower? How does one know if a tower operates under stable operation or near the best performance? What turnkey options are available to operators in order to simultaneously optimize product separation and energy use? How can the best performance be sustained? These are the questions that engineers constantly ask and these are the focus of this chapter.

As distillation is a complex topic, this chapter intends to cut through the maze straight to the most important matters such as the concept of feasible operating region, key operating parameters, and optimizing distillation from design and operation. For effective understanding, simplified explanation and symbolic terms are adopted without sacrificing fundamentals. This chapter is designed to focus on these aspects supported by comprehensive examples. The detailed and excellent discussions for all aspects related to column design and operations can be found in Kister (1990, 1992).

10.2 WHAT IS DISTILLATION?

A complex distillation column (Figure 10.1) is designed and operated for separating products from feed(s) and some of the products may need to go through further

Hydroprocessing for Clean Energy: Design, Operation, and Optimization, First Edition.
Frank (Xin X.) Zhu, Richard Hoehn, Vasant Thakkar, and Edwin Yuh.

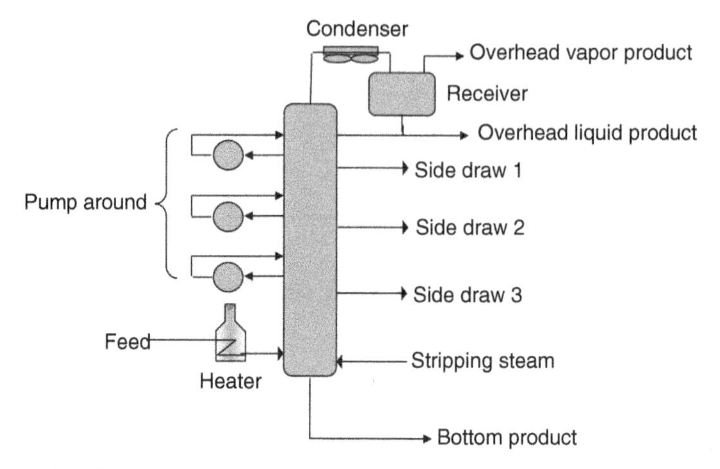

FIGURE 10.1. A complex configuration of a distillation column.

separation downstream to meet product specifications. Energy is provided in the form of reboiling or feed heating or steam injection as the driving force for desirable distillation. Pumparounds are used to remove excess heat within the column to achieve required vapor and liquid loading balances locally in the column. Although every distillation column is different in feeds/products hence resulting in different design and operation, the basic elements are common. Good understanding of major features of these elements will provide a sound basis to know what to watch (key indicators), what knobs to turn (operating parameters) and directional changes for improvements (effects on process and equipment performance) within operating limits (process and equipment constraints).

Distillation occurs in a tower due to the relative volatility between light and heavy components. Vapor flows upward at relatively higher temperatures and meets liquid that flows downward at relatively lower temperatures. On the tray, heat transfer takes place between vapor and liquid. As a result, vapor becomes cold, which makes heavy components separated out from vapor and join liquid traveling down. In contrast, liquid is heated up and thus light components are flashed out and join vapor traveling upward. This is what happens in mass and heat transfer between vapor and liquid on each tray. Consequently, two things happened. First, vapor and liquid approach equilibrium on each tray. Second, vapor becomes lighter and colder traveling upward while liquid becomes heavier and hotter traveling downward. The limit for the mass transfer between vapor and liquid on each tray is set by the composition equilibrium, which depends on the relative volatility. With a certain number of equilibrium stages, a distillation column accomplishes its operating objective that light and heavy products are separated and produced from the overhead and the bottoms, respectively. The ease or difficulty of using distillation to separate the more volatile components from the less volatile components in a mixture is denoted by the relative volatility. For a mixture with low (high) relative volatility, a tall (short) tower with many (fewer) number of stages is required.

It is often that distillation columns do not perform as they should, and operation performance deviates from design. Energy usage is often not optimized during operation when feed conditions and product specifications change due to operational requirements. In some cases, product quality could be better than specifications. We call this product quality giveaway as it consumes more energy than required. In other circumstances, distillation columns are operated in abnormal conditions such as flooding. Under abnormal operation, distillation efficiency suddenly drops, leading to little or virtually no distillation taking place. Therefore, the topic of distillation efficiency needs to be well understood before discussing the operating window.

10.3 WHY DISTILLATION IS THE MOST WIDELY USED?

Distillation technology is 100 years old as it was developed by Jean-Baptiste Cellier-Blumenthal in 1813. Distillation is widely used in process industry for over 90% of all separations. It is the technological and operation maturity as well as the economic viability that makes distillation a dominant separation method in the industry.

The technology is well understood by the industry. As expected, the technology maturity can be correlated well with commercial applications. Beyond technology maturity, distillation requires small equipment count and can effectively deal with large variations in feed rate, feed compositions, and product splits. Furthermore, distillation is easy and cheap to scale up. As a result, distillation processes can be readily designed, operated, and scaled up to commercial capacity from laboratory data. Therefore, distillation becomes an industrial benchmark for other separation methods to compare against.

When selecting among different technologies, the criteria used to make judgment include equipment count, scale up, design and operation reliability, process conditions requirement, physical size limitations, and energy and capital investment.

- *Small equipment count* Compared to other separation processes, distillation system is relatively simple. It requires one column, overhead condensing, and bottom reboiling. In contrast, processes based on mass separating agents (e.g., absorbent) require additional equipment for recovery of mass separating agents. If comparing with rate-govern separation processes such as crystallization, it involves solid phase and thus additional equipment are needed such as filtration, washing, and drying to obtain the final products. These additional equipment add process complexity and capital investment, which may well be justified due to the nature of feed composition, physical properties, and product specifications.
- *Economics of scale* Trays can be staged easily, and it is possible to have 100 trays in a distillation tower. The scaling factor for throughput increase is 0.6. This implies that doubling the column capacity increase the column capital cost only by a factor of 1.5. Thus, this makes distillation have strong economics of scale for large commercial applications.

  For most other separation processes, the scaling factor is higher than 0.6, which makes scaling up expensive. Use membrane module as an example. To increase capacity, more membrane modules must be added in parallel.
- *Design and operation reliability* The design maturity is one of the most important factors for successful commercial applications that concern about making

desirable products in all possible operating scenarios. Failure in achieving this goal would be fatal to business. Distillation can achieve this goal as distillation tower design is mature and design performance can be guaranteed in operation.

Furthermore, operation of distillation tower is better understood than other separation processes. With proper instrumentation and control systems in place, distillation tower can be operated to meet design performance and make on-spec products.

However, distillation has its own limitations due to the nature of distillation technology. Thus, the question becomes: what prevents use of distillation in some applications? The following discussions shed light on this direction.

- *Low relative volatility* Distillation is not effective to separate key components with very low relative volatility as it requires very large column with too many stages leading to prohibitive capital cost. When relative volatility of key components is less than 1.1, other separation technology should be considered such as extractive distillation, solvent extraction, and adsorption.
- *Extreme conditions* Temperature and pressure conditions affect distillation operation and capital costs significantly. For example, when a distillation column temperature is less than −30 °C, refrigerant is needed for overhead condensing and cost of a refrigerant system is very expensive. When the temperature is higher than 230 °C, fired heater may be required for bottom reboiling. In addition, under high temperature, chemical components become thermally unstable and undesirable reactions may occur. This could lead to formation of unwanted by-products and other operational problems.

  When column pressure is too low, the column dimension and overhead pump size increase rapidly. On the other hand, when column pressure is too high, column cost escalates significantly due to the column thickness required.
- *Feed compositions* Feed composition could also present problems for distillation. If only a small amount of component is to be separated from the feed, energy and capital costs for the distillation column could become excessive. For example, obtaining a product of a high boiling but with a low concentration less than 10% of the total feed, energy consumption could be excessive to boil up all the lighter components and capital cost increases rapidly as a large column is required to deal with high vapor traffic. Other separation processes such as stripping, extraction, and selective adsorption should be considered.

  Stripping is often used to remove volatile components via use of stripping gas or steam while extraction is applied for recovery of low volatility components. At very low concentration (<1000 ppm), extraction could be highly uneconomic as removal of solvent residues from the raffinate and recovery of solvent from the extract becomes expensive. The same is true for stripping. In these cases, selective adsorption could be the choice.
- *Energy cost* Distillation requires high energy cost when column is designed in isolation; it can be reduced by heat integration among distillation columns and between columns and background processes. The topic of column heat integration is discussed in detail in chapter 9.

Based on the above criteria, distillation is selected as the production fractionation method for hydroprocessing processes including hydrocracking and diesel hydrotreating.

10.4 DISTILLATION EFFICIENCY

The ultimate goal of a distillation tower is separation of products and naturally distillation efficiency becomes the key performance metric. A tower, if properly designed, can achieve 10% higher distillation efficiency. In operation, the tower operated with better distillation efficiency requires less energy use.

Many different measures of efficiency have been developed. Let us look at the two commonly used measures, which are the staged based Murphree tray efficiency (Murphree, 1925) and overall efficiency.

The Murphree tray efficiency is defined as

$$\eta_{\mathrm{M}} = \frac{\text{change in vapor for actual stage}}{\text{change in vapor for equilibrium stage}}. \tag{10.1}$$

The above expression can be better explained using Figure 10.2. The denominator in equation (10.1) is the vertical distance or vapor composition difference from the operating line to the equilibrium curve ($\overline{AC}$), while the numerator is the vertical distance or vapor composition differencc from the operating line to the actual concentration curve ($\overline{AB}$). For example, $\eta_{\mathrm{M2}} = \overline{AB}/\overline{AC}$ represents the efficiency for stage 2. The efficiency for other stages can be determined similarly.

The overall tower efficiency is defined as the ratio between the number of theoretical stages and the actual number of stages required for the separation as

$$\eta_{\mathrm{o}} = \frac{N_{\mathrm{eq}}}{N_{\mathrm{act}}}. \tag{10.2}$$

As an example for illustration, McCabe–Thiele diagram (McCabe and Thiele, 1925) in Figure 10.2 indicates 12 actual stages required in comparison with 8 theoretical stages in the tower. Partial condensers and partial reboilers are counted in both the theoretical stages and actual stages. Thus, the overall tower efficiency is 72%.

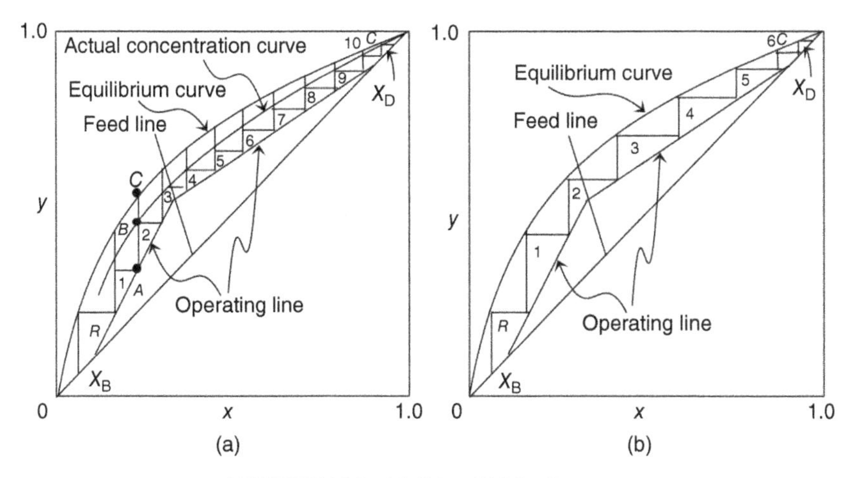

FIGURE 10.2. McCabe–Thiele diagram.

The overall efficiency lumps everything that happens in the column into one value. Based on the assumption of constant molar overflow and constant value of η_M for trays in each distinctive section in a tower, Lewis (1936) developed a relationship between the Murphree tray efficiency and overall efficiency, which is expressed as

$$\eta_o = \frac{\ln[1 + (\lambda - 1)\eta_M]}{\ln \lambda}, \tag{10.3}$$

where

$$\lambda = m\frac{V}{L}. \tag{10.4}$$

Equation (10.3) applies separately to rectifying and stripping sections as the G_V/G_L ratio is different between these two sections. In the rectifying section, vapor rate is higher while liquid rate is lower compared with those the in stripping section. But in each section, equation (10.3) is based on the assumption of straight operating and equilibrium lines and constant G_V/G_L ratio and η_M from tray to tray.

For complex towers with pumparounds, side condensers or side reboilers, side product draws, a tower can be divided into more than two sections, and overall efficiency in equation (10.3) is then applied to each section. This efficiency calculation method requires a tower simulation to give internal G_L/G_V distributions from stage to stage, which becomes the basis to determine the sections, each of which features near constant molar flows between stages. For the example in Figure 10.2, three actual stages are required in the stripping section versus two theoretical stages. Thus, the efficiency for this stripping section is 67%. In contrast, there need seven actual stages in the rectifying section versus four theoretical stages, which results in the section efficiency of 57%. The overall tower efficiency is 60%. For industrial towers, stage efficiency is typically around 60–75%; it is uncommon to have stage efficiency as low as 50%.

Many studies of efficiency have been conducted on both sieve and valve trays in industrial and academic laboratories. Most academic studies are performed in towers far too small in size to be useful for industrial applications. The industrial efficiency data are proprietary. The best example is the Fractionation Research Inc. (FRI) data, which are only available to FRI consortium members.

The simplest approach is to use a correlation to determine overall tower efficiency. The O'Connell (1946) correlation shown in Figure 10.3 is the standard of the industry (Kister, 1992) for industrial tower efficiency.

Figure 10.4 shows a typical trend of tower efficiency dependent on the balance of vapor and liquid rates. In the middle of the efficiency curve corresponding to stable operation, there is a relatively flat region although with marginal variation. Trays with good turndown features such as valve tray compared with sieve tray have wider flat or stable operating region. On the either side of the curve, efficiency drops off dramatically. Efficiency declines under low feed rate corresponding to turndown operation and falls off the cliff when liquid dumping occurs. On the other hand, efficiency reduces at excessive entrainment and plummets when flooding occurs. Optimization in design and operation tends to push the tower toward the boundary of stable operation. Understanding of these controlling mechanisms can provide insights into how to optimize tower design and operation while achieving stable operation.

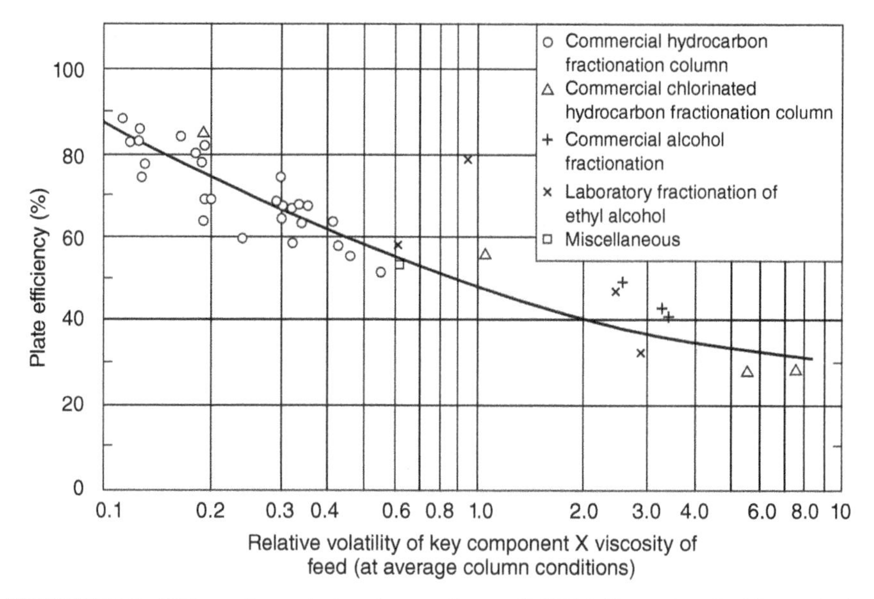

FIGURE 10.3. O'Connell correlation. Source: Oçonnell (1946). Reproduced with permission of AIChE.

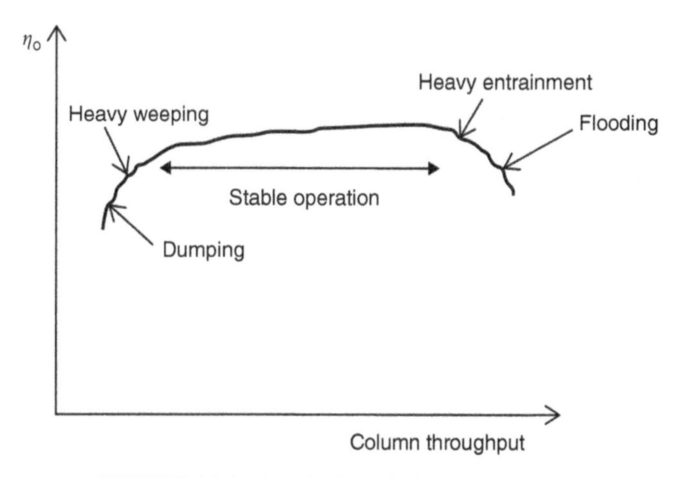

FIGURE 10.4. A typical trend of tower efficiency.

10.5 DEFINITION OF FEASIBLE OPERATING WINDOW

How to be sure if a tower is designed and can be operated to give a feasible operation with a satisfactory distillation? A valuable tool for this purpose is capability diagram or feasible operation window (Summers, 2004). This operation window is bounded by tray capability limits (Figure 10.5). Any specific operating scenario represents an operating point within this operating window. Any variation in tower feed rate, product rates, and quality and operating conditions such as feed temperature, tower

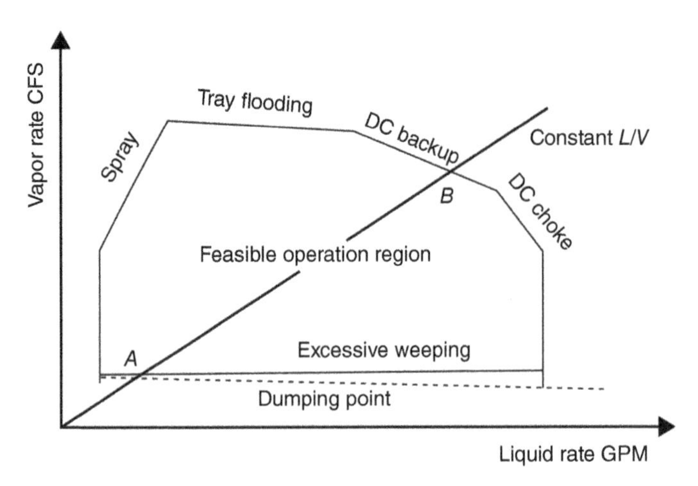

FIGURE 10.5. Capacity limits for distillation tower.

pressure, and reflux rate will move to a different operating point in the window. For simplicity of illustration, it can be said that the design and operation of a tower is all about *balancing between vapor and liquid loadings* with an utmost goal of achieving desirable separation at minimal operating cost.

10.6 UNDERSTANDING OPERATING WINDOW

The mechanisms of capability are described qualitatively here based on the concept of relative momentum between vapor and liquid while quantitative discussions will be given afterward. To understand the main modeling parameters of interest, Figure 10.6 is used for illustration purpose. Adjacent trays are apart with the tray spacing H_S. Vapor flows upward through the perforations with the hole diameter d_h. Froth is generated due to the momentum exchange between vapor and liquid with the froth height h_f. To measure the equivalent liquid height for the froth, the clear liquid height h_c is defined for modeling purpose, which is the height that the froth level h_f could collapse in the absence of vapor flow. From the froth level, vapor continues to flow upward and creates a two-phase layer with a height of h_v. In this vapor region, vapor is in a continuous phase but containing a small portion of liquid in the form of fine droplets. The concept of *relative momentum* between vapor and liquid is the key to understand the operating window (Lockett, 1986; Bennett and Kovak, 2000).

The liquid momentum is mainly generated by gravitational force represented by liquid height or liquid holdup, h_c, on tray deck, while the vapor momentum is generated by the buoyant force due to vapor flowing upward through perforations, which is indicated by $u_h d_h$. The relative momentum is largely defined by the ratio of $h_c/u_h d_h$. In the case of higher ratio of $h_c/u_h d_h$, small u_h and/or small d_h. In this case, vapor exchanges momentum sufficiently with liquid, which slows down the vapor momentum and prolongs the vapor residence time within the liquid zone. The consequence is better interactions between vapor and liquid, resulting in better distillation. As liquid loading is responsible for liquid height h_c while vapor loading for vapor velocity u_h,

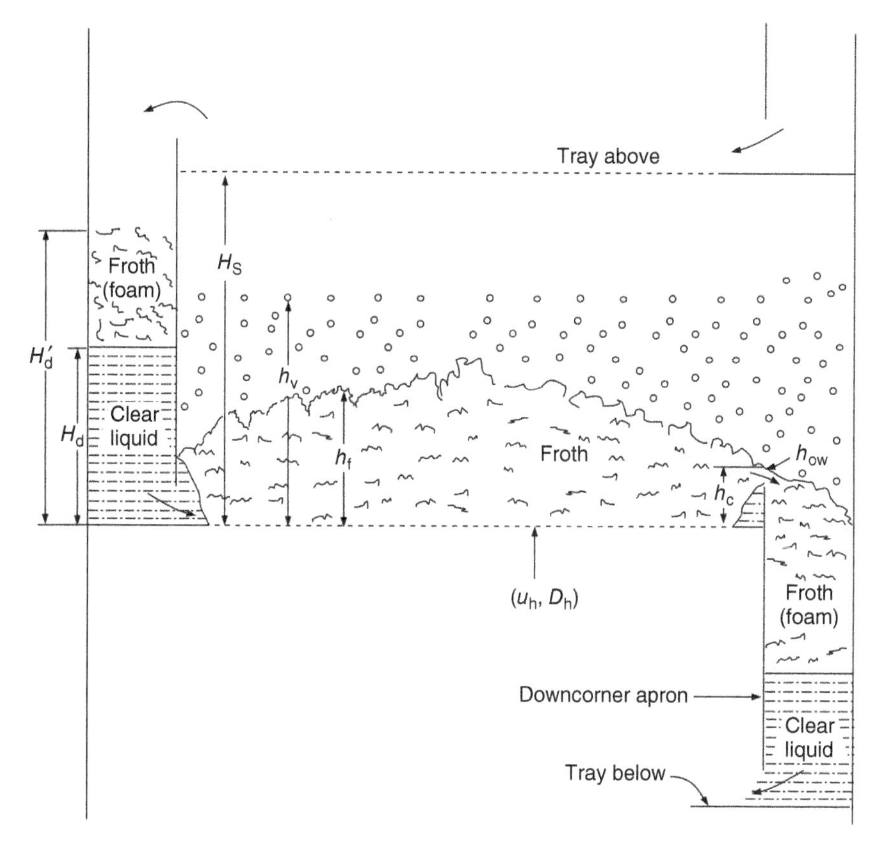

FIGURE 10.6. Vapor–liquid flow structure on tray deck.

thus a combination of high vapor and liquid loading with small perforation size d_h promotes mass transfer between vapor and liquid and enhances distillation. On the other hand, in cases when the combination of (h_c, u_h, d_h) is inappropriate, tower could be under unstable operation.

The first unstable operation of interest is *spray* flow, which occurs under a high vapor but low liquid loading on the tray. One could visualize what happens in spray operation: high vapor momentum causes liquid to blow off and reach the tray above, which contaminates the tray above and undoes separation. Under spray operation, with severe entrainment passing through the tray above, distillation efficiency reduces sharply and can drop half of the normal operation.

Spray flow occurs under very small $h_c/u_h d_h$ ratio when vapor momentum is much higher than liquid momentum and there is very thin layer of liquid h_c. In this case, vapor does not adequately exchange momentum with the liquid and maintains high momentum entering the two-phase region. The resulting spray flow carries fine liquid droplets to the tray above causing excessive entrainment (Figure 10.7). The spray flow can be analogous to fluidization or blowing effect. Spray flow should be avoided in design and operation. As preventing spray is equivalent to increase the ratio of $h_c/u_h d_h$, two effective ways of spray prevention can be devised. One is to increase liquid loading using picket fence weir. The other is to reduce the hole diameter d_h

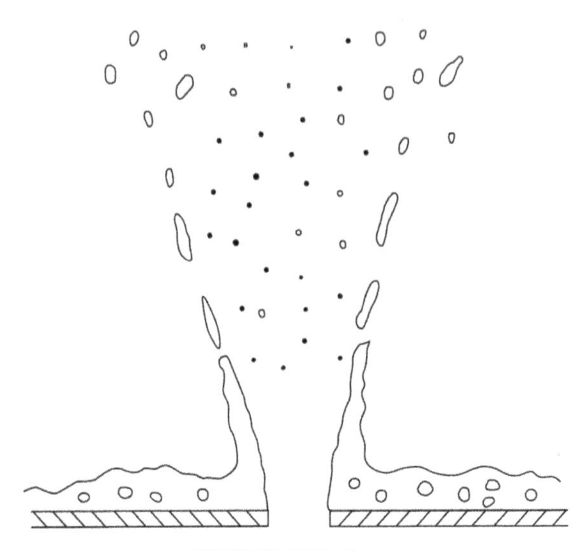

FIGURE 10.7. Spray.

while increasing the number of perforations for the same hole area. If spray flow cannot be sufficiently avoided by using these two methods for tray column, packed column could be the choice. Under spray regime, valve trays can perform better than sieve trays in both capacity and efficiency.

The second abnormal operation is *tray flood* when the froth level $h_f > H_S$ and thus liquid reaches the tray above. The tray flood occurs when both liquid and vapor loadings on the tray are high and thus sufficient vapor momentum generates higher froth level h_f. When $h_v > H_S$, entrainment occurs as vapor carries liquid droplets to the tray above. Tower can still operate in a stable manner under significant amount of entrainment if downcomer can handle the additional liquid loading. However, tray distillation efficiency drops depending on the amount of entrainment. If vapor rate further goes up and increased vapor momentum pushes the froth level to reach the tray above ($h_f > H_S$), excessive entrainment occurs and thus distillation efficiency falls off the cliff. As this happens on the tray deck, it is named tray flood in differentiation of downcomer flood. The tower may not be able to achieve stable operation under severe tray flood. In the literature, it is commonly called *jet flood*, which is a misconception as the name of jet flood is more suited for spray. To avoid misunderstanding, tray flood is used in place of jet flood in this book.

Clearly, tower diameter and tray spacing H_S are the two major parameters in design affecting tray flood. A large tower diameter reduces vapor velocity and liquid holdup (h_c), while high tray spacing increases liquid settling space. In operation, reboiling duty is the major parameter to be adjusted to avoid tray flood.

The third unstable operation is *downcomer flood* when the liquid stacks up in the downcomer and the liquid froth in the downcomer reaches the tray above ($H'_d > H_S$). Downcomer flood is caused by high hydraulic head loss along the liquid flow path. Downcomer flood could be *downcomer backup* and/or *choke*. Downcomer backup occurs when downcomer cannot allow additional liquid to freely flow out of the downcomer when there is high hydraulic resistance along the liquid pathway (Figure 10.8).

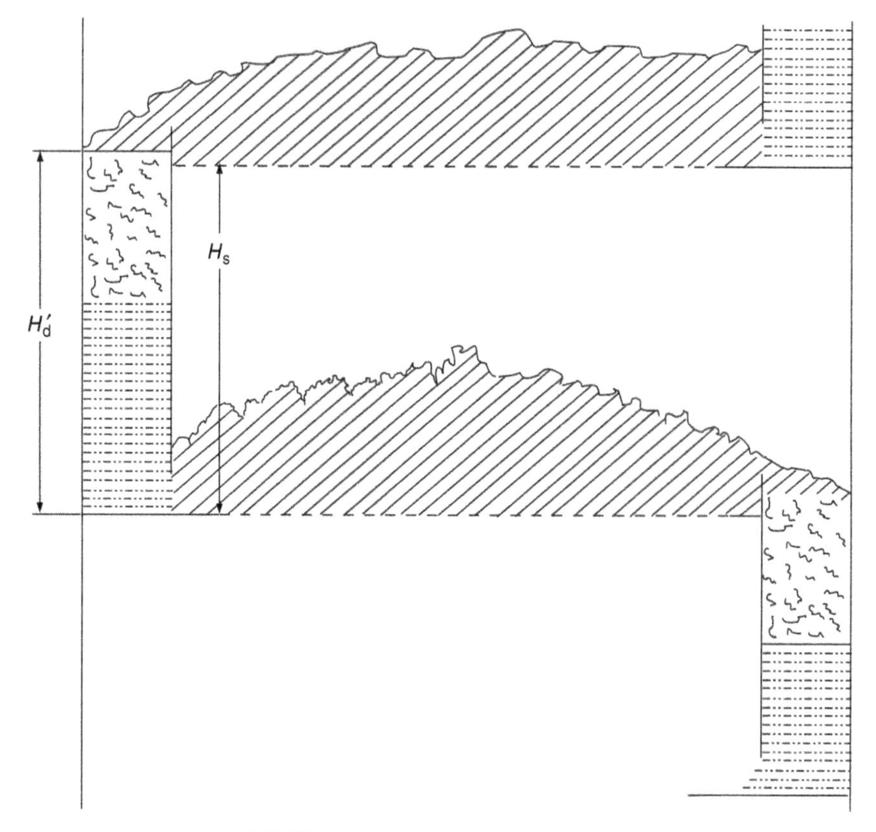

FIGURE 10.8. Downcomer backup flood.

In a different mechanism, downcomer choke happens when large froth crest gets stuck in the downcomer mouth and blocks the liquid to flow over the outlet weir (Figure 10.9).

In both downcomer backup and choke cases, downcomer liquid inventory increases and downcomer liquid backs up until the downcomer froth level reaches the tray above ($H_d' > H_S$). This phenomenon is called downcomer flood. When downcomer flood occurs to any tray, the whole tower will be flooded very quickly. A tower under downcomer flood provides virtually no distillation. In contrast, under tray flood, liquid can still leave the tower and tower could still operate if the control system allows although distillation efficiency suffers. Downcomer flood can be prevented in design by providing adequate downcomer area and clearance underneath the downcomer and minimizing tray pressure drop. Reducing reflux rate in operation could be effective in avoiding downcomer flood in operation.

The fourth unstable operation is *weeping*, which occurs at too low vapor loading when vapor momentum cannot hold liquid gravitational force. Under weeping, some of liquid falls down through perforations and short circuits the flow path. Thus, distillation suffers. In the worst case, all liquid flows through perforations; this is called *dumping*. A tower can still operate under a certain amount of weeping although at reduced distillation efficiency; but the operation is very unstable under dumping.

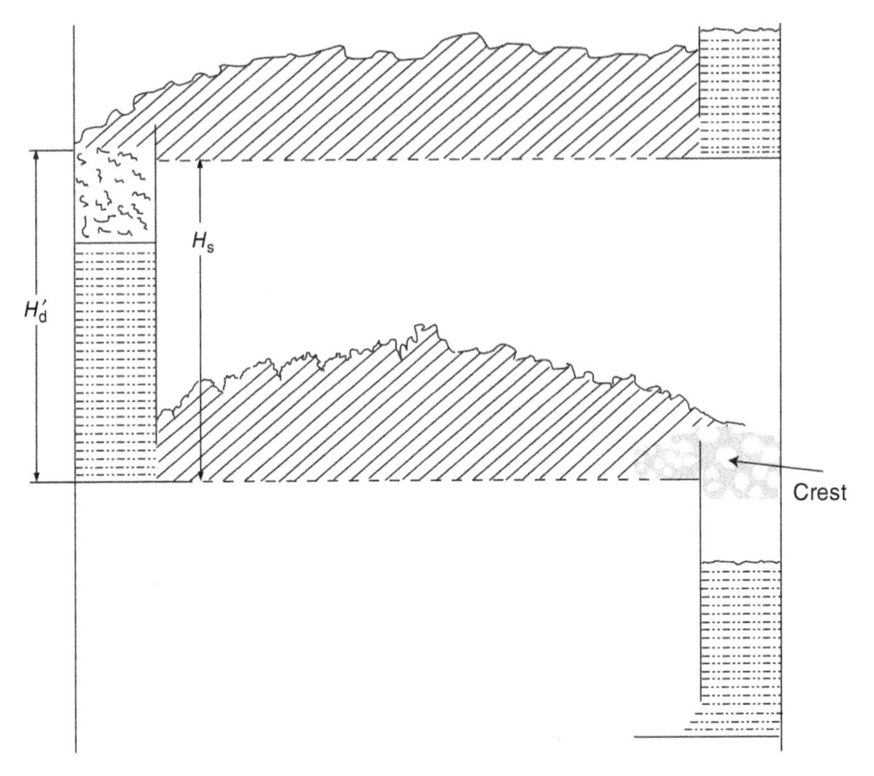

FIGURE 10.9. Downcomer choke.

Weeping usually happens at turndown operation when the feed rate is reduced significantly from the nominal rate. For the need of high turndown operation, selection of valve trays over sieve trays should be considered. Typically, valve trays can achieve 3:1 and higher turndown in comparison with 2:1 turndown by sieve trays.

Having discussed unstable operations, you may like to know what takes to achieve *stable operation*. Under stable operation, vapor can travel upward without excessive liquid entrainment while liquid travels down through the tray flow path effectively. This is achieved by two kinds of dimensions. The first one is the tray dimension that allows sufficient momentum exchange from vapor to liquid on tray deck. The tray dimension includes tower diameter, tray spacing, tray flow path, hole diameter, and area. The other is downcomer dimension that includes downcomer flow areas, clearance, weir type, and length. The tray and downcomer dimensions together are called tower dimensions, which are determined in design. In operation, a proper vapor and liquid loading balance is the basis for ensuring stable operation for a given tower design. A balanced vapor and liquid loading is indicated by the liquid and vapor (G_L/G_V) ratio. Feed rate and reboiling duty are the two major operating parameters affecting the G_L/G_V ratio. Reflux ratio is a dependent parameter on reboiling duty as reboiling generates reflux rate. A preferred stable operation is to locate the operating point near the upper right-hand side of the operating window (Figure 10.5). This is because such an operation features high vapor and liquid loadings resulting in sufficient encountering of vapor and thus high distillation efficiency. A tower designed based on this point could achieve a lower tower diameter and hence lower capital for a given separation objective.

But how does one design a column away from unstable conditions in the first place and then operate it within the operating window when process conditions vary?

10.6.1 Spray

As mentioned, spray occurs under large vapor momentum ($u_h d_h$) but small liquid gravitational force (h_c). By considering density difference in vapor and liquid, the equation for spray factor describing the relative momentum is developed by Lockett (1986) as

$$S_p = \frac{h_c}{u_h d_h}\left(\frac{\rho_l}{\rho_v}\right)^{0.5}. \tag{10.5}$$

Since the Lockett's correlation was developed for sieve tray, Summers and Sloley (2007) extended it for valve tray by introducing a constant K to the correlation

$$S_p = K\frac{h_c}{u_h d_h}\left(\frac{\rho_l}{\rho_v}\right)^{0.5}, \tag{10.6}$$

$K = 1$ for sieve trays and 2.5 for movable and fixed valve trays. Introduction of K to the spray equation manifests that valve tray has a better capability in suppressing entrainment than sieve trays. This is achieved by the mechanism of vapor entering the tray horizontally with valves, which reduces the entrainment significantly at low liquid loadings. According to Lockett (1986) and Summers and Sloley (2007), spray factor S_p in equation (10.6) must be larger than 2.78 to avoid spray regime.

The relationship of vapor and liquid under spray is observed by Sakata and Yanagi (1979) for sieve trays: as the liquid rate reduces beyond a certain amount corresponding to weir loading of 2 gpm/in. (gpm is gallon per minute), vapor rate must reduce to maintain the same entrainment rate. This reducing trend of both vapor and liquid rates under very small weir loading defines the spray phenomenon. This trend is different from the tray flood phenomenon under which vapor rate increases as liquid load reduces.

10.6.2 Tray Flooding

As tray flooding occurs when vapor loading is too high, let us start by defining the column vapor load as

$$V_L = \text{CFS}\sqrt{\frac{\rho_v}{\rho_l - \rho_v}}, \tag{10.7}$$

CFS is vapor flow in ft^3/s. ρ_v and ρ_l are densities for vapor and liquid, respectively. Vapor load represents the vapor flow under relative density difference of vapor and liquid, which contact each other on tray.

To describe the vapor load V_L over net area A_N ($=A_T - A_d$) on tray, let us define

$$C = \frac{V_L}{A_N} = \frac{\text{CFS}}{A_N}\sqrt{\frac{\rho_v}{\rho_l - \rho_v}}. \tag{10.8}$$

Let the vapor velocity u be expressed as

$$u = \frac{\text{CFS}}{A_N}. \tag{10.9}$$

Thus

$$C = u\sqrt{\frac{\rho_v}{\rho_l - \rho_v}}, \tag{10.10}$$

C defined in equation (10.10) is called C-factor in literature, which has the meaning of vapor capacity factor.

But how can we know the limit for vapor capacity at which column starts to flood? This question can be answered by making a force balance between vapor and liquid on tray. Assume the bulk of liquid with a volume v and the liquid gravitational force is $v(\rho_l - \rho_v)g$. At the same time, vapor flows upward with velocity of u and thus vapor momentum is $f(\rho_v u^2/2)$, where f is the friction coefficient between vapor and liquid.

When both forces are equal, the bulk of liquid stays floating on tray. If the vapor velocity is raised in any small amount above u, the bulk of liquid will be carried away by vapor. By denoting this threshold velocity as u_F, the force balance that defines the threshold condition for the occurrence of flooding can be expressed as

$$f\frac{\rho_v u_F^2}{2} = v(\rho_l - \rho_v)g. \tag{10.11}$$

Rearranging equation (10.11) gives

$$u_F = \sqrt{\frac{2vg}{f}}\sqrt{\frac{\rho_l - \rho_v}{\rho_v}}. \tag{10.12}$$

As it is difficult to measure bulk volume v and friction factor f in an actual column, C_F-factor (C-Factor at flood) is used to represent $\sqrt{2vg/f}$. Let

$$C_F = \sqrt{\frac{2vg}{f}}. \tag{10.13}$$

Thus, equation (10.12) can be expressed as

$$u_F = C_F\sqrt{\frac{\rho_l - \rho_v}{\rho_v}}. \tag{10.14}$$

To predict C_F, Fair (1961) developed a C_F correlation that can be plotted as Figure 10.10. Summers (2011) converted Fair's correlation into an equation form by ignoring the part with very low liquid loadings:

$$C_F = \left(\frac{H_S}{24}\right)^{0.5}\left(0.455 - 0.0055\rho_v^2\right). \tag{10.15}$$

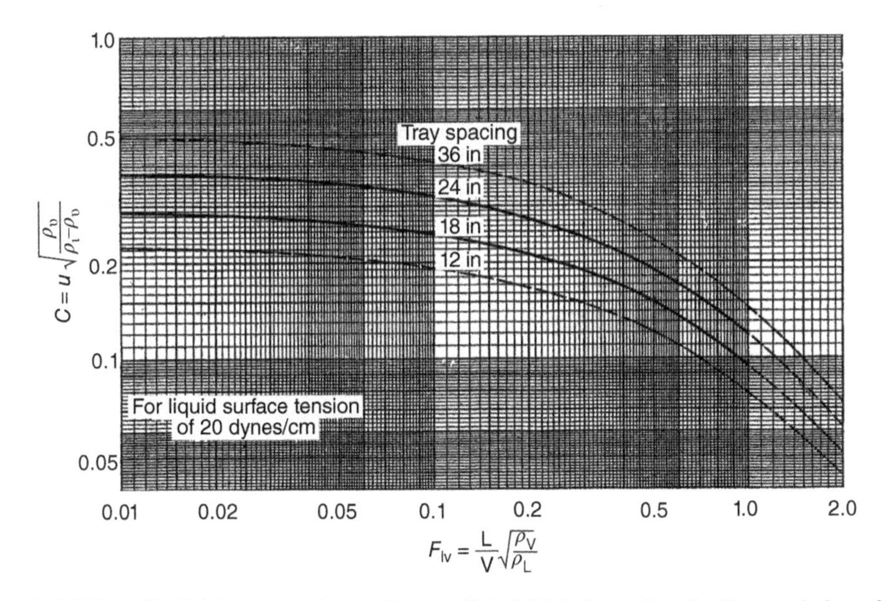

FIGURE 10.10. Fair's C_F correlation. Source: Fair (1961). Reproduced with permission of Hart Energy Publishing.

For foaming materials, a derating factor or system factor (SF) may apply to C_F as a multiplier. SF is generally related to the foaming tendency of the material. The higher the foaming tendency, the lower the SF value and vice versa. From practical point of view, the purpose that derating factors are used is to provide safety margin or overdesign in order to account for any inaccuracy associated with the empirical correlations. Excellent discussions related to derating factors can be found in Kister (1992) and typical SFs are given in Table 10.1 based on Glitsch (1974).

Replacing the C_F term in equation (10.14) with equation (10.15) yields with the expression for vapor velocity at flood as

$$u_F = \left(\frac{H_S}{24}\right)^{0.5} (0.455 - 0.0055\rho_v^2) \sqrt{\frac{\rho_l - \rho_v}{\rho_v}}. \tag{10.16}$$

With actual vapor velocity being calculated from equation (10.9), Fair's flooding limit is expressed as the ratio of actual vapor velocity and flooding vapor velocity, which

TABLE 10.1. System Factors

Nonfoaming, regular system	1.00
Fluorine systems, for example, BF_3, Freon	0.90
Moderate foaming, for example, oil absorbers, amine and glycol regeneration	0.85
Heavy foaming, for example, amine and glycol absorbers	0.73
Severe foaming	0.60
Foam-stable systems, for example, caustic regenerators	0.30–0.60

defines the close proximity of actual operation from tray flooding:

$$\%\text{flooding} = \frac{u}{u_\text{F}} \times 100\%. \tag{10.17}$$

Alternative to the above equation for flood% based on vapor velocity, Glitsch correlation (1974) based on both vapor and liquid loadings is arguably the most widely employed:

$$\%\text{flooding} = \frac{V_\text{L} + (\text{GPM} \times \text{FPL})/13{,}000}{C_\text{F} A_\text{a}}. \tag{10.18}$$

FPL is the liquid flow path length and A_a is the active area. Assume FPL $= A_\text{a}/L_\text{w}$, Weiland and Resetarits (2002) rearranged the second term in equation (10.13) algebraically to yield

$$\%\text{flooding} = \frac{V_\text{L}}{C_\text{F} A_\text{a}} + \frac{W_\text{L}}{13{,}000 C_\text{F}}, \tag{10.19}$$

where

$$W_\text{L} = \frac{\text{GPM}}{L_\text{w}}, \tag{10.20}$$

W_L is weir liquid loading in gpm/in., while L_w is the outlet weir length. Equation (10.19) states that both vapor and liquid loading contribute to vapor flood although vapor loading is much more dominant in tray flood. However, for downcomer flood, the weir loading plays a more dominant role, which will be discussed.

Clearly, a column is already in flooding when %flooding is near 100%. To prevent this from happening in the design phase, a certain safety margin for avoiding tray flooding is built into design. Hower and Kister (1991) recommend that large columns are typically designed with 80–85% flooding while 65–77% for small columns. For vacuum columns, %flooding is 75–77%.

The main factors affecting tray flooding include column diameter, tray spacing, and vapor and liquid loading. In operation, three key parameters can be used to control flooding, which are feed rate, column pressure, and reboiling duty (reflux rate depends on reboiling). When feed rate is raised or column pressure is lower or reboiling duty gets higher, more liquid is vaporized and hence vapor velocity is raised and hence the column will work closer to flooding. In contrast, reducing feed rate or raising the column pressure or lowering reboiling duty could suppress tray flooding. However, they come at a cost. Obviously, reduced feed rate will immediately impact column throughput and affects economic margin. Increasing pressure could make column more stable but have negative effect on relative volatility. To compensate with reduced distillation efficiency, higher reboiling duty is needed. On the other hand, increased reflux rate helps distillation because it reduces theoretical stages required and provides means for improved mass transfer in the column. But remember reflux rate is generated from reboiling and thus higher reflux rate requires extra reboiling duty leading to higher energy cost.

10.6.3 Downcomer Flooding

Downcomer flooding is caused by too much hydraulic resistance resulting in downcomer liquid backup. Figure 10.11 is used to show parameters of interest.

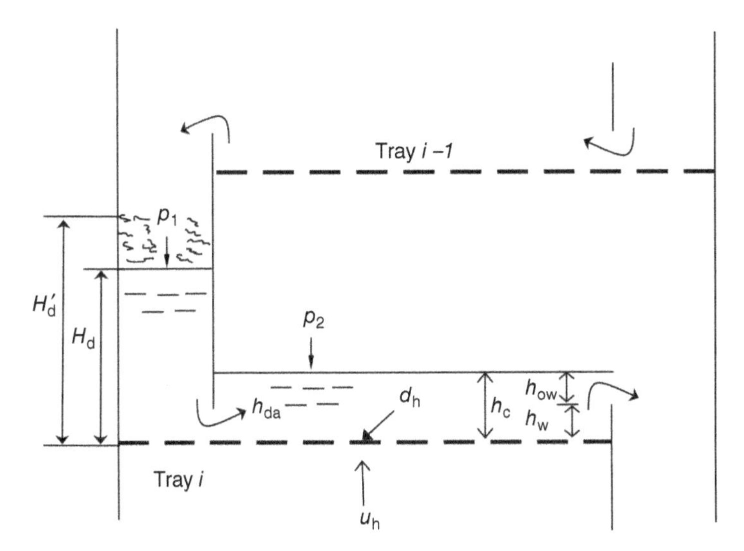

FIGURE 10.11. Key parameters for tray hydraulics (this graph is used for model illustration).

The downcomer liquid level depends on tray hydraulic balance that can be derived from the Bernoulli equation based on two liquid levels as

$$H_d + \frac{p_1}{\rho_L g} = h_c + \frac{p_2}{\rho_L g} + h_{da}, \tag{10.21}$$

where H_d is the liquid height in downcomer, h_c is clear liquid height or liquid holdup on tray, and h_{da} is the equivalent clear liquid height for pressure drop through downcomer apron. The unit for all these heights is inches.

By defining tray pressure drop as

$$h_t = \frac{\Delta P}{\rho_L g} = \frac{p_2 - p_1}{\rho_L g}.$$

Thus, equation (10.21) can be converted into

$$H_d = h_c + h_t + h_{da}, \tag{10.22}$$

$h_t = \Delta p/(\rho_L g)$ is the equivalent clear liquid height to tray pressure drop due to vapor pressure drop through holes, which is called dry pressure drop, h_d, as well as liquid pressure drop through liquid on tray, which is named as wet pressure drop, h_l. As defined in equation (10.22), H_d can also be called total hydraulic head of liquid going through downcomer and tray deck.

Does downcomer flooding occur when downcomer clear liquid level H_d reaches the tray above? As a matter of fact, the flooding occurs earlier than the point of clear liquid reaching the tray above. This is because there is a layer of liquid froth on top of clear liquid. As soon as the froth reaches the tray above, flooding happens as froth carries significant amount of liquid to the tray above. Thus, the total liquid height including froth level is $H'_d = H_d/\varphi_d$ ($\varphi_d \le 1$), where φ_d is the downcomer froth density. To be conservative, we assume the downcomer backup limit is 80%

and thus downcomer flood capacity is expressed as

$$H'_{\mathrm{d}} = \frac{H_{\mathrm{d}}}{\varphi_{\mathrm{d}}} = 80\%(H_{\mathrm{S}} + h_{\mathrm{w}}). \tag{10.23}$$

Combining equations (10.22) and (10.23) yields

$$\frac{h_{\mathrm{c}} + h_{\mathrm{t}} + h_{\mathrm{da}}}{\varphi_{\mathrm{d}}} = 80\%(H_{\mathrm{S}} + h_{\mathrm{w}}), \tag{10.24}$$

where h_{w} is the outlet weir height. φ_{d} is downcomer froth density and is also considered as a derating factor by which a safety margin can be built into. φ_{d} is 0.3–0.4 for high foaming tendency; 0.6–0.7 for low foaming tendency; 0.5 for average foaming tendency.

Clearly, downcomer backup can be estimated if we know the clear liquid height H_{d}. However, it is not straightforward in predicting H_{d}. As H_{d} consists of three components, namely clear liquid height h_{c}, tray pressure drop h_{t}, and downcomer apron pressure drop h_{da}, let us look into each individual component one at a time.

(a) *Clear liquid height on tray* h_{c} can be calculated by

$$h_{\mathrm{c}} = h_{\mathrm{w}} + h_{\mathrm{ow}} + h_{\mathrm{g}}/2, \tag{10.25}$$

where h_{ow} is the liquid height in clear head over outlet weir (Figure 10.11) and h_{g} is the hydraulic gradient, which liquid has to overcome along the tray flow path. Excessive gradient could cause weep at downcomer inlet where liquid level is low. However, hydraulic gradient is negligible for most trays, and it is common to omit this term from pressure drop calculations (Ludwig, 1979; Fair et al., 1984). Thus, equation (10.25) can be simplified as

$$h_{\mathrm{c}} = h_{\mathrm{w}} + h_{\mathrm{ow}}. \tag{10.26}$$

For segmental weir, h_{ow} can be calculated via Francis weir formula (Bolles, 1963):

$$h_{\mathrm{ow}} = 0.48 F_{\mathrm{w}} \left(\frac{\mathrm{GPM}}{L_{\mathrm{w}}} \right)^{2/3} \text{in.}, \tag{10.27}$$

where F_{w} is the weir constriction factor and it can be ignored in quick sizing as $F_{\mathrm{w}} \approx 1$ in most cases anyway.

Weir loading $W_{\mathrm{L}} = \mathrm{GPM}/L_{\mathrm{w}}$ is an important parameter. The lower limit of weir loading is 2–3 gpm/in., while the higher limit is 7–13 gpm/in. (Kister, 1992). When weir loading is less than the lower limit, weir constriction such as picket fence should be employed to increase weir loading artificially to avoid spray. On the other hand, when weir loading is larger than the upper limit, multiple flow passes must be considered to reduce weir loading in order to avoid both tray and downcomer flood.

h_{ow} calculated above could be used to determine the outlet weir height h_w. For most normal pressure applications, liquid level h_c is within 2–4 in. and thus h_w can be determined via

$$2 - h_{ow} \leq h_w \leq 4 - h_{ow}. \tag{10.28}$$

For vacuum conditions, h_c could be 1–1.2 in. Thus, we have

$$1.0 - h_{ow} \leq h_w \leq 1.2 - h_{ow}. \tag{10.29}$$

With relatively high outlet weir height h_w, the clear liquid height h_c increases and vapor and liquid contact time increases. This improves separation efficiency. However, a too high outlet weir height could effect on downcomer backup and tray capacity.

However, Bernard and Sargetnt (1966) show that the estimate of clear liquid height from pressure drop calculations could be unsatisfactory for some applications. Instead, h_c could be calculated from froth height (h_f) based on froth density φ_f as

$$h_c = \varphi_f h_f, \tag{10.30}$$

φ_f can be calculated from Colwell's correlation (1981) and the detailed discussions can be seen in Kister (1992).

(b) *Head loss under downcomer apron* h_{da} is calculated based on Bolles (1963):

$$h_{da} = 0.03\left(\frac{\text{GPM}}{100A_{da}}\right)^2, \tag{10.31}$$

A_{da} is the flow area at downcomer clearance and it is also called downcomer apron. A_{da} should take the most restrictive area in the downcomer.

(c) *Tray pressure drop* h_t includes dry and wet pressure drop as

$$h_t = h_d + h_l. \tag{10.32}$$

First let us look at the dry pressure, h_d, which is the pressure drop that vapor experienced when it flows through holes or valves. By definition, expression for h_d takes the format of the orifice correlation.

For sieve tray, the dry pressure drop h_d can be expressed as

$$h_d = K\frac{\rho_V}{\rho_L}u_h^2, \tag{10.33}$$

where

$$u_h = \frac{\text{CFS}}{A_h},$$

u_h is the hole vapor velocity in ft^3/s and A_h is the hole area. K is calculated based on orifice coefficient C_v as $K = 0.186/C_v$. C_v can be found in Lockett (1986).

For valve trays, dry pressure h_d varies with valve positions, namely partly open or fully open. At low and moderate vapor loading when valve partly opens, dry pressure is more depending on the valve weight than the vapor rate. In contrast, at high vapor rate when valve fully opens, the dry pressure follows the orifice correlation. Glitsch (1974) provides the following correlations of dry pressure for valve trays:

$$\text{when valve part opens,} \quad h_d = 1.35 t_m \frac{\rho_m}{\rho_L} + K_1 \frac{\rho_V}{\rho_L} u_h^2, \tag{10.34a}$$

$$\text{when valve fully opens,} \quad h_d = K_2 \frac{\rho_V}{\rho_L} u_h^2, \tag{10.34b}$$

where t_m is the valve thickness (in.) and ρ_m is the valve metal density (lb/ft^3). Pressure drop coefficients K_1 and K_2 depend on valve type and weight. For Koch–Glitsch A(V-1) type as an example, $K_1 = 0.20$ and $K_2 = 0.86$ for $t_m = 0.134$ in., respectively.

Now we turn our attention to wet pressure drop. As h_l represents the pressure drop that vapor goes through the aerated liquid on the tray, it can be modeled in proportion to the tray clear liquid height as

$$h_l = \beta h_c, \tag{10.35}$$

β is the tray aeration factor and has different correlations for sieve and valve trays.

h_l can also be calculated from the froth height by

$$h_l = \varphi_t h_f, \tag{10.36}$$

where φ_t is the relative froth density. The most known β and φ_t correlations are Smith correlation (1963) for sieve trays and Klein correlation (1982) for valve trays.

After calculating dry and wet pressure, we can determine the tray pressure drop as follows.

For sieve trays:

$$h_t = h_d + h_l = K \frac{\rho_v}{\rho_L} u_h^2 + \beta h_c. \tag{10.37}$$

For valve trays:

$$\text{when valve part opens,} \quad h_t = \left(1.35 t_m \frac{\rho_m}{\rho_L} + K_1 \frac{\rho_V}{\rho_L} u_h^2\right) + \beta h_c, \tag{10.38a}$$

$$\text{when valve fully opens,} \quad h_t = K_2 \frac{\rho_V}{\rho_L} u_h^2 + \beta h_c. \tag{10.38b}$$

(d) *Total hydraulic head* H_d

So far, we are ready to derive overall expressions for $H_d = h_t + h_c + h_{da}$, based on the above discussions for these three components.

H_d *for sieve trays*: combining equations (10.27), (10.31), and (10.37) yields

$$H_d = \left(K\frac{\rho_v}{\rho_L}u_h^2 + \beta h_c\right) + \left[h_w + 0.48F_w\left(\frac{GPM}{L_w}\right)^{2/3}\right] + 0.03\left(\frac{GPM}{100A_a}\right)^2. \quad (10.39)$$

Rearranging equation (10.39) gives a simplified H_d expression for sieve tray:

$$H_d = K\frac{\rho_v}{\rho_L}u_h^2 + (1+\beta)h_w + (1+\beta)0.48F_w(W_L)^{2/3} + 0.03\left(\frac{W_L}{100h_{cl}}\right)^2. \quad (10.40)$$

H_d *for valve part open*: combining equations (10.27), (10.31), and (10.38a) yields

$$H_d = \left(1.35t_m\frac{\rho_m}{\rho_L} + K_1\frac{\rho_V}{\rho_L}u_h^2 + \beta h_c\right) + \left[h_w + 0.48F_w\left(\frac{GPM}{L_w}\right)^{2/3}\right] + 0.03\left(\frac{W_L}{100h_{cl}}\right)^2. \quad (10.41)$$

Rearranging equation (10.41) gives a simplified H_d expression for partly open valve tray:

$$H_d = \left(1.35t_m\frac{\rho_m}{\rho_L} + K_1\frac{\rho_V}{\rho_L}u_h^2\right) + (1+\beta)h_w + (1+\beta)0.48F_w(W_L)^{2/3} + 0.03\left(\frac{W_L}{100h_{cl}}\right)^2. \quad (10.42)$$

H_d *for valve full open*: combining equations (10.27), (10.31), and (10.38b) yields

$$H_d = \left(K_2\frac{\rho_V}{\rho_L}u_h^2 + \beta h_c\right) + \left[h_w + 0.48F_w\left(\frac{GPM}{L_w}\right)^{2/3}\right] + 0.03\left(\frac{GPM}{100A_a}\right)^2. \quad (10.43)$$

Rearranging equation (10.43) gives a simplified H_d expression for fully open valve tray:

$$H_d = K_2\frac{\rho_V}{\rho_L}u_h^2 + (1+\beta)h_w + (1+\beta)0.48F_w(W_L)^{2/3} + 0.03\left(\frac{W_L}{100h_{cl}}\right)^2. \quad (10.44)$$

In column design, increasing tray spacing, downcomer clearance, and the number of tray flow passes can effectively reduce liquid loading and hence avoid downcomer backup. Increasing tray spacing is more economic because column tangent length is more expensive. In operation, reducing feed rate and reflux rate can avoid downcomer flooding.

10.6.4 Downcomer Choke

Downcomer choke occurs when large crests block downcomer entrance and thus choke the liquid flow in the downcomer (Figure 10.9), resulting in liquid backup onto the tray deck. The root cause is too high downcomer velocity that results in excessive frictional losses at downcomer entrance.

Maximum downcomer velocity defines the liquid capacity limit and it can be predicted by several different correlations. According to Kister (1992), the Glitsch correlation (1974) tends to predict the highest downcomer liquid load, the Koch correlation (1982) predicts the lowest downcomer liquid load, and the Nutter correlation (1981) provides the estimate in between. Let us focus on the Glitsch correlation.

The Glitsch correlation (1974) is expressed as

$$\left.\begin{aligned} Q^1_{\mathrm{D,max}} &= 250 \\ Q^2_{\mathrm{D,max}} &= 41\sqrt{\rho_L - \rho_G} \\ Q^3_{\mathrm{D,max}} &= 7.5\sqrt{H_S\left(\rho_L - \rho_G\right)} \\ Q_{\mathrm{D,max}} &= \left[\min\left(Q^1_{\mathrm{D,max}}, Q^2_{\mathrm{D,max}}, Q^3_{\mathrm{D,max}}\right)\right] \times \mathrm{SF} \end{aligned}\right\}, \tag{10.45}$$

where downcomer liquid load Q_D is gpm/ft^2 of the downcomer area A_D. The correlation for $Q^3_{\mathrm{D,max}}$ in equation (10.45) is plotted for different tray spacing in Figure 4 of Glitsch Ballast Tray Design Manual (1974), which is reproduced in Figure 10.12. SF is derating factor and SF = 1 for the case of no derating.

For quick tray sizing purpose, Summers (2011) developed a simplified version of the Glitsch correlation. First data points on different curves in Figure 10.12 are extracted to generate a single data plot on $V_{\mathrm{D,max}}$ (maximum downcomer entrance velocity) and $\Delta\rho$ axis (Figure 10.13). Then a correlation is developed based on the best fit method to give

$$V_{\mathrm{D,max}} = 0.1747 \ln\left(\rho_L - \rho_V\right) - 0.2536\,\mathrm{ft/s}. \tag{10.46}$$

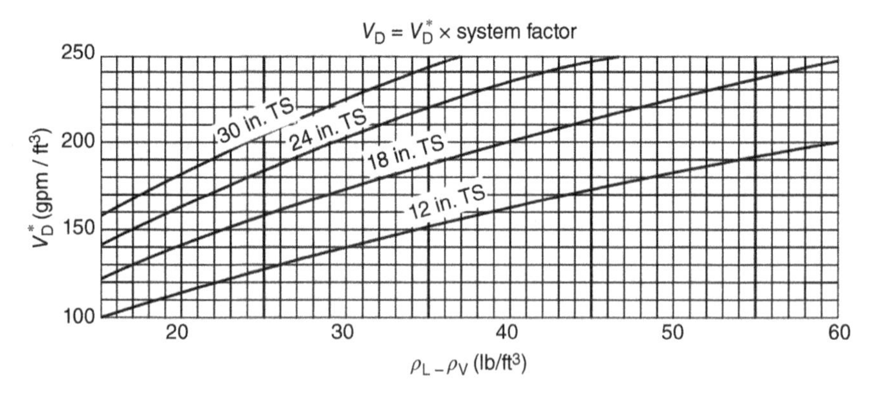

FIGURE 10.12. Downcomer design velocity curves in Figure 4 of Glitsch Design Manual (1974).

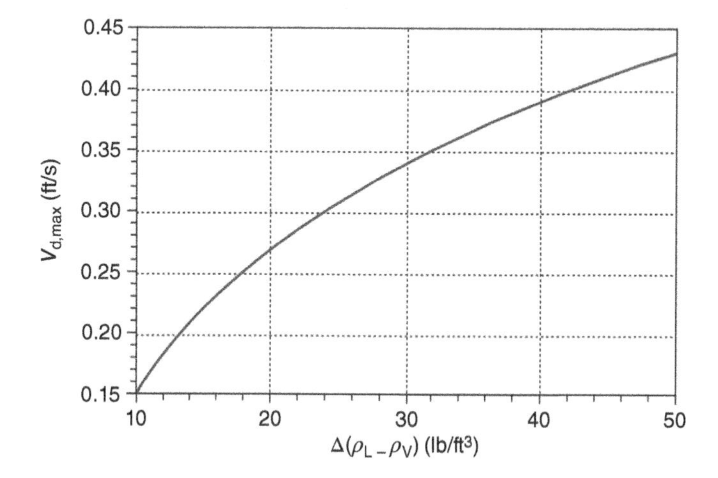

FIGURE 10.13. Maximum downcomer velocity correlation based on Glitsch's Figure 4 (1974). Source: Summers (2011). Reproduced with permission of AIChE.

For forming services, $V_{D,max}$ should be multiplied by a foaming factor. It must be noted that $V_{D,max}$ and $Q_{D,max}$ describe the same concept, namely maximum downcomer loading; but $V_{D,max}$ in ft/s while $Q_{D,max}$ in gpm/ft^2.

The Summers' correlation expressed in equation (10.46) is independent of tray spacing and it manifests the fact that greater density difference makes it easier for the vapor to escape from the froth in the downcomer. Summers (2011) indicated that this correlation works for nonfoaming services. For foaming services, a foaming factor must be applied. Since this correlation is independent of tray spacing and gives slightly conservative estimate of maximum downcomer velocity, Summers (2011) demonstrated that this correlation could be effectively applied for the purpose of quick tray sizing.

The Koch correlation (1982) and Nutter correlation (1981) are based on the minimum residence time criteria, which is converted to maximum downcomer liquid load criteria as

$$Q_{D,\max} = \mathrm{SF} \times 448.8 \frac{H_S}{12\tau_R}\ \mathrm{gpm/ft^2}. \tag{10.47}$$

10.6.5 Weir Loading Limits

Weir loading plays a critical role in stable operation. This is because spray could occur at low weir loading while downcomer flood could take place at high weir loading. For very low liquid loading, packing may be preferred to tray. The general guideline is consider packing when liquid loading is less than 4 gpm/ft^2 of column active area and use tray when liquid loading is larger than 14 gpm/ft^2 column active area. It is a judgment call for selecting either tray or packing for liquid loading in the range of 4–14 gpm/ft^2 column active area. The discussions for weir loading below are given with tray in mind.

10.6.5.1 Minimum Weir Loading A very low weir loading is manifested when the liquid crest or height over outlet weir, h_{ow}, is less than the limit of 0.25–0.5 in. (Chase, 1967; Davis and Gordon, 1961; Kister, 1992). A sufficient liquid crest over the weir maintained above the limit provides a stable liquid distribution. We can apply the Francis correlation in equation (10.27) to determine the minimum weir loading:

$$h_{ow,min} = 0.48F_w W_L^{2/3} = 0.25 \sim 0.5\,\text{in.}$$

Solving W_L yields

$$W_{L,min} = \left(\frac{h_{ow,min}}{0.48F_w}\right)^{3/2} = 0.4 \sim 1.0\,\text{gpm/in.} \tag{10.48}$$

The above $W_{L,min}$ is recommended by Glitsch (1974), Koch (1982), and Lockett (1986). When weir loading is close to $W_{L,min}$, spray could occur on tray deck. According to Summers and Sloley (2007), when actual liquid loading is small, the most effective way to avoid spray regime is using picket fence outlet weirs in order to artificially increase weir loading.

10.6.5.2 Maximum Weir Loading When weir loading is too high beyond the maximum limit, multiple passes are required. Kister (1990) suggested using 7–13 gpm/in. per single pass as the weir loading limit, while Summers (2011) recommended 12 gpm/in.

Resetarits and Ogundeji (2009) proposed that maximum weir loading should be determined according to tray spacing. In other words, different weir loading limits should be used for different tray spacing. For example, higher tray spacing can accommodate higher weir loading and vice versa. The intention of Resetarits and Ogundeji's argument is to minimize the number of tray flow passes as multiple passes feature lower distillation efficiency and higher tray cost than single pass tray. This argument is consistent with Nutter's data shown in Table 10.2.

10.6.5.3 Minimum Downcomer Residence Time Equation (10.47) can be converted to minimum residence time as

$$\tau_{R,min} = 448.8\frac{A_d H_S}{12\text{GPM}_{max}}. \tag{10.49}$$

TABLE 10.2. Relation of Weir Loading Limits and Tray Spacing (Nutter Engineering, 1981)

Tray Spacing (in.)	$W_{L,max}$ for Single Pass (gpm/in.)
12	3
15	5
18	8
21	10
24	13

Source: Resetarits (2009). Reproduced with permission by AIChE.

As can be observed from equation (10.49), A_d should be large enough to allow liquid have sufficient residence time so that vapor can disengage from descending liquid in the downcomer. Liquid can be relatively vapor free when it enters the tray below. Otherwise, too short residence time could lead to inadequate disengaging of vapor from liquid and then choke the downcomer. For low foaming tendency, $\tau_{R,min} = 3.0$ s; for high foaming tendency, $\tau_{R,min} = 6\text{–}7$ s; for average foaming tendency, $\tau_{R,min} = 4\text{–}5$ s.

10.6.6 Excessive Weeping

Excessive weeping usually occurs at turndown operation when vapor loading is very low and liquid short circuits the flow path and falls through holes directly. A small amount of weeping can be tolerated. But when the fraction of liquid falling through holes increases and exceeds a certain percent, dumping occurs and distillation efficiency falls off the cliff. As a guideline, a weeping rate of 25% is corresponding to a 10% loss in distillation efficiency. If vapor rate reduces until all liquid flows through the holes and does not reach the downcomer, this condition is called dumping. The weep point can be set in the range of 25–30%. Although there are no reliable methods available to predict the weep rate in accuracy, trays can be designed to operate above the weep point with confidence. There is a gray area between weep point and dumping point and the mechanism for this area is not well understood.

The mechanism for weeping is the force balance on tray between the vapor momentum upward and the liquid gravitational force downward. Under normal operation, vapor keeps liquid on tray when the vapor momentum is higher than the gravitational force. If the vapor momentum is too high, vapor carries liquid to the tray above and causes either spray flow and tray flood. In the opposite extreme where vapor momentum is too weak, liquid leaks through tray holes and weeping occurs. The force balance for weeping is established when vapor momentum is equal to liquid gravity as

$$h_d + h_\sigma = h_w + h_{ow}. \tag{10.50}$$

In the right-hand side, liquid force is formed by clear liquid height $h_c = h_w + h_{ow}$. On the left-hand side, the vapor force consists of h_d, the dry pressure drop required for vapor to go through the tray holes and h_σ, the surface tension associated with bubble formation. h_σ is given by Fair et al. (1984):

$$h_\sigma = \frac{0.04}{\rho_L d_h}\sigma, \tag{10.51}$$

where d_h is the hole diameter (in.) and σ is surface tension (dyne/in.).

The weep point check is important if the flexibility of large turndown operation is desired. The vapor and liquid rates under turndown are used to calculate h_d (using equation (10.34a) for sieve tray and equation (10.34b) for valve tray) and h_{ow} based on Francis' equation (10.27).

The weep point corresponding to $(h_d + h_\sigma)_{min}$ can be obtained using Fair's correlation (1963) via

$$\%\text{weep} = \frac{h_d + h_\sigma}{(h_d + h_\sigma)_{min}}. \tag{10.52}$$

The above Fair weep point method is applied to sieve trays. Bolles (1976) extended it to valve trays and details can be found in the Bolles' article. The weep rate is affected by minimum vapor loading in relation with weir loading, the number of holes for sieve tray and the number of valves and weight of valves. The experience shows that a well-designed valve tray must maintain the weep point below the vapor loads at which valve opens. To do this, a valve tray design should avoid use of too many valves.

To achieve an effective turndown operation, a manufacturer usually specifies two valve weights. When light valves open, the heavy ones still closed, which reduces active bubble area and thus avoids weeping. Furthermore, weeping usually occurs at the exit of the downcomer apron area. Thus, it is critical to maintain the level of tray.

For safety, 80% weep point from equation (10.52) could be used as the weep point limit. If the %weep is larger than 80%, weep rate evaluation is warranted, which is discussed as follows.

Lockett and Banik (1986) developed the correlation for weep rate W as

$$\frac{W}{A_\text{h}} = \frac{29.45}{\sqrt{Fr_\text{h}}} - 44.18, \tag{10.53}$$

where Fr_h is the Froude number and it is expressed as

$$Fr_\text{h} = 0.373\frac{u_\text{h}^2}{h_\text{L}}\frac{\rho_\text{V}}{\rho_\text{L} - \rho_\text{V}}. \tag{10.54}$$

Colwell and O'Bara (1989) suggested an alternative weep rate correlation as

$$\frac{W}{A_\text{h}} = \frac{1841}{Fr_\text{h}^{1.533}}. \tag{10.55}$$

Colwell and O'Bara recommended applying equation (7.53) for cases with Froude number less than 0.2 and equation (10.55) for large Froude number. For high pressure towers (>165 psia), the Hsieh and McNulty (1986) correlation is recommended and details are provided there.

The weep ratio is then defined as

$$w = \frac{W}{\text{GPM}}. \tag{10.56}$$

Above the weep point, $w=0$ while $w=1$ at the dumping point. Weeping occurs between the weep point and dumping point. Turndown operation could be still acceptable even if it is below the weep point, but the weep ratio w is less than 0.1. This is because tray efficiency is not affected too severely when w is less than 0.1. Increasing vapor load and reducing the clear liquid height could help to avoid weeping.

10.6.7 Constant G_L/G_V (L/V) Operating Line

The G_L/G_V ratio represents the internal liquid to vapor mole ratio in the top tray of the tower. The guideline range of G_L/G_V mole ratio is 0.3–3.0. G_L/G_V ratios outside this range may give sloppy or too easy distillation. Determination of G_L/G_V ratio is a major part of energy optimization, which is based on trade-off between the reboiling duty and the number of trays. The general trend is that the reboiling duty can be lowered at the increase of the number of trays required resulting in the lower reflux rate. Detailed discussions are given in this chapter.

When a distillation column operates under a constant G_L/G_V ratio, the operating region is reduced to a straight line as shown in Figure 10.5 with G_L/G_V as the slope and A and B as two limits. When the same column is operated under a different G_L/G_V ratio, the slope of the operating line varies. There are process variations such as changes to feed rate and compositions, and product rates and specifications as well as tower operating conditions such as pressure, temperature, and reflux rate. These variations will cause changes to the operation points within the operating window. The effects of changes are discussed in this chapter.

10.7 TYPICAL CAPACITY LIMITS

Typical capacity limits are given here as general guidelines. Specific limits need to be developed based on specific conditions. There are three capacity limits related to vapor loading, which are spray, jet flood, and weeping. The spray limit is set as 2.78 and the flow regime is spray when spray factor S_p is less than 2.78 as calculated by equation (10.6). The tray flood limit is 75–85%, where the lower limit is used for vacuum and near atmospheric towers while the higher limit should be adopted for moderate and high-pressure towers (>165 psia). For low-pressure towers, valve trays should be considered because they are better in suppressing spray than sieve trays. A tower should be designed above 50% flood to avoid too much spare capacity with unnecessarily large diameter and height. Thus, to get these two parameters right is the paramount of tower design. Weeping limit is set at 90% of Fair weep point or at 10% weeping rate.

There are two capacity limits relating to liquid loading, which are the downcomer backup limit and downcomer velocity limit. The downcomer backup limit is set at 80% of liquid settling height based on the froth level. The downcomer velocity limit is 75% of maximum velocity allowed in order to avoid downcomer choke. The number of tray passes is the most important parameter affecting the downcomer loading and thus these two downcomer limits.

Lastly, there are two capacity limits based on weir loading. For stable operation, minimum weir loading of 2 gpm/in. is required. If weir loading is less than this limit, a picket fence weir should be considered to increase the weir loading artificially. On the other hand, when weir loading is larger than the maximum limit per pass corresponding to tray spacing (Table 10.2), multiple passes should be considered.

Based on the capacity guidelines, a typical capacity diagram or operating window is illustrated in Figure 10.14 although specific applications warrant different capacity limits.

10.8 EFFECTS OF DESIGN PARAMETERS

There are three common flow regimes for industrial towers, namely spray, froth, and emulsion, which are the results of vapor and liquid loadings (Figure 10.15).

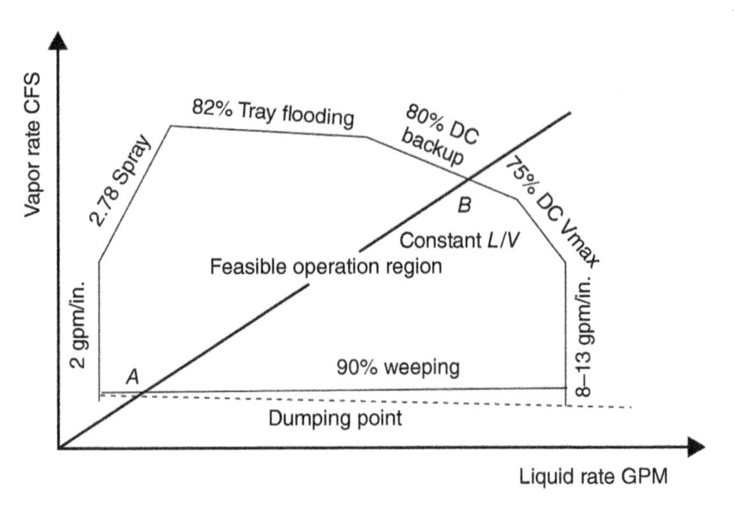

FIGURE 10.14. Typical capacity diagram.

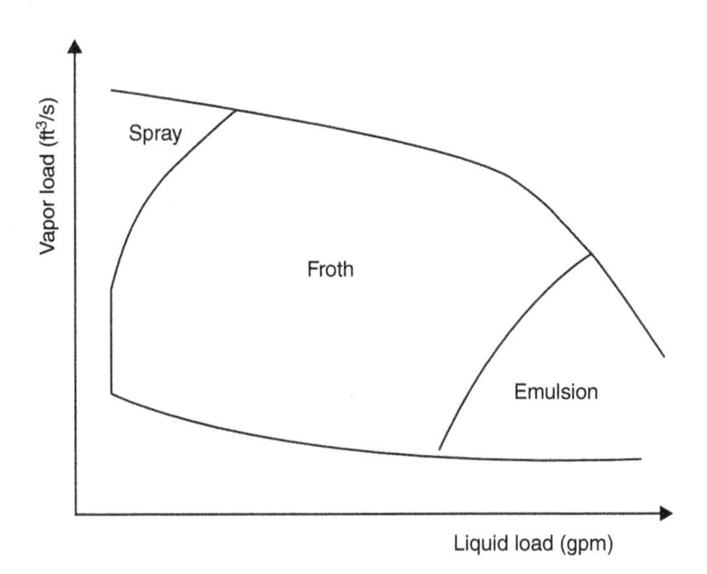

FIGURE 10.15. The flow regimes in distillation.

Froth regime is favored among these flow regimes as it enhances distillation efficiency and promotes stable operation. The fundamental reason why froth regime features higher distillation efficiency is that liquid penetrates into continuous vapor phase in the form of fine droplets while vapor penetrates into continuous liquid phase in the form of bubbles. The feature of the highest vapor/liquid interface area from the froth regime enhances mass transfer between vapor and liquid. In contrast, spray regime features too high upward vapor momentum, resulting in excessive entrainment of liquid. On the other hand, emulsion regime has too low vapor upward momentum, resulting in no penetration of liquid into vapor phase and thus poor mass transfer. Thus, avoiding spray, reducing emulsion, and enhancing froth flow is the focus of design considerations.

The effects of different parameters can be better explained based on one simple concept, which is momentum exchange, which is captured the best by Lockett (1986) and modified by Summers and Sloley (2007):

$$S_P = K\frac{h_L}{u_h d_h}\left(\frac{\rho_L}{\rho_V}\right)^{0.5}, \quad (10.57)$$

where h_c represents liquid gravitational force and $u_h d_h$ vapor momentum multiplied by density. The magic number for S_p is 2.78 below which spray flow occurs. Thus, design effort strives to increase S_p above 2.78 at all cost.

10.8.1 Effects of Pressure

Vacuum and low pressure condition enhance vapor loading at the expense of liquid loading resulting a low S_p and thus promotes spray flow. Valve trays have greater tendency to operate under froth regimes than sieve trays, which is indicated by a higher K value. Thus, it is common practice that valve trays are used for vacuum and low-pressure applications. At the same time, small valves (for small d_h) with increased number of valves (for small u_h) are used in order to increase S_p.

In contrast to low-pressure towers, high-pressure towers (>200 psi) are characterized with low vapor loading but high liquid loading and thus have a high tendency of downcomer flood. Multipass trays or counterflow trays such as UOP MD/ECMD or Shell HiFi or Sulzer high capacity or Koch–Glitsch high-performance trays can be applied to increase weir length and reduce weir loading as a result.

10.8.2 Column Diameter and Tray Spacing

Larger diameter (D) and lower tray spacing (H_S) will increase the capability of liquid loading more than vapor loading, resulting in a higher S_p. In other words, a fat and short tower (low H_S/D ratio) suppresses spray and shifts the operating point toward froth and emulsion. On the other hand, a tall and thin tower (high H_S/D ratio) can prohibit emulsion and shift the operating point to the left toward spray and froth. In addition, multipass trays also shift operating point to the left with reduced liquid loading in comparison with single pass trays. It is common to see fat and short towers for low pressure and vacuum applications while long and thin towers for high-pressure applications.

10.8.3 G_L/G_V Ratio

The rectifying section of a tower has a greater tendency to operate under spray, while the stripping section has a greater tendency to operate toward emulsion due to the characteristics of liquid and vapor loadings in these sections. That is, the rectifying section has a relatively higher vapor loading and lower liquid loading than in the stripping section.

10.8.4 Hole Diameter and Fractional Hole Area

Large hole diameter promotes spray. In contrast, small hole diameter reduces d_h and thus increases S_p and suppresses spray. Therefore, it is beneficial to use the smallest practical hole diameter possible considering fouling, tray deck material, thickness, and standard size. In contrast, small fractional hole area promotes spray as it increases u_h and thus reduces S_p. Therefore, for a flow regime leaning toward spray, one resolution is to design tray decks with greater number of holes and smaller hole diameter. For conditions promoting emulsion regime, lower fractional hole area can increase vapor and liquid contact via increased vapor velocity. Lower fractional area can extend operating window for sieve trays.

10.8.5 Weir Height

There is also an optimal weir height. High outlet weir promotes froth regime as it increases h_c and thus S_p, but it is at the expense of higher tray pressure drop and low tray capacity. The pressure drop increases at a higher rate than efficiency. Thus, it is not beneficial to have too high outlet weir. On the other hand, at very low weir height, liquid inventory is low, resulting in low mass transfer and affecting tray efficiency.

10.8.6 Tray Type

Sieve trays have wide applications due to lower capital cost and less maintenance required; but valve trays can promote froth regime and thus achieve distillation efficiency up to 20% higher than sieve trays (Anderson et al., 1976). Furthermore, valve trays can achieve 3:1 turndown compared to sieve trays in 2:1. But the disadvantages of valve trays are higher pressure drop and cost.

10.8.7 Summary

From the discussions, the main causes for spray regime are (1) low operating pressure; (2) low weir loading; (3) small column diameter; (4) high tray spacing; (5) large hole diameter; and (6) low fractional area. In short, a tall and thin sieve tray column under low pressure with multipass and large hole diameter has every tendency of spray. Another observation is that valve trays provide a higher distillation efficiency (Anderson et al., 1976) and greater turndown ratio than sieve trays although valve trays are more expensive.

10.9 DESIGN CHECKLIST

A design checklist could act as a safeguard for review of a tower design to avoid any unexpected issues to occur in late stage. Engineering companies usually have a design checklist. Table 10.3 gives an example although this example checklist is not extensive. Extra items can be added depending on applications. As mentioned, a column is divided into sections based on feed and draw locations. Thus, the check should be performed for each section.

TABLE 10.3. Example Tower Design Check List

Check Items	Typical Limits	Calculation Basis	Comments	Recommended Values	Potential Cures
Internal L/V ratio (%)	Lower limit = 0.3; Upper limit = 3	Mole percent	Sloppy or easy fractionation if outside of this range. If unexpected L/V is obtained, review the vapor and liquid loading	Within the range	
Spray factor S_p (s/ft)	2.78	Equation (12.6) (Lockett, 1986; Summers and Sloley, 2007)	Valve tray is better in suppressing spray than sieve tray	> 2.78	Increase tower diameter and open area; reduce hole diameter; consider valve tray
Tray flood (%)	75–85%	Equation (12.18) (Glitsch, 1974)	75–77% for vacuum towers; 80–85% for normal towers; 65–77% for small tower	> 50% for new design; > 85% for revamp	Increase tower diameter and tray spacing
Downcomer froth backup (%)	80%	Equation (12.24)	It is the height of froth instead of clear liquid height	< 80%	Increase tray spacing; consider multiple tray passes; increase downcomer clearance
Downcomer velocity limit (%)	75%	Equation (12.46) (Glitsch, 1974; Summers, 2011)		> 65%	Consider multiple passes and sloped downcomer if larger than the limit

(continued)

TABLE 10.3. ***(Continued)***

Check Items	Typical Limits	Calculation Basis	Comments	Recommended Values	Potential Cures
Weir loading (gpm/in.)	Lower limit = 2; upper limit = 8–13	Equation (12.48) for min limit (Glitsch, 1974; Koch, 1982; Lockett, 1986). Table 12.2 for max limit (Nutter, 1981)	The limits for single pass and straight weir	> 4 and < max limit in Table 12.2	Use multiple passes when larger than upper limit. Consider fewer passes if less than 4 gpm/in. Use blocked weir or baffled weir when less than 2 gpm/in.
Weeping	90% Fair weep point or 10% weep rate	Equation (12.52) (Fair, 1963)		> 10% weep rate	Reduce fractional hole area and outlet weir height. Choose right hole diameter based on liquid loading
Fractional hole area (%)	5%		Reduce hole diameter and fractional hole area if possible to enhance fractionation efficiency	7–15%	Use of fewer passes and short FPL if less than 5%
Flow path length (FPL) (in.)	18		For easy access in inspection and maintenance	> 18	
Tray spacing (in.)	15–24		For easy access in inspection and maintenance	> 15	
Weir length (in.)	5		For hand access when installing downcomer	> 5	

10.10 EXAMPLE CALCULATIONS FOR DEVELOPING OPERATING WINDOW

The following example is designed with the purpose of enhancing the understanding of column tray design and the capacity limits or operating window.

The overall tray design procedural guideline is to get tray diameter and spacing right first considering spray and tray flooding and then size downcomer adequately and then taking into account downcomer backup and choke. The guideline for an "optimal" design is that the least expensive design is usually the best one when comparing alternative designs.

The calculation method adopted here is based on Summers (2011) in which the tower diameter is determined based on liquid load entirely and weir loading is used as an iteration parameter to determine the number of tray passes. The method is straightforward to use and effective for a quick assessment.

Example 10.1

A stripper column is required to remove H_2S in the overhead in the hydrocracking unit. The task in hand here is to generate column tray design based on the column heat and mass balance simulation as shown in Table 10.4. Following the tray design philosophy as discussed, the data in the table are compiled as the combination of the most constrained vapor and liquid loadings. In other words, the most constrained vapor loading for one tray is selected in combination with the most constrained liquid loading from the another tray.

Working guide for diameter calculations

Step 1: Assume TS and W_{load}. Adjust them based on jet flood and weir loading criteria.

Step 2: Calculate maximum downcomer velocity $V_{\text{D,max}}$ based on equation (10.46), which determines the downcomer area A_{d}.

Step 3: Calculate tower active area A_{a} using equation (10.58).

Step 4: Calculate total tower area $A_{\text{t}} = A_{\text{a}} + 2A_{\text{d}}$ and then tower diameter.

Step 5: Calculate downcomer width w_{d} and weir length L_{w} based on geometry.

The calculation procedure can be summarized in Figure 10.16.

TABLE 10.4. Column Overhead Conditions

Vapor				Liquid				
T (°F)	V (lb/h)	CFS (ft³/s)	ρ_{v} (lb/ft³)	L (lb/h)	GPM (gal/m)	ρ_{L} (lb/ft³)	σ (mN/m)	μ (cP)
377	94,912	25.80	1.022	77,512	253.64	38.10	9.2	0.2

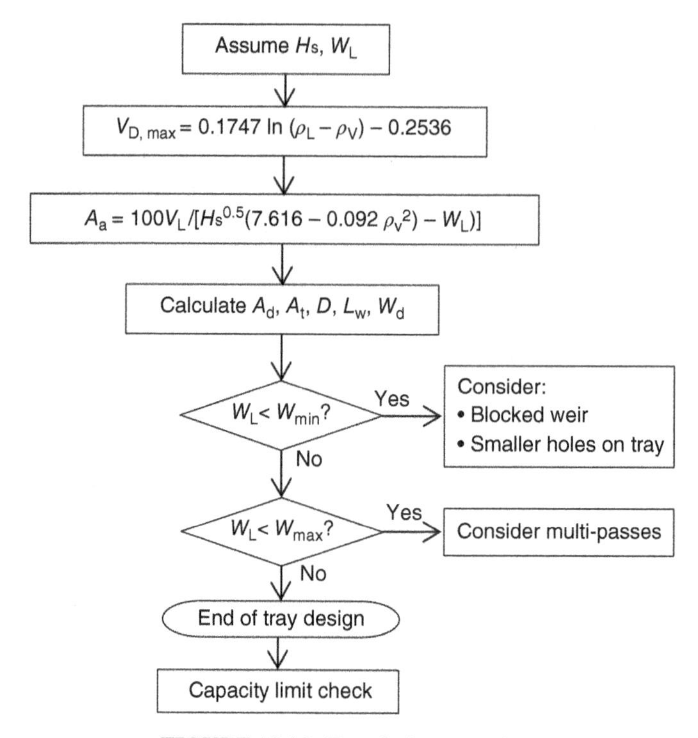

FIGURE 10.16. Tray design procedure.

(1) Determination of tower diameter

First let us assume *tray spacing and weir loading*

$$H_S = 24\,\text{in.};\quad W_L = 5\,\text{gpm/in.}$$

Downcomer area: Applying equation (10.46) with a foaming factor $F = 0.9$ yields

$$V_{D,\max} = F\left[0.1747\ln\left(\rho_L - \rho_V\right) - 0.2536\right] = 0.34\,\text{ft/s}.$$

As liquid rate is 253.6 gpm (0.565 ft^3/s), downcomer area becomes

$$A_d = \frac{L}{V_{D,\max}} = \frac{0.565}{0.34} = 1.66\,\text{ft}^2.$$

Tray active area: Vapor loading can be calculated based on equation (10.7)

$$V_L = \text{CFS}\sqrt{\frac{\rho_V}{\rho_L - \rho_V}} = 25.8\sqrt{\frac{1.022}{38.1 - 1.022}} = 4.28\,\text{ft}^3/\text{s}.$$

For quick tray sizing purpose, Summers (2011) derived a correlation mainly based on equation (10.18) (Glitsch flood correlation 13, 1974), which is expressed as

$$A_{\mathrm{a}} = \frac{100V_{\mathrm{L}}}{H_{\mathrm{S}}^{0.5}(7.616 - 0.0092\rho_{\mathrm{V}}^2) - 1.1W_{\mathrm{L}}}. \tag{10.58}$$

Applying equation (10.58) yields

$$A_{\mathrm{a}} = \frac{100 \times 4.28}{24^{0.5}(7.616 - 0.0092 \times 1.022^2) - 1.1 \times 5} = 13.67\,\mathrm{ft}^2.$$

Thus, total tray area and tray diameter are calculated by

$$A_{\mathrm{t}} = A_{\mathrm{a}} + 2A_{\mathrm{d}} = 13.67 + 2 \times 1.66 = 17.0\,\mathrm{ft}^2.$$

Then tower diameter can be calculated as

$$D = \sqrt{\frac{4A_{\mathrm{t}}}{\pi}} = \sqrt{\frac{4 \times 17.0}{3.14}} = 4.65\,\mathrm{ft}.$$

Rounding up the tower diameter to 5 ft. Weir length L_{w} is the cord length in a circle, while weir width w_{d} is the rise as shown in Figure 10.17. From geometry, there is a relationship between arc area (A_{d}), rise (the same as downcomer width w_{d}), and radius ($r = D/2$):

$$A_{\mathrm{d}} = r^2 \cos^{-1}\frac{r - w_{\mathrm{d}}}{r} - (r - w_{\mathrm{d}})\sqrt{2rw_{\mathrm{d}} - w_{\mathrm{d}}^2}. \tag{10.59}$$

For $A_{\mathrm{d}} = 1.66\,\mathrm{ft}^2$ and $r = D/2 = 2.5$ ft, equation (10.59) becomes

$$1.66 = 2.5^2 \cos^{-1}\frac{2.5 - w_{\mathrm{d}}}{2.5} - (2.5 - w_{\mathrm{d}})\sqrt{2 \times 2.5w_{\mathrm{d}} - w_{\mathrm{d}}^2}.$$

Solving this equation gives 8.4 in. for w_{d}. This w_{d} is larger than 5 in. (the minimum downcomer width) but smaller than 12 in. (the maximum downcomer width). The reason for the minimum is to allow easy access by hand in installing a downcomer.

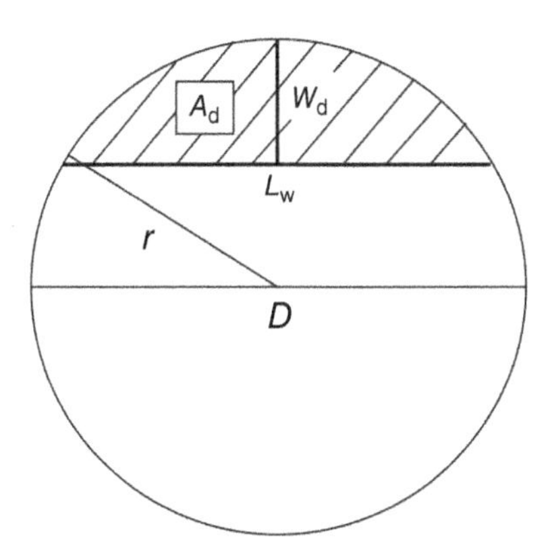

FIGURE 10.17. Derive A_{d} from w_{d} and r.

L_w can be calculated as

$$L_w = 2\sqrt{r^2 - (r - w_d)^2} = 2\sqrt{2.5^2 - \left(2.5 - \frac{8.4}{12}\right)^2} = 3.47\,\text{ft}.$$

Flow path length (FPL) is

$$\text{FPL} = D - 2w_d = 5 - 2 \times \frac{8.4}{12} = 3.6\,\text{ft}.$$

Weir loading can then be calculated as

$$W_L = \frac{\text{GPM}}{L_W} = \frac{253.6}{3.6 \times 12} = 6.29\,\text{gpm/in.}$$

The weir loading of 6.1 gpm/in. calculated is different from 5 gpm/in. assumed. Thus, the whole calculation repeats at the second trial based on $W_L = 6.1$ gpm/in. The procedure repeats until W_L converges as shown in Table 10.5. As a result, $D = 5$ ft is determined.

Tray deck layout is shown in Figure 10.18.

(2) Tray layout

The following calculations are based on the converged tray layout as shown in Table 10.5.

Number of passes: Single flow path is acceptable for weir loading $W_L =$ 6.1 gpm/in. as it is less than the weir loading limit of 8 gpm/in. for a tray spacing of 18 in. based on Table 10.2.

Downcomer and weir type: Straight downcomer and weir are selected for simplicity.

Selection of valve or sieve tray: Valve tray is selected mainly for fouling consideration as we turn down operation. For this application, Koch–Glitsch A (V-1), which is float valve, is selected.

Number of valves: The standard valve density for KG A(V-1) type is 12 ft^2. With active area A_a being 17.14 ft as recalculated based on $W_L = 6.29$ gpm/in., thus

TABLE 10.5. Trials of Calculating Tray Diameter

Tray Layout	First trial	Second trial	Third trial
Weir loading W_L (gpm/in.)	5.00	6.10	6.10
DC top area A_d (ft^2)	1.66	1.66	1.66
DC bottom area = DC top (ft^2)	1.66	1.66	1.66
Active area A_a (equation 10.58), (ft^2)	13.67	14.21	14.21
Total area $A_T = A_A + 2A_d$ (ft^2)	16.99	17.54	17.54
Net area $A_N = A_T - A_d$ (ft^2)	15.33	15.88	15.88
Diameter $D = (4A_T/\pi)^{1/2}$ (ft)	5.00	5.00	5.00
Weir length L_w (ft)	3.47	3.47	3.47
Weir width w_d (in.)	8.37	8.38	8.38
Flow path length (FPL) (ft)	3.60	3.60	3.60

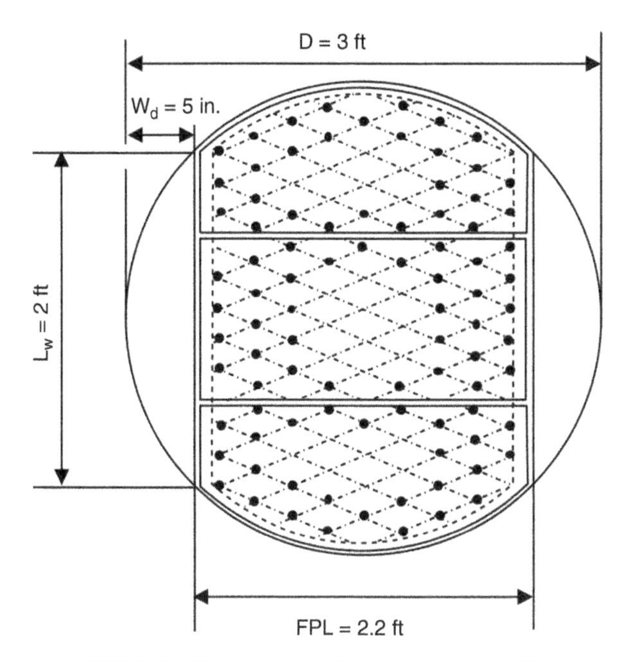

FIGURE 10.18. Tray layout for the example problem.

the number of valves is calculated by

$$N = 12A_a = 12 \times 14.2 = 171.$$

Hole diameter: The hole diameter is calculated based on hole factor as

$$F_{hc} = u_{hc}\sqrt{\rho_V}. \tag{10.60a}$$

Assume the selected valve will open when $F_{hc} = 9.5$. Thus, the critical hole vapor velocity is

$$u_{hc} = \frac{F_{hc}}{\sqrt{\rho_V}} = \frac{9.5}{\sqrt{1.022}} = 9.4\,\text{ft/s}.$$

According to the relationship of u_h and d_h as

$$u_h = \frac{\text{CFS}}{N\frac{\pi}{4}d_h^2}. \tag{10.60b}$$

Thus, for $u_{hc} = 9.4$ ft/s, the maximum hole diameter becomes

$$d_{h,max} = \sqrt{\frac{\text{CFS}}{N\frac{\pi}{4}u_{hc}}} = \sqrt{\frac{25.8}{171 \times \frac{3.14}{4} \times 9.4}} = 0.14\,\text{ft} = 1.7\,\text{in}.$$

Thus, by setting the actual hole diameter as 1.5 in., the actual vapor hole velocity is

$$u_h = \frac{25.8}{171 \times \frac{3.14}{4}\left(\frac{1.5}{12}\right)^2} = 12.3\,\text{ft/s} > u_{hc} = 9.4.$$

Valve will open since the actual hole velocity u_h is greater than the critical hole velocity u_{hc}. At the same time, we can derive the valve open percentage over the column diameter:

$$\psi_h = N\left(\frac{d_h}{D}\right)^2 \times 100\% = 171 \times \left(\frac{1.5/12}{5.0}\right)^2 \times 100\% = 11\%.$$

Liquid height over weir h_{ow}: For the purpose of quick calculation, assume weir correction factor $F_w = 1$. In most cases, F_w is close to one. Applying the Francis correlation (equation (10.27)) to the weir loading of 6.1 gpm/in. yields

$$h_{ow} = 0.48F_w W_L^{2/3} = 0.48 \times 1 \times 6.1^{2/3} = 1.6\,\text{in.}$$

Weir height h_w: By ignoring the liquid height gradient on tray and using $h_{ow} = 1.6$, applying equation (10.28) yields

$$0.4 \le h_w \le 2.4\,\text{in.}$$

Set $h_w = 2$ in. for being on conservative side due to relatively large liquid loading. Then clear liquid height becomes

$$h_c = h_w + h_{ow} = 2 + 1.6 = 3.6\,\text{in.}$$

Downcomer clearness h_{cl}: Assume liquid seal $h_s = 0.5$ in., which implies that weir height h_w exceeds downcomer clearness h_{cl} by 0.5 in. Thus,

$$h_{cl} = h_w - 0.5 = 2.0 - 0.5 = 1.5\,\text{in.}$$

Selection of downcomer clearness h_{cl} needs to consider a good downcomer seal but also needs to avoid too high pressure drop at the downcomer exit. Too small downcomer clearness could cause downcomer backup.

The downcomer layout is listed in Table 10.6. Until now, we have conducted a preliminary column design that provides the column geometry and tray layout.

(3) Hydraulic performance evaluations

It is only half of the job done when you performed tray sizing and layout design. The remaining questions are what is the capacity or operating window and can the tower give satisfactory performance. The questions can only be answered by conducting hydraulic check for spray regime, tray flooding, downcomer backup, downcomer choke, and weeping.

TABLE 10.6. Downcomer Layout for Example Problem

Downcomer (DC) Layout	
Weir length L_w (ft)	3.47
Weir height h_w (in.)	2.0
Liquid height over weir h_{ow} (in.)	1.6
Clear liquid height $h_L = h_{ow} + h_w$ (in.)	3.6
DC seal h_s (in.)	0.5
DC clearance $h_{cl} = h_w - h_s$ (in.)	1.5
DC clearance area $A_{da} = h_{cl} L_w$ (ft^2)	0.43
Flow passes	1
Downcomer type	Straight
Valve type	Koch–Glitsch Valve Type A (V-1)
Number of valves N; 12 valves/ft^2	171
Hole diameter d_h (in.)	1.5

(a) Spray flow check

Summers and Sloley (2007) spray factor correlation in equation (10.6) is employed:

$$S_p = K \frac{h_c}{u_h d_h} \left(\frac{\rho_L}{\rho_V} \right)^{0.5} = 2.5 \frac{3.6}{12.3 \times 1.5} \left(\frac{38.1}{1.022} \right)^{0.5} = 2.97,$$

$K = 2.5$ is used for valve trays in this case. The tray design point will be away from spray regime as the spray factor is greater than 2.78.

(b) Tray flood check

The Fair C_F-Factor correlation generalized in equation (10.15) by Summers (2011) is employed to calculate C_F-Factor:

$$C_F = \left(\frac{24}{24} \right)^{0.5} \times (0.455 - 0.0055 \times 1.022^2) = 0.449.$$

Then applying equation (10.14) to calculate vapor velocity at flood based on a tray spacing of 24 in. and assuming SF = 0.9 give

$$u_F = \text{SF} \times C_F \sqrt{\frac{\rho_L - \rho_V}{\rho_V}} = 0.9 \times 0.449 \times \sqrt{\frac{38.1 - 1.022}{1.022}} = 2.44\,\text{ft/s}.$$

Applying equation (10.9) for actual vapor velocity yields

$$u_N = \frac{\text{CFS}}{A_N} = \frac{25.8}{15.88} = 1.62\,\text{ft/s}.$$

Fair flooding% based on equation (10.17) becomes

$$\text{flood\%} = \frac{u_N}{u_F} = \frac{1.62}{2.44} = 67\%.$$

As an alternative calculation, Glitsch flood correlation 13 (1974) expressed in equation (10.18) is applied. In this case, the Glitsch flooding level can be evaluated using equation (10.18):

$$\text{Food\%} = \frac{V_L + \dfrac{\text{GPM} \times \text{FPL}}{13{,}000}}{C_F A_a} = \frac{4.28 + \dfrac{253.6 \times 43.25}{13{,}000}}{0.449 \times 14.21} = 80\%.$$

Both flooding calculations give flood% estimates less than 82% in normal operation for this example. Thus, it could be expected that the tray design will ensure the liquid entrainment less than 0.1 lb liquid/lb vapor. It should be noted that Glitsch flood correlation 13 (1974) is the most commonly used flood correlation for industrial distillation.

(c) Downcomer backup%

According to equation (10.22), total clear liquid height in downcomer is

$$H_d = h_c + h_t + h_{da}.$$

The height of froth in the downcomer is H_d/φ, where φ is froth density. Assume the target value of 80% for the froth height in relation to the liquid setting height. For a stable operation, the liquid capacity must satisfy the condition defined in equation (10.23):

$$\frac{H_d/\varphi}{H_S + h_w} \leq 80\%.$$

In the following, we calculate individual hydraulic head losses.

Clear liquid height $h_c = h_w + h_{ow} = 2 + 1.6 = 3.6$ in.

Dry pressure drop h_d is the pressure drop that vapor goes through holes. For KG valve type A(V-1), $K = 0.86$. Based on $u_h = 12.3$ ft/s and $d_h = 1.5$ in. applying equation (10.34b) yields

$$h_d = K_2 \frac{\rho_V}{\rho_L} u_h^2 = 0.86 \times \frac{1.022}{38.1} 12.3^2 = 3.5 \text{ in.}$$

Wet pressure drop h_l is the pressure drop that vapor goes through the aerated liquid on the tray. h_l can be calculated by using equation (10.35):

$$h_l = \beta h_c = 0.58 \times 3.6 = 2.1 \text{ in.}$$

β is the tray aeration factor that is found to be 0.58 based on the correlation for valve trays (Klein, 1982).

Thus, *tray pressure drop* can be calculated as

$$h_t = h_d + h_l = 3.5 + 2.1 = 5.6 \text{in.}$$

Downcomer apron pressure drop based on equation (10.31) is

$$h_{da} = 0.03\left(\frac{\text{GPM}}{100A_{da}}\right)^2 = 0.03\left(\frac{253.6}{100 \times 0.43}\right)^2 = 1.03 \text{in.}$$

Total hydraulic height in the downcomer or downcomer backup is

$$H_d = h_c + h_t + h_{da} = 3.6 + 5.6 + 1.03 = 10.23\,\text{in.}$$

Froth density $\varphi = 0.6$ based on the Glitsch correlation (1974). There will be no downcomer backup flood because

$$\frac{H_d/\varphi}{H_S + h_w} = \frac{10.23/0.6}{24+2} = 65.6\% < 80\%(\text{DC backup limit}).$$

(d) Downcomer choke

The downcomer liquid capacity check is conducted based on minimum residence time. The actual liquid residence time in downcomer can be calculated using equation (10.49):

$$\tau_R = 448.8\frac{A_d H_S}{12\text{GPM}} = 448.83\frac{1.66\times 24}{12\times 253.6} = 5.9 > \tau_{min} = 5\,s.$$

Thus, there will be no downcomer choke in normal operation.

(4) Tray design summary

So far, we have obtained tray layout and conducted hydraulic evaluations. The results can be summarized in Tables 10.7–10.9.

TABLE 10.7. Tray Design Overall Summary

Overall Summary	
Tower diameter D (ft)	5.0
Tray spacing H_s (in.)	24
Number of passes	1
Type of tray	Valve
Tray thickness (in.)	0.134
Number of valves; $12/\text{ft}^2$	171

TABLE 10.8. Tray Layout Summary

Layout Summary	
Column total area A (ft^2)	17.54
Column active flow area $A_a = A - 2A_d$ (ft^2)	14.21
Downcomer type	Straight
Weir shape	Straight
Downcomer area A_d (top) (ft^2)	1.66
Outlet weir length L_w (ft)	3.47
Downcomer width w_d (in.)	8.38
Liquid height over outlet weir h_{ow} (equation (4.20)) (in.)	1.60
Outlet weir height h_w (in.)	2.00
Downcomer clearance area $A_{da} = h_{cl}L_w$ (ft^2)	0.43
Downcomer clearance $h_{cl} = h_w - 0.5$ (in.)	1.50
Flow path length, FPL $= D - 2w_d$ (in.)	3.60

TABLE 10.9. Hydraulic Performance Summary

Hydraulic Performance	
1. Spray factor	2.97
2. Tray flood	
Based on Fair correlation	67%
Glitsch flood correlation	80%
3. Downcomer capacity limits	
Downcomer backup in froth	66%
Residence time (s)	5.9
4. Tray pressure drop	
Dry pressure drop h_d (in.)	3.51
Wet pressure drop h_l (in.)	2.09
Total pressure drop h_t (in.)	5.60
5. Downcomer hydraulics	
Downcomer apron head loss h_{da} (in.)	1.03
Downcomer backup, in clear liquid (in.)	10.23
Downcomer backup, in froth liquid (in.)	17.05
Downcomer froth density (φ)	0.60

(5) Feasible operation window

It seems we have completed the task of tower layout design at hand. You may still not be fully satisfied with the question in mind: how good is this design in the context of the tower capacity limits? Indeed, it would be very helpful if one can visualize the tower capability diagram or operating window. This could help to know where the current tower design stands and what flexibility the tower can offer in operation. The following discussions are used to show the procedure of generating the operating window for this example problem.

Working guide for obtaining the operating window

Step 1: Assume capacity limits for spray, tray flooding, downcomer backup, downcomer velocity, weeping and liquid rates, and so on.

Step 2: Apply the limits to each capacity equation.

Step 3: Derive relationships between vapor and liquid rates based on capacity equations under these limits.

Step 4: Plot these relationships onto a diagram of vapor and liquid rates.

Solutions

(a) Spray limit

Use the spray factor limit as 2.78 and applying equation (10.6) yields

$$S_p = K\frac{h_L}{u_h d_h}\left(\frac{\rho_l}{\rho_v}\right)^{0.5} = 2.5 \times \frac{2 + h_{ow}}{1.5u_h}\left(\frac{38.1}{1.022}\right)^{0.5} = 2.78, \qquad (10.61)$$

where liquid height over weir h_{ow} is expressed as

$$h_{ow} = 0.48\left(\frac{\text{GPM}}{L_w}\right)^{2/3} = 0.48\left(\frac{\text{GPM}}{3.47 \times 12}\right)^{2/3} = 0.04 \times \text{GPM}^{2/3}, \quad (10.62)$$

and the hole velocity expression for u_h

$$u_h = \frac{\text{CFS}}{A_h} = \frac{\text{CFS}}{N_h \frac{\pi}{4} d_h^2} = \frac{\text{CFS}}{171 \times \frac{3.14}{4} \times \left(\frac{1.5}{12}\right)^2} = 0.477 \times \text{CFS}. \quad (10.63)$$

For simplicity of mathematical expressions, let $G_V = \text{CFS}$ and $G_L = \text{GPM}$. Replacing h_{ow} and u_h in equation (10.61) with equations (10.62) and (10.63) yields

$$2.5 \times \frac{2 + 0.04 G_L^{2/3}}{1.5 \times 0.477 G_V}\left(\frac{38.1}{1.022}\right)^{0.5} = 2.78.$$

Solving G_V gives

$$G_V = 0.31 G_L^{2/3} + 15.3. \quad (10.64)$$

Equation (10.64) forms the spray limit at 2.78, which will be shown on a $G_V - G_L$ plot together with other operating limits.

(b) Vapor flooding limit

Assume the flooding limit at 82% vapor capacity. Application of equation (10.18) with V_L replaced via equation (10.7) gives

$$\text{Flood\%} = \frac{\text{CFS}\sqrt{\frac{\rho_V}{\rho_L - \rho_V}} + \frac{\text{GPM} \times \text{FPL}}{13{,}000}}{C_F \times A_a} = 82\%,$$

where C_F is calculated by equation (10.15) as

$$C_F = \left(\frac{H_S}{24}\right)^{0.5}\left(0.455 - 0.0055\rho_V^2\right) = \left(\frac{24}{24}\right)^{0.5}(0.455 - 0.0055 \times (1.022)^2)$$
$$= 0.449.$$

Let $G_V = \text{CFS}$ and $G_L = \text{GPM}$, thus, the above flood% can be expressed as

$$\frac{G_V\sqrt{\frac{1.022}{38.1 - 1.022}} + G_L\frac{(3.6 \times 12)}{13000}}{0.449 \times 14.2} = 82\%.$$

Solving for G_V leads to the linear flooding limit as

$$G_V = 31.54 - 0.02 G_L. \quad (10.65)$$

Equation (10.65) forms the vapor capacity limit at 82% flooding.

(c) Downcomer backup limit

The maximum liquid height allowable for downcomer backup is set by the tray spacing and outlet weir height. Downcomer backup flooding occurs when the liquid froth in downcomer reaches the tray above. Assume the maximum backup limit as 80% and the froth density $\varphi = 0.6$ (Kister, 1992). Applying equation (10.23) yields

$$H_d = 0.8\varphi(H_S + h_w). \tag{10.66a}$$

RHS of equation (10.66a) becomes

$$0.8\varphi(H_S + h_w) = 0.8 \times 0.6 \times (24 + 2) = 12.48\,\text{in}. \tag{10.66b}$$

LHS of equation (10.66a) is

$$H_d = h_t + h_c + h_{da} = (h_d + h_l) + (h_w + h_{ow}) + h_{da}.$$

By letting $h_l = \beta h_c = \beta(h_w + h_{ow})$ and setting RHS = LHS, equation (10.66b) becomes

$$h_d + \beta(h_w + h_{ow}) + h_w + h_{ow} + h_{da} = 12.48. \tag{10.67}$$

Assume $\beta = 0.58$ (Kister, 1992). For given $h_w = 2$ in., rearranging equation (10.67) yields

$$h_d + 1.58h_{ow} + h_{da} = 9.32. \tag{10.68}$$

In order to express equation (10.68) in G_V and G_L, h_d is expressed as a function of vapor flow rate G_V (CFS) according to equation (10.34b), while h_{ow} and h_{da} are functions of liquid flow rate G_L (GPM) based on equations (10.27) and (10.31). Furthermore, u_h is expressed in equation (10.60b). Thus, equation (10.68) becomes

$$\left.\begin{aligned} h_d &= 0.86\frac{\rho_V}{\rho_L}u_h^2 = 0.86 \times \frac{1.022}{38.1} \times \left(\frac{G_V}{171 \times \pi/4 \times (1.5/12)^2}\right)^2 = 0.0052G_V^2 \\ h_{ow} &= 0.48F_W\left(\frac{G_L}{L_W}\right)^{2/3} = 0.48\left(\frac{G_L}{3.47 \times 12}\right)^{2/3} = 0.04(G_L)^{2/3} \\ h_{da} &= 0.03\left(\frac{G_L}{100A_{da}}\right)^2 = 0.03\left(\frac{G_L}{100 \times 0.43}\right)^2 = 1.6 \times 10^{-5}G_L^2 \end{aligned}\right\}. \tag{10.69}$$

Replacing h_d, h_{ow}, and h_{da} in equation (10.68) with the corresponding expressions in equation (10.69) and then solving for G_V give

$$G_V^2 = 1768 - 11.99G_L^{2/3} - 0.003G_L^2. \tag{10.70}$$

Equation (10.70) forms the downcomer liquid capacity limit at 80% downcomer flood.

(d) Maximum liquid loading limit

The minimum residence time $\tau_{\min}=3\text{–}5$ s was suggested by Kister (1992). When liquid flows faster than $\tau_{\min}$, there is no sufficient time to separate vapor from liquid in downcomer and large froth crest forms and blocks the flow path at the downcomer entrance resulting in downcomer choke. For being conservative, $\tau_{\min}=5$ s is selected.

Based on equation (10.49), the minimum residence time can be calculated as

$$\tau_{\min}=\frac{448.8A_{\mathrm{d}}(H_{\mathrm{S}}/12)}{G_{\mathrm{L,max}}}=\frac{448.8\times 1.66\times(24/12)}{G_{\mathrm{L,max}}}=5\,s.$$

Thus,

$$G_{\mathrm{L,max}}=\frac{448.8\times 1.66\times(24/12)}{5}=298.5\,\mathrm{gpm}. \tag{10.71}$$

(e) Minimum liquid loading limit

According to Kister (1990), liquid height over outlet weir, h_{ow}, must be maintained larger than 0.25–0.5 in. in order to keep steady flow over the outlet weir and cover it completely during operation. Assume $h_{\mathrm{ow,min}}=0.5$ in. for conservative consideration. By applying equation (10.27) for this case, we have

$$h_{\mathrm{ow,min}}=0.48\left(\frac{G_{\mathrm{L,min}}}{L_{\mathrm{W}}}\right)^{2/3}=0.5\,\mathrm{in}.$$

Rearranging the above equation yields

$$G_{\mathrm{L,min}}=L_{\mathrm{w,e}}\left(\frac{h_{\mathrm{ow,min}}}{0.48}\right)^{3/2}=(3.47\times 12)\left(\frac{0.5}{0.48}\right)^{3/2}=44.2\,\mathrm{gpm}. \tag{10.72}$$

Minimum weir loading limit becomes

$$W_{\mathrm{L,min}}=\frac{G_{\mathrm{L,min}}}{L_{\mathrm{w,e}}}=\frac{44.2}{3.47\times 12}=1.06\,\mathrm{gpm/in}.$$

(f) Minimum vapor loading limit

Assume 60% turndown flexibility for this example tower operation. For valve trays, the minimum vapor momentum to open the valve should be considered for turndown operation. Since this minimum momentum corresponding to F-Factor = 4 at 60% turndown, applying equation (10.60) for F-Factor gives

$$F_{\min}=u_{\mathrm{v,min}}\sqrt{\rho_{\mathrm{V}}}=4.$$

Thus,

$$u_{\mathrm{v,min}}=\frac{F_{\min}}{\sqrt{\rho_{\mathrm{V}}}}=\frac{4}{\sqrt{1.022}}=3.956\,\mathrm{ft/s},$$

and

$$G_{\mathrm{V,min}}=u_{\mathrm{v,min}}N\frac{\pi}{4}d_{\mathrm{H}}^{2}=3.956\times 171\times\frac{3.14}{4}\times\left(\frac{1.5}{12}\right)^{2}=8.3\,\mathrm{ft/s}. \tag{10.73}$$

(g) The design point and constant G_V/G_L line

The basis used for the tower layout design is 25.8 CFS and 253.6 GPM (Table 10.4), which gives

$$G_L/G_V = 253.6/25.8 = 9.8. \tag{10.74}$$

This G_L/G_V ratio corresponds to molar flows for vapor and liquid as V= 1461 lb mol/h and L = 633 lb mol/h, respectively, which leads to L/V ratio as 0.433. This falls in the recommended range of (0.3, 3.0) as shown in Table 10.3, which indicates a good distillation. Operation outside of this range may imply sloppy or too easy distillation.

(h) Put all capacity limits together to generating capacity diagram

Based on the capacity limits derived, we are ready to generate a capacity diagram based on the G_V–G_L relationships:

$$\left.\begin{array}{l}
\text{2.78 Spray factor:} \quad G_V = 15.3 + 0.31G_L^{2/3} \quad (12.64)\\
\text{82\% Tray flood:} \quad G_V = 31.54 - 0.02G_L \quad (12.65)\\
\text{80\% DC backup:} \quad G_V^2 = 1768 - 11.99G_L^{2/3} - 0.003G_L^2 \quad (12.70)\\
\text{Maximum weir loading at } \tau_{min} = 5\,\text{s:} \quad G_{L,max} = 398.5 \quad (12.71)\\
\text{Minimum weir loading at 1.1 gpm/in.:} \quad G_{L,min} = 44.2 \quad (12.72)\\
\text{Vapor loading when valve close:} \quad G_{V,min} = 8.3 \quad (12.73)\\
\text{Design point:} \quad G_L/G_V = 9.8 \quad (12.74)
\end{array}\right\}. \tag{10.75}$$

By plotting the set of equations (10.75) onto G_V (CFS)–G_L (GPM) axis, we derive the capacity diagram for this example problem as shown in Figure 10.19. This diagram defines the feasible operating region for the example problem at hand. It must be emphasized that *capability diagram depends on the physical conditions of specific distillation.*

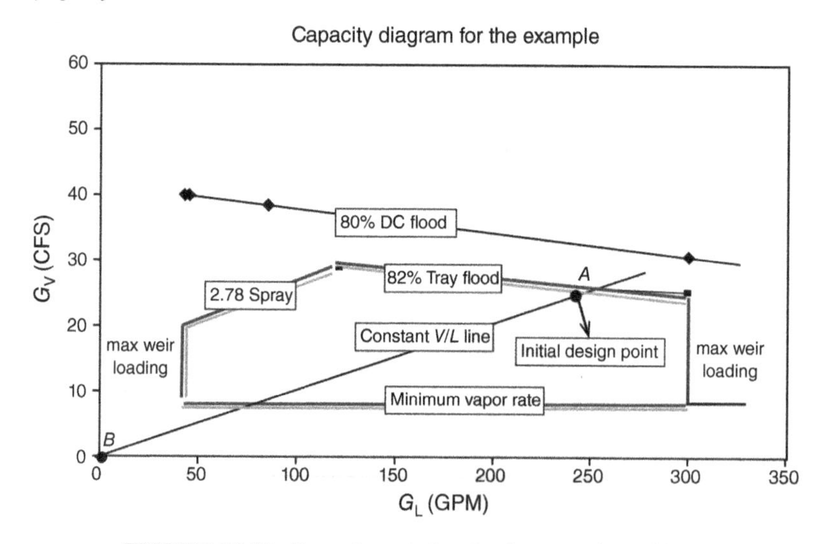

FIGURE 10.19. Operating window for the example problem.

Several observations can be obtained from the capacity diagram or the operating window as shown in Figure 10.19 and summarized as follows:

- The upper vapor capacity boundary of the operating region is controlled by both spray and tray flood. Spray is dominant under low liquid loadings, while tray flood becomes dominant at moderate and large liquid loadings.
- Downcomer flood does not impose a constraint in this example because the downcomer area is provided generously.
- Based on the G_L/G_V ratio of 9.8, the operating line (*AB*) can be determined. In the operating line, the upper limit is the *A* point ($G_L = 259$ GPM and $G_V = 26.3$ CFS), which is the intersection between the constant G_V/G_L line and the tray flood line. The lower limit is the *B* point ($G_L = 81.4$ GPM and $G_V = 8.3$ CFS), which is the intersection between the constant G_V/G_L line and the minimum vapor limit. The operating flexibility is 3.2 ($G_{V(A)}/G_{V(B)} = 26.3/8.3$).
- At the design point, $G_L = 253.6$ GPM (=633 lb mol/h) and $G_V = 25.8$ CFS (=1461 lb mol/h). The G_L/G_V molar ratio is 0.433 at this top tray, which falls in the recommended range of 0.3–3. The G_L/G_V molar ratio will gradually increase from top to bottom of the column.
- 82% flooding and 80% downcomer backup are used as the design limits for vapor flooding and liquid backup. The gap between these limits and 100% is the design margins, which allows the column to deal with variations in feed rate and composition. In other words, operating the column at 90% flooding and backup could be acceptable as long as it makes on-spec products.
- Although the design tray flooding is 80% that is close to 82% limit, there is a large spare capacity regarding the downcomer backup as the downcomer backup is 66% by design versus 80% as the design limit. This spare capacity can be removed if tray spacing is reduced from 24 to 18 in. By doing so, downcomer backup by design will be close to 80%. As a result, the column will become shorter in height, leading to low capital for the column. Similarly, the spare capacity for tray flooding can be removed by reducing the column diameter, which could make the column thinner. Thus, spare capacity costs money but provide flexibility when it is needed.
- If the feed rate increases beyond the design rate, the column could operate beyond 82% tray flooding and 80% downcomer backup as long as the column is able to make on-spec products.

The million dollar question is: does the tower design above is optimal? The answer is "probably not" as there are other alternative tray designs available, which are yet to be determined. The same design procedure in Figure 10.16 can be employed for generating alternative designs. As mentioned, the first thing to be determined in tray design is tower diameter, tray spacing, and downcomer layout. The general trend of changes between them is that tower diameter reduces (increases) when tray spacing is increased (reduced), thus resulting in a tall and thin (short and fat) tower. On the other hand, different tray designs provide different capacity limits, resulting in different operating windows. Consequently, distillation efficiency will be different, energy for reboiling and condensing will be different, and capital cost will be different.

These differences occur in operation for a given tray design when process and operating conditions vary. In all cases, it is important to know where the design and operating point in relation to the corresponding operating window. This could give you a good understanding if the column is operating under healthy conditions. There are two kinds of operating policy: pushing maximum throughput and meeting product quality. The former can be achieved by reducing reflux ratio when the reboiler size is constrained, while the latter is achieved by increasing reflux ratio and reducing feed rate. That tends to be common practice in the industry to push more throughput with minimum extra reboiling duty. However, if product specification cannot be met under increased feed rate, the reboiler duty must be increased for obtaining higher return ratio. At the same time, the capacity limits or the operating window will change when operating conditions change.

In general, when comparing alternative design options, the least cost including capital and utility costs usually is the choice by engineers who constantly seek economic solutions unless special circumstances justify a more expensive design such as in revamp for capacity expansion. In operation, you should seek the minimum energy cost operation while satisfying throughput and product quality. In some cases, column energy use in terms of feed heating, stripping steam, and reboiling are not adjusted in proportion to process variations such as feed rate, product rates, and quality. This could result in wasted energy.

Thus, it is challenging to design a tower with the lowest overall cost (capital and utility) and simultaneously provide sufficient operation flexibility to deal with process variations. In contrast, the challenge for operation is to determine the optimal operating conditions within the feasible region under which the column is operated in the most efficient manner. Discussions and guidelines for tower operation optimization are given in Chapter 13.

10.11 CONCLUDING REMARKS

What have we learned from operating window discussions above? It might be helpful for us to pause a little and generalize some of the key points and stock them in our memory.

- Momentum exchange and balance between vapor and liquid is the key concept in understanding the mechanism of stable operation. A tower comes with a capacity diagram that defines the operating window. Too high vapor load relative to liquid loading could cause tray flood. Conversely, too high liquid loading is the main cause for downcomer flood. On the other hand, too low vapor loading in relation to liquid could lead to weeping, while too low liquid loading and high vapor loading could result in spray flow.
- Spray mechanism can be best remembered by $h_c/u_h d_h$, the momentum ratio of liquid and vapor. Spray factor is defined in equation (10.6) based on Lockett (1986) with the value of 2.78 as the spray limit. To avoid spray, one needs to increase weir loading, reducing vapor loading and/or hole diameter.
- Tray flood is better represented by equation (10.18) (Glitsch correlation 13, 1974). Both vapor load and weir loading contribute to tray flood in relation

to tower diameter defining vapor open area and tray spacing defining settling space. Thus, tower diameter and tray spacing are the two major design parameters, while feed rate, column pressure, and reboiling duty are major operating parameters for avoiding tray flood.

- Downcomer flood is controlled by combination of tray pressure drop, weir loading, and downcomer apron. Remember that it is the liquid froth level instead of clear liquid height that determines the downcomer flood. This fact is clearly stated in equation (10.24). To prevent downcomer flood in design, downcomer size must be provided generously. From equations (10.40), (10.42), and (10.44), feed rate and reboiling duty are the major operating parameters to avoid downcomer flood.
- Weeping capacity is more related to turndown operation. Typically, 80% weep defined by equation (10.52) is used as the weep limit. Valve trays provide better turndown capacity in 3:1 versus sieve trays in 2:1. In other words, valve trays could operate in 30% of normal load with weep rate less than 10%.
- For towers operating at near atmospheric or under vacuum, it is recommended to use valve tray as it is more resistant to spray than sieve trays. For low liquid loading applications, tray with picket fence weir could be used to increase the liquid loading artificially. But for much lower liquid loading where picket fence could fail, packing tower should be used. As a rule of thumb, use packing when liquid loading is less than 4 gpm/ft^2 column active area and use tray for liquid loading higher than 14 gpm/ft^2 column active area. It is a judgment call as to what trays to use for liquid loading in between.

NOMENCLATURE

Variables

A_a active or bubble area (ft^2)
A_d downcomer top area (ft^2)
A_{da} area under downcomer apron (ft^2)
A_h tray hole area (ft^2)
A_N column net area (=column cross-sectional area less downcomer top area) (ft^2)
A_T total column cross-sectional area (ft^2)
C C-factor (describing vapor load) (ft/s)
CFS vapor flow (ft^3/s)
C_F C-Factor at flood (ft/s)
d_h tray hole diameter (in.)
D column diameter (ft)
f friction coefficient between vapor and liquid (dimensionless)
F_{hc} hole factor
FPL flow path length (distance from the inlet downcomer edge to outlet weir) (in.)
Fr Froude number (dimensionless)
F_w weir correction factor (dimensionless)

g acceleration due to gravity (32.2 ft/s^2)
GPM gallon per minute (gal/min)
G_L GPM
G_V CFS
h_c liquid height or liquid holdup (in. of liquid)
h_{cl} clearance under downcomer (in.)
h_d dry pressure drop (in. of liquid)
h_{da} clear liquid height for pressure drop through downcomer apron (in. of liquid)
h_f froth height on tray (in. of froth)
h_g hydraulic gradient on tray (high minus low clear liquid height) (in. of liquid)
h_l pressure drop through aerated liquid on tray (in. of liquid)
h_{ow} liquid head over the outlet weir (in. of liquid)
h_s downcomer seal (in.)
h_t tray pressure drop (in. of liquid)
h_v height of two-phase (froth and vapor) layer (in. of froth plus vapor)
h_w outlet weir height (in.)
h_σ head loss due to surface intension associated with bubble formation (in. of liquid)
C_v orifice coefficient (in dry pressure h_d calculation) (dimensionless)
H_S tray spacing (in.)
H_d downcomer liquid height (in. of liquid)
H'_d downcomer froth head (in. of froth)
L_w outlet weir length (in.)
N number of valve per square feet (number/ft^2)
p pressure (atm)
ΔP pressure drop (in. of liquid)
Q_D downcomer liquid load (gpm/ft^2)
$Q_{D,max}$ maximum downcomer liquid load (gpm/ft^2)
SF derating factor (dimensionless)
S_p spray factor (s/ft)
t_m valve thickness (in.)
u_h vapor hole velocity (ft/s)
u velocity (ft/s)
u_F vapor velocity at flood (ft/s)
v volume (ft^3)
$V_{D,max}$ maximum downcomer entrainment velocity (ft/s)
V_L vapor load (lb/ft^3)
w weep ratio (dimensionless)
W weep rate (gpm)
w_d downcomer width (in.)
W_L weir liquid loading (gpm/in.)

Greek letters

α relative volatility (dimensionless)
β tray aeration factor (dimensionless)

λ ratio of the slope of the equilibrium curve to the slope of the component balance line
ρ density (lb/ft^3)
ρ_m valve metal density (lb/ft^3)
σ surface tension (dyne/in.)
μ liquid viscosity (cP)
τ_R downcomer residence time (s)
φ_d downcomer froth density (dimensionless)
φ_f froth density (dimensionless)
φ_t relative froth density (dimensionless)
ψ valve open percentage (%)
η_M Murphree tray efficiency (dimensionless)
η_o Column overall efficiency (dimensionless)

Subscripts

l (L) = liquid
v (V) = vapor

REFERENCES

Anderson, RH, Garnett G, Winkle, MV (1976), Efficiency comparison of valve and sieve trays in distillation columns. *Ind. Eng. Chem. Process Des. Dev.*, 96–100, **15**, 1.

Bennett BL, Kovak KW (2000), *Optimize Distillation Columns*, Chem. Engng. Prog, May, pp19.

Bernard, JDT, Sargetnt, RWH (1966), *Trans. Inst. Chem. Engrs (London)*, 44, p.T314.

Bolles WL (1963), in Smith BD (ed), *Design of Equilibrium Stage Process*, McGraw-Hill

Bolles WL (1976), *Chem. Eng. Progr.* **72**(9), 43.

Chase JD (1967), Sieve Tray Design, Chem. Eng., July 31, p. 105.

Colwell, CJ (1981), *Ind, Eng, Chem, Proc Des, Dev,*.

Colwell CJ, O'Bara JT, (1989), Paper presented at AIChE Spring Meeting, Huston, April.

Davis JA, Gordon KF (1961), What to Consider in Your Tray Design, Petro/Chem. Eng. Oct., p. 230.

Fair JR (1961), *Petro/Chem Eng.* **33** (10), 45

Fair JR (1963), in Smith BD, *Design of Equilibrium Stage Processes*, McGraw-Hill.

Fair JR, Steinmeyer DE, Penney WR, Crocker BB (1984), in Perry RH and Green DW, *Chemical Engineering's Handbook*, 6th ed., Chapter 18, McGraw-Hill.

Glitsch Inc. (1974), *Ballast Tray Design Manual*, 3rd ed, Bulletin No. 4900, Texas.

Hower, TC, Kister, HZ (1991), *Solve process column problems, Parts 1&2*, Hydrocarbon Processing, May and June.

Hsieh CL, McNulty, KJ (1986), Paper presented at the AIChE Annual Meeting, Miami Beach, Florida, November.

Kister, HZ (1990), Distillation Operation, *McGraw-Hill.*

Kister, HZ (1992), Distillation Design, *McGraw-Hill.*

Klein, GF (1982), the correlation for valve tray, Chemical Engineering, May 3, p81.

Koch Engineering Company Inc. (1982), *Design Manual – Flexibility*, Bulletin 960-1, Kansas.

Lewis, WK (1936), Ind. Eng. Chem., 28 pp 399.

Lockett MJ (1986), *Distillation Tray Fundamentals*, pp35, Cambridge University Press, Cambridge, England.

Lockett, MJ, Banik, S (1986), *Ind. Eng. Proc. Des. Dev.* **25**, 561.

Ludwig, FE (1979), *Applied Process Design for Chemical and Petrochemical Plants*, 2nd ed, vol.**2**, Gulf Publishing.

McCabe, WL, Thiele EW (1925), Graphic design of fractionation columns, *Industrial and Engineering Chemistry*, **17**:605–611.

Murphree, EV (1925), Graphical rectifying column calculations, *Industrial and Engineering Chemistry*, **17**, 960.

Nutter Engineering (1981), Float Valve Design Manual, Rev. 1, pp 10, Tulsa, Oklahoma.

O'Connell, HE (1946) *Trans. AIChE*, **42**, 741.

Resetarits, MR, Ogundeji, AY (2009), *On Distillation tray weir loadings*, AIChE Spring Meeting, Florida, April.

Sakata M, Yanagi T (1979), Performance of a commercial-scale sieve tray, I. Chem. E Symposium Series, No. 56.

Smith BD (1963), *Design of Equilibrium Stage Processes*, McGraw-Hill, Inc.

Summers DR (2004), Performance Diagrams – All Your Tray Hydraulics in One Place, AIChE Annual Meeting: Distillation Symposium – Paper 228f, November 9, Austin.

Summers DR (2011), A novel approach to quick sizing trayed towers, AIChE Spring Meeting, Chicago, March.

Summers DR, Sloley A (2007), How to handle low-liquid loadings, *Hydrocarbon Processing*, 67, January.

Weiland RH, Resetarits MR (2002), New Uses for Old Distillation Equations, AIChE Spring Meeting, New Orleans, LA, March 11–14.

PART 4

PROCESS EQUIPMENT ASSESSMENT

11

FIRED HEATER ASSESSMENT

11.1 INTRODUCTION

Fired heaters are used to provide high temperature heating when high pressure steam is unable to satisfy process heating demand in terms of temperature. The primary role of industrial fired heaters is to provide heat required for reaction and separation processes. In a fired heater as shown in Figure 11.1, the process fluid enters the tubes at the top of the convection section and flows down countercurrent to the flue gas flow. The fuel mixes with the combustion air in the burner and provides the heat to heat up the process steam. The hot combustion gases need residence time to transfer the heat to the tubes. The shock tubes are often the hottest tube in the fired heater. The shock tubes receive the full radiant heat transfer in the range of 10,000 Btu/h ft^2, plus the hot gases flowing over the tubes result in an additional convective heat transfer rate in the range of 5000 Btu/h ft^2. Since the firebox operates at very high temperatures, refractory lining is required to prevent heat loss to the atmosphere.

Because fired heaters operate under severe conditions, fired heaters are designed with careful considerations of high temperature characteristics of the alloy. With proper maintenance and operation, a fired heater can have a long operating life. However, the life of a fired heater can be greatly shortened due to creep, fatigue, corrosions, and erosion by lack of maintenance and reliability considerations. Fired heater failure could not only result in significant production loss. In the worst case, it could cause damage to human life.

Therefore, maintaining fired heater in reliable operation is the highest priority. With this priority in place, process plants strive to maximize fired heater efficiency and hence reduce its operating cost. This is because of a simple fact: fired heaters

Hydroprocessing for Clean Energy: Design, Operation, and Optimization, First Edition.
Frank (Xin X.) Zhu, Richard Hoehn, Vasant Thakkar, and Edwin Yuh.

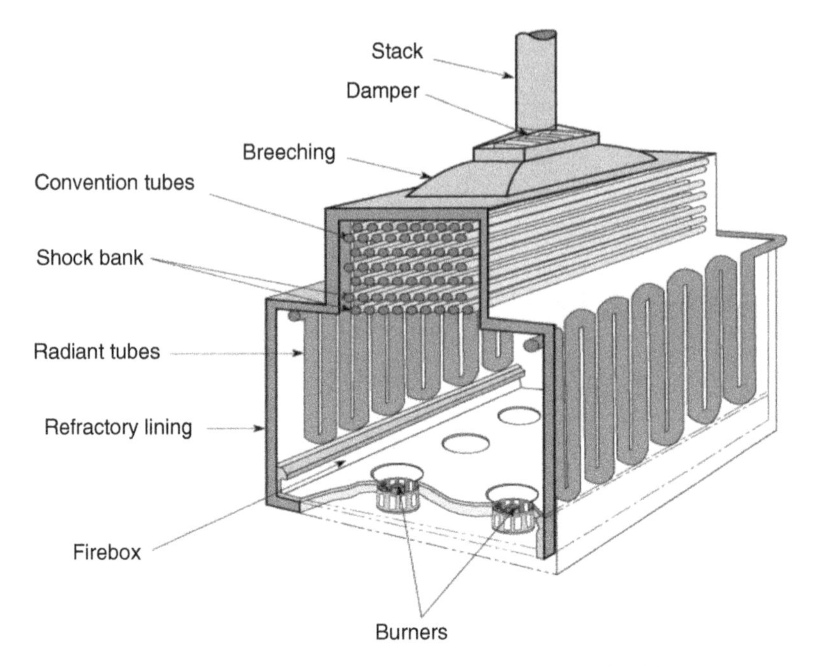

FIGURE 11.1. Schematic view of a typical process fired heater.

are the largest energy consumers in process plants and accounts for majority of total energy use.

11.2 FIRED HEATER DESIGN FOR HIGH RELIABILITY

The eventual measure of fired heater reliability is availability and the goal is that a fired heater needs to be online almost 100% of time. What does it take for a fired heater to achieve this high reliability from design point of view?

In this section, we discuss the critical issues that a highly reliable fired heater must acquire and shed light on fundamentals for these features so that they can be used as the benchmark for assessing a fired heater. The critical reliability issues are discussed as follows:

- Flux rate
- Burner to tube clearance
- Burner selection
- Fuel conditioning system.

11.2.1 Flux Rate

Radiant heat flux is defined as heat intensity on a specific tube surface. Thus, heat flux represents the combustion intensity and is analogous to "how hard a fired heater

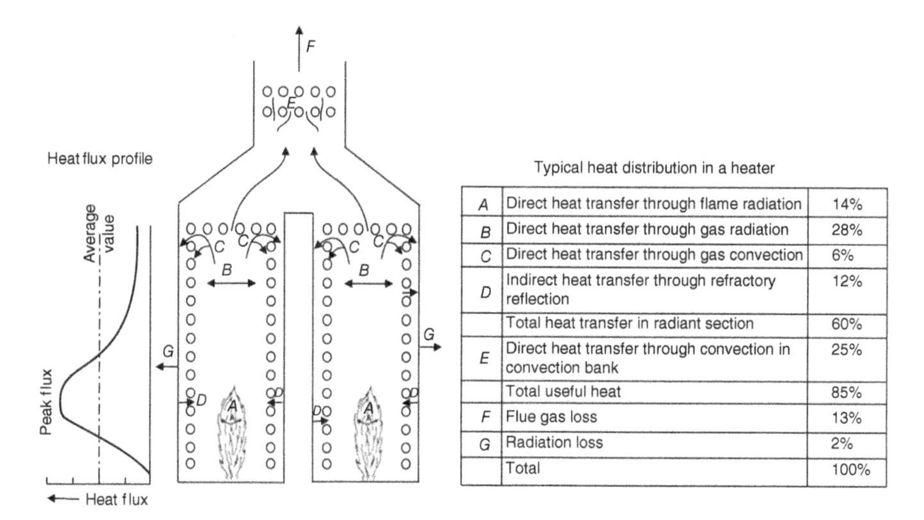

A	Direct heat transfer through flame radiation	14%
B	Direct heat transfer through gas radiation	28%
C	Direct heat transfer through gas convection	6%
D	Indirect heat transfer through refractory reflection	12%
	Total heat transfer in radiant section	60%
E	Direct heat transfer through convection in convection bank	25%
	Total useful heat	85%
F	Flue gas loss	13%
G	Radiation loss	2%
	Total	100%

FIGURE 11.2. Flux profile and heat distribution in a heater.

is run." More specifically, to keep firing rate within the safe limit is equivalent to maintain the peak heat flux being less than the design limit because high firebox temperatures could cause tubes, tube-sheet support, and refractory failures. What is the peak flux and why it is so important to keep it within the limit? These questions are answered as follows.

Flux rate is influenced by combustion characteristics and heat distribution. While the combustion characteristics can be described by combustion intensity, the heat distribution is explained by heat flux. The heat flux is defined as the heat transferred to the process feed, while combustion intensity is the heat released from flame divided by the flame's external surface area. Clearly, combustion intensity is related to combustion flame while heat flux to process. Another difference is that combustion intensity is inevitably an average value while heat flux is either average or local value. Local heat flux requires more attention in design and operation.

Figure 11.2 shows a typical pattern of a heat flux profile. It can be observed that the flux distribution is not uniform. The heat intensity nearest flames is the highest (peak flux) and declines away from flames. The peak flux can exceed twice the average value and should be maintained below the flux limit all the time for safe operation. The nonuniformity is described by the ratio of peak flux to average flux. A good heater design and operation should have a heat flux profile featuring a high average and low peak flux values.

Jenkins and Boothman (1996) reported an operating case with average flux at only 700 Btu/h ft^2, but the peak flux was over 20,000 Btu/h ft^2 nearly three times of the average. The peak flux exceeded the safe limit of the process, and residual oil in the tubes in the peak flux area was cracked and coke was deposited in the inner tube surface. As a consequence, the coke acted as an insulating layer and caused the tubes to overheat, which was measured by tube wall temperature (TWT). This was dangerous as the tubes in the peak flux area could rapture. To mitigate the safety risk, the plant operators had to reduce the process feed rate and hence cut down the firing rate in order to keep the peak flux below the safe limit. Furthermore, the heater had to be shut

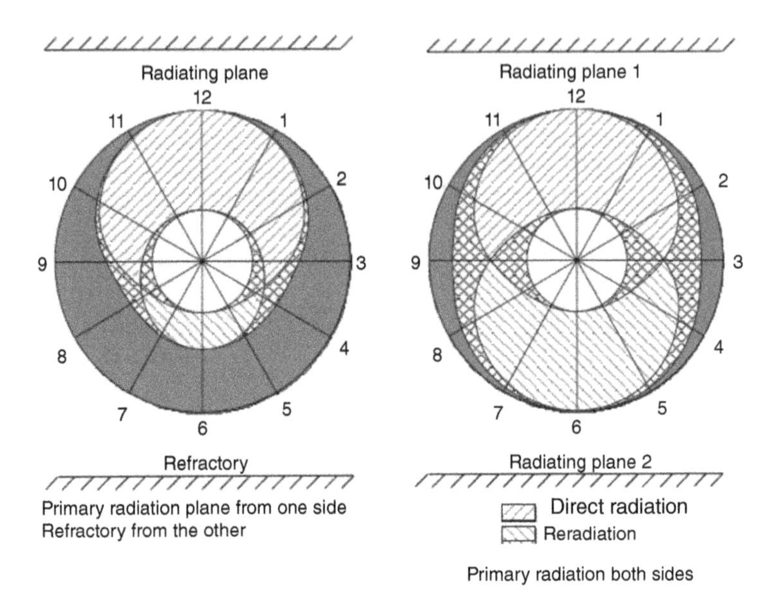

FIGURE 11.3. Flux distribution around fired heater tube.

down at regular intervals so that tubes could be cleaned to remove the carbon deposits. These mitigation actions resulted in significant cost to the plant. The problem was solved fundamentally by improving air distribution between the burners together with changes to the burner gas nozzle and flame stabilizer. These changes result in a lower air pressure loss and improved fuel and air mixing. After these changes, the burners produced a more even flux profile: the average flux was increased from 700 to 10,000 Btu/h ft^2, while the peak flux reduced to 18,000 Btu/h ft^2. Coking inside the tubes was eliminated, which allowed the feed rate to increase by 4% and the heater could run continuously between scheduled outages.

The flux distribution around the tube is not uniform as well. As indicated in Figure 11.3, the radiating plane is the flame. The diagram on the left side shows the flux profile for a single fired heater. The front of the tube facing the fire picks up most of the heat. The diagram on the right side shows the profile for a double-fired heater with flames on both sides of the tubes. The flux pattern is close to uniform.

The nonuniformity is described by the circumferential flux factor, which is the ratio of peak flux to average flux. Peak flux determines the maximum TWT. The peak flux is typically 1.5–1.8 times of the average for a single fired heater, while it is 1.2 times of the average for a double-fired heater. That explains why double-fired heater has a longer run length as it has a lower flux rate and hence lower TWT compared to the single fired heater.

The tube thinning follows the same pattern of flux distribution. Figure 11.4 shows a fired heater tube with severe thinning creep caused by internal coking especially on the fireside of the tube. The internal coking follows the same pattern with much greater coke thickness at the front face facing the flame. This is why inspectors concentrate their tube inspections on the fireside of the tube. As a reference, Table 11.1 gives the typical maximum heat flux.

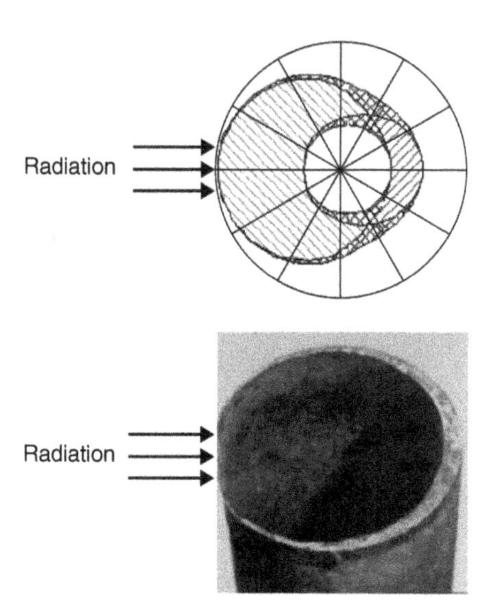

FIGURE 11.4. Tube thinning follows the flux distribution.

TABLE 11.1. Maximum Flux Rate Used in an Operating Company, Btu/h ft^2

Vertical cylindrical with tube length 20–30 ft	12,000
Vertical cylindrical with tube length >30 ft	13,000
Cabin	14,000
Double-fired U-tube	22,000

Heat distribution throughout the fired heater is not even. Radiation section makes up 70–75% of total process heat transfer, while convection section accounts for 25–30%, which can be observed in Figure 11.2. Different fuels have different heat distribution. For a gas fired heater, one-third of the heat transfer in the radiant section is flame radiation and two-thirds is hot gas radiation. If the flame height is too high, there is not enough residence time for the hot gas cloud, represented as "B" in Figure 11.2, to transfer heat to the tubes. This situation occurs when a long flame burner is placed in a short firebox. Oil firing is different. The oil flame has very high flame radiation, so approximately two-thirds of the heat transfer in the radiant section is flame radiation and one-third is hot gas radiation.

Oil and gas firing have different combustion characteristics. Oil firing is governed by flame radiation with the presence of visible flame light waves. In contrast, hot gas radiation produced by combustion is governed by gas firing. Oil has high emissivity close to one and thus to be able to drive the heat through the ash resistance.

11.2.2 Burner to Tube Clearance

Burner to tube clearance is very important in heater design because flame radiation is directly proportional to the square of the distance to the tube. Small burner to tube clearance can result in flame impingement, hot spots, and tube failure. That is why most heater failures can be traced to flame impingement due to close placement of burners to the tubes. For example, consider a 5 ft–0 in. burner to tube clearance versus 3 ft–0 in. spacing; the smaller spacing case results in 2.8 [$=(5/3)^2$] times of the flame radiation as the larger spacing.

11.2.3 Burner Selection

There are four types of burners, namely standard, premixed, staged air/fuel (low NO_x), and next generation (ultra-low NO_x). Standard gas and premixed burners have luminous flames. The combustion reaction occurs within the visible flame boundaries. Ultra-low NO_x and next-generation burners have nonluminous flames and much of the combustion reaction is not visible.

11.2.3.1 NO_x Emission NO_x emission is an important environmental issue for the process industry today. The NO_x is formed by nitrogen and oxygen reacting at the peak temperatures of the flames. A standard gas burner produces 100 ppm NO_x; staged air gas burner 80 ppm; staged gas burner 40 ppm; ultra-low NO_x gas burner 30 ppm; and the latest-generation ultra-low NO_x gas burner produces 8–15 ppm NO_x. SO_x is controlled by the sulfur in the fuel. Many plants have sulfur limits that require burning low-sulfur fuel oil. CO should be less than 20 ppm.

11.2.3.2 Objective of Burner Selection The objective of burner selection is to determine the burner type and configuration in order to obtain the desired heat flux profile to meet process heating demand. The combustion space and shape may be determined by physical, mechanical, or structural factors, but that space must be able to accommodate efficient aerodynamic mixing and combustion of the fuel and generate the desired heat flux profile for the product. The heat release and hence heat flux generated from burner flames are not even. The heat flux is generally high in the region near the burner port, where fuel and air are plentiful, and reduces as the flame develops, owing to the depleting fuel content, and loss of heat to its surroundings. The burner designer can adjust this profile from burner type and configuration and flame envelope although it never achieves uniform flux distribution.

11.2.3.3 Flame Envelope The flame envelope is defined as the visible combustion length and diameter. The flame length should be 1/3–1/2 of the firebox height. The hot combustion gases need residence time to transfer the heat to the tubes. Many burners have flame diameters that are between 1 and 1.5 times the diameter of the burner tile. Since the tile diameters are often larger for ultra-low NO_x and latest-generation burners, the flame diameters at the base of the flame may be slightly larger. The flame diameter often expands, giving a wider flame at the top.

Ultra-low NO_x and latest-generation burners have longer flame lengths than that of conventional burners. Longer flame lengths change the heat transfer profile in the firebox and can result in flame impingement on the tubes.

11.2.3.4 Physical Dimension of Firebox Optimized designs have burner spacing that is designed to have gaps between the flame envelopes. Since the tile diameters are often larger for ultra-low NO_x and latest-generation burners, retrofits can result in closer burner-to-burner spacing and flame interaction. Flame interaction can produce longer flames and higher NO_x. Flame interaction can interrupt the flue gas convection currents in the firebox, reducing the amount of entrained flue gas in the flame envelope. This condition increases the NO_x levels. Ultra-low NO_x and latest-generation burners should be spaced far enough apart to allow even flue gas recirculation currents to the burners.

The burner centerline to burner centerline dimension is one of the most important dimensions in the firebox tube. Many tube failures are caused by flame and hot gas impingement. When ultra-low NO_x and latest-generation burners are being retrofitted, the larger size of the flame envelope must be evaluated. Firebox convection currents can push the slow burning flames into the tubes.

Flame impingement on refractory often causes damage. When ultra-low NO_x and latest-generation burners are being retrofitted, the larger burner diameter may result in the burners being spaced closer to the refractory. Unshielded refractory may require hot face protection.

Many heaters are designed for flame lengths that are 1/3 to 1/2 the firebox height. Ultra-low NO_x and latest-generation burners typically have flame heights of 2–2.5 ft/million Btu (2–2.5 m/MW). Longer flame heights from ultra-low NO_x and latest-generation burners may change the heat transfer profile in the firebox. The longer flames may result in flame or hot gas impingement on the roof and shock tubes. In this case, the solution is to change burners. Some older heaters have very short firebox heights and may not be suitable for retrofits to ultra-low NO_x and latest-generation burners.

11.2.3.5 Process-Related Parameters Ultra-low NO_x and latest-generation burners have longer flames that change the heat flux profile. This is especially important on thermal cracking heaters such as cokers and visbreakers in oil refineries. The longer flames may increase the bridge wall temperature (BWT) and change the duty split between the radiant section and convection section.

The location of the maximum tube metal temperature (TMT) changes as the heat flux profile changes. Retrofitting ultra-low NO_x and latest-generation burners in short fireboxes can result in high metal temperatures for roof and shock tubes.

Ultra-low NO_x and latest-generation burners may have less turndown capability than conventional burners. High CO levels can occur when firebox temperatures are below 1240 °F. Flame instability and flameout can occur when firebox temperatures are below 1200 °F. Since ultra-low NO_x and latest-generation burners are often designed at the limit of stability, a fuel composition change may cause a stability problem.

The proper design basis for the burner selection is extremely important. Sometimes, the process requirements have changed significantly since the fired heater was designed. Important design basis factors include (a) emission requirements, (b) process duty requirements, (c) turndown requirements, (d) fuel composition ranges, (e) fuel pressure, and (f) startup considerations.

The guideline for burner selection is to select the most appropriate burner technology while meeting the NO_x emission limit. Reliability should be placed as higher priority than cost in burner selection because industrial applications show that 90% of fired heater problems come from poorly maintained and operated burners.

Although it could be more expensive with the best burner technology, the money spent is worthwhile as burners cost only 5–10% of fired heater overall cost, but it could avoid 90% of fired heater problems.

11.2.4 Fuel Conditioning System

Poor fuel conditioning could cause problems in burners and combustion. While many conventional burners have orifices 1/8 in. (3 mm) and larger, ultra-low NO_x and latest-generation burners often have tip drillings of 1/16 in. (1.5 mm). These small orifices are extremely prone to plugging and require special protection. Most fuel systems are designed with carbon steel piping. Pipe scale forms from corrosion products and plugs the burner tips. Although tip plugging is unacceptable for any burner, it is even more important not to have plugged tips on ultra-low NO_x and latest-generation burners because plugged tips can result in stability problems and higher emissions.

Many companies have installed austenitic piping downstream of the fuel coalescer/filter to prevent scale plugging problems:

- Coalescers or fuel filters are required on all ultra-low NO_x and latest-generation burner installations to prevent tip plugging problems. The coalescers are often designed to remove liquid aerosol particles down to 0.3–0.6 μm. Some companies install pipe strainers upstream of the coalescer to prevent particulate fouling of the coalescing elements.
- Piping insulation and tracing are required on fuel piping downstream of the coalescer/fuel filter to prevent condensation in fuel piping. Some companies have used a fuel gas heater to superheat the fuel gas in place of pipe tracing. Unsaturated hydrocarbons can quickly plug the smaller burner tip holes on ultra-low NO_x and latest-generation burners.

11.3 FIRED HEATER OPERATION FOR HIGH RELIABILITY

Fired heater capacity for critical processes is usually pushed hard for more production and thus the fired heaters are operated near or at the operation limits. It is essential to make sure the fired heater is running in a safe and reliable manner with the following key operating reliability parameters within acceptable limits:

- Draft control: Avoid positive pressure to prevent safety hazards and provide sufficient primary air for burners.
- High BWT: BWT directly relates to flux rate and indicates how hard a heater is running.
- TWT or TMT: Identify root cause for high TWT operation.
- Flame impingement: The most common reliability hazard for fired heaters.
- Excess air or O_2 content: Optimal O_2% is the balance between reliability and efficiency.
- Flame pattern: Visualize the flame shape, height, and color to identify abnormal combustion problems.

11.3.1 Draft

There are two types of drafts: natural draft and forced draft. For natural draft, draft depends on the density difference between hot flue gas and ambient air. Thus, stack height must be sufficient in order to provide adequate draft while stack damper opening must be adjusted properly in operation at the same time. For forced draft, stack height can be short as fan is used for providing air. Thus, stack height is only set based on dispersion requirements. Similar to natural draft, stack opening must be adjusted properly in operation. The key objective of draft control for both natural and forced draft is to avoid positive pressure inside the heater to prevent damage or safety hazards and provide sufficient combustion air (primary air) at the same time. A proper draft control is to maintain the draft at the range of 0.1–0.2 in. water column (WC) vacuum measured underneath the convection tubes or at the bridge wall (line *Y*; Figure 11.5a). This can be achieved by adjustment of both stack damper and air register. With this draft, sufficient air can be drawn in through the burners as primary air to obtain flame stability, while secondary air is provided by air register for O_2 control. However, too high or low draft must be avoided. A too high draft could occur when the damper is widely open and register fully closed. This could result in a too high vacuum in the stack and could increase cold air leakage into the heater (Figure 11.5b). Excessive draft could cause flame lift off the burners touching the tubes, and this could lead to serious damage to the heater. A too low draft corresponds to the case when the damper is almost closed and the air register is widely open. In this case, positive pressure at the bridge wall could be developed, which forces hot flue gases flowing outward through leaks in the convection section (Figure 11.5c). This could lead to serious structural damage. The draft profiles for these three cases are provided in Figure 11.5d.

Weather change could cause draft fluctuation. For example, when strong winds occur and cause draft fluctuation, the damper opening should be increased gradually to maintain flame stability. On the other hand, in a windy weather, if the heater faces toward the wind with the highest static atmospheric pressure, this may result in a too high draft. In this case, the damper should be closed slightly.

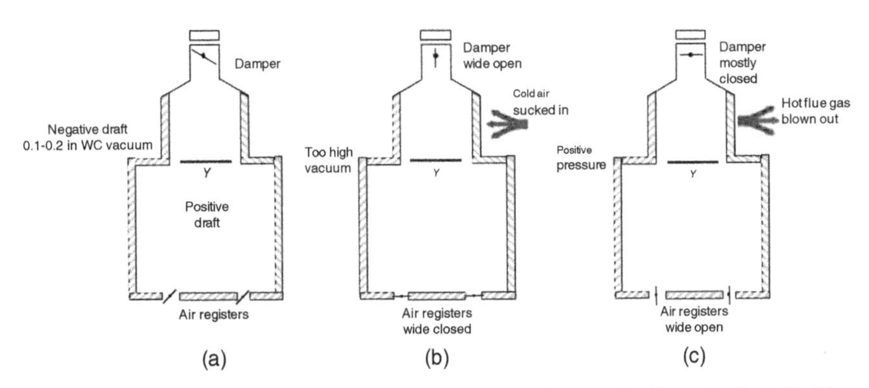

FIGURE 11.5. Correct and incorrect draft. (a) Proper draft control; (b) too high draft; (c) too low draft; (d) draft representation.

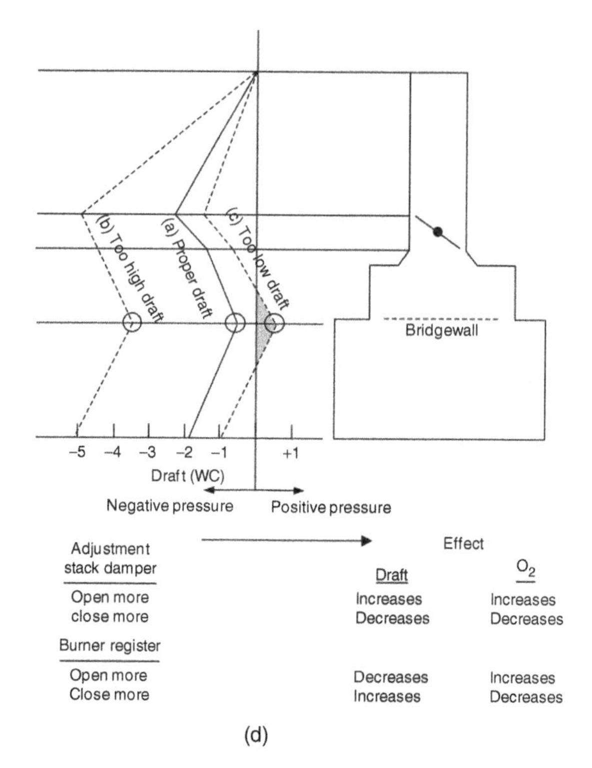

FIGURE 11.5. *(Continued)*

11.3.2 Bridge Wall Temperature (BWT)

A high BWT measurement indicates the heater operates at high radiant flux rates. BWT is the key reliability parameter for fired heaters as high BWT or TWT can cause mechanical failures on tube sheet supports and refractory. Majority of heater failures are accompanied with high BWT. The general guideline for BWT is not to exceed the mechanical design limit that tube sheet supports and the BWT limit depends on design. High BWT could be caused by long flames, inadequate flue gas residence time, and external fouling on the tubes.

11.3.3 Tube Wall Temperature

The skin temperature is the process temperature inside the tube, plus the temperature differences across the film and metal resistances. The film resistance is usually larger than the metal resistance. It is calculated by taking the peak flux and dividing by the heat transfer coefficient. The heat transfer coefficient is usually 200–500 Btu/(h ft^2 °F), which provides a typical film resistance of 45–80 °F. The metal resistance is much smaller than the film resistance. It is calculated by taking the peak flux and dividing by the thermal conductivity of the metal. The thermal conductivity is usually 12–16 Btu/(h ft °F), which results in a typical metal resistance of 15–20 °F (8–11 °C). The exception is for thick walled tubes, which could have a metal resistance as high as 80 °F (44 °C).

Tube wall or skin temperature is an important reliability parameter and should be closely monitored and guidelines for tube life can be developed. Guidelines should be effectively communicated to operators so that appropriate tube temperature can be determined that could meet the production requirement while minimizing the risk of tube damage. It is important for operators to know that overfiring is the main cause of tube damage. Process plants use skin thermocouples and infrared pyrometers to monitor TWT.

It is very important to monitor the amount of scale on tubes in order to measure coking/fouling/corrosion rates. This can be achieved by thermocouple and infrared pyrometer monitoring program. The scales on tube increase TWT or skin temperature. Ten mils (0.01 in.) scale on tube could raise the tube surface temperature by 100 °F. The common way to get rid of scale is to sandblast the scale off the tubes while ceramic coating on tubes is a preventive measure; but it is expensive.

11.3.4 Flame Impingement

Flame impingement could be caused by low air as well as burner tip fouling, which could be avoided by adjusting excessive air and fuel pressure. Figure 11.6 shows a fired heater operating with severe flame impingement in which a long flame reaches tubes and the tube front receives almost six times as much heat as the back side of tube does. The best way to know if hard flame impingement is formed is to view the firebox using the glasses especially for that purpose. These glasses eliminate the glare and bright haze and make it possible to view real flame positions.

The following guidelines for better mixing could be used to determine the root causes for flame impingement:

FIGURE 11.6. An example of flame impingement.

- Primary air is used for achieving flame stability while secondary air for O_2/NO_x control. Thus, primary air should be increased via damper opening to a limit beyond which the flame will lift off the burners. Excess air is provided by adjusting secondary air via register. Too much and too little secondary air gives poor combustion. This is because a minimum excess air is required for flame stability and too much excess air reduces flame temperature and hence efficiency drops and NO_x increases as a result.
- Close ignition ports, peep doors, and other holes around burners. Combustion air only mixes well with fuel gas when it flows through the air registers.
- At turndown operation, some of the burners may be blanketed off and do not forget to close the air registers for the idle burners. Burners work more satisfactorily close to design capacity.
- Plugged burners require more excess air for combustion but too much excess air could lift off flame. Sulfur deposits is the common cause of burner plugging and a solution is to prevent oxygen from entering the fuel gas system as it could combine with hydrogen sulfide in the fuel gas to form NH_3Cl.

11.3.5 Tube Life

Realistic average tube life can be assessed based on creep measurement and metallurgic examination. The guidelines derived from assessment should be illustrated to operators for the serious damage that could occur by operating a fired heater over the TWT limit. In general, 18 °F increase over the TWT limit could half the life of a heater. 30 °F over the TWT limit could shorten a heater's life substantially and cause rapid failure when a heater is in the creep range. It is important to know that it is the peak TWT that should not exceed the limit instead of the average TWT.

A fired heater is not operated uniformly over the entire run as it could run light in turndown operation and harder in full capacity and toward the end of run for reaction heaters. To estimate the effects of changing tube wall temperature, corrosion rates, and pressure, metallurgic examination can be applied to estimate the remaining life of tubes. Knowing the tube life not only prevents premature tube failure but also identifies the need of metal upgrade if operating skin temperature increases over time.

11.3.6 Excess Air or O_2 Content

It must be stated that optimal O_2% is the balance between safety and efficiency. There are several signs visible when a firebox is short of combustion air: a hazy flame, regular thumping sound, and long flame touching the tubes.

One of the reasons causing insufficient air is aggressive O_2% management regardless of burner conditions. Another root cause of insufficient air is the O_2 measurement based on the flue gas sample taken from the stack. This measurement is not an accurate representative of the oxygen available in the firebox. Leaks in the convection section allow air to bypass the firebox and exit in the stack and contribute to the O_2% measured in the stack. When air registers are adjusted based on the oxygen level measured from stack, the firebox could be in short of air. On the other hand, air leak is wasteful of hot flue gas for heating up cold air that is sucked into the convection section.

The cost-effective activities include seal welding of casing, mudding up header boxes, using high temperature sealants. Leaks through roof penetration are also a major source of air leak, which should be inspected during turnaround. These activities are especially important for NO_x control.

11.3.7 Flame Pattern

Proper control of combustion air is the key to make complete combustion and stable flame and thus avoid flame impingement. Lower fuel pressure also helps avoid flame impingement. When the amount of excess air is appropriate, flame is orange and flue gas from stack is light gray. With sufficient air, if flame is long with much smoke, burners may have problems.

Figure 11.7 shows a good combustion with orange color and a proper flame height about 1/3–1/2 of the firebox height. In contrast, Figure 11.8 displays a poor combustion with plugged gas tips on first burner. There is a strong haze from the flame of the first burner indicating incomplete combustion. The burner tip plugging could be reduced by using fuel gas coalescer and steam heater.

11.4 EFFICIENT FIRED HEATER OPERATION

Operators understand the importance of maintaining fired heaters in safe and reliable operation. The response from operators to this priority could go to another extreme: run fired heaters with too much excess air. The result of much excess air is much reduced flame length and thus the risk of flame impingement is minimized. However, the price for too much excess air is the higher operating cost from burning extra fuel. Therefore, there is an optimization need for excess air.

FIGURE 11.7. Good flame color and height.

FIGURE 11.8. Poor flame pattern from the first burner.

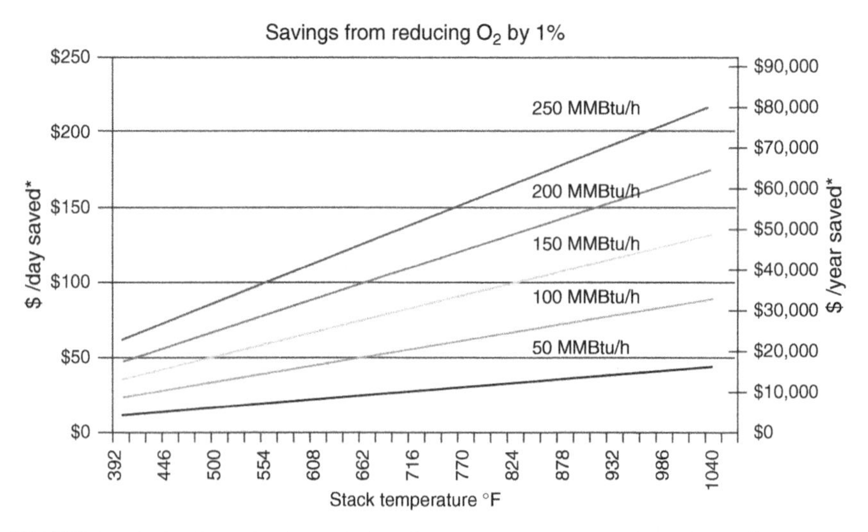

FIGURE 11.9. Dollar value for reducing O_2% by 1%. *Based on fuel price at $3/MMBtu.

Too much excess air is waste of fuel as cold air needs to be heated up from ambient to stack temperature. Figure 11.9 shows the fuel saving by dropping 1% of O_2 from reduced excess air. For example, for a heater with operating duty of 200 MMBtu/h with stack temperature at 500 °F, a reduction of 1% oxygen saves 1 MMBtu/h of fuel, which is worth $72/day or $26,280/year for fuel price at $3/MMBtu. Reducing O_2% from 7% to 3%, the saving could worth around $100,000/year. If fuel price is at $6/MMBtu, dropping 4% of O_2 could save $400,000/year. Three percent of O_2

is used as the basis for benefit calculation here as 3% is a typical limit for industrial fired heaters. However, do not start O_2% reduction before burners are in good working condition and O_2 analyzers are installed and calibrated with corrected readings.

Similarly, reducing stack temperature could improve heater efficiency more than O_2% optimization. Every 40 °F increase in stack temperature is equivalent to 1% fuel efficiency improvement. For example, a small heater with duty of 50 MMBtu/h does not have a convection section and stack temperature is at 1250 °F. If the flue gas is routed to the convection section of a large heater in a close location and thus the stack temperature could be reduced to 500 °F. Capture of this waste heat could worth $60/day and $220,000/year for fuel price at $6/MMBtu. In general, reducing stack temperature is more of a design issue, for example, installing steam generator and economizer in the convection section to recover waste heat. In contrast, O_2 level is an operation issue that can be controlled by adjusting secondary air via air register.

11.4.1 O_2 Analyzer

Fired heaters have either forced draft fans or induced draft fans to control air to the burners. This allows control of oxygen amount by direct measurement of air and fuel flow rates. Large and efficient process fired heaters with natural draft burners usually have induced draft fans. It is desirable to have control systems devised to maintain the desired amount of excess air. With O_2 analyzers, the control system adjusts damper openings automatically to control O_2 subject to a limitation on absolute draft level. Relatively small fired heaters can also justify O_2 analyzers for energy saving.

To obtain more uniform O_2 reading, every 30 ft should have one sample point and sample points should be installed downstream from the convection section. The requirement is that there should be minimum air leakage into the convection section to avoid false O_2 readings. In general, sample points should not be located in the radiation section for the reason that flue gas from different burners are not well mixed. Otherwise, the O_2 reading would mainly reflect the operation of the burners close to the sample points. The exception to placing the oxygen analyzer downstream of the convection section is for fired heaters with high tube temperatures in the convection section. This is because it is desirable to monitor radiant section oxygen to avoid afterburning.

11.4.2 Why Need to Optimize Excess Air

In an ideal combustion of fuel purely based on stoichiometric conversion, fuel is burnt to CO_2 and H_2O 100% with 0% excess air so that there is no oxygen left in the combustion flue gas. However, in reality, industrial fired heaters require excess air. To achieve complete combustion, minimum 10–15% excess air (2–3% O_2 in flue gas) is required for fuel gas. Otherwise, carbon monoxide and unburned hydrocarbon could appear in flue gas leaving stack. Fuel oil usually requires 5–10% higher excess air than fuel gas. In other words, minimum 15–20% excess air (3–4% O_2) is required for fuel oil for complete combustion.

Older heaters with poor burner conditions could have O_2% higher than 5%. This is because many older heaters are not designed for low O_2 operations. The burner flame will be very poor below the 0.15 in. H_2O. High excess air is required for these operations. For fuel oil used as fuel, black smoke is visible from stack under incomplete combustion. For fuel gas and natural gas, smoke is not visible from stack, but incomplete combustion can be measured by CO concentration in the flue gas.

Typically, 1% CO measure in the stack flue gas implies that 3–4% of fuel is wasted. Because O_2% is measured online, thus, efficient and reliable operation of heaters should maintain O_2% as close (but not less than) to the limit as possible. It is important to make sure that O_2% is not a false indication as air coming from leaking could contribute to the O_2% measured. Too little excess air available for combustion could cause flame impingement to tubes and cause local hot spots and coking on tubes eventually cause severe tube damage. Another consequence of too little excess air is afterburning in the convection section, which could result in elevated tube temperature, which is the root cause for premature tube failure and sagging of horizontal tubes. This is because the fired heater undergoes incomplete combustion and thus the combustible or CO in the flue gas increases. The incomplete combustion makes lazy flames. These long flames can reach tubes in the radiation section and even convection section. In the worst case, flames could reach the exit of stack.

So what is optimal excess air or O_2%? The basis is to achieve complete combustion. For reliability considerations, optimal O_2% should be determined with a safety margin on top of minimum excess air when burners are under good conditions. The safety margin depends on specific technology, design, and conditions for each heater as well as measurement. Figure 11.10 is commonly used to explain qualitatively the existence of optimal excess air.

The more rigorous way than O_2 measurement is to measure CO in the flue gas. This can be accomplished by measuring combustibles in the flue gas. Combustibles here refer to the products of incomplete combustion including carbon monoxide (CO), hydrogen, and trace hydrocarbons while CO accounts for the majority of combustibles. For consistency with O_2 measurement, the combustibles measurement should be taken in the same location as the O_2 analyzer. With reliable combustibles measurement available for ppm concentration, it allows the O_2% level to be reduced safely (safety margin) until the combustibles start to increase (Figure 11.11). This is the optimal O_2% for the heater.

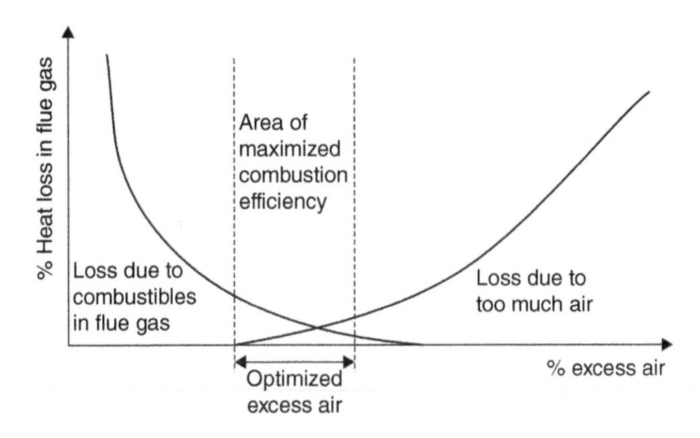

FIGURE 11.10. Optimizing excess air.

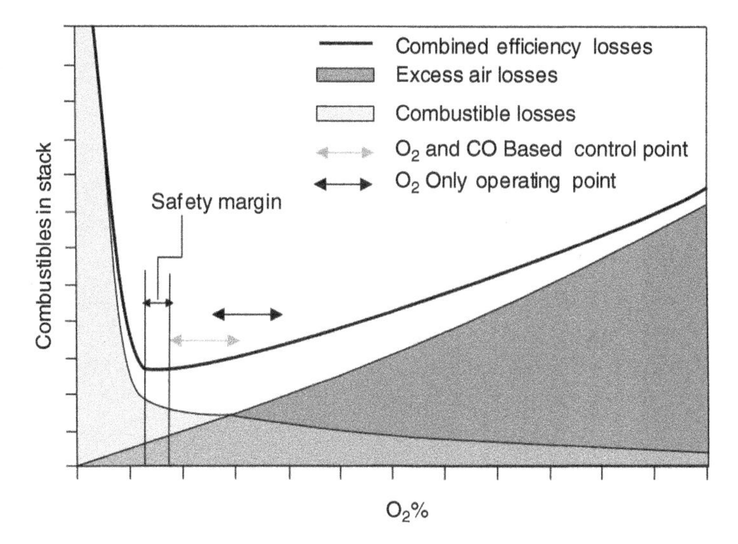

FIGURE 11.11. Determining optimal O_2% level.

11.4.3 Draft Effects

Efficient heater operation requires that excess air entering the convection section be minimized, which is indicated by a very small negative pressure at the convection section inlet. To achieve this, it should have a well-balanced draft pressure profile between firebox and stack. The hot gas pushes so that the pressure is always greatest at the firewall while the stack draft pulls. "When this draft is correctly balanced, the pressure at the bridge wall should be around 0.1–0.2 WG (water gauge)." Too much draft allows cold air leakage into fired box resulting in wasted fuel.

11.4.4 Air Preheat Effects

Air preheating is a classical example of upgrading low-valued heat. This is done by providing heat to raise the combustion air temperature from the ambient temperature using waste heat. Air preheat can be accomplished via low pressure steam or flue gas. Typically, air preheat can increase fired heater efficiency up to 5%, which is more significant than reducing O_2%.

11.4.5 Too Little Excess Air and Reliability

Too little excess air could result in flame impingement and afterburn in the convection section, which impose reliability risks. With too little excess air, incomplete combustion occurs and reduces flame temperature, which might encourage operators to increase fuel flow in order to increase heater duty. Increased fuel with too little excess air enhances afterburn and could be dangerous.

11.4.6 Too Much Excess Air

This is inefficient operation and should be avoided. According to Kenney (1984), the common causes of too much excess air are as follows:

- Improper draft control
- Air leakage into the convection section
- Improper calibration of O_2 analyzer
- Faulty burner operation: (a) dirt burners, (b) poor maintenance on air doors, and (c) dual fuel burners needed.

11.4.7 Availability and Efficiency

Making fired heater in high availability is desirable for continuous production without interruption. As a consequence, a plant can achieve high profit and high energy efficiency at the same time. Experience from the industry indicates that high availability is the major contributor to improved energy efficiency.

11.4.8 Guidelines for Fired Heater Reliability and Efficient Operation

Draft and excess air control should be considered together in operation. This is because draft provides primary air while air register delivers secondary air for burners. As discussed before, both air supplies could affect reliability and efficiency. The systematical method for optimizing draft and excess air together is proposed in Figure 11.12.

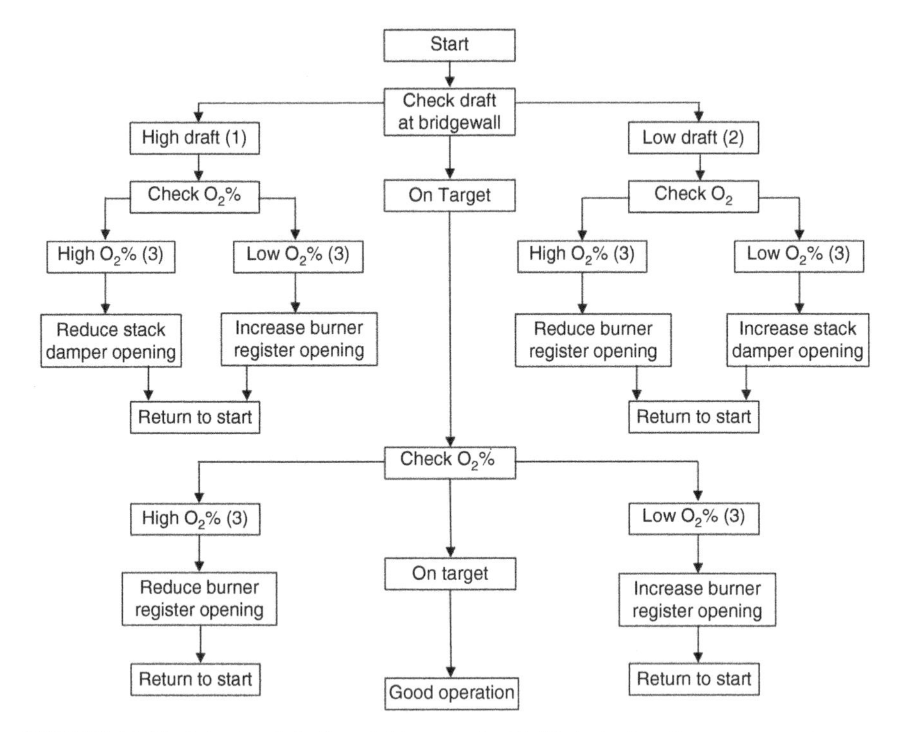

FIGURE 11.12. Integrated draft and O_2 control. (1) High draft – fire box pressure more negative; (2) low draft – fire box pressure more positive; (3) low or high $O_2\%$ – $O_2\%$ is above or below target.

I feel the need to provide additional comments on excess air as many plants have an O_2 reduction program. O_2 reduction (or minimum excess air) must be built upon the basis of proper draft control. Minimum excess air for the fired heater can be obtained when it is reduced to the point where combustibles begin to appear in the stack. For modern fired heaters, this occurs at 8% excess air equivalent to 1.8% of oxygen level in the flue gas. However, practical constraints prevent achieving this minimum excess air in operation, and these constraints include variations in fuel quality, feed rates, and other process variables. Thus, operation without flame impingement sets the limit for practical minimum excess air. The optimal flue gas O_2 concentration depends on the heater duty, burner design, types of fuel, and burner performance.

To achieve the limit, the first step is monitoring O_2%. O_2 measurement must reflect the true amount of excess air and air leaks must be eliminated. The following guidelines can be used for operation reference:

- O_2 analyzers should be installed below the convection section instead of stack. If not, a correction factor must be developed for the readings with a portable analyzer. O_2 analyzers should be calibrated once per week.
- Efficiency based on stack temperature and corrected O_2% should be reported daily.
- Draft should be monitored and maintained as required for the specific fired heater design. Even fired heaters without draft control should be periodically checked.
- Convective section air leakage should be measured once per shift and determined as the difference in convective inlet and outlet O_2. The source of leakage should be identified via inspection and eliminated. Ideally, all oxygen should enter the fired heater through the burners.
- Coil flow paths should be balanced within ±5% accuracy once per shift in order to obtain equal outlet temperatures. On large fired heaters, this may be as often as every 2 h, or continuously with control systems.
- In cases of turnarounds and large load changes, flue-gas parameters (draft, O_2%, etc.) should be checked and adjusted as necessary.
- Soot blowers on oil-fired fired heaters and boilers should be activated once a shift. The operator should observe which ones actually rotate and report (in writing) those soot blowers that have failed. Where operability of soot blowers is less than 70%, an alternative plan for cleaning should be prepared and executed. This may include on-stream water-washing.
- The need for on-stream cleaning of outside tube surface should be evaluated. This may include water-washing of both convective and radiant sections.

11.5 FIRED HEATER REVAMP

In general, fired heaters are revamped for capacity expansion, process conversion changes, energy efficiency, and NO_x reduction. For capacity-expansion revamps, the type of limitation for the revamp is usually the same as for the original design. In conversion revamps, one type of process technology is converted to another. Thus, in conversion revamps, a heater designed for one service may be used in a new service. Therefore, the type of heater limitation may be different for the new service.

Heaters encounter with four major design limitations: heat flux, process pressure drop, TWT, and BWT. Heat flux limited heaters are usually characterized with high pressure ΔP (>20 psi) and most general service heaters fall into this flux-limited category. Typically, the flux limit for single fired heaters is around 10,000 Btu/ft^2 h. Small heaters (<10 MM Btu/h process duty) have lower flux rates. For revamps, the heat flux limit can go up to 12,000 Btu/ft^2 h.

Double-fired heaters are usually TWT-limited. However, flux limits are specified for revamps, which depend on specific services. The limits are provided by heater specialists.

TWT-limited heaters are characterized by low process ΔP (2–6 psi). Because of low ΔP, the heaters have low tube mass velocities, which result in low heat transfer coefficients, and thus high tube wall temperatures. TWT-limited heaters usually occur in high temperature processes. For example, TWT of 800 °F is used for killed carbon steel heaters. The chrome (Cr) limits are based on inhibiting tube oxidation and use a limit of 1075–1100 °F per recent data. For stainless steel (SS), process temperature limits usually occur before reaching the TWT limit. There can be exception for high-pressure heaters.

On the other hand, BWT-limited heaters are usually encountered when a heater with conventional burners is replaced with low NO_x/ultra-low NO_x/new generation burners or the revamp requires higher turndown ability. Flame instability and flame-out can occur at low BWT. For fuel gas firing with ultra-low NO_x and next-generation burners, BWT should be greater than 1200 °F. For low NO_x burners, the BWT should be greater than 1000 °F. For oil firing in combination burners, the BWT shall be greater than 1200 °F. When determining the BWT limit, the burner spacing and BWTs at normal and turndown operations must be investigated on the stability and operation of burners.

There is much more to discuss about heater revamp, which is beyond the scope of this chapter. As a general recommendation, heater revamp projects should be conducted by heater specialists who not only have good knowledge of the heaters but also process that the heaters serve.

NOMENCLATURE

BWT bridge wall temperature
TMT tube metal temperature
TWT tube wall temperature

REFERENCES

Jenkins BG, Boothman M (1996) Combustion science: a contradiction in terms, *Petroleum Technology Quarterly*, **Autumn**, 71–76.

Kenney, WF (1984) *Energy Conservation in the Process Industries*, Academic Press, Inc.

12

PUMP ASSESSMENT

12.1 INTRODUCTION

The pump's role is to move the fluid through the system at desired flow rate and pressure. Pumps include centrifugal type and positive displacement types. The latter usually includes reciprocating and rotary pumps. The following discussions focus on the centrifugal pump as it is the most common type used in the process industry.

There are two basic tasks for pump selection. The first is to determine the pump head required for given process requirement, while the second is to select the pump that can deliver the desired flow rate through a pump system under required head. The required head depends solely on the process characteristics (suction pressure, discharge pressure, and liquid density), while the flow rate relies on the pump characteristics (impeller size and speed). As a process engineer, your role is to find the best match between these two characteristics and make sure the pump selected can satisfy process requirement in the most reliable and efficient manner.

In selecting and operating a pump, it is essential that you have good knowledge of the process requirement in terms of pump head and flow rate required, how the pump works, and guidelines for pump selection and operation. In particular, adequate knowledge of the process conditions and stream compositions is the most important aspect for optimizing pump selection. Process constraints must be considered such as operation flexibility, turndown, startup, and shutdown. With good understanding of process conditions and constraints as well as pump characteristics, application of API Standard 610 for centrifugal pumps and Standard 682 for mechanical seals will result in improved reliability and extended on-stream operation.

Hydroprocessing for Clean Energy: Design, Operation, and Optimization, First Edition.
Frank (Xin X.) Zhu, Richard Hoehn, Vasant Thakkar, and Edwin Yuh.

12.2 UNDERSTANDING PUMP HEAD

Why is pump performance always measured in flow versus head rather than flow versus pressure? This is the myth we want to clarify through discussions here. When the pump operates in a process unit, the process requires the liquid at a desired flow rate to be delivered from the suction pressure (P_S) to discharge pressure (P_D). If the pump is capable, it will move fluid forward at the required discharge pressure for given suction pressure. Otherwise, if the pump cannot create enough head to move the fluid forward at the discharge pressure, it will operate at no flow.

The relationship between the pump head and differential pressure is described by the following formula:

$$H_T = \frac{2.31(P_D - P_S)}{SG}, \tag{12.1}$$

where H_T = pump head, ft; P_D = discharge pressure, psig; P_S = suction pressure, psig; specific gravity $SG = \rho/\rho_{H_2O}$ and ρ is the density of the liquid; standard water density $\rho_{H_2O} = 62.4\,\text{lb/ft}^3$ ($1000\,\text{kg/m}^3$) at temperature 4 °C (39.2 °F); 2.31 is the conversion factor. Note that equation (12.1) will be derived later.

By measuring pump head the density or specific gravity of the fluid is already accounted for. Consider Figure 12.1 for a visual interpretation where three identical pumps are pumping three fluids of different specific gravity or density under the same suction atmospheric pressure. The head or height of the fluids is the same for three cases even though the discharge pressures and power requirements are different. Thus, by using pump head, the pump performance depends on the pump mechanical design characteristics instead of types of liquids. As a result, the pump performance curve based on pump head versus flow rate will remain constant for any liquids. This explains the reason why the pump performance is described by pump head instead of pressure.

Otherwise, use of discharge pressure could be problematic for specifying a pump because the discharge pressure depends upon the suction pressure and the specific gravity of the liquid being pumped. The specific gravity changes with temperature,

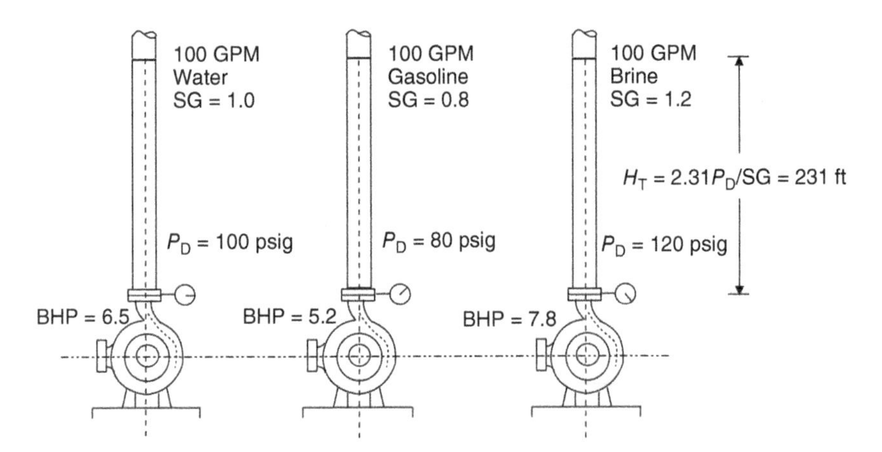

FIGURE 12.1. Pump head applies to any liquid (pump operating under no flow condition).

type of fluid, and fluid composition. But the pump manufacturer does not know these process parameters in prior and use of pump head for pump selection can avoid these uncertainties.

12.3 DEFINE PUMP HEAD – BERNOULLI EQUATION

After you have some understanding about the pump head, you may want to know how the head is defined and calculated. This is the question we focus in this and the following sections.

Let us consider Figure 12.2. The principle of Energy Conservation states that energy is neither created nor destroyed but is simply converted from one form of energy to another. Bernoulli equation is the most well-known expression for this principle.

Thus, applying the Bernoulli equation to the liquid at points A and B in Figure 12.2 gives

$$\frac{P_A}{\rho} + \frac{u_A^2}{2g} + Z_A = \frac{P_B}{\rho} + \frac{u_B^2}{2g} + Z_B, \tag{12.2}$$

where three energy components are involved as follows:

- Potential energy from elevation: Z in ft using the pump suction centerline as the datum
- Pressure energy: P/ρ in ft as P in lb/ft^2 and density ρ in lb/ft^3; P is the surface pressure acting upon the liquid surface
- Kinetic energy from flow velocity: $u^2/2g$ in ft as velocity u in ft/s and acceleration g is 32.2 ft/s^2.

If a pump is placed between A and B as shown in Figure 12.3, the net liquid column in terms of pump head (H_T) should be added to the LHS of equation (12.2) to account for the net energy input by the pump on the suction side, which yields

$$\frac{P_A}{\rho} + \frac{u_A^2}{2g} + Z_A + H_T = \frac{P_B}{\rho} + \frac{u_B^2}{2g} + Z_B. \tag{12.3}$$

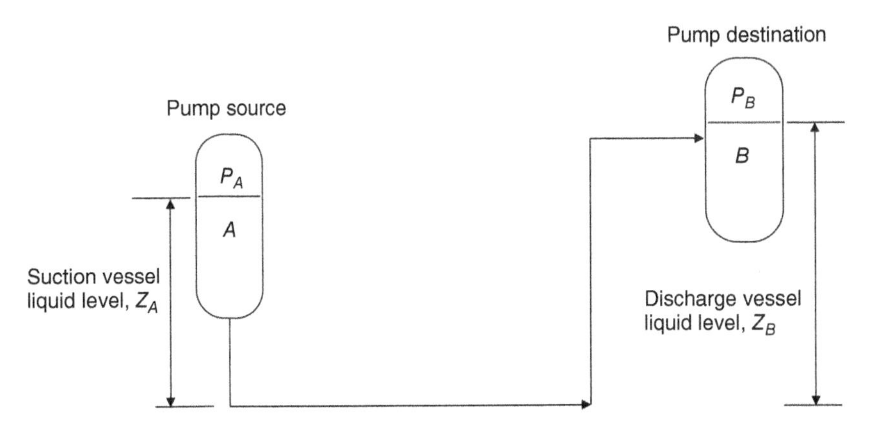

FIGURE 12.2. A simple process system.

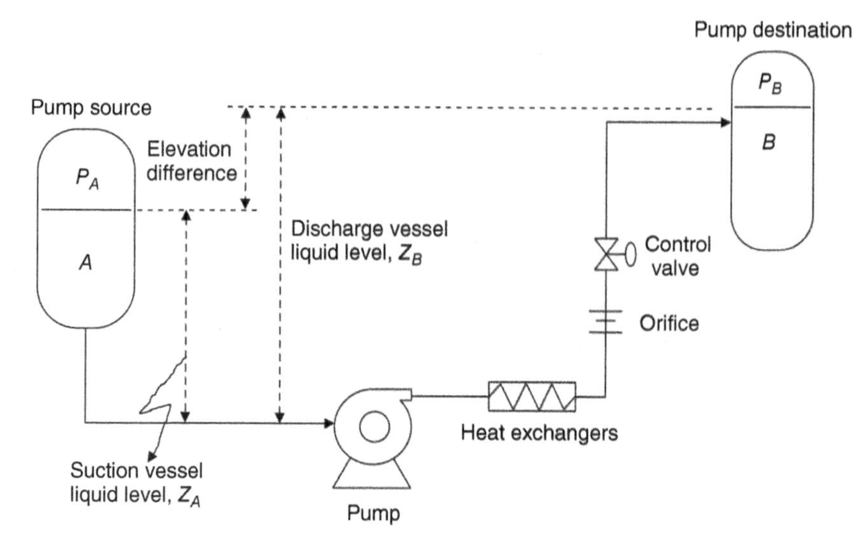

FIGURE 12.3. A practical process system.

In reality, a process will have piping and fittings, heat exchangers, and control valves between the source and destination. Thus, friction losses occur to both sides of the pump: $H_{A,\mathrm{f}}$ represents the friction loss in the suction side while $H_{B,\mathrm{f}}$ is the friction loss in the discharge side. To count for these losses, $H_{A,\mathrm{f}}$ must be deducted from the LHS (left-hand side) of equation (12.3) as it makes negative contribution to the suction energy. In contrast, the friction loss $H_{B,\mathrm{f}}$ must be added to the RHS (right-hand side) of equation (12.3) as it increases the energy requirement in the discharge side. Thus, equation (12.3) becomes

$$\frac{P_A}{\rho} + \frac{u_A^2}{2g} + Z_A - H_{A,\mathrm{f}} + H_\mathrm{T} = \frac{P_B}{\rho} + \frac{u_B^2}{2g} + Z_B + H_{B,\mathrm{f}}. \tag{12.4}$$

The pump head is derived as

$$H_\mathrm{T} = \frac{P_B - P_A}{\rho} + (H_{B,\mathrm{f}} + H_{A,\mathrm{f}}) + \frac{u_B^2 - u_A^2}{2g} + (Z_B - Z_A). \tag{12.5}$$

If the surface pressure P_A, P_B are expressed in psi instead of lb/ft^2, the conversion of pressure to liquid head is

$$H = \frac{144(P_B - P_A)}{\rho} = \frac{2.31(P_B - P_A)}{\mathrm{SG}}. \tag{12.6}$$

Thus, equation (12.5) can be expressed as

$$H_\mathrm{T} = \frac{2.31(P_B - P_A)}{\mathrm{SG}} + (H_{A,\mathrm{f}} + H_{B,\mathrm{f}}) + \frac{u_B^2 - u_A^2}{2g} + (Z_B - Z_A) \tag{12.7a}$$

or

$$H_T = \frac{2.31(P_B - P_A)}{SG} + H_f + \frac{u_B^2 - u_A^2}{2g} + (Z_B - Z_A), \tag{12.7b}$$

H_T is called total dynamic head required from a pump while the terms in the order of sequence in RHS in equation (12.7b) are static surface pressure head, friction head, velocity head, and static elevation head, respectively. Equation (12.7b) states that a pump must overcome the total head H_T to deliver a desired process fluid rate in order to satisfy the process requirement.

Let us consider the following example to illustrate the Bernoulli equation as it is always beneficial to walk through theory with practical examples.

Example 12.1 Pump water to a higher location

According to equation (12.7b), the total pump head is

$$H_T = \frac{2.31(P_2 - P_1)}{SG} + H_f + \frac{u_2^2 - u_1^2}{2g} + (Z_2 - Z_1),$$

where the differential static head $(Z_2 - Z_1)$ is 31 m. P_1 and $P_2 = 0$ because the water surfaces open to atmosphere (Figure 12.4). $u_1 = 0$ because the inlet reference is the reservoir water surface (the decline in the large reservoir surface is negligible). u_2 can be calculated as

$$u_2 = \frac{Q}{A} = \frac{4 \times (0.102)}{\pi (0.295)^2} = 1.49\,\mathrm{m/s}.$$

Thus, the velocity head $u_2^2/2g$ is 0.11 m.

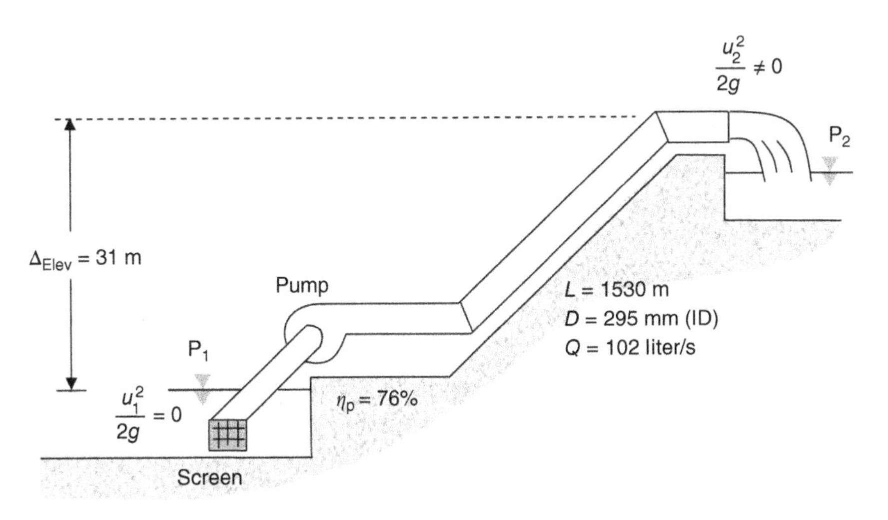

FIGURE 12.4. Illustration of Bernoulli equation (12.7).

The flow resistances occur in screen, three elbows, and piping, resulting in total friction loss as

$$H_{\mathrm{f}} = H_{\mathrm{f,screen}} + 3H_{\mathrm{f,elbow}} + H_{\mathrm{f,pipe}}.$$

For the 12 in. pipe made of PVC, the friction coefficient can be found in engineering design manual to be 0.0141. The piping friction loss can be calculated as

$$H_{\mathrm{f,pipe}} = f\frac{L}{D}\frac{u^2}{2g} = 0.0141\frac{1530}{0.295} \times 0.11 = 8.04\,\mathrm{m}.$$

For a 12-in. pipe and 45° flanged elbow, the friction coefficient is 0.15. Thus, the elbow friction loss can be calculated via

$$H_{\mathrm{f,elbow}} = K\frac{u^2}{2g} = 0.15 \times 0.11 = 0.0165\,\mathrm{m}.$$

By assuming the screen friction loss as 0.2 m, the total friction loss becomes

$$H_{\mathrm{f}} = H_{\mathrm{f,screen}} + 3H_{\mathrm{f,elbow}} + H_{\mathrm{f,pipie}} = 0.2 + 3 \times 0.0165 + 8.04 = 8.26\,\mathrm{m}.$$

Thus,

$$\begin{aligned} H_{\mathrm{T}} &= \frac{2.31(P_2 - P_1)}{\mathrm{SG}} + H_{\mathrm{f}} + \frac{u_2^2 - u_1^2}{2g} + (Z_2 - Z_1) \\ &= 0 + 8.26 + 0.11 + 31 = 39.37\,\mathrm{m}. \end{aligned}$$

The above calculations give one total pump head required for delivering the flow rate of 102 lb/s.

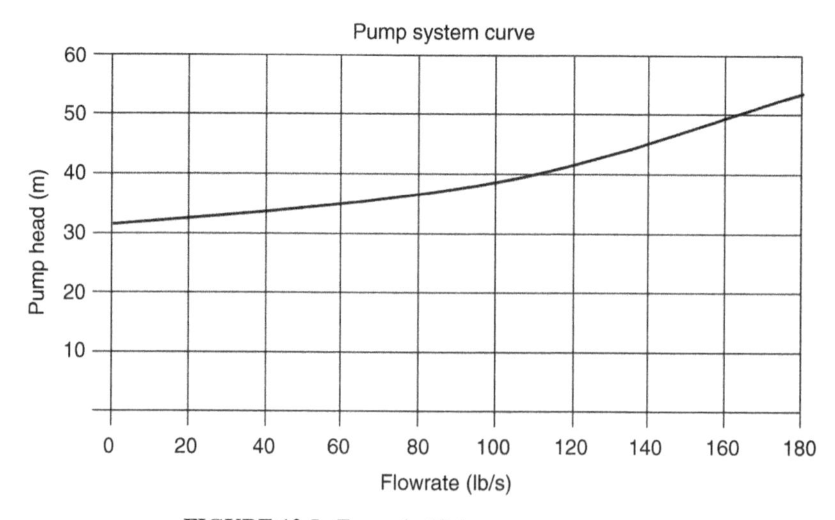

FIGURE 12.5. Example 12.1 pump system curve.

When flow rate varies, the calculation of the pump head can be repeated as above. Thus, the relationship of pump head and flow rate can be developed to form the system curve as shown in Figure 12.5 and the characteristics of the system curve will be discussed later. In conclusion, total head is a function of the process as it is comprised of suction pressure, discharge pressure, liquid specific gravity, friction losses, and elevation. All of these are process conditions.

12.4 CALCULATE PUMP HEAD

Let us take another look at equation (12.7a) and it can be expressed as

$$H_{\mathrm{T}} = \left[\frac{2.31P_B}{\mathrm{SG}} + H_{B,\mathrm{f}} + \frac{u_B^2}{2g} + Z_B\right] - \left[\frac{2.31P_A}{\mathrm{SG}} - H_{A,\mathrm{f}} + \frac{u_A^2}{2g} + Z_A\right]. \tag{12.8}$$

Or simply (if use D denoting discharge or the destination and S for suction or the source)

$$H_{\mathrm{T}} = H_{\mathrm{D}} - H_{\mathrm{S}}, \tag{12.9a}$$

where

$$H_{\mathrm{D}} = \frac{2.31P_{\mathrm{D,SP}}}{\mathrm{SG}} + H_{\mathrm{D,f}} + \frac{u_{\mathrm{D}}^2}{2g} + Z_{\mathrm{D}}, \tag{12.9b}$$

$$H_{\mathrm{S}} = \frac{2.31P_{\mathrm{S,SP}}}{\mathrm{SG}} - H_{\mathrm{S,f}} + \frac{u_{\mathrm{S}}^2}{2g} + Z_{\mathrm{S}}, \tag{12.9c}$$

where SP stands for surface pressure.

Following equation (12.9b), total discharge head (H_{D}) can be expressed in head as

$$H_{\mathrm{D}} = H_{\mathrm{D,SP}} + H_{\mathrm{D,f}} + H_{\mathrm{D,v}} + H_{\mathrm{D,E}}, \tag{12.10}$$

where

$$H_{\mathrm{D,SP}} = 2.31\frac{P_{\mathrm{D,SP}}}{\mathrm{SG}},$$

$$H_{\mathrm{D,v}} = \frac{u_{\mathrm{D}}^2}{2g},$$

$$H_{\mathrm{D,E}} = Z_{\mathrm{D}},$$

and subscript D stands for discharge, v for velocity head, and E for elevation head.

Similarly, total suction head (H_S) can be expressed in head based on equation (12.9c) as

$$H_S = H_{S,SP} - H_{S,f} + H_{S,v} + H_{S,E}, \tag{12.11}$$

where

$$H_{S,SP} = 2.31\frac{P_{S,SP}}{SG},$$

$$H_{S,v} = \frac{u_S^2}{2g},$$

$$H_{S,E} = Z_S,$$

and subscript S stands for suction, v for velocity head, and E for elevation head.

In simplicity, you can think of total suction head as if you stand right at the pump suction flange and look back from the pump toward the suction tank. With this perspective, you can understand why the suction friction head takes a negative sign because the suction friction head negatively contributes to the total suction head. Similarly, you can think of total discharge head as if you stand right at the pump discharge flange and look forward from the pump toward the discharge terminal. Therefore, the pump must overcome the surface pressure, elevation, and friction losses in the discharge side.

By expanding equation (12.9a) based on equations (12.10) and (12.11), we have

$$H_T = (H_{D,SP} - H_{S,SP}) + (H_{D,f} + H_{S,f}) + (H_{D,v} - H_{S,v}) + (H_{D,E} - H_{S,E}). \tag{12.12}$$

Equation (12.12) indicates that we calculate the total head based on the individual energy differences between the suction and discharge sides.

12.5 TOTAL HEAD CALCULATION EXAMPLES

The following examples show how the total head is determined based on the individual energy differences defined in equation (12.12). It can be demonstrated from the following examples that with this decomposition, the calculations of the total head can be simplified while better understanding of the key components contributing to the total head can be obtained.

Example 12.2 Calculate total head

The process conditions and equipment elevations are shown in Figure 12.6. The maximum and minimum liquid levels in the column on the suction side are 32 and 22 ft, respectively. The specific gravity of the liquid is 0.488. The elevation of the pump is 3 ft while the liquid level in the column on the discharge side is 72 ft. The pressure drop in the suction piping is 1 psi. In the discharge side, the pressure drop through 6 in. pipe at 500 gpm is 10 psi while the pressure drops through orifice and control valve

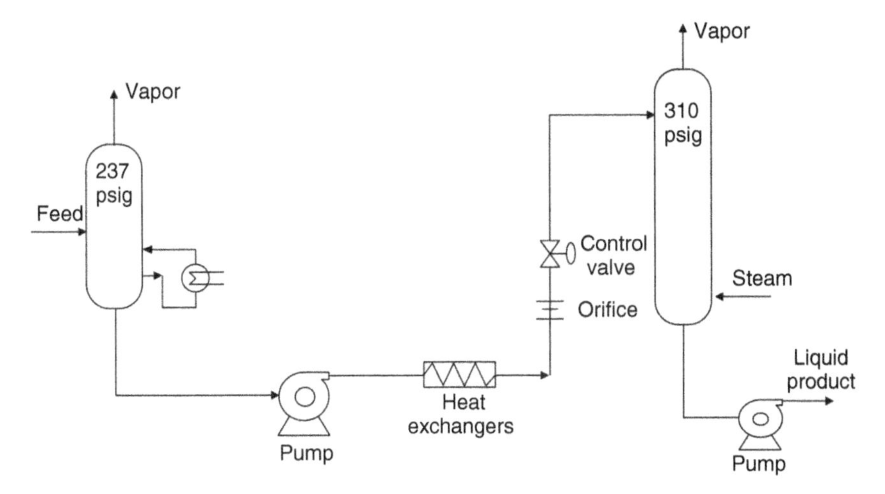

FIGURE 12.6. Process system.

are 2 and 30 psi, respectively. The pressure drop through the exchangers is 15 psi. Calculate the pump head.

Solution:

(i) Total suction head:

- The suction surface pressure is 237 psig. Thus, $H_{S,SP} = 2.31 \times 237/0.488 =$ 1121.86 ft
- The pressure drop in suction piping due to friction loss at 500 gpm is 1 psi. Thus, $H_{S,f} = 2.31 \times 1/0.488 = 4.73$ ft
- The velocity head $H_{S,v}$ is almost zero as the suction reference in this case is the surface of the suction tank.
- The suction elevation head $H_{S,E} = 22 - 3 = 19$ ft, where the pump is 3 ft above suction centerline. For being conservative, the minimum suction vessel level of 22 ft is used.

  Therefore, the total discharge head is

$$H_S = H_{S,SP} - H_{S,f} + H_{S,v} + H_{S,E} = 1121.8 - 4.7 + 0 + 19 = 1136.1\ \text{ft}.$$

(ii) Total discharge head:

- Discharge surface pressure =310 psig. Thus, $H_{D,SP} = 2.31 \times 310/0.488 =$ 1467.4 ft
- The total discharge friction head ($H_{D,f}$) is the sum of all the friction losses in the suction line:
  - Pressure drop through 6 in. pipe at 500 gpm is 10 psi.
  - Pressure drops through orifice and control valve are 2 and 30 psi, respectively.

  - Pressure drop through exchangers is 15 psi.
  - Thus, the total discharge friction head is the sum of the above losses, $H_{D,f} = 2.31 \times (10 + 2 + 30 + 15)/0.488 = 269.8$ ft at 500 gpm.
- The velocity head $H_{D,v}$ is almost zero as the discharge reference in this case is the surface of the discharge tank.
- The discharge elevation head $H_{D,E} = 72 - 3 = 69$ ft.

  Therefore, the total discharge head is

$$H_D = H_{D,SP} + H_{D,f} + H_{D,v} + H_{D,E}$$
$$= 1467.4 + 269.8 + 0 + 69 = 1806.2 \text{ ft gauge.}$$

(iii) Thus, the total system head:

$$H_T = H_D - H_S = 1806.2 - 1136.1 = 670.1 \text{ ft for } 500 \text{ gpm.}$$

(iv) Alternatively, since the pressure drops for the piping, fittings, and exchangers are given in this example, the pump head required can be calculated based on the total pressure drop between the discharge and suction flanges via

$$H_T = \frac{2.31(P_D - P_S)|_{\text{at flanges}}}{SG}. \tag{12.1}$$

The pressure at the suction flange is

$$P_S = P_{S,SP} - \Delta P_{S,f} + P_{S,E} = 237 - 1 + \frac{(22 - 3) \times 0.488}{2.31} = 240 \text{ psi.}$$

The pressure at the discharge flange is

$$P_D = P_{D,SP} + \sum \Delta P_{D,f} + P_{D,E} = 310 + (10 + 2 + 30 + 15)$$
$$+ \frac{(72 - 3) \times 0.488}{2.31} = 381.6 \text{ psi.}$$

Thus, the total pressure drop between the discharge and suction flanges is

$$(P_D - P_S)|_{\text{at the flanges}} = 381.6 - 240 = 141.6 \text{ psi.}$$

Total pump head required can then be calculated based on equation (12.1) as

$$H_T = \frac{2.31 \times 141.6}{0.488} = 670.1 \text{ ft for } 500 \text{ gpm.}$$

12.6 PUMP SYSTEM CHARACTERISTICS – SYSTEM CURVE

Bernoulli equation (12.7) defines the total head requirement and describes the pump system characteristics. The equation can be represented graphically by the system curve as shown in Figure 12.7, which describes the relationship between the system

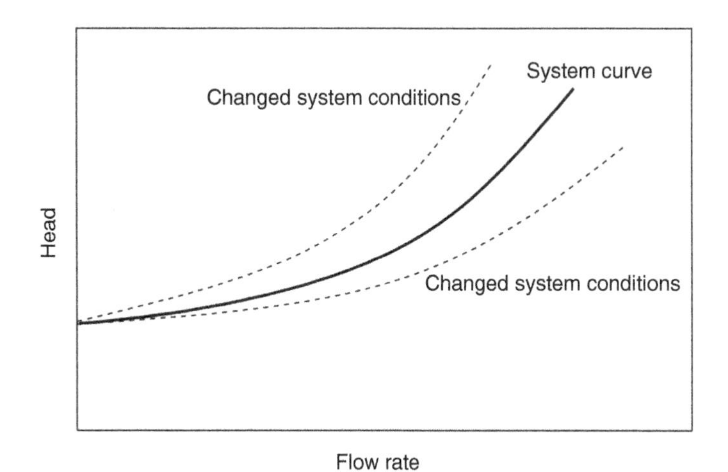

FIGURE 12.7. Pump system curve.

total head and flow rate for a given pump system. A few important facts for the pump system curve are summarized as follows:

- There is only one system curve for a given pump system design and control valve openings.
- The system curve changes if the pump system conditions vary (e.g., degree of opening varies in control valves).
- The shape of the system curve is parabolic because the terms for friction losses and velocity head take the exponent of 2.
- The system curve is independent of the pump mechanical characteristics.
- At zero flow, the curve will be vertically offset due to static head or elevation difference. At the zero flow, equation (12.12) becomes

$$H_T|_{Q=0} = (H_{D,E} - H_{S,E}). \tag{12.13}$$

12.6.1 Examples of System Curves

For illustration purposes, system curves for different pump system designs are provided as follows. You may need to think why these curves look like as shown.

Case 1: All friction losses and no static lift (Figure 12.8)
Case 2: More static lift and less friction losses (Figure 12.9)
Case 3: Negative static lift (Figure 12.10)
Case 4: Two different static lifts in a branching pipe system (Figure 12.11).

12.7 PUMP CHARACTERISTICS – PUMP CURVE

A pump curve indicates the pump capability to overcome the system head, which is different from the system curve that indicates the system requirement of pump head

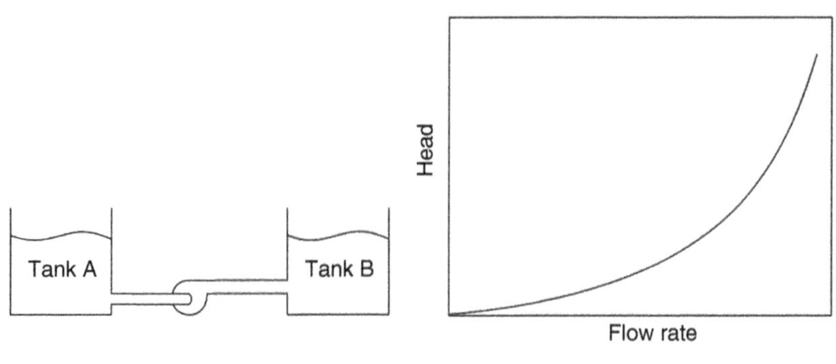

FIGURE 12.8. System curve for no static lift.

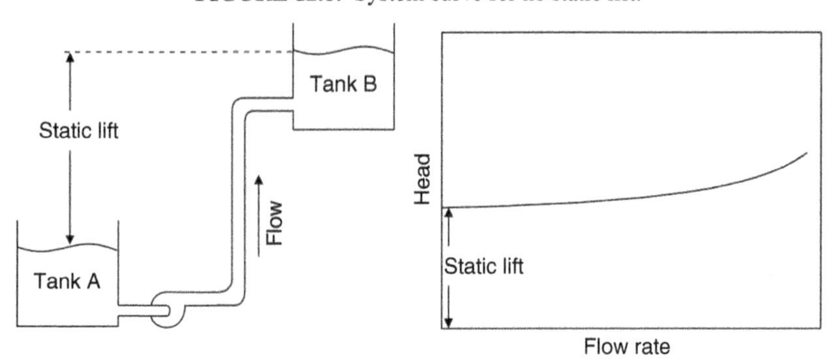

FIGURE 12.9. System curve for small friction losses.

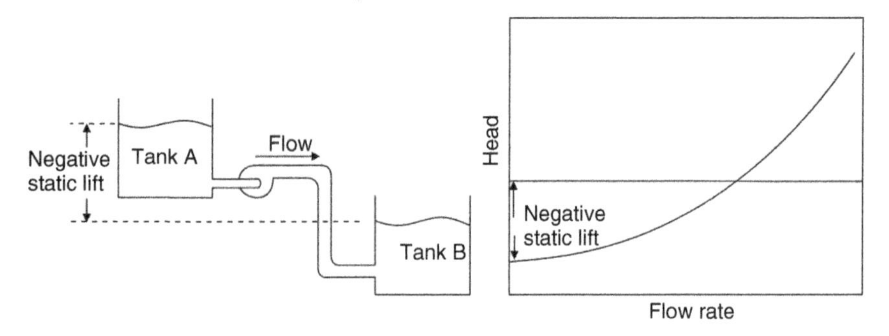

FIGURE 12.10. System curve for negative static lift.

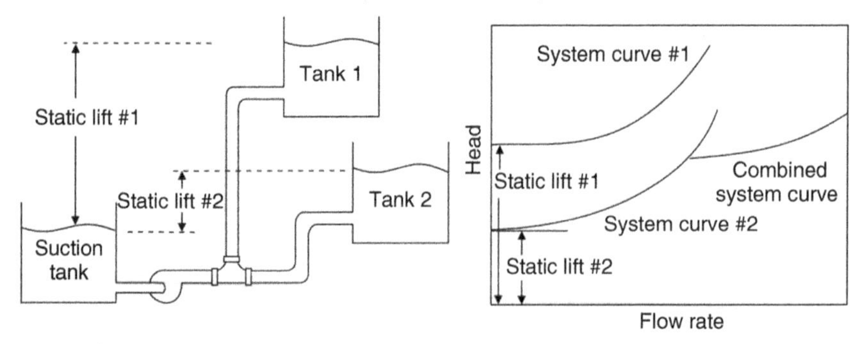

FIGURE 12.11. System curve for double discharges.

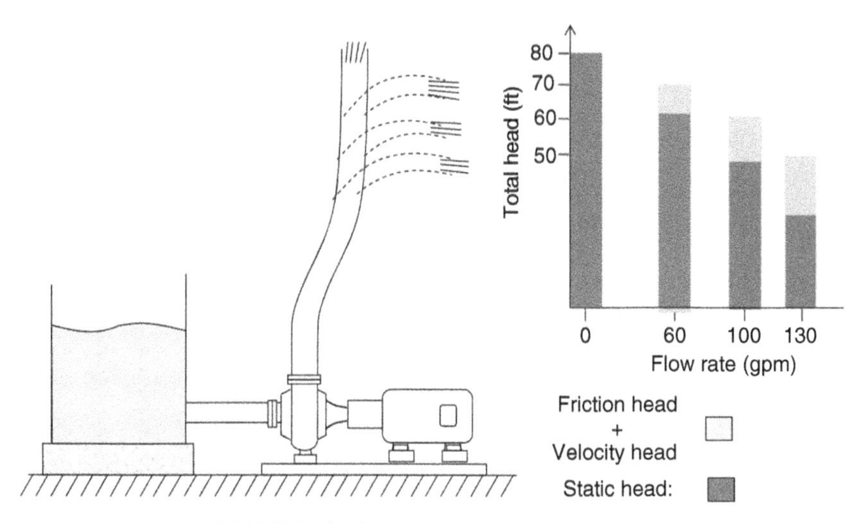

FIGURE 12.12. Pump head versus flow rate.

in terms of elevations and friction losses. Thus, while the system curve is a function of the process, the pump curve is the function of the pump. You may ask how a pump curve is generated and how to interpret it.

To answer this question, let us conduct a simple experiment Figure 12.12 with the pump suction at atmospheric condition: raise the discharge pipe end of the pump vertically until the flow stops. This means that the pump cannot raise the fluid higher than this point. Thus, we generate a zero flow point with the maximal pump head (H_T). For illustration, let us assume $H_{T,max} = 80$ ft. The maximum liquid height corresponds to the maximum discharge pressure as

$$H_{T,max}|_{Q=0} = \frac{2.31 P_{D,max}}{SG}. \tag{12.14}$$

At the zero flow when the pump is still running, the friction head is zero since there is no flow. Thus, the pump head is equal to the static head with zero flow. The pump head at zero flow is called shut-off head when the discharge valve is closed. Of course, a pump should not run under this condition continuously as the liquid would rapidly heat up to a temperature greater than what the seal can tolerate and the pump could be severely damaged.

If the discharge pipe is now cut at a slightly lower height, say 70 ft, there is a certain amount of flow out of the pipe, which can be measured and assumed to be 60 GPM for this example. Now we have a second point at 60 GPM flow on the flow (Q) – head (H) diagram.

If we continue to cut the pipe at several descending heights and measure the flow rates, more Q–H points can be generated as shown in Figure 12.12. By connecting these points, a pump performance curve is obtained as in Figure 12.13. This may explain how a pump curve is generated. However, use of cutting the pipe here is just for a symbolic illustration. In reality, it relies on flow measuring and pressure reading devices.

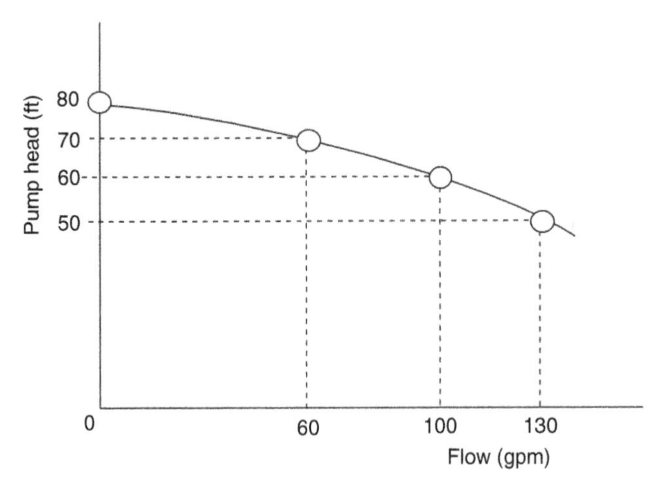

FIGURE 12.13. Pump curve for Figure 12.12.

Key pump characteristics can be summarized as follows:

- Head and flow: The performance curve indicates the range of pump head that a pump is capable of providing over a range of flow rates. The process conditions dictate the pump head that the pump selected will operate at, while the pump curve determines the flow rate according to the head. If the head goes down, the flow rate goes up. If the head goes up, the flow rate goes down.
- Design point: The system curve defines the process requirement, which becomes the basis for choosing a pump. The pump selected must satisfy the process requirement of flow rate and head. Thus, by plotting the selected pump curve on top of the system curve, the intersection of these curves as shown in Figure 12.14 is known as the design point.

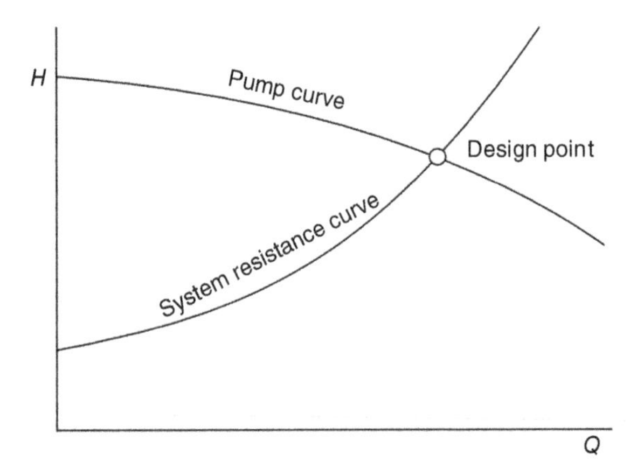

FIGURE 12.14. Pump normal operating point.

- Impeller size: Increasing the impeller diameter changes the performance capability. As a result, the pump curve moves upward. As an example, Figure 12.15 shows pump curves for pump impellers with different diameters under a given operating speed. Consider Figure 12.16 and assume the process head is 2400 ft. If the pump has a 9 ½ in. impeller, the flow rate will be 600 gpm. If the pump has a 10 ½ in. impeller, the flow rate will be slightly over 1000 gpm.

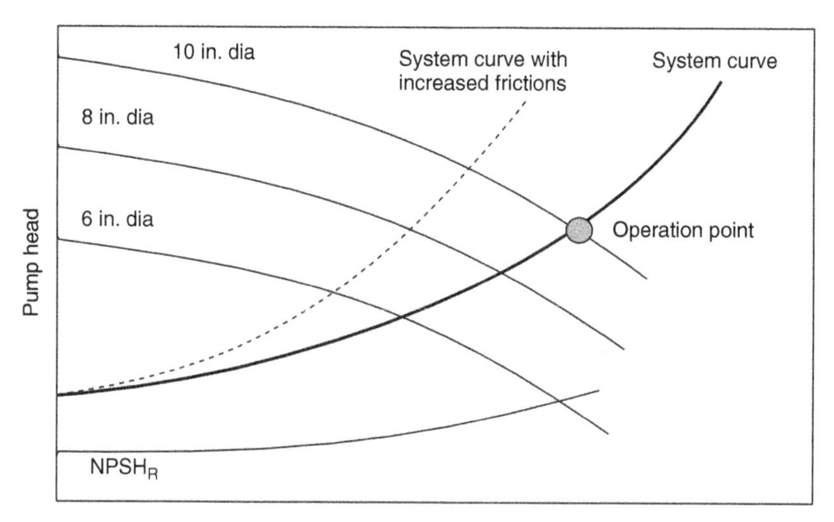

FIGURE 12.15. Pump curve and system curve could change.

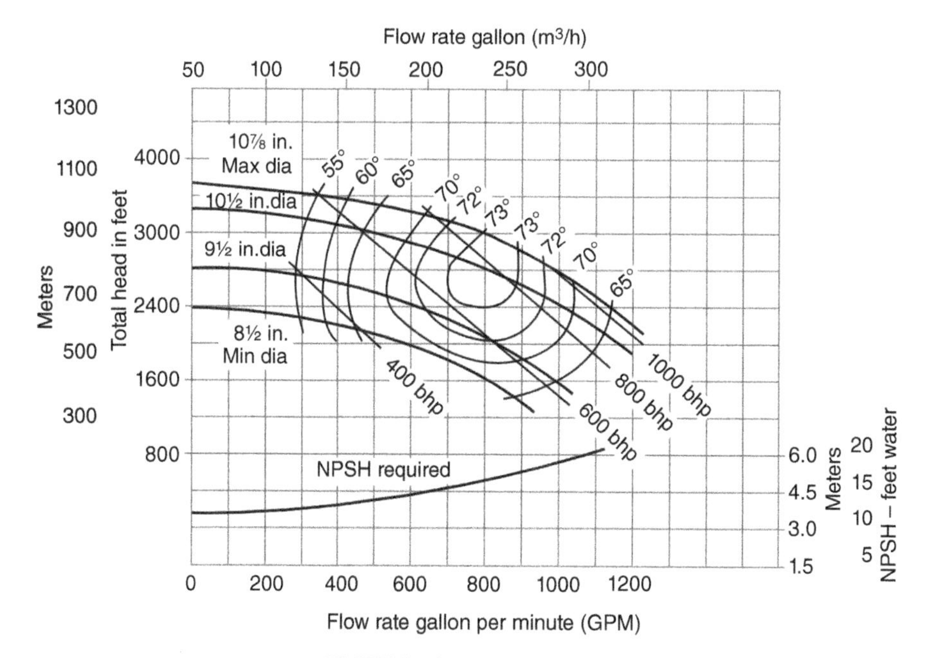

FIGURE 12.16. Pump curve.

- Pump speed: The pump curve indicates the performance at a certain speed (e.g., 3450 RPM, a common electric motor speed in 60 Hz countries). Increasing pump speed has similar effect as increasing impeller size.
- Brake power: The pump power in brake horsepower (BHP) can be calculated as

$$\text{BHP} = \frac{(\text{GPM} \times \text{SG})H_{\text{T}}}{3960\eta}. \tag{12.15}$$

where flow rate is in GPM (gallon per minute), total head H_{T} in ft, and η is the pump efficiency in %.

In summary, increasing the impeller diameter and/or speed will raise the pump flow-head curve. Pump flow rate will vary with the suction and discharge pipe diameter and length as the system friction drop changes. A system with a long and narrow discharge pipe will lead to high friction loss and thus lower flow rate. In this case, the pump system curve will move upward. The pump is designed to produce a certain nominal flow rate for the piping system sized accordingly. The impeller size and its speed dictate the pump to deliver the nominal flow rate. To change the flow rate in operation, appropriate valves must be adjusted.

12.8 BEST EFFICIENCY POINT (BEP)

The best efficiency point (BEP) indicates that the pump will operate most efficiently and reliably at the BEP point. Per API 610, the rated flow (process condition) shall be 80–110% of the BEP. At the BEP, there will be minimal amount of vibration and noise. This is because at BEP the impeller is balanced radially. At flows higher and lower than the BEP flow, there is a radial force on the impeller. Of course, the pump can operate at other flow rates, higher or lower than the BEP flow.

12.9 PUMP CURVES FOR DIFFERENT PUMP ARRANGEMENT

Series arrangement: This arrangement may be required for higher head applications. Centrifugal pumps are connected in series if the discharge of the first pump is connected to the suction side of the second pump. Two similar pumps, in series, operate in the same manner as a two-stage centrifugal pump. Each of the pumps is putting energy into the pumping fluid, so the resultant head is the sum of the individual heads. If two of the same pumps are in series, the combined performance curve will have double the head of a single pump for a given flow rate (Figure 12.17). For two different pumps, the head will still be added together on the combined pump curve, but the curve will most likely have a piecewise discontinuity. At the same flow rate, the heads of the two pumps are added together. For example, if a single pump operating at 50 gpm at 70 ft of head and 3 bhp is put in series with an identical pump, the two pumps will provide 140 ft of head and require 6 bhp total. Each pump supplies the required flow at one-half the required head.

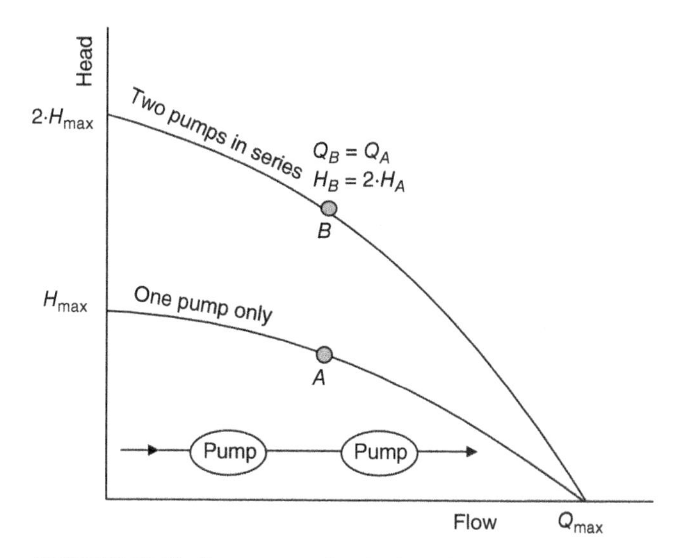

FIGURE 12.17. Pump curves for single and two pumps in series.

Some things to consider when you connect pumps in series:

- Both pumps must have the same width impeller. Otherwise, the difference in capacities could cause a cavitation problem if the first pump cannot supply enough liquid to the second pump. For the same reason, both pumps must run at the same speed.
- When pumps are operating in series, if either the low pressure or the high pressure booster pump fails, the remaining pump will operate at zero flow! Therefore, the second pump in series should be automatically shut down at low flow rate.
- Be sure the casing of the second pump is strong enough to sustain the higher pressure. Higher strength material, ribbing, or extra bolting may be required.
- Be sure both pumps are filled with liquid during startup and operation.
- Start the second pump after the first pump is running.

Parallel arrangement: This arrangement may be selected for large flow rate. Pumps are operated in parallel when two pumps are connected to a common discharge line and share the same suction conditions. The pump curves for single and two pumps in parallel are given in Figure 12.18. For example, a single pump operating at 50 gpm at 70 ft of head and 3 bhp is put in parallel with an identical pump. The result is a total flow of 100 gpm, requiring 6 bhp and operating at 70 ft of head. Each pump supplies one-half the required flow at the required head. For the same head, the flow rates are added together.

Some things to consider when pumps are operated in parallel:

- Both pumps must produce the same head, which is resulted from the same speed and the same diameter impeller.

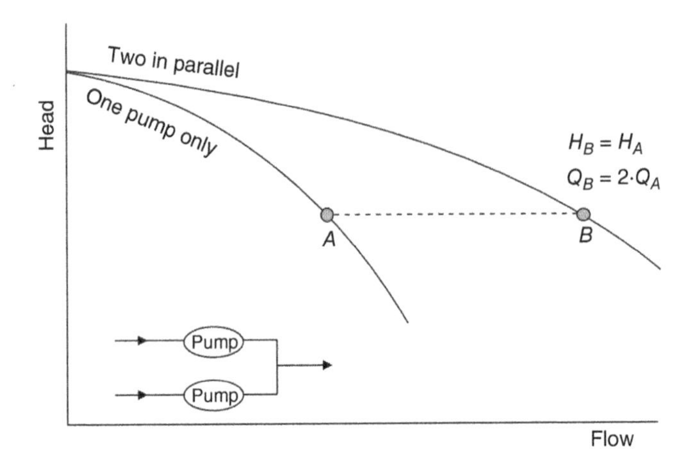

FIGURE 12.18. Pump curves for single and two pumps in parallel.

- When pumps are operating in parallel, if the internal clearances in one pump deteriorate substantially more than the other, the stronger pump may force the other to operate below its minimum continuous capacity.
- Two pumps in parallel will deliver less than twice the flow rate of a single pump in the system because of the increased friction in the piping. If there is additional friction in the system from throttling, two pumps in parallel may deliver a flow only slightly more than a single pump.
- Most plants read only total flow and cannot see the differences in individual pump performance.
- Parallel pumps are notorious for operating at different flows. Often a weaker pump is operating close to its shut-off point, while a stronger pump is operating to the far right of its curve and running out of net positive suction head available ($NPSH_A$). This is why it is important to have similar curves that rise to shutoff.

12.10 NPSH

Cavitation occurs when there is presence of vapors in the impeller. As the fluid enters the pump and impeller there is a pressure drop. This pressure drop is caused due to entrance losses as the liquid enters the pump and friction losses in the pump nozzle. There are also friction and turbulence losses as the liquid enters the impeller. If this pressure drop causes the fluid to drop below its vapor pressure or bubble point, the fluid will start to boil and vapor bubbles will occur. When vapor bubbles occur, there is expansion. One cubic feet of water at room temperature will create 200 ft^3 of vapor.

Pump cavitation occurs when the vapor bubbles develop within the pump casing. When the fluid pressure rises as the fluid leaves the impeller these vapor bubbles collapse. The liquid strikes the impeller and casing at the speed of sound. The noise generated from these collisions of vapor bubbles sounds like pumping marble stones. Cavitation could possibly stop flow altogether and damage the impeller. Thus, noise and capacity loss are the major indicators of cavitation.

Under cavitation, the first pump seal will be damaged due to high vibration. In the propane service, running with too low a liquid level for a few hours will often damage the mechanical seal sufficiently to require taking the pump out of service. Passing vapor through the impeller causes rapid changes in the density of the fluid pumped. This uneven operation forces the impeller and shaft to shake, and the vibration is transmitted to the seal. A mechanical seal consists of a ring of soft carbon and a ring of hard metal pressed together. Their smooth, polished surfaces rotate past each other. When either surface is chipped or marred, the seal leaks. Continued operation of a cavitating pump will damage its bearing and eventually the impeller wear ring and shaft.

To avoid cavitation, the liquid pressure within the pump should never fall below the vapor pressure of the liquid at the pumping temperature. There must be enough pressure at the pump suction and thus not vaporize the fluid. This pressure available at the pump suction over the fluid vapor pressure is called as NPSH_A, where A stands for "Available", NPSH_A is a function of the pumping system.

12.10.1 Calculation of NPSH_A

For better understanding of NPSH_A, consider a typical pump suction system shown in Figure 12.19 and NPSH_A can be defined as

$$\text{NPSH}_\text{A} = P_{\text{S,A}} - P_{\text{S,V}}, \tag{12.16}$$

where

$$P_{\text{S,A}} = P_{\text{S,SP}} + \frac{H_{\text{S,E}} \times \text{SG}}{2.31} - \Delta P_{\text{S,f}},$$

$P_{\text{S,A}}$ = actual suction pressure, psia; $P_{\text{S,SP}}$ = surface pressure at the liquid free surface in the suction, psia; $P_{\text{S,V}}$ = vapor pressure of liquid under suction temperature, psia; $\Delta P_{\text{S,f}}$ = pressure drop due to suction friction losses, psia.

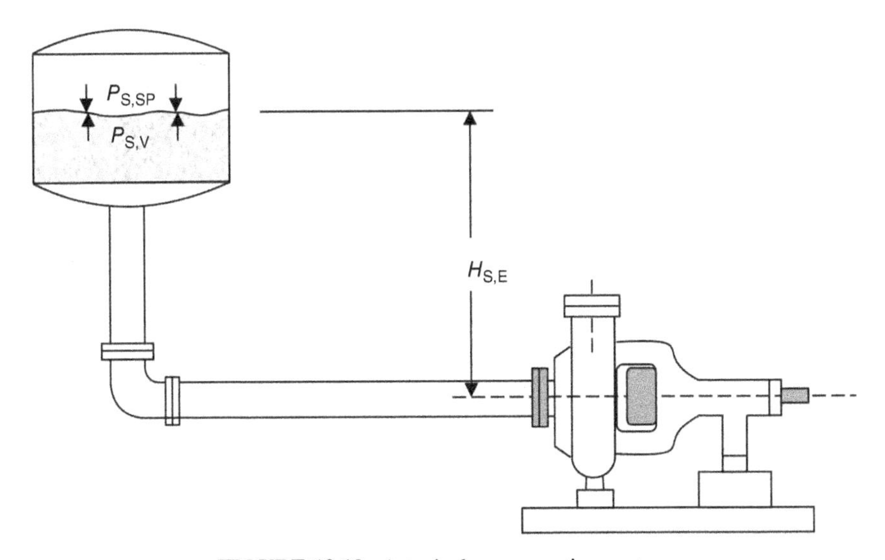

FIGURE 12.19. A typical pump suction system.

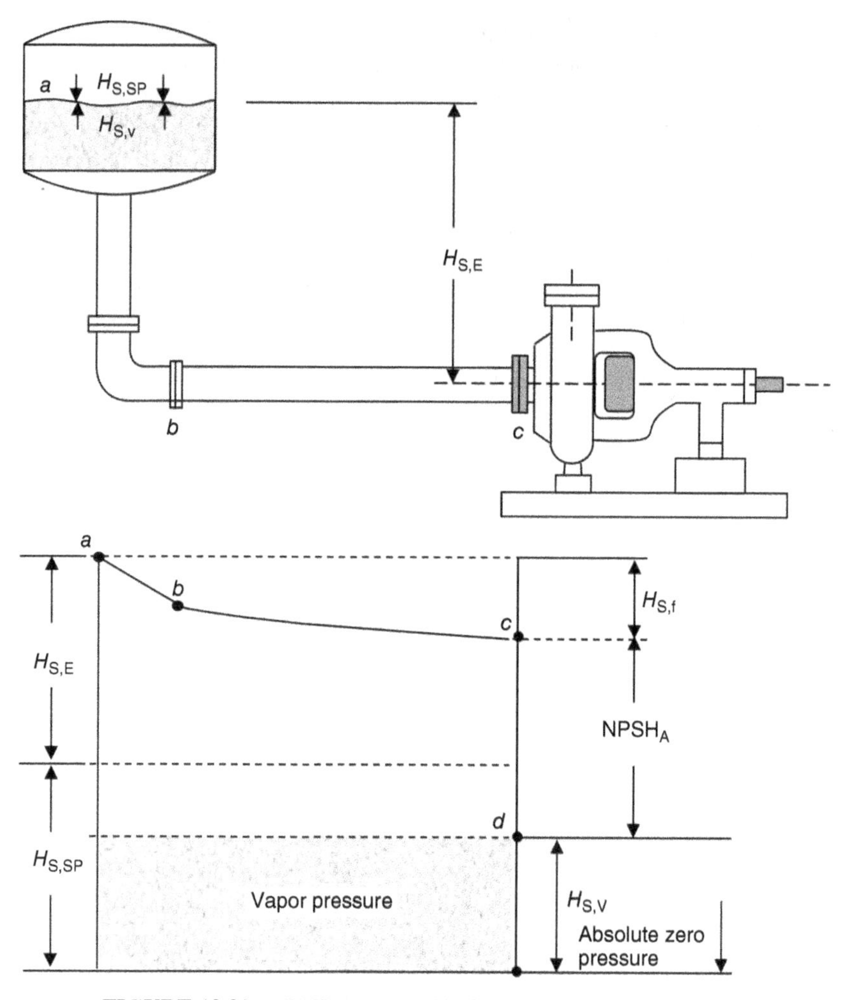

FIGURE 12.20. NPSH$_A$ expressed in feet for typical pump suction.

If all pressures are converted to feet of liquid as shown in Figure 12.20, available NPSH can be calculated via

$$\text{NPSH}_A = (H_{S,SP} \pm H_{S,E} - H_{S,f}) - H_{S,V}, \tag{12.17}$$

where

$H_{S,P} = 2.31 P_{S,SP}/\text{SG}$; absolute pressure at the liquid free surface converted to feet of liquid.

$H_{S,E}$ = suction head (takes "+") or lift (takes "−") with liquid specific gravity considered, in feet of liquid; make sure to use the lowest liquid level allowed in the tank.

$H_{S,V} = 2.31 P_{S,V}/\text{SG}$, vapor pressure of liquid (P_V) at pumping temperature converted to feet of liquid.

$H_{S,f}$ = friction loss through suction line, fitting and entrance, in feet of liquid.

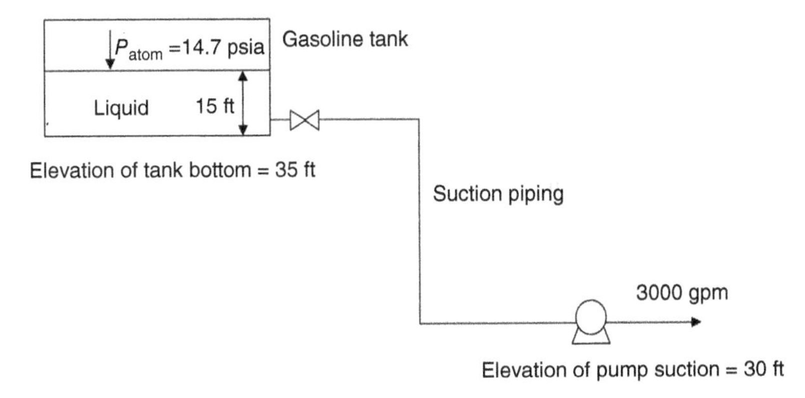

FIGURE 12.21. Pump suction for Example 12.3.

If the surface pressure is atmospheric pressure, 34 ft ($H_{S,P}$) is the value of atmospheric pressure at sea level. If the suction head ($H_{S,E}$) is 10 ft, $NPSH_A$ is 44 ft minus a small quantity of friction loss and vapor pressure based on equation (12.17). This $NPSH_A$ should be sufficient.

One should start to have concern when the $NPSH_A$ falls to within 4 ft of the net positive suction head required ($NPSH_R$). But how could this happen? This is possible if the pipe diameter is small and pipe length is long plus a lot of plugging, which increases friction in the suction line. Also during startup, the suction strainer is prone to plugging if the suction piping is not sufficiently cleaned.

Note that the above calculation does not include the velocity head, which is common for $NPSH_A$ as velocity head is small relatively. But velocity head is included in the $NPSH_R$ curves provided by the manufacturer.

Example 12.3

Gasoline is stored in an open tank and the tank is piped to a centrifugal pump. The pump suction system is depicted in Figure 12.21. The pressure drop through the suction piping is = 0.186 psi. Vapor pressure of gasoline at pumping temperature = 5.0 psia. SG of gasoline = 0.74. Calculate the available NPSH at a flow rate of 3000 gpm. If $NPSH_R$ is 22 ft, will the pump cavitate?

Solution:

$$H_{S,SP} = 14.7 \times 2.31/0.74 = 45.89\,\text{ft}; \quad H_{S,E} = 15 + 35–30 = 20\,\text{ft};$$

$$H_{S,f} = 0.186 \times 2.31/0.74 = 0.58\,\text{ft}; \quad H_{S,V} = 5 \times 2.31/0.74 = 15.61\,\text{ft};$$

$$NPSH_A = H_{S,SP} + H_{S,E} - H_{S,f} - H_{S,\ V} = 45.89 + 20–0.58–15.61 = 49.7\,\text{ft};$$

Thus, the pump will not cavitate as $NPSH_A$ = 49.70 ft is much larger than $NPSH_R$ = 22 ft.

12.10.2 NPSH Margin

$NPSH_R$ is a characteristic of the pump and it is provided by pump vendor. Per API 610, the pump vendor can report NPSH required when there is a 3% loss of head due to cavitation. So if NPSH available is equal to NPSH required, there is cavitation. Therefore, there must be a margin of NPSH available over NPSH required. For example, a margin of 4 ft for hydrocarbon liquids (including low S.G.) and 10 ft for boiling water.

During initial system design, one variable is suction vessel height. For example, the minimum liquid level in the suction vessel must be at a high enough elevation to provide the required margin.

12.10.3 Measuring NPSHA for Existing Pumps

NPSH calculations must be conducted for every pump installation. It is also recommended to do the same for existing pumps. To know the true NPSHA for existing pumps, first install a compound gauge at the pump suction that can measure both vacuum pressures as well as positive gauge pressures. When the pump is running, the reading from this gauge will indicate the suction pressure. For given vapor pressure at the pump temperature, NPSHA can be calculated via equation (13.17) and the safety margin should be provided, which depends on the company's best practice. If this NPSHA value calculated is less than the pump's NPSHR provided by the manufacturer, this pump is under cavitation.

12.10.4 Potential Causes and Mitigation

There are various causes for a low NPSH, and major causes are summarized as follows. If changes to the system are not adequate to increase $NPSH_A$, you may need to consult the pump manufacturer about reducing $NPSH_R$.

(i) Low P_S due to reduced pressure at the suction nozzle

This is the basic cause when the fluid at pump suction is not available sufficiently above the vapor pressure of liquid at operating conditions.

(ii) Low P_S due to low density of the liquid

Light hydrocarbon and vacuum services encounter low NPSH more frequently.

(iii) Low P_S due to low liquid level

The problem may be a low liquid level in the suction side vessel. The level controller on the pump discharge line may have malfunctioned. To check this, blow out the taps on the vessel gauge glass and verify that there is a good level in the vessel.

(iv) Low P_S due to increase in the fluid velocity at pump suction

Higher liquid velocity leads to lower suction pressure. At the same time, higher velocity results in higher friction losses in the piping and fittings in pump inlet system. The above consequence is the lower pressure available at the pump suction and thus cavitation has a greater chance to occur.

(v) Low P_S due to plugged suction line

A suction-line restriction will also cause a pump to cavitate. To verify this, run the pump with its discharge valve pinched back just enough to suppress cavitation. Then, measure the pressures at the upstream vessel and the pump suction (use the same gauge) to calculate the pressure difference ΔP. Then, ΔP is converted to the liquid height as $2.31\Delta P/SG$ of liquid. Finally, subtract the vertical distance between the two gauges from the liquid height to give the pump head. The resulting head at the pump suction should only be 1–2 ft less than the head of liquid in the vessel. If the head difference is quite a bit more than 2 ft, there is probably a plugged suction line.

(vi) High P_V due to increase in the pumping temperature

Vapor pressure is a function of temperature only. Increase in liquid temperature at the pump suction increases the vapor pressure of the liquid. It becomes more likely for operating pressure to fall below this vapor pressure. In some cases, a slight warming of the fluid at the pump suction could promote flashing.

(vii) Reduction of the flow at pump suction

A certain minimum flow as indicated by the pump curves is required to keep the pump from running dry. If liquid flow falls below this limit, it has greater possibility of developing vapor within the pump and the likelihood of cavitation increases.

(viii) The pump is not selected correctly

Every centrifugal pump has a certain requirement of positive suction head ($NPSH_R$). If the pump is not selected properly, $NPSH_A$ might fall below this $NPSH_R$ limit, causing cavitation.

According to Fernandez et al. (2002), increasing static head is the most viable approach. There are three basic methods to raise the static head:

- Lower the pump elevation. However, this could prove to be less practical since pumps are typically located just above the ground level. Lowering the pump suction may require the suction nozzle to be below grade, which usually results in a more expensive pump.
- Raise the level of fluid in the suction tank.
- Increase the suction piping diameter and remove elbows.

Reduction in friction losses through suction piping and fittings can also mitigate the risk of low NPSH. Reducing friction is more appealing in the existing plants where throughput is usually increased above the design throughput.

There are also other options available, which include using a larger but slower speed pump, a double-suction impeller, and a larger impeller inlet area.

12.11 SPILLBACK

A "spillback" is a jargon for a partial flow recycle as a small percentage of a pump's discharge flow is routed back to the suction of the pump to ensure that the pump has a sufficient continuous flow. Spillback may be required for preventing cavitation and failures in bearing and seals if the process flow falls below the pump minimum continuous flow for a prolonged duration. In some cases, spillback may be required to support off-design operation requirements, particularly start-up. A proper spillback system must be selected for a specific application.

There are three kinds of minimum flow limits. The objective of defining minimum flow limits is to prevent undue wear and tear in bearings and seals. In the real environment of a process plant, a pump is operated at just about any condition demanded by the situation at hand. Thus, these different pump minimum flows are used for different application purposes. In other words, for a specific application, the pump may be governed by a certain minimum flow limit.

The minimum continuous stable flow (MCSF) is the lowest flow at which the pump may operate without exceeding vibration limits imposed by API Standard 610 for hydrocarbon process industry. It is the flow below which the pump should not be operated continuously. If the minimum continuous flow through a pump is insufficient, the resultant pump damage takes place over the long term and usually results in bearing or seal failure.

Generally speaking, single stage, overhung process pumps with typically low power requirements are less susceptible to damage and cheaper to repair when subjected to minimum flow limits. Multistage pumps are much more susceptible to damage when subjected to these situations and are much costlier to repair. Therefore, multistage pumps may require a spillback system and an automatic low flow shutdown of the driver. The automatic low flow shutdown removes energy input in the upset event and limits the duration/severity of dry running condition when the suction vessel liquid level is lost.

Most Sundyne pumps require spillback systems because of their limited turndown capability, high temperature rise, and drooping head curves. Specify a spillback for the pump turndown requirement is 60% or lower of the rated pump capacity.

12.12 RELIABILITY OPERATING ENVELOPE (ROE)

From the pump curve provided by manufacturing, we can determine the preferred operation reliability operating envelope (ROE) in the flow range of 70–110% of BEP as shown in Figure 12.22. Outside of the allowable operation region is unstable operation (Figure 12.22). When the flow is too high, the pump may suffer high velocity cavitation where too large friction losses could cause too low suction pressure. On the other hand, when the flow is too low, the pump will suffer cavitation and failures in bearing and seals.

The bad-actor pump (more than one failure per year) should be checked with ROE (Forsthoffer, 2011).

To determine if a pump is operating within ROE, calculate the pump head required and measure the operating flow. Plot the flow-head point on the pump curve. If the bad-actor pump is operating outside the ROE, the equipment engineer must discuss with the process engineer to determine the operating target ranges for flow, motor amps, control position, and temperature difference between the suction and discharge.

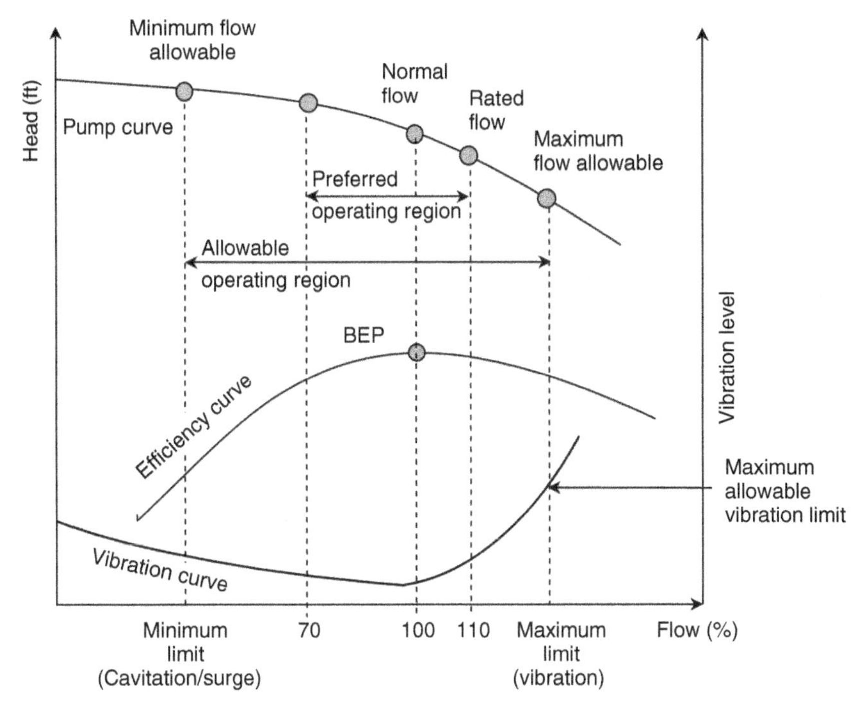

FIGURE 12.22. Reliability operating envelope.

12.13 PUMP CONTROL

In operation, the pump head will change due to variations in throughput to meet processing objectives. There are two control options available for the pump to meet varied throughput (Figure 12.23). If the pump is driven by a steam turbine or variable speed motor, the driver speed can be adjusted to vary the pump head produced, which is essentially to change the pump performance. The other option is to adjust discharge control valve to vary the pump head required, which is essential to change the system characteristics (change the system friction losses). It can be harmful to restrict a pump's flow by putting a valve on the suction line. This can cause pump cavitation because the suction pressure will be reduced.

12.14 PUMP SELECTION AND SIZING

Both pump curves and the system curve must be evaluated to choose the right pump for a specific application. As a general guideline, the following steps are taken for selecting a centrifugal pump:

Step 1: *Determine the process conditions*

Correctly specifying process conditions are essential for defining the operating requirements leading to proper pump selection. A process flow diagram for the pump system should be generated, which covers all the

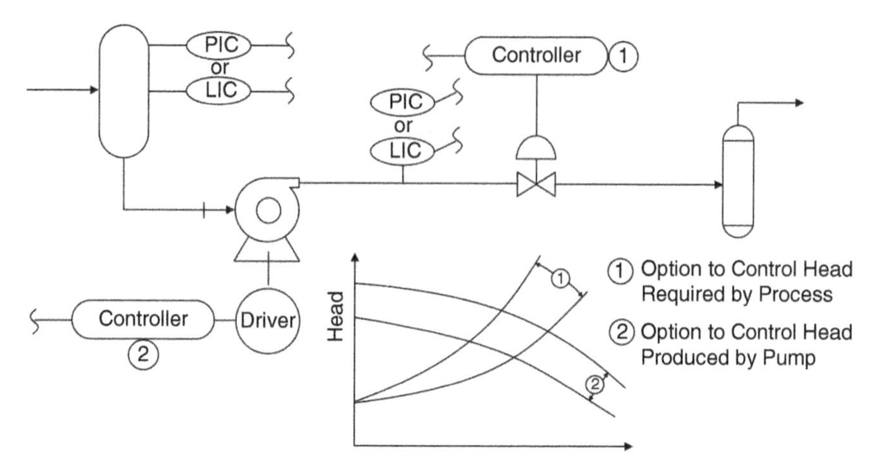

FIGURE 12.23. Two flow control options.

components to be included in the system. The following factors must be considered in generating the pump system flow diagram.

- *Flow rate:* Flow rates including minimum, normal, and rated should be specified in the datasheet. Normal flow is to achieve a specific process operation while the rated flow is typically 10% over the normal flow depending on company practices to accommodate process variation and pump wear. The minimum flow rate for process turndown operation must be provided in order to establish if a flow bypass line is required in process design.
- *Liquid properties:* Viscosity, vapor pressure, and specific gravity are important parameters in achieving the required reliability of a pump. The viscosity affects pump performance. Since the performance of most centrifugal pumps is determined from water, the procedure developed by the Hydraulic Institute is adopted to correct the performance curves when pumping viscous fluids. Vapor pressure of the process liquid at the suction temperature is an important property when determining whether there is a sufficient NPSH. Specific gravity is used to calculate the pump head required to overcome the resistance of the suction and discharge systems. The process engineer must specify not only the normal values for density and viscosity but also the maximal and minimal values that pump may encounter in abnormal operations such as startup, shutdown, turndown, and process upset. In addition, the engineer must specify the maximum operating temperature.

Step 2: *Determine the total head required*

The pump head at the design point (Figure 12.14) is determined based on the normal process conditions while rated pump head is based on consideration of design margin, which could be 10–20% to account for variations in physical properties and process conditions.

When operation is toward the end of run, pressure drops become much higher in heat exchangers and heaters due to fouling accumulation. This requires higher pump head, which is considered in determining rated discharge pressure.

Step 3: *Select the pump*
The pump is selected based on rated flow and head. Adequate knowledge of the process conditions and compositions is the most important aspect for optimizing pump selection. Process constraints must be considered such as operation flexibility, turndown, startup, and shutdown.

Since centrifugal pumps are not normally custom designed, it is important to ensure that each vendor will provide quotes for similar pump configurations for the specific operating conditions. Often, there could be a group of pumps available for selection. This provides opportunity for good pump selection to achieve optimal trade-off between pricing, efficiency, and reliability. For given rated pump head and flow, the preliminary pump selection can be made based on the normal operating ranges.

Figure 12.16 shows that there are a number of different impeller diameters available for each pump. Selection of geometry and type are governed by the operating conditions and properties and compositions of the liquid. Select an impeller that allows for future changes in the diameter. Pumps are rarely operated at their exact designed point. Therefore, the flow or head may need to be changed to increase the pump efficiency or to accommodate changes in process requirements. API 610 requires that 5% higher head must be achieved with a larger impeller, so the largest impeller for a casing size should not be selected.

The rated flow should be no greater than 10% to the right of the BEP. This will result in both rated and normal operation close to the BEP (see Figure 12.24). The pump selected must ensure reliability, which is discussed in more detail previously.

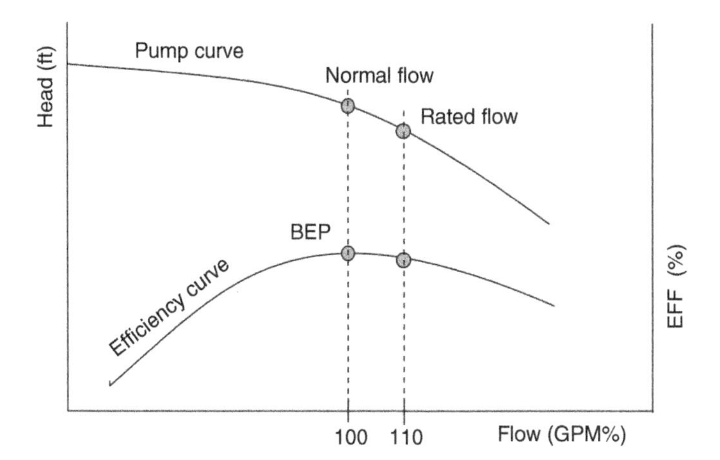

FIGURE 12.24. Optimal pump selection.

The next task is to match $NPSH_A$ and $NPSH_R$. $NPSH_A$ is calculated based on equation (12.17). It is prudent to incorporate a margin of safety for $NPSH_A$ above $NPSH_R$ to effectively prevent potential cavitation. The actual margin will vary from company to company. Some use the normal liquid level as the datum point, while others use the bottom of the vessel. Typical margins are 4 ft for hydrocarbon liquids (including low S.G.) and 10 ft for boiling water.

Step 4: *Select the driver*

When sizing a motor driver to fit an application, it is necessary to consider whether the pump will ever be required to operate at a flow higher than the rated point. The motor will need to be sized accordingly. If the pump may flow out to the end of the curve (it occurs if someone opens the restriction valve all the way, for example), it is important that the motor does not become overloaded as a result. Therefore, it is normal practice to size the motor based on the end-of-curve (EOC) horsepower requirements. Figure 12.25 shows an example where a 7.5 hp motor would adequately power the pump at a duty point of 120 gpm at 150 ft head. But note that the EOC BHP requires that a 10 hp motor be used. Note that at the bottom of the pump performance curve in Figure 12.25, BHP lines slope upward from left to right. These BHP lines are developed based on water with SG = 1. These lines correspond to the pump performance curves above them (the top performance curve corresponds to the top BHP line, and so on). These BHP lines indicate the amount of driver BHP required at different points of the performance curve.

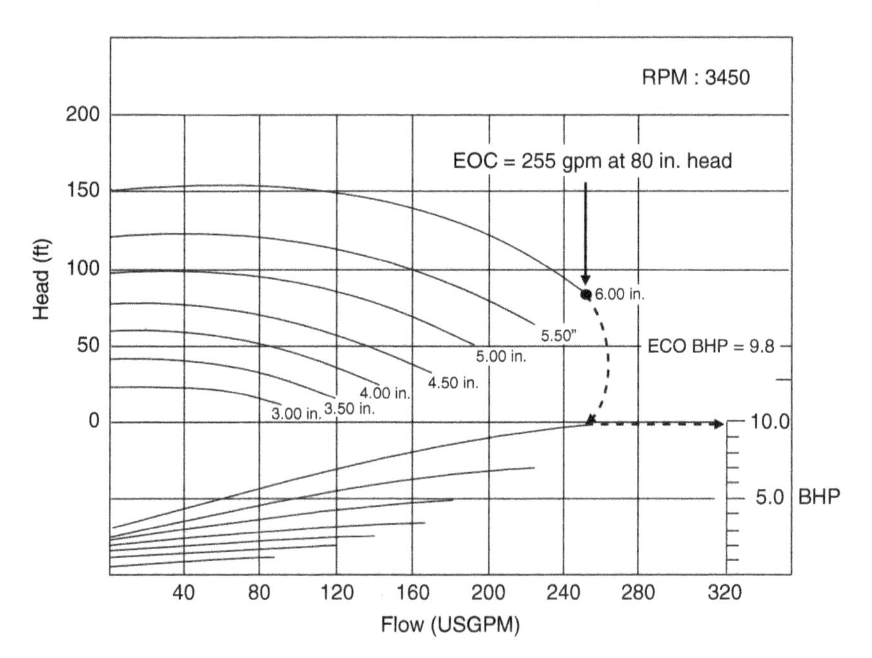

FIGURE 12.25. Pump curves with corresponding impeller diameters and BHP curves.

NOMENCLATURE

BHP	brake horsepower, hp
BEP	best efficiency point
GPM	gallon per minute, g/m
H_E	static head or elevation head, ft
H_f	friction head, ft
H_{SP}	surface pressure head, ft
H_T	pump head, ft
H_v	velocity head, ft
H_V	vapor pressure head, ft
$NPSH_A$	Net Positive Suction Head Available, psi or ft
$NPSH_R$	Net Positive Suction Head Required, psi or ft
P_D	pump discharge pressure, psi
P_S	pump suction pressure. Psi
P_V	vapor pressure of liquid, psi
Q	flow rate, liter/s
SG	specific gravity, dimensionless
u	fluid velocity, ft/s
Z	elevation, ft

Greek letters

η pump efficiency, %

REFERENCES

Fernandez K, Pyzdrowski B, Schiller D, Smith MB (2002) Understand the basics of centrifugal pump operation, May issue, *Chemical Engineering Progress, AICHE*, 52–56.

Forsthoffer WE (2011) *Forsthoffer's Best Practice Handbook for Rotating Machinery*, Butterworth-Heinemann.

13

COMPRESSOR ASSESSMENT

13.1 INTRODUCTION

Compressors are important equipment in the process industries. Their primary purpose is to compress air or gas into a smaller volume and thus simultaneously raise the pressure as well as the temperature. The basic principles of compression are summarized in Figure 13.1.

Compressors are very expensive, and they account for major part of capital costs in the overall process. Selecting the right type of compressor for a specific application is important for both cost and reliability considerations. It has been realized that one of the most common reliability issues for compressor is caused by improper selection of a compressor.

There are numerous types of compressors. They can be categorized under two basic types: positive displacement and dynamic. Positive displacement compressors include piston or reciprocating, screw, vane, and lobe compressors. Axial and radial (or centrifugal) compressors belong to the dynamic type as the required pressure rise and flow are imparted to the fluid by transferring kinetic energy to the process gas.

Positive displacement compressors feature a constant volume flow where the volumetric flow rate is not affected by changes in gas characterizations (pressure, temperature, or molecular weight (MW)). In contrast, with dynamic compressors, volumetric flow rate is affected by changes in gas characterizations.

Positive displacement compressors are generally suitable for gases with low flow, low MW gases, and requiring high compression ratios. Centrifugal compressors can handle higher flow rates although head limited. Head is a function of compression

Hydroprocessing for Clean Energy: Design, Operation, and Optimization, First Edition.
Frank (Xin X.) Zhu, Richard Hoehn, Vasant Thakkar, and Edwin Yuh.

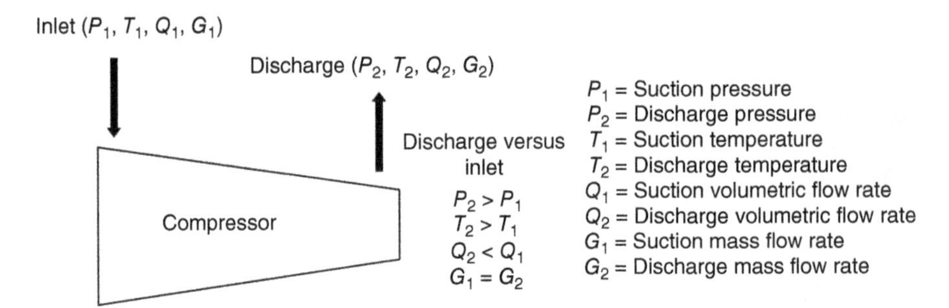

FIGURE 13.1. Basic principles of compressor.

ratio and MW. The gas MW will define the allowable compression ratio per centrifugal casing. Axial flow type compressors are used for high flow and low head applications.

Multistage centrifugal and reciprocating compressors are commonly used in the refinery for recycle gas, net gas, and hydrogen makeup services. The other types of compressors previously mentioned are used for more specialty applications. Multistage centrifugal compressors are the most common type used since they have wide operating range, are reliable and efficient, and are less affected by performance degradation due to fouling compared to reciprocating type. Furthermore, reciprocating compressors have moving and wearing parts, so they are typically spared, which raises the cost compared with centrifugal compressors. Therefore, in this chapter, multistage centrifugal compressor is the focus of discussions. For readers who like to dig deeper into the subject, detailed discussions can be found in Sorokes (2013).

13.2 TYPES OF COMPRESSORS

Centrifugal compressors can be beam type or integrally geared and both types of compressors could be single stage or multistage.

13.2.1 Multistage Beam Type Compressor

For volumetric flow rates between 1000 and 100,000 actual cubic feet per minute (ACFM) and polytropic heads under 120,000 ft (recycle gas, wet gas, and net gas applications), a multistage beam type centrifugal compressor is used and is typically unspared.

Beam type compressor casings can be horizontally or vertically split. A horizontally split compressor has a casing that is divided into upper and lower halves along the horizontal centerline. With this arrangement, all that is necessary is to lift the upper casing and gain access to the internal components without disturbing the rotor to casing clearance or bearing alignment (Figure 13.2).

In vertically split compressor, the casings are formed by a cylinder closed by two end covers: hence, the name "barrel" compressor is termed. With one end cover being removable, it allows access to the inner casing with the internal components

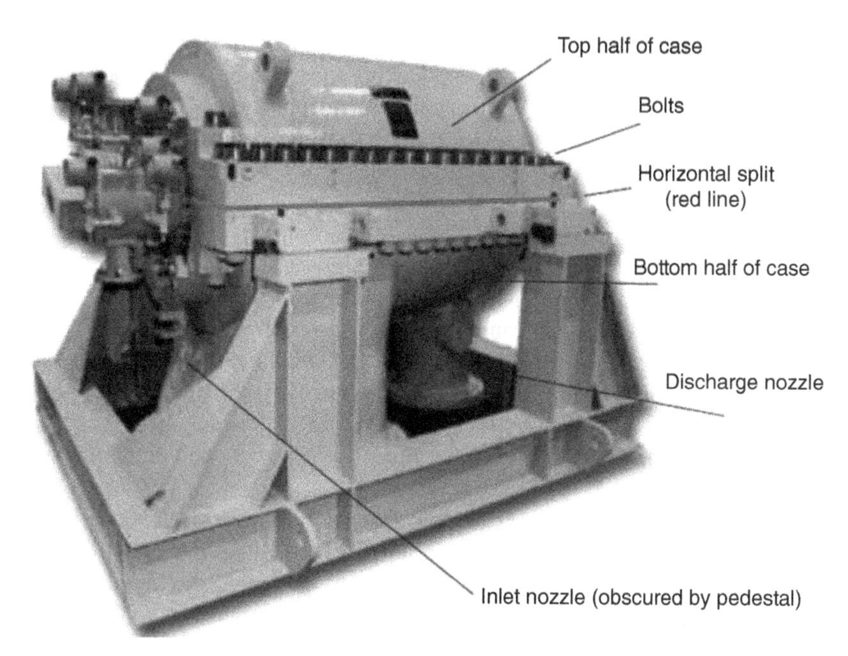

FIGURE 13.2. Centrifugal multistage horizontal split. Source: Sorokes (2013). Reproduced with permission of AIChE.

FIGURE 13.3. Centrifugal multistage radially split compressor. Source: Sorokes (2013). Reproduced with permission of AIChE.

(Figure 13.3). Inside the casings, the rotor and diaphragms are essentially the same as that of horizontally split compressors.

API 617 (2012) "Axial and Centrifugal Compressors and Expander-compressors" states that when the partial pressure of hydrogen is above 200 psig, casings with vertical or radial splits shall be used. This is to prevent leakage of light gases along the casing split.

Effective sealing is important to prevent leakages. In particular, it is the shaft end seals, which keep the process gases from leaking to atmosphere. Dry gas seals have gained wide acceptance and are the choice for most applications.

13.2.2 Multistage Integrally Geared Compressors

This type of compressors has a low-speed bull gear that drives multiple high-speed gears (pinions), which connect impellers. Integrally geared compressors achieve the required head with smaller and fewer number of higher speed impellers compared to a beam type compressor, which can have up to 10 impellers. Impellers are mounted at one or both ends of each pinion (Figure 13.4). Each impeller has its own casing that is bolted to the gear casing. The gear casing is usually horizontally split to allow access to the gears. These are used for some lower flow applications and may be less expensive than a multistage beam type compressor. They are used predominantly in gas processing plants.

In the integrally geared compressor, an impeller receives the gas from the first-stage inlet nozzle, compresses it, and discharges it to the diffuser where the velocity is converted into pressure. Then, the gas exits the first stage via discharge nozzle and enters an intercooler, if required, and is then piped to the second stage. The discharge from the second stage enters an intercooler, if required, to keep gas temperatures within limits, and it enters the third stage, and so on.

Integrally geared compressor can be single staged as well, which is mainly used for relative low pressure ratio applications. Single-stage integral gear compressor is also called "Sundyne" compressor. For a Sundyne type compressor, it is an in-line type similar to a pump, usually driven by motor through an integrally mounted gear

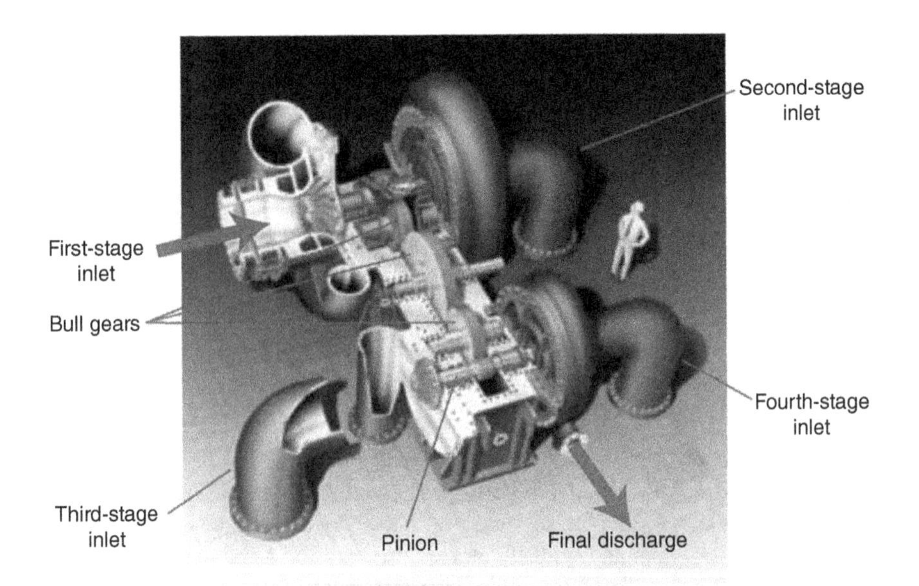

FIGURE 13.4. Integrally geared centrifugal compressor. Source: Sorokes (2013). Reproduced with permission of AIChE.

box. The Sundyne compressors feature low flow and high head and are used in many applications that used to be served by positive displacement compressors.

13.3 IMPELLER CONFIGURATIONS

The most critical component in a centrifugal compressor is rotor, which is shaft plus impellers. Impellers provide 100% kinetic energy to the gas, which is responsible for around 70% of static pressure rise in a compression stage. Well-designed impellers are very energy efficient and only 4% energy expanded is lost. The losses in stationary parts in a compressor reduce overall energy efficiency. The type of impellers chosen depends on required pressure ratio, gas compositions, operating speed, equipment cost, and so on.

Multistage centrifugal compressors have two types of impeller configurations: between-bearing (for beam type) and integrally geared.

13.3.1 Between-Bearing Configuration

Impellers in the between-bearing compressor are mounted on a single shaft. Between-bearing compressors are available with horizontally or vertically split. A driver rotates the shaft and impellers at a common speed. Between-bearing compressors come with two categories of configurations: straight-through and back-to-back. The straight-through arrangement is typically used for vertically split compressors in which the fluid flows at the one end and exits at the opposite end of the compressor (Figure 13.5). A balance drum or balance piston is required to absorb axial thrust.

In the back-to-back arrangement that is typically used for horizontally split compressors, impellers are arranged back to back with the exit flow at the center. In other words, the impellers are allocated in opposite directions (Figure 13.6). In this design, the main inlet is at the both ends of the rotor and the impellers guide the flow toward the center of the compressor. In this configuration, the axial thrust is self-correcting and balanced so that the force on the thrust bearings is reduced.

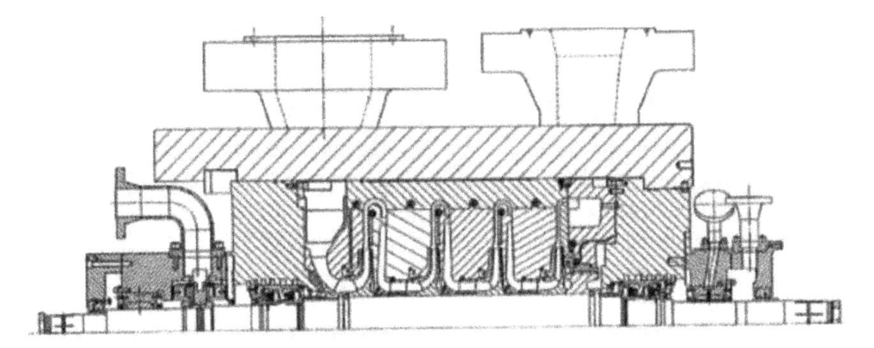

FIGURE 13.5. Straight-through compressor. Source: Sorokes (2013). Reproduced with permission of AIChE.

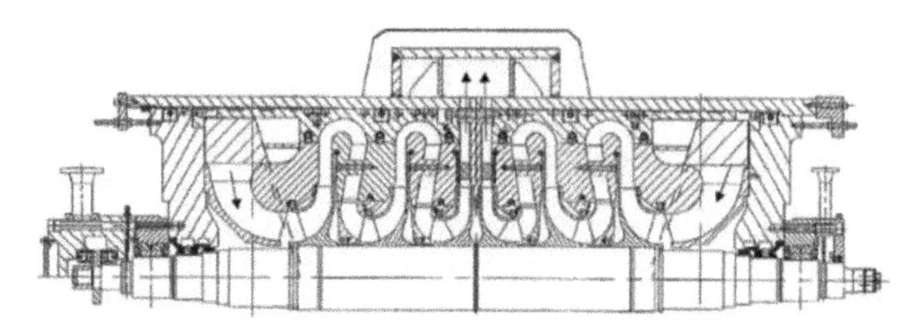

FIGURE 13.6. Back-to-back compressor with double flow inlet. Source: Sorokes (2013). Reproduced with permission of AIChE.

13.3.2 Integrally Geared Configuration

In an integrally geared configuration, multiple shafts may be used in which the impellers are mounted at the ends of multiple pinions that can rotate at different speeds depending on the gear ratio between the individual pinions and the bull gears. The number of impellers and the number of pinions vary depending on applications. Typical integrally geared compressors have two to four pinions with one or two impellers mounted at the ends of each pinion.

Instead of circumferential arrangement as in the between-bearing compressors, axial flow inlet arrangement is obtained in the integrally geared compressors. This is achieved by flow entering the first impeller via an axial or straight run of pipe and the flow at the volute (or collector) is piped to the axial inlet for the next impeller. This eliminates the flow inefficiency incurred from the inlet bend, return bend, and return channel used in the between-bearing design.

13.4 TYPE OF BLADES

The blades can be two-dimensional (2D) or three-dimensional (3D) (Figure 13.7). 3D impellers have better aerodynamic efficiency than 2D impellers but more expensive. Selection of the type of impellers depends on flow coefficient (FC), operating speed, desire pressure ratio, efficiency, and equipment cost.

13.5 HOW A COMPRESSOR WORKS

A centrifugal compressor (Figure 13.8a) consists of four basic parts: inlet nozzle for inlet flow guidance, impeller for increasing gas velocity, diffuser for converting velocity energy to pressure energy, and volute (or scroll) for existing flow guidance.

In the impeller, the gas velocity is increased due to the centrifugal action of the rotating blades. Then, this velocity is converted to pressure in the diffuser. In this mechanism, the principle of centrifugal compressor is very similar to that of a centrifugal pump.

FIGURE 13.7. 2D blades with circular arc shape (a) or 3D blades with complex shape (b). Source: Sorokes (2013). Reproduced with permission of AIChE.

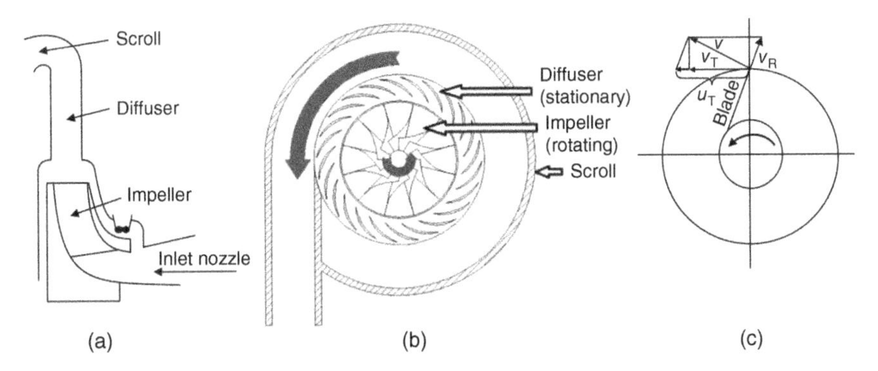

FIGURE 13.8. Key components of centrifugal compressor.

An impeller consists of a hub and a number of rotating blades that impart mechanical energy to the gas via increasing the velocity of the gas (Figure 13.8b). The gas leaves the impeller with increased velocity and enters the diffuser. The diffuser gradually reduces the velocity of the gas in order to increase gas pressure. In this manner, the diffuser converts the velocity energy to a higher pressure. In a single-stage compressor, the gas leaves the diffuser and enters a volute before it exits the compressor through the discharge nozzle. The volute collects the exiting gas and reduces the gas velocity further through increased cross-sectional area. Thus, it gives additional pressure rise. In a multistage compressor, the gas leaves the diffuser and enters return vanes that direct the gas into the impeller of the next stage.

The velocity of the gas is the key to understand how a centrifugal compressor can compress the gas dynamically. When the gas enters the impeller, it flows into the narrowed passage between blades. The gas velocity in the passage relative to the blades is called relative velocity (V_R) and also called radial velocity as it occurs in a radial direction. The gas has a higher relative velocity in a long and narrow flow passage. The opposite is true that the gas's relative velocity is lower in a short and wide flow passage. As soon as the gas flows out of the flow channel between blades, the tangent velocity of the gas increases from V_T to the blade tip speed U_T as shown in Figure 13.8c. The blade tip speed is the product of blade tip diameter and RPM. Due to increased tangent velocity, the gas energy increases in proportion to the net velocity (V) (also called exit velocity) that is the vector sum of the relative velocity (V_R) and the blade tip speed (U_T). Therefore, the blade tip speed and the relative velocity determine the pressure ratio, that is, the performance of the compressor.

13.6 FUNDAMENTALS OF CENTRIFUGAL COMPRESSORS

Flow coefficient, φ_G, is the most important parameter for selection of centrifugal compressor types. φ_G relates flow rate with both impeller size and speed and is defined as

$$\varphi_{\mathrm{G}} = \frac{Q}{ND^3}, \tag{13.1}$$

where Q is impeller's volumetric flow rate, N impeller's speed, and D impeller's exit diameter.

Centrifugal impellers can be classified into two categories based on FC: low FC and high FC impellers. The former features long and narrow passages with higher pressure ratio. In contrast, the latter has much wider flow passages to accommodate higher flows but with lower pressure ratio. Figure 13.9 shows impellers with different FCs. The impeller with the highest FC is located at the right end of the rotor with remaining impellers being progressively narrower in flow passages from the right end toward the center. In contrast, the impellers with the low FC are located at the left end of the rotor with the lowest FC impeller located closest to the center from the left end. As what can be observed, these low FC impellers have much narrow flow passages than the high FC impellers on the right. In this configuration, fluid pressure increases beginning with the high FC impellers on the right with relative lower pressure ratio and then to the FC impeller in the center on the right-hand side with medium pressure ratio. Lastly, the fluid pressure continues to increase with relative higher pressure ratio from the low FC impellers on the far left toward the highest pressure in the center on the left-hand side.

Low FC impellers have 2D blades (simpler blade design). In contrast, high FC impellers have 3D blades (complex blade design), which can be observed in

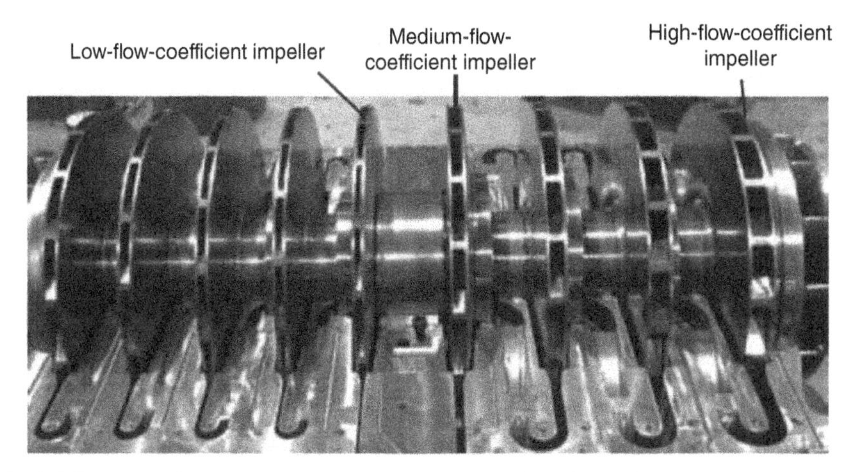

FIGURE 13.9. Different FC impellers: from low at the left to high at the right. Source: Sorokes (2013). Reproduced with permission of AIChE.

Figure 13.7. Due to their narrow passages and simpler blades, low FC impellers have lower aerodynamic efficiency than high FC ones.

Both the straight-through and back-to-back designs can allow intercooling, which keeps the temperature of the material below the strength limit as well as reduces the shaft power requirement. If intercooling is required, the gas is discharged from the compressor after traveling through half the impellers, cooled and injected back into the compressor before traveling through the remaining impellers.

The integrally geared design has several advantages over the between-bearing design. The most important is the aerodynamic advantage due to axial flow arrangement, which eliminates flow turning and thus reduces pressure losses. Furthermore, impellers can have different speeds or diameters. These two features make integrally geared design more energy efficient than the between-bearing design.

However, the disadvantage of the integrally geared design is the complex mechanical arrangement because it contains a large number of bearings and seals. Vibration could be a problem in operation for integrally geared design due to the fact that impellers are located outside the bearing supports (overhung) and they can vibrate if not mounted properly.

To relate the pressure head with impeller geometry and speed, use head or pressure coefficient μ_p, defined as follows:

$$\mu_p = K\frac{H_p}{N^2D^2}, \tag{13.2}$$

where K is constant and it can be calculated as

$$K = \frac{g}{\left(\frac{\pi}{720}\right)^2}, \tag{13.3}$$

where g is gravitational constant.

High head impellers have a narrower flow range than those of low head impellers. High head impellers are also more prone to surge than those of low head impellers. Compressor surge occurs when the compressor cannot overcome the discharge pressure. This causes reverse flow through the compressor and damage to the seals, bearings, and rotor. To prevent surge, user often specifies an antisurge spillback.

13.7 PERFORMANCE CURVES

Figure 13.10 shows a typical compressor performance curve at a design speed. The curve shows that the centrifugal compressor has limited head capability, with variable volume characteristic.

Head is a process condition and is a function of compression ratio and MW. The curve sets the flow rate that a compressor needs to deliver. If process conditions do not change, raising the compressor speed will raise the flow rate. The required flow rate is obtained by changing the speed of a variable speed compressor or throttling the suction valve for a constant speed compressor.

Design point (D): The design point for the compressor at the given speed. The section between "*S*" and "*D*" is the normal operating range. The curve shows that the compressor discharge pressure is balanced with relatively large changes in volume flow. The compressor can operate at the rated condition represented by point "*R*"; but the region between points "*R*" and "*C*" is unstable. For optimum compressor size selection, the compressor should be selected to have the design point close to the right-hand side of the curve, but not too close to the stonewall or choke point "*C*." If the design point is located too far to the left the compressor is sized too large, the compressor is too small if farther to the right.

Surge (S): Surge is characterized by intense and rapid flow and pressure fluctuations throughout the compressor and is associated with stall involving one or more

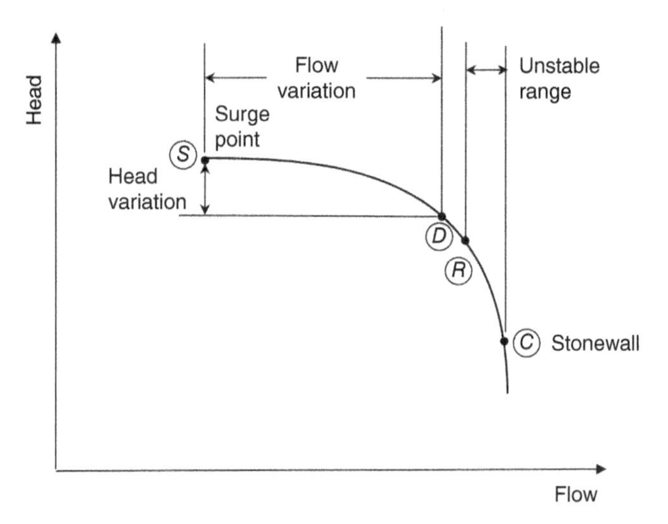

FIGURE 13.10. Performance curve for a centrifugal compressor.

compressor stages. It is accompanied by strong noise and violent vibration, which can damage the compressor. Surging occurs at a minimum suction flow with point "*S*" in Figure 13.10. When the discharge pressure in the centrifugal compressor increases, the mass flow rate decreases. There is a minimum flow limit. Below this limit, the compressor operation becomes unstable. When the compressor cannot overcome the discharge pressure, the easiest path for the gas is back through the compressor. After the back flow slug has been discharged, the compressor still faces the problem of insufficient gas flow and the back flow reoccurs. This unstable operation manifests itself in the forms of pressure and flow oscillation.

If the compressor is to be operated below the minimum flow, the compressor must be equipped with a low flow spillback. The details for partial control can be seen as follows.

Choking (C): Choking is the opposite of surge in the centrifugal compressor. It occurs at point "*C*" under which the gas flow is too large and that is more than what the impeller can handle. At the choking point "*C*," the fluid reaches sonic conditions and there will be an abrupt decrease in the compressor performance. The occurrence of choking depends not only on the high flow condition but also on the fluid thermodynamic properties. For example, choking can occur to compressors operating with fluids of high MW.

Many industrial compressors normally operate at conditions far away from choking. For these compressors, the maximum flow limit is usually defined as the flow corresponding to a sharp reduction in efficiency based on company's best practice.

If variation of impeller speed is considered, there will be a family of compressor performance curves as shown in Figure 13.11. An increase in rotor speed (n) or RPM increases the compressor flow rate. At a particular rotor speed, a decrease in flow rate can be obtained by increasing the compression ratio. Minimum and maximum permissible flow rates at constant RPM are termed surge and choke (stonewall) limits. A line tracing the stall points of all the constant RPM lines is called a surge line.

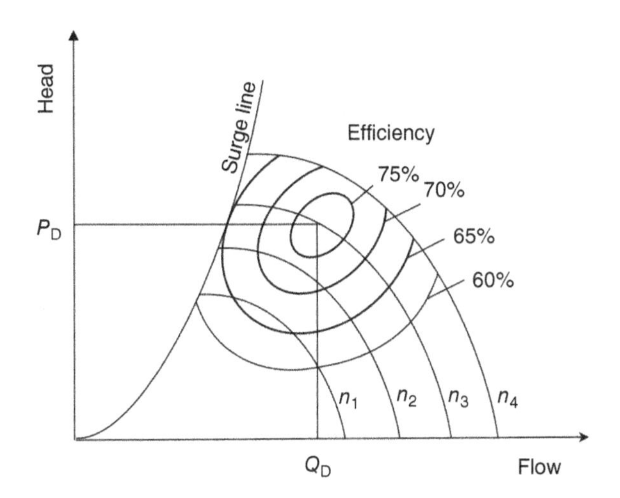

FIGURE 13.11. Compressor performance curves.

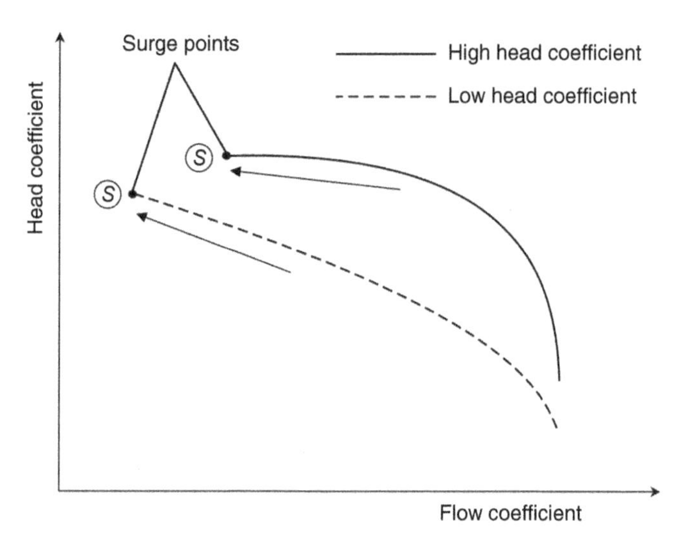

FIGURE 13.12. Impeller with higher head coefficient has a smaller rise-to-surge.

Flow coefficients and pressure coefficients can be used to determine various design characteristics. Use of flow coefficient enables selection of impeller type while knowing head coefficient helps to determine the hydraulic performance. For example, Figure 13.12 is a simplified performance curve for low and high head coefficient impellers. As can be observed, the low head coefficient impeller has a steeper rise-to-surge slope than the high head coefficient impeller. Therefore, comparing with the high head coefficient impellers, the low head coefficient impellers has much higher pressure sensitivity from the flow change and thus is easier to measure the pressure condition to detect how close the compressor to the surge limit.

A performance curve is obtained under new and clean conditions. After operating for a period of time, however, a compressor will deteriorate in performance. Even after a full maintenance, a compressor will rarely retain its original performance.

13.8 PARTIAL LOAD CONTROL

As mentioned, surge could lead to violent flow oscillation and thus cause damage to the compressor. Hence, it should be avoided by all means. The surge line is established during manufacturing shop performance testing. Instrumentation will open a recycle valve before the compressor goes into the surge region.

Recycle or surge control valve: At constant speed, the head–flow relationship will vary in accordance with the performance curve (see Figure 13.11). For a constant compressor speed, a recycle valve is needed for surge control if process conditions (rising compression ratio and dropping MW) push the operating point into surge.

Variable speed control: A performance curve is established for each speed, as shown in Figure 13.13. If the compressor-driver system (compressor, driver, and gear) is designed for 90–110% speed variation as shown in Figure 13.13, by varying the compressor speed, the centrifugal compressor can be operated at any partial load point on the right-hand side of the surge line. The speed control is actually shifting

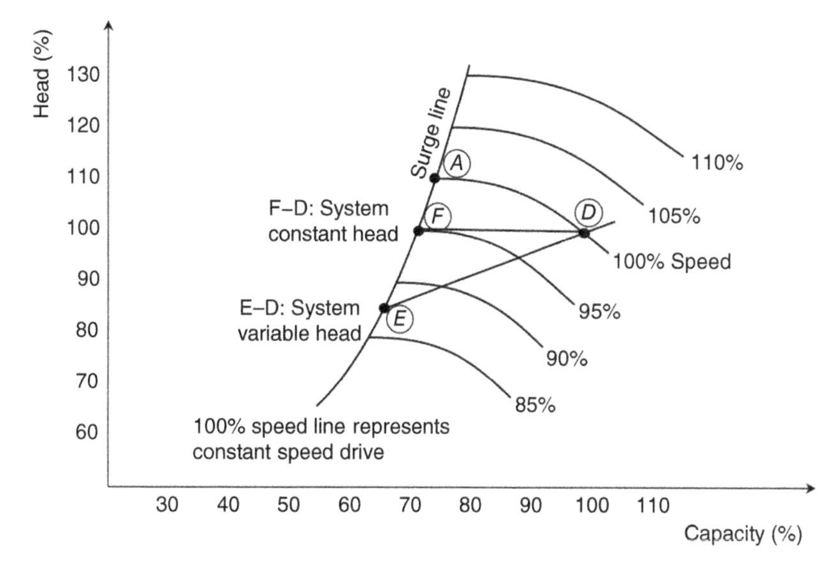

FIGURE 13.13. Typical variable speed control compressor performance curves.

the compressor curve until it reaches the new head and flow requirement of the compressor under the partial load conditions. The compressor can even handle the head higher than the design head by increasing the speed of the compressor. However, the compressor will surge if the flow is less than the surge point, unless it is equipped with recycle control valve.

The partial load performance can be presented by a horizontal line *D–F* in Figure 13.13 if the compressor is operated under constant head mode. The point *D* is design point. The minimum partial load is about 73% at the surge point of "*F*." If the compressor is operating under decreasing head mode, then the performance line is *D–E*. The minimum partial load without surge is about 67% at the surge point of "*E*."

Inlet guide vane (pre-rotation vane): Inlet guide vanes are primarily used for integrally geared compressors. Located at the compressor inlet, the guide vanes change the direction of the velocity entering the first-stage impeller. By changing the angle of flow, these vanes direct the flow into the impeller, and consequently the shape of the performance curve is changed. With velocity change to the inlet gas by the guide vanes, the performance curve steepens with very little efficiency loss.

Figure 13.14 is the typical performance curve for the centrifugal compressor equipped with inlet guide vane control for constant speed driver. The performance line of *D–A–G–H* is the maximum head capability of the compressor. The inlet guide vane allows the centrifugal compressor to operate any points below envelope of *D–A–G–H* without surge. Compared to variable speed control as shown in Figure 13.13, the surge limits are extended and the partial load capability of the compressor is greatly increased without change in compressor speed.

Figure 13.15 is used to explain the performance curves with inlet guide vane openings at constant speed. The compressor is operating with the vane fully open on the line of *B–D–A*. The inlet guide vane is adjusting and changing its position for the compressor to operate in other areas away from full load. The compressor will exceed

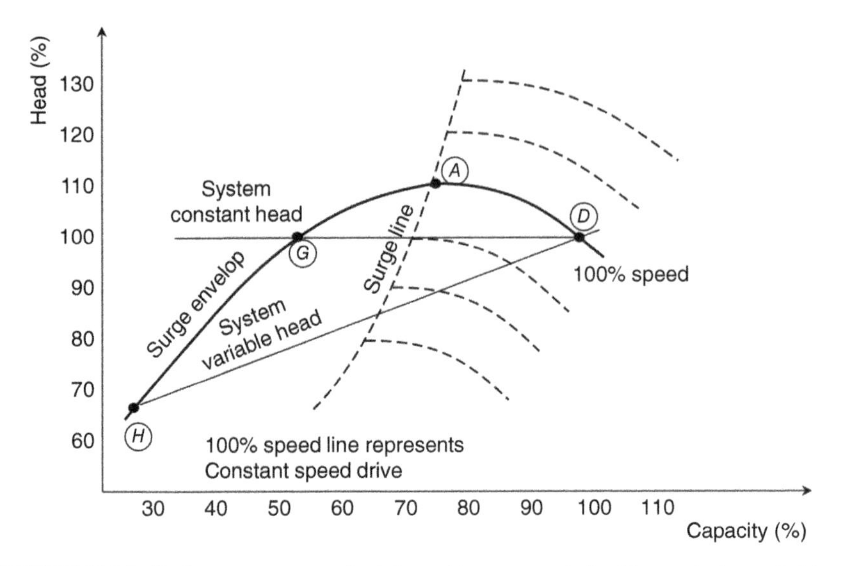

FIGURE 13.14. Performance curves for inlet guide vane control with constant speed driver.

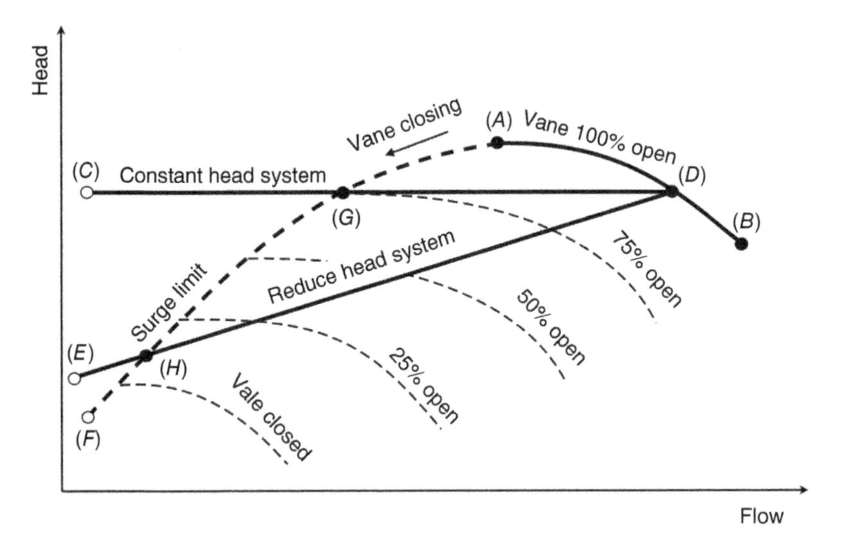

FIGURE 13.15. Inlet guide vane control – constant speed driver.

its head capability if it is operating above the performance curve of *D–A*; hot gas bypass is required if the operating conditions of the compressor is located in the area of the left-hand side of the surge limit of *A–G–H*.

13.9 INLET THROTTLE VALVE

Between-bearing compressors with constant speed electric motor drivers are controlled with suction throttle valves. Throttling at the suction instead of the discharge takes advantage of volume reduction, so less pressure is throttled and less energy is used when operating at off-design conditions.

When throttling at the discharge, the suction pressure does not change, so mass flow rate (G) is proportional to volumetric flow rate (ACFM). When throttling at the suction, the compressor suction pressure drops and the same G is obtained at a higher ACFM (see equation (13.4)). Thus, less throttling is required and less energy is used by using inlet throttle valve.

$$\text{ACFM} \approx \frac{G}{P_1}. \tag{13.4}$$

13.10 PROCESS CONTEXT FOR A CENTRIFUGAL COMPRESSOR

The objective of using a compressor is to compress a certain amount of gas to a desired pressure. A typical process involving a compressor is shown in Figure 13.16. The responsibility of an engineer is to provide process conditions as the basis for determining the type of compressor and the number of compression stages. The process data that the engineer will provide include mass flow, inlet and discharge pressure, temperature, and gas compositions. Then, the designer will calculate actual flow, the type of compressor, horsepower, and efficiency.

Firstly, the gas flow range that the compressor must handle should be provided. The minimum flow corresponds to turndown operation, while maximum flow is for rated operation. Secondly, the fluid conditions (temperature and pressure) and property (composition) should be specified for several process scenarios so that the selected compressor can handle all the variations. Next, inlet and discharge pressure must be specified.

We need to differentiate volumetric flow, mass flow, and standard flow. The compressor is sized on volumetric flow rate. For a compressor with compression ratio of two, one actual cubic feet of gas per minute will be compressed to a discharge volume of exactly one half of a cubic feet per minute, assuming no increase in temperature in the compressor and the gas is dry because the gas pressure is doubling.

Standard volume has only one volume referenced to the same pressure and temperature. At default, the standard conditions are defined at 14.7 psia (atmospheric pressure) and 60 °F. A measurement in standard cubic feet is the ratio of the actual pressure to the referenced standard pressure multiplied by the actual volume.

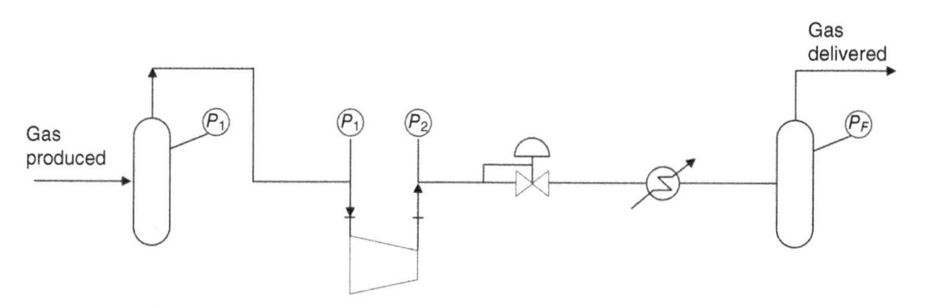

FIGURE 13.16. Typical process involving a compressor.

Referring back to the previous example of a compressor with compression ratio of two and no change in temperature, the standard volume from this compressor would remain the same because the pressure remains constant at 14.7 psia. Even though the actual volume of the gas decreases by one half, the discharge standard volume is the ratio of the discharge pressure to the standard atmospheric pressure multiplied by the discharge volume. This will result in the same discharge standard volume as the inlet one.

Mass flow is the product of the actual volume flow and the density of the specific gas. Same as the standard volume, mass flow through the aforementioned compressor will remain the same provided the gas is dry.

Both standard volume and mass flow are used to describe process capacity and in horsepower calculations. The gas price is based on the standard volume or mass flow.

13.11 COMPRESSOR SELECTION

Based on the actual flow rates provided, flow coefficient (FC) as defined in equation (13.1) can be calculated, which is used to select impeller blade type: low FC or high FC impellers. As mentioned, the major difference between these types is that low FC impellers are characterized by long and narrow passages, while high FC impellers are featured with wide passages to accommodate high flow rate.

The pressure coefficient defined in equation (13.2) can be used to determine either low or high head impellers. To prevent surge, user often specifies a minimum rise-to-surge limit for compressor selection.

Based on the discharge pressure specified, the pressure head (H_p) raised by the compressor can be determined by

$$H_p = \frac{\gamma}{\gamma - 1} ZRT_1 \left[\left(\frac{P_2}{P_1} \right)^{\frac{\gamma - 1}{\gamma}} - 1 \right], \tag{13.5}$$

where $\gamma = C_p/C_v$ in which C_p is the specific heat capacity at constant pressure and C_v is the specific heat capacity under constant volume, Z is the compressibility of the gas, R is the gas constant in ft lb$_f$/(lb mol) (°R), T_1 is the inlet temperature in °R, P_1 is the inlet pressure in psia, and P_2 is the discharge pressure in psia.

Knowing the mass flow (G) and pressure increase that the compressor must deliver, polytropical efficiency can be determined as follows:

$$\eta_P = \frac{\text{Work out}}{\text{Work in}} = \frac{\gamma - 1}{\gamma} \times \frac{\ln(P_2/P_1)}{\ln(T_2/T_1)}. \tag{13.6}$$

Then, horsepower requirement can then be calculated by

$$W_{hp} = \frac{G \times H_p}{\eta_P}. \tag{13.7}$$

Obviously, the compressor with highest efficiency will require least power and hence lowest operating cost. Horsepower is linearly proportional to flow rate and head increase.

When selecting a compressor, the best practice for ensuring reliability is to preselect compressor casing type (horizontally split, vertically split, or integrally geared), impeller type (open or closed), the number of impellers allowed in each casing based on head per stage limits and shaft stiffness. One should request vendors' experience and references for installed compressor with similar design parameters.

NOMENCLATURE

Variables

C_p the specific heat capacity at constant pressure, kJ/kg.K
C_v the specific heat capacity under constant volume, kJ/kg.K
D impeller's exit diameter, ft^3 or m^3
H hydraulic head, ft
Q impeller's volumetric flow rate, GPM or ft^3/m or m^3/s
G impeller's mass flow rate, lb/s or kg/s
N impeller's speed, RPM
P pressure, psia or psig
R the gas constant in ft-lb_f/(lb-mol) (°R)
T temperature, °C or °F
W horsepower, hp
Z gas compressibility factor, dimensionless

Greek letters

φ_G flow coefficient, dimensionless
μ_p head or pressure coefficient, dimensionless
γ heat capacity ratio, dimensionless
η compressor efficiency, %

REFERENCES

API STANDARD 617 (2012) *Axial and Centrifugal Compressors and Expander-compressors for Petroleum, Chemical and Gas Industry Services*, 7th edition, American Petroleum Institute.

Sorokes, J (2013) Selecting a centrifugal compressor, *Chemical Engineering Progress (CEP)*, 44–51, June, AIChE, New York.

14

HEAT EXCHANGER ASSESSMENT

14.1 INTRODUCTION

Often exchangers do not perform as they should and their performance deviates from optimum. Sometimes, they do not accomplish what they are capable of and other times they are expected to perform what they are not capable of. The primary purpose of heat exchanger assessment is to identify the root causes if it is due to poor design, excessive fouling, or mechanical failure, and determine the required actions to improve the performance.

This chapter provides the basic understanding of heat exchange assessment supported with examples considering whether the exchanger is designed correctly, evaluation of operating performance, evaluation of fouling, and its effect on heat transfer and pressure drop. On this basis, methods for improving exchanger performance are provided using examples associated with different application scenarios. The methods discussed in this chapter focus on shell-and-tube exchangers as they are the most commonly used in the process industry although the assessment methodology can be applied to other types of heat exchangers. Detailed assessment of heat exchange performance may be conducted using commercial software.

14.2 BASIC CONCEPTS AND CALCULATIONS

As it is well known, the primary equation for heat exchange between two fluids is the Fourier equation expressed as

$$Q = UA\Delta T_{M}, \tag{14.1}$$

Hydroprocessing for Clean Energy: Design, Operation, and Optimization, First Edition.
Frank (Xin X.) Zhu, Richard Hoehn, Vasant Thakkar, and Edwin Yuh.

where

Q = heat duty, MMBtu/h
A = heat transfer surface area, ft^2
U = overall heat transfer coefficient, Btu/(ft^2 °F h)
ΔT_M = effective mean temperature difference (EMTD), °F.

Let us define U value first based on Figure 14.1, where h_i *and* h_o are film coefficients for fluids inside and outside of the tube and they can be calculated from the physical form of heat exchanger, physical properties of streams, and process conditions of streams. Thus, clean overall heat transfer coefficient (U_C) can be determined based on

$$\frac{1}{U_C} = \frac{1}{h_o} + \frac{1}{h_i}\left(\frac{A_o}{A_i}\right) + r_w, \quad (14.2)$$

where r_w is the conductive resistance of the tube wall. A_o and A_i are outside and inside tube surface area with subscripts i and o denoting inside and outside of the tube.

In reality, heat exchangers operate under fouled conditions with dirt, scale, and particulates deposit on the inside and outside of tubes. Allowance for the fouling must be given in calculating overall heat transfer coefficient. The graphical description of fouling resistances (R_o, R_i) and film coefficients (h_o, h_i) inside and outside of tube is provided in Figure 14.1. Conceptually, R_i and h_i are equivalent to R_t and h_t (t for tube side) while R_o and h_o are for R_s and h_s (s for shell side).

The overall fouling resistance is then defined as

$$R_f = R_o + R_i\left(\frac{A_o}{A_i}\right). \quad (14.3)$$

By adding the overall fouling resistance to U_C, actual U_A is defined as

$$\frac{1}{U_A} = \frac{1}{U_C} + R_f. \quad (14.4)$$

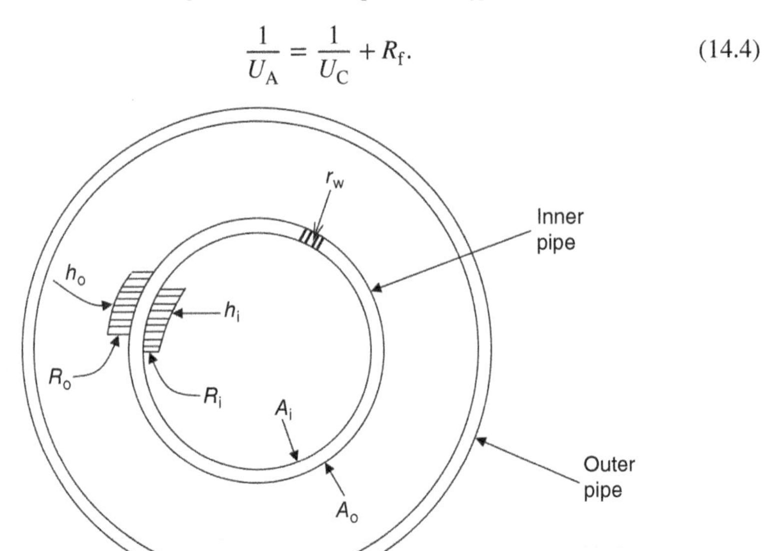

FIGURE 14.1. Location of h's and R's.

More detailed discussions for U values are provided later in this chapter. Now let us turn our attention to ΔT_M or EMTD. Several temperature differences can be used to calculate ΔT_M including inlet temperature difference, arithmetic temperature difference, and logarithmic mean temperature difference. Figure 14.2 is used for illustration.

Inlet temperature difference can be expressed as

$$\Delta T_1 = T_1 - t_2 \quad \text{(for countercurrent)}, \tag{14.5a}$$

$$\Delta T_1 = T_1 - t_1 \quad \text{(for cocurrent)}. \tag{14.5b}$$

This temperature difference could lead to a gross error in estimating true temperature difference over the entire pipe length.

Arithmetic mean temperature difference is defined as

$$\Delta T_A = \frac{\Delta T_1 + \Delta T_2}{2} = \frac{(T_1 - t_2) + (T_2 - t_1)}{2} \quad \text{(for countercurrent)}, \tag{14.6a}$$

$$\Delta T_A = \frac{\Delta T_1 + \Delta T_2}{2} = \frac{(T_1 - t_1) + (T_2 - t_2)}{2} \quad \text{(for cocurrent)}. \tag{14.6b}$$

This temperature difference could give erroneous estimate of true temperature difference when ΔT_1 (hot end approach) and ΔT_2 (cold end approach) differ significantly.

The logarithmic temperature difference (LMTD) is defined as

$$\Delta T_{LM} = \frac{\Delta T_1 - \Delta T_2}{\ln(\Delta T_1 / \Delta T_2)}, \tag{14.7}$$

ΔT_{LM} represents a true temperature difference for a perfect countercurrent as well as cocurrent heat exchange (Figure 14.2).

At this point, the first question is: why the countercurrent pattern is widely adopted in shell-and-tube exchangers? The answer is that the countercurrent LMTD is always

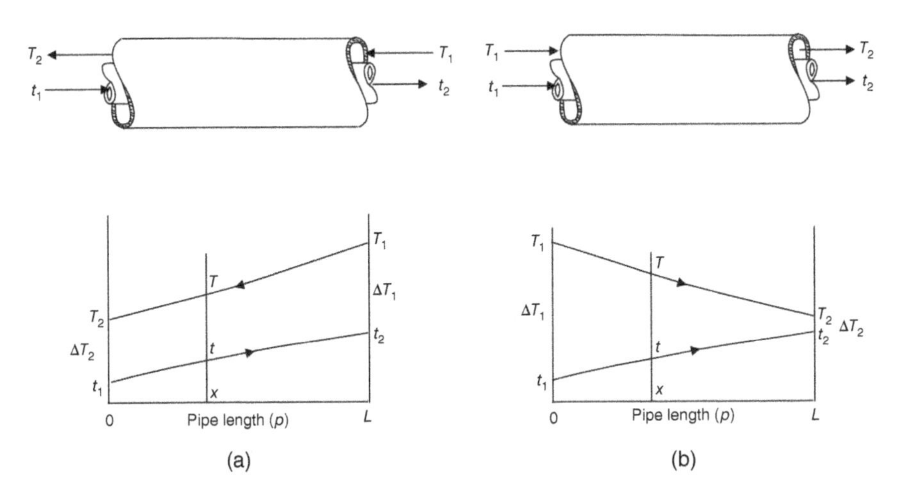

FIGURE 14.2. (a) Countercurrent and (b) cocurrent flows.

greater than the cocurrent LMTD. An example corresponding to Figure 14.2 is shown as follows.

Countercurrent			**Cocurrent**		
Hot fluid	Cold fluid		Hot fluid	Cold fluid	
$T_1 = 350°$	$t_2 = 230°$	$\Delta T_1 = 120°$	$T_1 = 350°$	$t_1 = 150°$	$\Delta T_1 = 200°$
$T_2 = 250°$	$t_1 = 150°$	$\Delta T_2 = 100°$	$T_2 = 250°$	$t_2 = 230°$	$\Delta T_2 = 20°$
		LMTD = 109.7°			LMTD = 78.2°

The second question is: what should be done if a heat exchange is not a perfect countercurrent? In fact, the flow pattern in most shell and tube exchangers is a mixture of cocurrent, countercurrent, and cross-flow. In these cases, EMTD ≤ LMTD. Thus, an LMTD correction factor F_t must be introduced.

$$\text{EMTD} = F_t \times \text{LMTD}, \tag{14.8}$$

$F_t = 1$ for a true countercurrent heat exchange; otherwise, $F_t < 1$.

As a short summary, EMTD is obtained by calculating LMTD based on equation (14.7) first and then applying F_t to account for nonperfect countercurrent flow.

F_t can be obtained via equations or charts (Shah and Sekulić, 2003). For example, F_t for 1–2 heat exchangers can be numerically calculated by

$$F_t = \frac{\sqrt{R^2+1}\ln[(1-P)/(1-R\times P)]}{(R-1)\ln\dfrac{2-P(R+1-\sqrt{R^2+1})}{2-P(R+1+\sqrt{R^2+1})}}, \tag{14.9}$$

where P is temperature efficiency and R is the ratio of heat flow, which are defined as

$$P = \frac{t_2 - t_1}{T_1 - t_1}, \tag{14.10}$$

$$R = \frac{T_1 - T_2}{t_2 - t_1} = \frac{m \times c_p}{M \times C_p}, \tag{14.11}$$

F_t for 1–2 type (one shell pass and two tube passes) exchangers can also be found in the chart as shown in Figure 14.3. It can be observed from the figure that F_t values drop off rapidly below 0.8. Consequently, if a design indicates an F_t less than 0.8, it probably needs to redesign to get a better approximation of countercurrent flow and thus higher F_t value. Different F_t charts are available for each exchanger layout (TEMA 1–2, 1–4, etc.).

14.3 UNDERSTAND PERFORMANCE CRITERION – *U* VALUES

The critical question for operation assessment is: what is the "performance" indicator for heat exchanger? As a heat exchanger is used to transfer heat, someone may naturally consider heat exchanger duty as the performance indicator. For verification, let

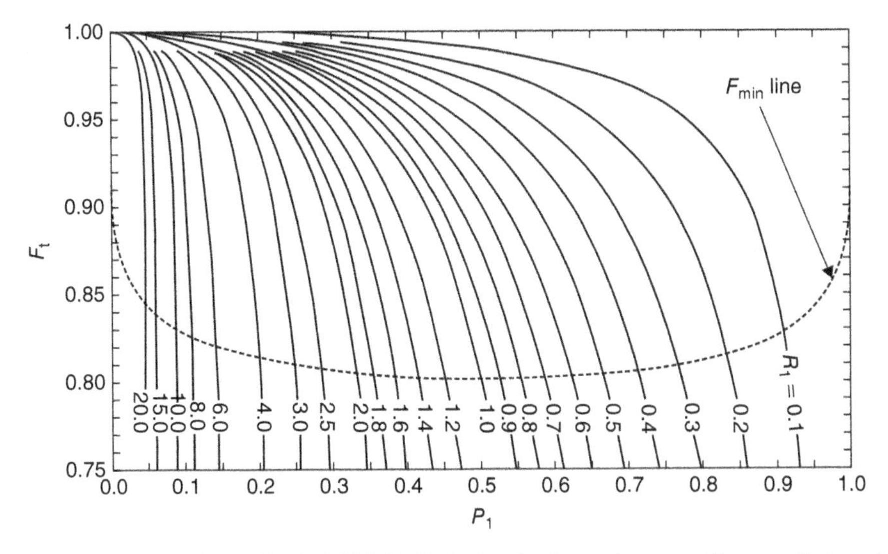

FIGURE 14.3. F_t factor for 1–2 TEMA E shell-and-tube exchangers. (Source: Shah and Sekulić (2003), Reproduced with permission of John Wiley & Sons.)

TABLE 14.1. Gathered Data for a Reaction Air Cooler

	Design	Operation
Q (MMBtu/h)	16.3	18.0
$M_{(effluent)}$ (lb/h)	154,447	162,919
T_1 (°F)	296.6	341.6
T_2 (°F)	105.8	149.0
t_1 (°F)	89.6	81.0
t_2(°F)	104.0	95.0
$\Delta T(T_1 - T_2)$(°F)	190.8	192.6
$\Delta t(t_2 - t_1)$(°F)	14.4	14.0

us look at an example of a reaction effluent cooler. The operating data was obtained, which is shown against the design data (Table 14.1). Operating temperatures can be measured from instrumentation. Then LMTD is calculated by equation (14.7) and F_t factor is obtained from F_t charts. Then U_A is calculated through equation (14.1) and U is derived for a given surface area A of the cooler. The calculation results for operating data are shown in Table 14.2. The design data were obtained from the exchanger datasheet.

As can be observed in Table 14.2, the heat exchanger duty in operation is little higher and the temperature changes are similar for both operation and design. What do you think of the performance of this exchanger in operation? If based on the exchanger duty, we could conclude that it is performing well or at least not worse than design performance. However, if comparing overall heat transfer coefficient, surprisingly, the operation U value is only half of the design U value, although the heat duty in operation is 10% higher. If the operation U value could maintain similar to design U value, the heat duty could be increased much higher than 10%!

TABLE 14.2. Calculation Results for a Reaction Air Cooler

	Design	Operation
Q (MMBtu/h)	16.3	18.0
M (effluent) (lb/h)	154,447	162,919
T_1 (°F)	296.6	341.6
T_2 (°F)	105.8	149.0
t_1 (°F)	89.6	81.0
t_2 (°F)	104.0	95.0
ΔT_1 (°F)	192.6	246.6
ΔT_2 (°F)	16.2	68.0
ΔT_{lm} (°F)	71.3	138.7
F_{t}	0.85	0.97
U_{A} (Btu/°F h)	269,896	133,759
A (ft²)	5167	5167
U (Btu ft²/°F h)	52	26
$U_{\text{operation}}/U_{\text{design}}$		0.50

T_1 (Effluent in)
Q
T_2 (Effluent out)
ΔT_1 (Hot end approach)
ΔT_2 (Cold end approach)
t_2 (air out)
t_1 (air in)

This example concludes that U value is a true performance indicator for heat exchanger under any process conditions. *The higher the U value, the better performance that a heat exchanger achieves.*

Clearly, good understanding of U value is of paramount importance for appropriate assessment of heat exchanger performance as it is the most important characteristic of heat exchanger representing its heat transfer capability. In view of the fact that many engineers are confused about the terminologies related to U, it is essential to get the basic understanding right before diverging into details of assessment methods.

14.3.1 Required U Value (U_{R})

The need of heat exchanger is to satisfy process requirement in terms of heat duty (Q) and temperatures (LMTD or ΔT_{LM}). Thus, heat exchanger is designed to have a certain surface area (A) in order to fulfill the process requirement. Based on process temperature requirement for reaction and fractionation, a certain amount of heat duty must be transferred. Under the basis of heat transfer duty and process temperatures, required U value can be calculated as

$$U_{\text{R}} = \frac{Q}{AF_{\text{t}}\Delta T_{\text{LM}}}.$$

14.3.2 Clean U Value (U_{C})

Independent of the required U based on thermodynamics stated as in equation (14.1), U value can be calculated based on transport considerations without taking into account fouling resistances. In other words, transport-based U is a function of film coefficients (h_{t} for tube side and h_{s} for shell side in Btu/(h ft² °F)) as expressed in equation (14.2):

$$\frac{1}{U_{\text{C}}} = \frac{1}{h_{\text{o}}} + \frac{1}{h_{\text{i}}}\left(\frac{A_{\text{o}}}{A_{\text{i}}}\right) + r_{\text{w}}. \qquad (14.2)$$

This U value is called as clean U value as fouling resistances (R_i, R_o) are not taken into account in equation (14.2). The film coefficients (h_t for tube side and h_s for shell side) can be calculated based on fluids' physical properties and geometry of heat exchanger. For example, for U-tube exchangers with streams all liquid or all vapor (no boiling and condensing), the correlation (Dittus and Boelter, 1930) is used to estimate the tube-side Nusselt number, Nu_t, and then tube-side film coefficient, h_t:

$$Nu_t = 0.027\left(\frac{C_p\mu}{k}\right)^{\frac{1}{3}}\left(\frac{\rho u d_i}{\mu}\right)^{0.8}, \tag{14.12}$$

$$h_t = \frac{k}{d_i}Nu_t = 0.027\frac{(C_p k^2)^{\frac{1}{3}}(\rho u)^{0.8}}{d_i^{0.2}\mu^{\frac{7}{15}}}, \tag{14.13}$$

where

C_p = fluid heat capacity, Btu/(lb °F)
d_i = inner diameter of the tube, ft
k = fluid thermal conductivity, Btu/(h ft °F)
u = fluid velocity, ft/h
ρ = fluid density, lb/ft^3
μ = fluid viscosity, lb/(ft h)

Equation (14.13) states that physical properties of tube-side stream (namely conductivity k, specific heat capacity C_p) and mass velocity u have positive effect on tube-side film coefficient h_t. In contrast, viscosity μ and tube inside diameter d_i have negative effect.

The Kern's correlation (1950) is used to estimate the shell-side Nusselt number, Nu_S, and then shell-side film coefficient, h_s:

$$Nu_S = 0.36\left(\frac{C_p\mu}{k}\right)^{\frac{1}{3}}\left(\frac{D_e\rho u}{\mu}\right)^{0.55}\left(\frac{\mu}{\mu_w}\right)^{0.14}, \tag{14.14}$$

$$h_S = \frac{k}{D_e}Nu_S = 0.36\frac{(C_p k^2)^{\frac{1}{3}}(\rho u)^{0.55}}{D_e^{0.45}\mu^{0.08}\mu_w{}^{0.14}}, \tag{14.15}$$

where

$$D_e = \frac{4\left(p^2 - \frac{\pi d_o^2}{4}\right)}{\pi d_o} \quad \text{(for spare tube pitch)}, \tag{14.16}$$

where

μ_w = water viscosity, lb/(ft h)
D_e = shell-side equivalent diameter, ft
d_o = outer diameter of the tube, ft
p = tube pitch, ft

Equation (14.15) states that physical properties of shell-side stream (namely conductivity k and specific heat capacity C_p) and velocity u and tube outside diameter d_o have positive effect on shell-side film coefficient h_t. In contrast, viscosity μ and tube pitch p have negative effect.

The above heat transfer equations provided the well-known observations: heat transfer coefficient on tube side is proportional to the 0.8 power of velocity, the 0.67 power of thermoconductivity, and the −0.47 power of viscosity,

$$h_t \propto u^{0.8}, \tag{14.17a}$$

$$h_t \propto k^{0.67}, \tag{14.17b}$$

$$h_t \propto \mu^{-0.47}. \tag{14.17c}$$

That is the reason why cooling water has a very high heat transfer coefficient, followed by hydrocarbon and then hydrocarbon gases because of the values of thermoconductivities for these fluids. Hydrogen is an unusual gas due to its extremely high thermoconductivity (greater than that of hydrocarbon liquids). Thus, its heat transfer coefficient is toward the upper limit of the range for the hydrocarbon liquid. The heat transfer coefficients for hydrocarbon liquids vary in a large range due to the large variations in viscosity, from less than 1 cP for ethylene to more than 1000 cP for bitumen. Heat transfer coefficients for hydrocarbon gases are proportional to pressure because higher pressure generates higher gas density resulting in higher gas velocity.

14.3.3 Actual U Value (U_A)

In reality, heat exchangers operate under fouled conditions with dirt, scale, and particulates deposit on the inside and outside of tubes. The overall fouling resistance is defined in equation (14.3) as

$$R_f = R_o + R_i \left(\frac{A_o}{A_i} \right). \tag{14.3}$$

By adding the overall fouling resistance to U_C, actual U_A is defined in equation (14.4) as

$$\frac{1}{U_A} = \frac{1}{U_C} + R_f. \tag{14.4}$$

Clearly, U_C is the heat transfer capability that the exchanger can deliver when no fouling is included while U_A takes into account fouling resistances. U_A can be thought of predicted or expected overall coefficient for actual heat transfer including the design fouling resistances.

Fouling resistances for streams are based on the physical properties of the streams and the average fouling factors are documented in TEMA (2007). For illustration purposes, Table 14.3 shows typical overall fouling resistances for hydrocarbon liquids based on the API gravity of the streams.

The U values should follow the order: $U_C \geq U_A \geq U_R$. The main reasons for the inequality are the practical considerations of fouling, process variations as well as inaccuracy in physical properties estimates and heat transfer calculations.

TABLE 14.3. Liquid Fouling Factors

API Gravity	R_f (ft^2 h °F/Btu)
>40	0.002
<40	0.003
<15	0.004
<5	0.005

14.3.4 Overdesign (OD_A)

For a heat exchanger to satisfy process requirement under changing process conditions, U_A must be greater than or equal to U_R. Actual overdesign or design margin can be defined as

$$\%OD_A = \left(\frac{U_A}{U_R} - 1\right) \times 100. \tag{14.18}$$

Overdesign is provided in the design stage beyond fouling factors in order to account for operation variations in fluid rates and properties as well as calculation inaccuracy for heat transfer and pressure drops. Some designers may use 5–10% overdesign for new heat exchangers if the designers have confidence in fluid properties and heat transfer calculation accuracy. Otherwise, 10–20% or higher overdesign might be used. In contrast, near-zero overdesign could be used for services with well-known fluid properties and accurate heat transfer calculations.

Statistically, heat exchangers are often designed with large overdesign intentionally because the designer wants to make sure it will satisfy process demand no matter whatever occurs in operation. There are several uncertain factors that the designer has to consider in design stage (Bennett et al., 2007).

First, uncertainty is accuracy in estimating fouling resistances to reflect the actual fouling. Furthermore, fouling resistances are static values, which are used in computation. In reality, fouling is a dynamic mechanism. The designer uses overdesign to account for this fouling dynamics based on his/her experience or company's best practices so that the exchanger can still satisfy the process demand under more severe fouling scenarios than estimated fouling resistances. The second factor is variations in process conditions. In particular, increasing feed rate is common as companies want to generate additional revenue using existing equipment. The designer provides overdesign to accommodate operating scenarios with increased feed rate. Third, the designer uses overdesign to account for the effects of inaccuracy in fluid properties and heat transfer calculations. These uncertainties become the basis for the designer to provide overdesign.

However, excessive overdesign can cause fouling and other problems with the exchanger. When too much overdesign in surface area is added, velocity reduces, which makes it easier for fouling deposits to accumulate. In some cases, a temperature-controlled by-pass line may be required for critical services to avoid too much heat transfer than process requirements in the start of run. Bypass operation could increase fouling as fluid velocity reduces.

14.3.5 Controlling Resistance

If the actual film coefficient of one side is much larger than the other, this side is referred to controlling side of resistance. In design and operation, special attention is devoted to this controlling resistance as any incremental decrease to this controlling resistance will greatly increase the overall U value. On the other hand, incremental change to the noncontrolling side film coefficient has very little effect on the overall U value.

One way to minimize the adverse effect of controlling resistance is to use extended surface area to offset the effect. Another way is to increase the velocity on the controlling side. Furthermore, the most heavily fouling stream should be placed on the tube side for ease of cleaning. Use of fouling mitigation methods such as fluid treatment, antifouling additive, and regular cleaning to prolong the "clean" operation can help maintain high U value.

14.4 UNDERSTAND FOULING

Fouling is accumulation of undesirable materials as deposits on heat exchanger surfaces. Fouling deposits come in many different causes and forms. Irrespective of the material contained in heat exchanger fouling deposits, it leads to similar consequences, which are reduction in thermal performance and an increase in pressure drop. There are complex factors causing heat exchanger fouling such as physical and chemical properties of process streams, operating conditions, heat exchanger design, and operation. Since complex factors affect the choice of methods to reduce and prevent fouling, identification of root causes could derive more effective solutions.

14.4.1 Root Causes of Fouling

Since there are a great variety of fouling phenomena, it is useful to group them into six types of fouling mechanisms for better understanding (Melo et al., 1988):

(1) Crystallization fouling: precipitation and deposition of dissolved salts, which are supersaturated at the heat transfer surface. Supersaturation may be caused by the following:
   - Evaporation of solvent.
   - Cooling below the solubility limit for normal solubility (increasing solubility with decreasing temperature, such as wax deposits, gas hydrates, and freezing of water/water vapor). The precipitation fouling occurs on the cold surface (i.e., by cooling the solution).
   - Heating above the solubility limit for inverse solubility (increasing solubility with increasing temperature, such as calcium and magnesium salts). The precipitation of salt occurs with heating the solution.
   - Mixing of streams with incompatible compositions.
   - Variation of pH that affects the solubility of CO_2 in water.

(2) Particulate fouling: accumulation of particles from heat exchanger working fluids (liquids and/or gaseous suspensions) on the heat transfer surface. Most often, this type of fouling involves deposition of corrosion products dispersed

in fluids, clay, and mineral particles in river water, suspended solids in cooling water, soot particles of incomplete combustion, magnetic particles in economizers, deposition of salts in desalination systems, deposition of dust particles in air coolers, particulates partially present in fire-side (gas-side) fouling of boiler, and so on. If particular fouling is of gravitational settling of relative large particles onto horizontal surfaces, this phenomenon is also called sedimentation fouling.

(3) Chemical reaction fouling: deposit formation (fouling precursors) at the heat transfer surface by unwanted chemical reaction (such as polymerization, coking) within the process fluid, but heat transfer surface material itself does not involve in the chemical reaction. Thermal instability of chemical components such as asphaltenes and proteins can become fouling precursors. Usually, this type of fouling starts to form at local hot spots in a heat exchanger. It can occur over a wide temperature range from ambient to over 1000 °C (1832 °F) but is more pronounced at higher temperatures.

(4) Corrosion fouling: the heat transfer surface itself reacts with chemical species present in the process fluid. Its trace materials are carried by the fluid in the exchanger, and it produces corrosion products that deposit on the surface. The thermal resistance of corrosion layers is low due to high thermal conductivity of oxides.

(5) Biological fouling: deposition and growth of macro- and microorganisms on the heat transfer surface. It usually happens in water streams.

(6) Freezing fouling: it is also called solidification fouling that occurs due to freezing of a liquid or some of its constituents to form deposition of solids on a subcooled heat transfer surface. For example, formation of ice on a heat transfer surface during chilled water production or cooling of moist air, deposits formed in phenol coolers, and deposits formed during cooling of mixtures of substances such as paraffin are some examples of solidification fouling. This fouling mechanism occurs at low temperatures, usually ambient and below.

There is no single unified theory to model the fouling process because combined fouling occurs in many applications and no single solution exists for fouling control. Appropriate theories and methods must be selected to tackle fouling issues for each application.

14.4.2 Estimate Fouling Factor R_f

The fouling factor has to be determined from actual heat exchanger performance based on online measurement taken from a process unit test run. Heat exchanger clean performance is obtained from process flowsheet simulation software (e.g., Hysys by Aspen Tech or Unisim by Honeywell) while dirt performance from exchanger rating software (e.g., HTRI by Heat Transfer Research Institute).

First, heat exchanger heat balance calculations are conducted in a flowsheet simulation software that has adequate thermal data and can describe process streams according to their physical properties and operating conditions. By providing measured temperatures, the simulation can determine heat transfer duty from $Q = mC_p\Delta T$ under design mode. At the same time, the simulation calculates heat transfer capability $U.A$ as the product of overall heat transfer coefficient and surface area as $U.A=Q/\Delta T_{LM}$. $U.A$ is also called effective surface area.

Second, heat exchanger performance calculations are performed in exchanger rating software. The thermal and physical property data for process streams are transferred from the flowsheet simulation and the dimensions and geometry of the heat exchanger are entered into the rating software based on the manufacturing datasheet.

The rating software calculates two U values, namely required and actual U. For given surface area (A), process heat duty (Q) and temperatures, the required U value is obtained according to the equation (14.1): $U_R = Q/(A\Delta T_{LM})$. At the same time, the software calculates actual U value. Then $(U_A - U_R)\cdot A$ indicates the loss of effective area due to fouling. The fouling factor can then be calculated by equation (14.4).

14.4.3 Determine Additional Pressure Drop Due to Fouling

As the tube wall thickness increases with fouling deposits, pressure drop measurement must be conducted and used as the basis for pressure drop rating calculations. In doing so, the tube wall thickness including fouling deposits are assumed and iterated until the calculated pressure drops from the rating software converges with measured ones.

Typical fouled exchanger pressure drops are 1.3–2 times that of clean exchangers (Barletta, 1998). For extreme cases, fouled exchanger pressure drops are much higher than that of clean exchangers.

It is recommended that hydraulic calculations be conducted in exchanger rating software (e.g., HTRI) as the rating software is more rigorous in pressure drop calculations than flowsheet simulation software.

14.5 UNDERSTAND PRESSURE DROP

In technical discussions on heat exchangers, pressure drop will naturally become an important topic. Process engineers usually prefer to keep pressure drop as low as possible in order to maintain sufficient suction pressure of the compressor or pumps downstream to reduce power consumption and avoid process issues. For example, high pressure drops could cause feed flashing before fired heaters downstream. In contrast, reliability and design engineers would like to maintain pressure drop as high as possible in order to reduce fouling and improve film coefficients. This helps to avoid operation issues and minimize overdesign.

Basically, heat exchanger pressure drop is a function of velocity, that is, tube velocity for tube-side pressure drop and bundle velocity for shell-side pressure drop.

14.5.1 Tube-Side Pressure Drop

Pressure drop for tube side can be expressed as

$$\Delta P_t = \frac{1}{2}\rho(u_t)^2\frac{4L}{d_t}f_t\{f_t = f(Re)\}, \tag{14.19}$$

where u_t = tube velocity, ft/h; f_t = tube-side friction factor, (ft^2 °F h)/Btu.

From equation (14.19), we can observe that the major parameters affecting the tube-side pressure drop include tube diameter and length, fluid density, viscosity, and velocity.

$$\Delta P_t \propto u_t^2, \tag{14.20a}$$

$$\Delta P_t \propto \rho, \tag{14.20b}$$

$$\Delta P_t \propto L, \tag{14.20c}$$

$$\Delta P_t \propto f_t, \tag{14.20d}$$

$$\Delta P_t \propto d_t^{-1}. \tag{14.20e}$$

14.5.2 Shell-Side Pressure Drop

The shell-side flow path is more complex than that for tube; hence, the calculation of shell-side pressure drop is more difficult. More accurate calculation of shell-side pressure drop could be obtained by the Bell–Delaware method (1973). For the purpose of providing explanation of shell pressure drop conceptually, Kern's correlation (1950) is used here. Based on bundle velocity, Kern's correlation for shell-side pressure drop (equation (14.21)) mirrors equation (14.19) for tube-side pressure drop:

$$\Delta P_s = \frac{1}{2}u_s^2 \frac{4D_s(N_B + 1)}{\rho D_e} f_s \{f_s = f(Re)\}, \tag{14.21}$$

where

u_s = shell-side cross-flow velocity, ft/h
D_s = shell diameter, ft
D_e = equivalent shell diameter, ft
N_B = number of baffles
f_s = shell-side friction factor, (ft^2 °F h)/Btu. f_s is a function of Reynolds number and f_s charts are available in Hewitt et al. (1994).

To transform the friction factor to a shell-side pressure drop, the number of the fluid crossing the tube bundle should be given. As the fluid crosses between baffles, so the number of "crosses" will be one more than the number of baffles, N_B. If the number of baffles is unknown, it can be determined using the baffle spacing P_B and tube length L:

$$N_B + 1 = \frac{L}{P_B}. \tag{14.22}$$

Equation (14.21) is then reduced to

$$\Delta P_s = \frac{1}{2}u_s^2 \frac{4D_s L}{\rho D_e P_B} f_s. \tag{14.23}$$

Clearly, equation (14.23) indicates major parameters affecting shell-side pressure drop, which include baffle spacing, tube length, fluid density, velocity, and viscosity.

Some of the important observations are

$$\Delta P_s \propto u_s^2, \tag{14.24a}$$

$$\Delta P_s \propto L, \tag{14.24b}$$

$$\Delta P_s \propto N_B, \tag{14.24c}$$

$$\Delta P_s \propto D_S^{-1}. \tag{14.24d}$$

14.6 EFFECTS OF VELOCITY ON HEAT TRANSFER, PRESSURE DROP, AND FOULING

Examination of equations (14.13) and (14.15) for heat transfer, and equations (14.19) and (14.23) for pressure drop, indicates that for given heat exchanger and fluids, fluid velocity is the most important parameter effecting pressure drop on both tube and shell sides. Thus, with increasing velocity, both pressure drop and heat transfer coefficient increase. The rate of pressure drop increase is faster than that of heat transfer coefficient. Since pressure drop is supplied by pumping (for liquid) or compression (for gas), higher pressure drop is at the expense of extra power cost while increased heat coefficient results in smaller surface area.

Learning from the above equations can lead to the conclusion that a short and wide heat exchanger could have a low pressure drop but a low heat transfer coefficient for both tube and shell sides. Clearly, higher pressure drop (ΔP value) forces the fluids through the heat exchanger at higher velocity leading to higher overall heat transfer coefficient (U value). But this is at the cost of high pump power. On the other hand, for a large surface area, the U and ΔP do not need to be so high; but this is at the expense of a larger heat exchanger. Therefore, there is an optimal velocity for each side in a heat exchanger, which can be obtained from the trade-off between the capital cost of a heat exchanger in terms of size and the operating cost in terms of power.

One common case is that actual pressure drop could be less than allowable pressure drop on either tube or shell side. This opportunity may be used to enhance the U value via increasing the fluid velocity. Velocity increase can be achieved by increasing flow passes on either tube side or shell side depending on which side is the controlling side on U value. Due to the fact that tube-side pressure drop rises steeply with increase in tube passes, it often happens that pressure drop is much lower than allowable value for a given number of tubes and two tube passes, but it exceeds the allowable value with four passes. In this case, the tube diameter and length could be varied to increase pressure drop with the result of a higher tube-side velocity obtained.

Another common scenario is that hydraulics can impose constraints when a heat recovery opportunity is implemented. In this case, the fluid velocity could be reduced via parallel arrangement of new and existing heat exchangers by splitting a total flow into two flows. Assuming that the flow split is equal, the fluid velocity for each branch flow is reduced by half while pressure drops on both sides are reduced by four times.

Fouling has to be addressed in heat exchanger design and operation. When heat exchanger is fouling, the fouling deposits build up additional resistance to heat transfer. At the same time, fouling deposits reduce the cross-sectional flow area and increase pressure drop. Plugging could also reduce the cross-sectional flow area and it could be treated the same as fouling in its effect on pressure drop. Fouling in liquids reduces heat transfer coefficient more rapidly than increase in pumping power. In contrast, fouling in gases reduces heat transfer in the range of 5–10%, but it increases pressure drop and fluid pumping power more steeply.

Increasing fluid velocity also reduces fouling tendency. Bennett et al. (2007) provided design guidelines for heavy fouling services with fluid velocity for shell-and-tube exchangers: tube-side velocity ≥ 2 m/s (6.5 ft/s) and shell-side B-stream (the main cross-flow stream through the bundle) ≥ 0.6 m/s (2 ft/s).

14.7 HEAT EXCHANGER RATING ASSESSMENT

When an evaluation is performed to assess the suitability of an existing heat exchanger for given process conditions or for new conditions, this exercise is called heat exchanger rating. Applications of rating can be for operational performance, for changes in process conditions or in process design. There are three fundamental points in determining if a heat exchanger performs well for given operating conditions or for a new service:

(1) What actual coefficient U_A value can be "performed" by the two fluids as a result of their flow rates, individual film coefficients h_t and h_s, and fouling resistance?
(2) From the heat balance: $Q = MC_p(T_1 - T_2) = mC_p(t_2 - t_1)$, known area A, and actual temperatures, required U value (U_R) can be calculated based on the Fourier equation (14.1).
(3) The operating pressure drops for the two streams passing through the existing heat exchanger.

The criteria can be established for the suitability of an existing exchanger for given or new services as two necessary and sufficient conditions:

(a) U_A must exceed U_R to give desired overdesign (%OD) so that the heat exchanger can meet changing process conditions for a reasonable period of service continuously.
(b) Operating pressure drops on both sides must be less than allowable pressure drops.

When these two conditions are fulfilled, an existing exchanger is suitable for the process conditions for which it was rated. When the process conditions undergo significant changes, a rating should be performed to make sure the exchanger can perform the task satisfactorily under the new conditions.

14.7.1 Assess the Suitability of an Existing Exchanger for Changing Conditions

When it is considered to use an existing exchanger for changing conditions or new services, rating assessment must be conducted well in advance for the suitability of existing exchangers for such services.

Example 14.1

Rating of an existing naphtha–heavier naphtha exchanger to operate under small changes in flow rates. 124,600 lb/h (vs 122,500 in design) of a 56.3 °API heavy naphtha leaves the naphtha splitter tower at 276 °F and is cooled to 174 °F by 193,000 lb/h (vs 188,000 in design) of 69 °API naphtha feed at 116 °F and heated to 170 °F. There is 6.3% vapor in the naphtha at 170 °F. 10 and 5 psi pressure drops are permissible on tube and shell sides, respectively. Can this exchanger operate satisfactorily under new conditions?

The exchanger is TEMA type AES as shown in Figure 14.4 with 21 in. shell ID having 268 tubes with 3/4 in. tube OD, 14 BWG thickness and 20 ft long, which are laid out on 1 in. triangle pitch. There are four tube passes and one shell pass with baffles spaced 11 ¼ in. apart and baffle cut 32% of shell diameter. The hot heavy naphtha is on the tube side.

Solution:

(1) Heat balance:

For naphtha feed,

$$Q = M \times C \times \Delta T + q \times M \times \text{vapor}\%$$
$$= 193{,}000 \times [0.53 \times (170 - 116) + 141.6 \times 6.3\%] = 7.23\,\text{MMBtu/h}$$

For heavy naphtha,

$$Q = m \times c \times \Delta t = 124{,}600 \times 0.57 \times (276 - 174)/10^6 = 7.23\,\text{MMBtu/h}$$

(2) ΔT_{lm} and F_t:

Tube Side		Shell Side	Differences	
Hot Stream		Cold Stream		
276	Higher temperature	170	106	ΔT_1
174	Lower temperature	116	58	ΔT_2
102	Differences	54	48	$\Delta T_1 - \Delta T_2$
ΔT		Δt		

$$\Delta T_{\text{lm}} = (\Delta T_1 - \Delta T_2)/\ln(\Delta T_1/\Delta T_2) = 48/\ln(106/58) = 79.6$$
$$R = \Delta T/\Delta t = 102/54 = 1.89$$
$$P = \Delta t/(T_1 - t_1) = 54/(276 - 116) = 0.34$$
$$F_t = 0.83$$
$$F_t \Delta T_{\text{lm}} = 66.1\,°\text{F}$$

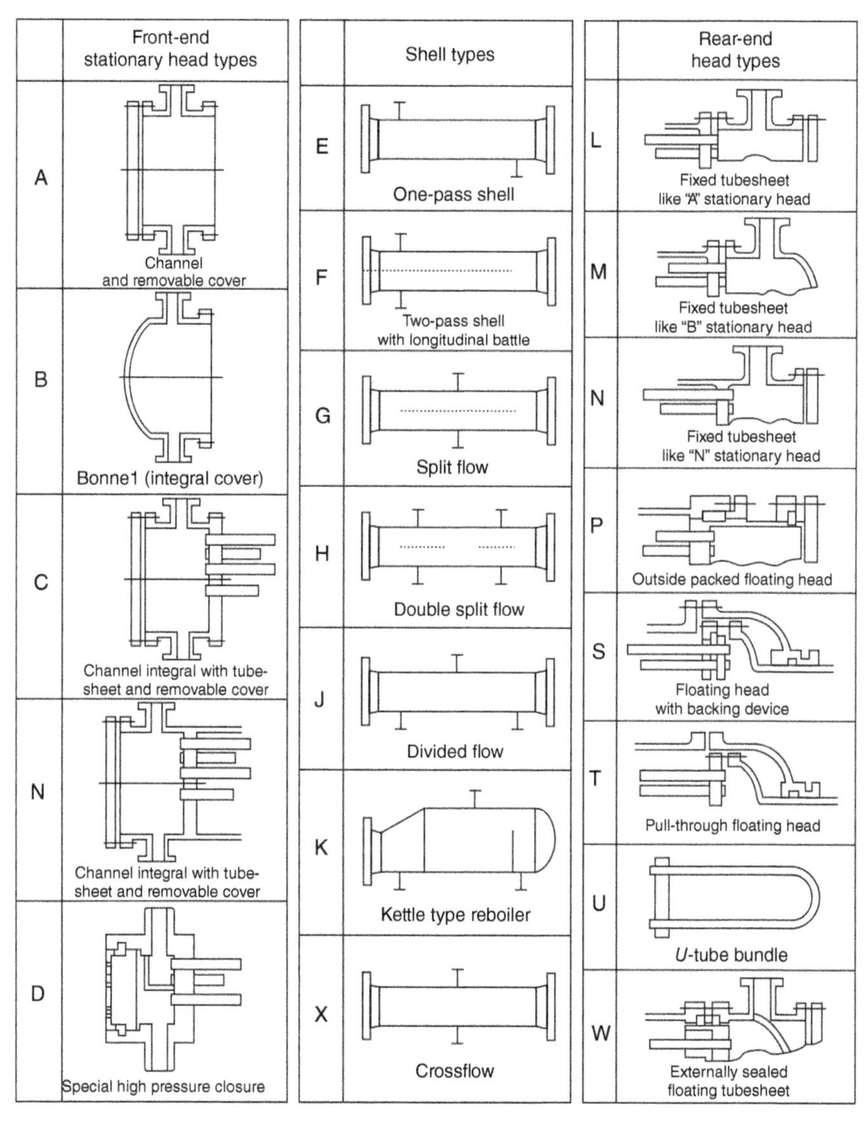

FIGURE 14.4. TEMA standard shell types and front and rear-end head types. (Source: TEMA, 1999.)

(3) Rating Summary

U_C	166.1
U_A	124.7
U_R	115.1
Overdesign	8%
R_f calculated	0.003
R_f required	0.002
ΔP_s calculated	3.8
ΔP_s allowable	5
ΔP_t calculated	9.9
ΔP_t allowable	10

The allowable fouling factor of 0.002 is assumed based on Table 14.3. U_A *is calculated by taking fouling into account. The heat exchanger has 8% of overdesign over normal fouling conditions. Pressure drops on both sides of the exchanger are less than allowable pressure drops. Thus, this exchanger meets the two criteria. Therefore, it can operate satisfactorily to fulfill the new flow conditions.*

From time to time, a process plant wishes to increase feed rate and/or make different product yields due to economic drivers. In feasibility evaluation, it is essential to assess the suitability of existing heat exchangers for new process conditions and find the most economic ways to handle significant changes.

Example 14.2

Rating of an existing naphtha–diesel exchanger to handle large increase in flow rate. A refinery plant plans to increase diesel production by 20% via revamp the hydrocracking unit. This is because diesel is highly desirable commodity in today's energy market. Currently, the naphtha–diesel exchanger is located downstream of naphtha–heavy naphtha exchanger, which is discussed in Example 14.1. Via the naphtha–diesel exchanger under the scenario of increased diesel production, 193,000 lb/h naphtha feed to the tower will increase vaporization up to 29.8% from 6.3% by 121,500 lb/h diesel product at 351 °F cooled to 260 °F. A 5 psi pressure drop is permissible on both sides. Can the current exchanger operate satisfactorily under new conditions?

The exchanger is TEMA type AES with 16-in. shell ID having 130 tubes with ¾-in. tube OD, 14 BWG wall thickness and 16 ft long, which are laid out on 1 in. triangle pitch. There are two tube passes and one shell pass with single segmental baffles spaced 14 in. apart and baffle cut 40% of shell diameter. The hot diesel is on the tube side.

Solution:

(1) Heat duty:

Naphtha feed,

$$Q = W \times q \times \Delta\text{vapor\%} = 193{,}000 \times 141.7 \times (29.8\% - 6.3\%)$$
$$= 6.4\,\text{MMBtu/h}$$

Diesel product,

$$Q = w \times c \times \Delta t = 121{,}500 \times 0.587 \times (351 - 260)/10^6$$
$$= 6.4\,\text{MMBtu/h}$$

(2) ΔT_{lm} and F_t:

Tube Side		Shell Side	Differences	
Hot Stream		Cold Stream		
351	Higher temperature	171	180	ΔT_1
260	Lower temperature	170	90	ΔT_2
91	Differences	1	90	$\Delta T_1 - \Delta T_2$
ΔT		Δt		

$$\Delta T_{\text{lm}} = (\Delta T_1 - \Delta T_2)/\ln(\Delta T_1/\Delta T_2) = 129.8$$
$$R = \Delta T/\Delta t = 91$$
$$P = \Delta t/(T_1 - t_1) = 1/(351 - 170) = 0.01$$
$$F_t = 0.99$$
$$F_t \Delta T_{\text{lm}} = 128.5\,°\text{F}$$

(3) Rating summary

U_C	135.4
U_A	106.6
U_R	123.5
Overdesign	−14%
ΔP_s calculated	10.7
ΔP_s allowable	5
ΔP_t calculated	5.6
ΔP_t allowable	5

The above rating calculations show that the existing exchanger alone cannot handle 20% increase in diesel flow rate. This is because surface area is not sufficient as well as the shell-side pressure drop in particular is too large to be allowed. Thus, it violates the criteria for the suitability of an existing exchanger to fulfill changing process conditions. The following discussions will show how to assess practical solutions by use of spare heat exchangers.

14.7.2 Determine Arrangement of Heat Exchangers in Series or Parallel

In some plants where a large number of exchangers are used, certain size standards are usually established in-house for 1–2 type of exchangers so that future services can be satisfied by making arrangement of standard exchangers in series or in parallel. Use of standard exchangers could come at a price because of impossibility of utilizing the standard equipment in the most efficient manner. However, it does offer a great advantage of reducing spare parts, tubes, and tools for replacement. When tube bundles are retubed, the standard exchangers can provide services as new ones to meet process conditions.

There are two basic arrangements of exchangers, namely series and parallel arrangements. When use of a single 1–2 exchanger could not satisfy new process conditions or lead to a severe temperature cross-signaled by a low F_t factor, it may be necessary to use two 1–2 exchangers in series. On the other hand, when hydraulic limitation could be an issue for a 1–2 exchanger, placing multiple 1–2 exchangers in parallel could resolve the issue.

Example 14.3 (continuing from Example 14.2)

The above rating assessment for the existing naphtha–diesel exchanger showed that the single 1–2 exchanger is not sufficient to meet 20% increase in diesel flow rate. A spare 1–2 exchanger was considered to add to the existing naphtha–diesel exchanger. What is the proper arrangement of this spare 1–2 exchanger in relation to the existing naphtha–diesel exchanger to handle the large increase in diesel flow rate?

Under increased diesel production, 193,000 lb/h naphtha feed (placed on the shell side of the exchanger) to the tower will increase vaporization up to 29.8% from 6.3% by 121,500 lb/h diesel product at 351 °F cooled to 260 °F. 5 psi pressure drop is permissible on both sides.

Based on the rating assessment in Example 14.2, it is observed that pressure drop on the shell side is too large to be allowed. Thus, a parallel arrangement is considered in this assessment as shown in Figure 14.5. The two exchangers are accounted as one exchanger unit for the rating calculations as follows.

(1) Heat duty:

Naphtha feed:

$$Q = W \times q \times \Delta\text{vapor\%} = 193{,}000 \times 141.7 \times (29.8\% - 6.3\%)$$
$$= 6.4\,\text{MMBtu/h}$$

Diesel:

$$Q = w \times c \times \Delta t = 121{,}500 \times 0.587 \times (351 - 260)/10^6 = 6.4\,\text{MMBtu/h}$$

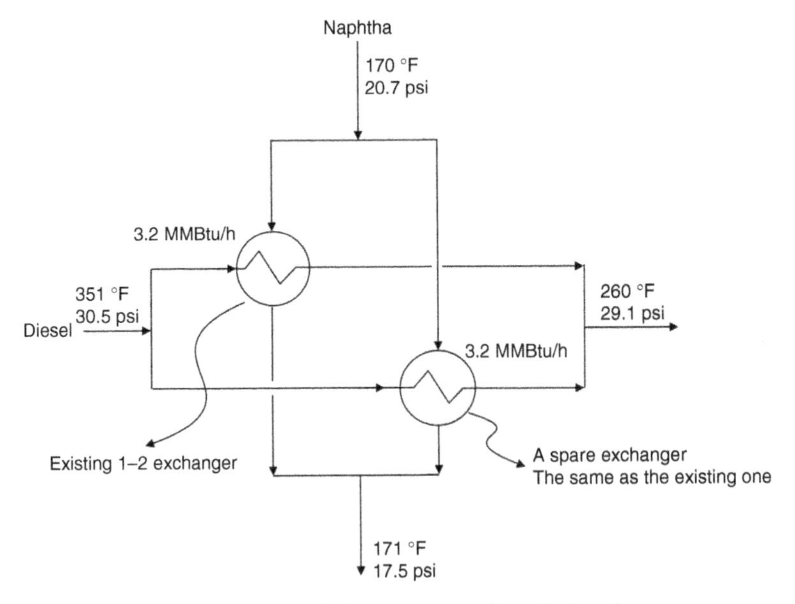

FIGURE 14.5. A parallel arrangement of two 1–2 exchangers.

(2) ΔT_{lm} and F_t:

Tube Side		Shell Side	Differences	
Hot Stream		Cold Stream		
351	Higher temperature	171	180	ΔT_1
260	Lower temperature	170	90	ΔT_2
91	Differences	1	90	$\Delta T_1 - \Delta T_2$
ΔT		Δt		

$$\Delta T_{lm} = (\Delta T_1 - \Delta T_2)/\ln(\Delta T_1/\Delta T_2) = 129.8$$

$$R = \Delta T/\Delta t = 91$$

$$P = \Delta_t/(T_1 - t_1) = 0.01$$

$$F_t = 0.99$$

$$F_t \Delta T_{lm} = 128.5\,°\text{F}$$

(3) Summary

U_C	78.7
U_A	68.0
U_R	61.8
Overdesign	10%
R_f calculated	0.003
R_f required	0.002
ΔP_s calculated	3.2
ΔP_s allowable	5
ΔP_t calculated	1.4
ΔP_t allowable	5

Two 1–2 heat exchangers in parallel are adequate to satisfy process heat transfer requirement with 10% overdesign. Pressure drops on both sides of the exchanger are less than allowable pressure drops.

Example 14.4

Use of spare exchangers in series to an existing acetone–acetic acid exchanger. Acetone at 250 °F is to be sent to storage at 100 °F and at a rate of 60,000 lb/h. The heat will be received by 185,000 lb/h of 100% acetic acid coming from storage at 90 °F and heated to 150 °F. Pressure drops of 10.0 psi are available for both fluids, and an overall fouling factor of 0.004 should be provided.

Available for the service are several 1–2 exchangers having 21 ¼ in. shell ID, having 270 tubes with ¾ in. tube OD, 14 BWG, 16 ft 0 in. long and laid out on 1 in. square pitch. The bundles are arranged for two tube passes with segmental baffle spaced 5 in. apart. Determine the suitability of these 1–2 exchangers for the specific service.

Solution:

(1) Exchanger data:

Shell Side	Tube Side
ID = 21 ¼ in.	Number and length = 270, 16 ft 0 in.
Baffle spacing = 5in.	OD / BWG / pitch = 3/4 in. / 14 BWG / 1 in. square
Shell passes = 1	Tube passes = 2

(2) Heat balances:

Acetone $Q = 60{,}000 \times 0.57(250\text{–}100) = 5{,}130{,}000\,\text{Btu/h}$

Acetic acid $Q = 168{,}000 \times 0.51(15\text{–}90) = 5{,}130{,}000\,\text{Btu/h}$

(3) F_t factor:

Shell Side		Tube Side	Differences	
Hot Stream		Cold Stream		
250	Higher temperature	150	100	ΔT_1
100	Lower temperature	90	10	ΔT_2
150	Differences	60	90	$\Delta T_1 - \Delta T_2$
ΔT		Δt		

$$\Delta T_{lm} = 39.1\,°F$$
$$R = 150/60 = 2.5$$
$$P = 60/(250 - 90) = 0.375$$

Thus,
one 1–2 exchanger, F_t is not on F_t charts
two 1–2 exchangers, $F_t = 0.57$ (too small)
three 1–2 exchangers, $F_t = 0.86$ (OK)
$F_t \times \Delta T_{lm} = 33.6\,°F$

To permit the heat transfer with the temperatures given by the process, a minimum of three 1–2 exchangers is required. If the sum of the surface area in three exchangers is insufficient, a greater number of 1–2 exchangers are required.

(4) U_c: Calculated from heat exchanger rating software:

$$h_t = 194\,\text{Btu/h ft}^2\,°F \quad \text{and} \quad h_s = 242\,\text{Btu/h ft}^2\,°F$$
$$U_c = h_t \times h_s/(h_t + h_s) = 194 \times 242/(194 + 242)$$
$$= 107.7\,\text{Btu/h ft}^2\,°F$$

(5)

$$U_R : U_R = Q/(AF_t \text{ LMTD}) = 5{,}130{,}000/(2540 \times 34.4) = 58.8\,\text{Btu/h ft}^2\,°F$$

(6)

$$R_f : R_f = (U_c – U_R)/U_c U_R = (107.5–58.8)/(107.5 \times 58.8)$$
$$= 0.0077\,\text{h ft}^2\,°F/\text{Btu}$$

(7) ΔP_s and ΔP_t: Calculated from heat exchanger rating software:

$$\Delta P_s = 10.4\,\text{psi (allowable } \Delta P_s = 10.0\,\text{psi) and } \Delta P_t = 5.2\,\text{psi (allowable } \Delta P_t = 10.0\,\text{psi)}$$

(8) Rating summary:

U_C	107.5
U_A	75.2
U_R	58.8
OD%	28%
R_f calculated	0.0077
R_f required	0.004
ΔP_s calculated	10.4
ΔP_s allowable	10
ΔP_t calculated	5.2
ΔP_t allowable	10

Conclusion: Three 1–2 exchangers are more than adequate for heat transfer even though the pressure drop on the shell side is slightly higher than allowable. Fewer exchangers cannot fulfill the process requirement.

14.7.3 Assess Heat Exchanger Fouling

Heat exchanger fouling occurs during operation and cause exchanger performance to deteriorate over time. In some cases, it requires cleaning several times before the entire process is shut down for turnaround maintenance. In extreme cases, it may not be possible to remove tube bundles that must be replaced. It is an economic decision for selecting fouling mitigation methods and when to apply. Heat exchanger rating can determine the level of fouling and if the heat exchanger in question requires attention.

Example 14.5

Calculation of heat transfer performance for an existing vacuum residue–crude exchanger. 710,000 lb/h of a 31.1°API crude oil going through the tube side of the exchanger is heated from 359 to 375 °F by 213,500 lb/h of 11.1°API vacuum residue entering at 503 °F and cooled to 449 °F. How is this heat exchanger performing?

The exchanger is TEMA Type AES with 48 in. ID shell having 964 tubes with 1 in. OD, 12 BWG thickness, and 24 ft length, which are laid out on 1 ¼ in. square pitch. There are four tube passes and one shell pass with baffles spaced 9.5 in. apart and baffle cut 15%. Fouling factors of 0.003 and 0.01 are provided for crude and vacuum residue, respectively.

(1) Heat duty:
Crude oil $Q = 730{,}000 \times 0.60 \times (375\text{–}359) = 7.0\,\text{MMBtu/h}$
Vacuum residue $Q = 213{,}500 \times 0.61 \times (503\text{–}449) = 7.0\,\text{MMBtu/h}$

(2) ΔT_{lm} and F_t:

Tube Side		Shell Side	Differences	
Hot Stream		Cold Stream		
503	Higher temperature	375	128	ΔT_1
449	Lower temperature	359	90	ΔT_2
54	Differences	16	38	$\Delta T_1 - \Delta T_2$
ΔT		Δt		

$$\Delta T_{\text{lm}} = (\Delta T_1 - \Delta T_2)/\ln(\Delta T_1 - \Delta T_2) = 107.9$$

$$R = \Delta T/\Delta t = 3.38$$

$$F_t = 0.98$$

$$F_t \Delta T_{\text{lm}} = 105.8\,^\circ\text{F}$$

(3) Initial assessment:

U_C	49.6
U_A	30.2
U_R	11.1
OD%	171%
R_f calculated	0.070
R_f required	0.013

The required U value to achieve 7.0 MMBtu of heat transfer is only 11.1 in comparison with the actual U value of 30.2 based on fouling factors of 0.01 for vacuum residue and 0.003 for crude. In other words, the heat exchanger only accomplishes one-third of the heat transfer capability offered by the heat exchanger, which warrants a more detailed investigation.

(4) More detailed assessment:

As a follow-up, engineers conducted the performance comparison between operation and design and the results are given in the following table. It can be observed that the flow rates in operation are higher than those in design. The higher flow rates should have corresponded to a higher U value in operation. However, in this case, the U value in design is 41% higher than in operation.

	Design	Operation
Q (MMBtu/h)	8.9	7.0
W (resid) (lb/h)	167,250	213,500
T_1 (°F)	505	503
T_2 (°F)	416	449
w(crude) (lb/h)	665,000	710,000
t_1 (°F)	360	359
t_2 (°F)	382	375
ΔT_1 (°F)	123	128
ΔT_2 (°F)	56	90
ΔT_{1m} (°C)	85	108
F_t	0.93	0.98
U_A	112,658	66,074
A	5,936	5,936
U	18.98	11.13
$U_{operation}/U_{design}$		0.59

Field inspection was performed and pressure drop was measured. It was found the pressure drop on the crude (tube side) was around 60 psi versus 6.8 psi under normal fouling conditions. It was concluded that the heat exchanger suffers severe fouling with loss of more than half the heat transfer capability. In addition, the much higher pressure drop caused crude feed flashing before the charge heater, which could be the potential safety issue for the heater. Thus, it was decided to clean the exchanger immediately online by means of by-pass arrangement. After cleaning, dedicated investigation was conducted to identify the root causes of this fouling.

It was found from rating assessment that the tube-side crude velocity is a bit too low, which is 5.3 ft/s. It should be 7 ft/s for this hot crude heating service because precipitation fouling becomes more active under high temperature. The change was made to the number of tube passes from four to six. As a result, the tube velocity was increased to 7.9 ft/s, but at the expense of higher pressure drop on the tube side. The tube-side pressure drop was increased to 21 psi versus 6.8 psi with four tube passes. This change helped reducing tube fouling and prolonged the operation of the exchanger between cleanings.

The lessons learnt from this investigation indicate that fluid flow rate affects fouling behavior significantly. Flow rates much lower than design result in lower velocity, which can promote accumulation of fouling deposits. High temperature is another major cause for promoting fouling. The heat exchangers in the high temperature region are more prone to be fouled due to inherent thermal coking tendency. Threshold conditions in terms of velocity and temperature should be identified beyond which fouling occurs in a faster pace.

14.8 IMPROVING HEAT EXCHANGER PERFORMANCE

The objective of heat exchanger operation management is to maintain good performance to fulfill process requirements for desirable periods of time. Basically, there are three major reasons why exchanger operation could deviate from design: poor design, excessive fouling, and mechanical failure. In any event, heat exchangers can deliver trouble-free services while meeting process requirements if the heat exchanger is designed well thermally and mechanically, stored carefully before use, installed correctly, operating within its design limits, and cleaned periodically depending on fouling formation. In contrast, it can be stressful if heat exchangers do not perform as expected in meeting process requirements. In the worst case, mechanical and performance failure of heat exchangers could cause undesirable unit shutdowns.

The methods for monitoring and troubleshooting are provided here with the focus on thermal and hydraulic performance. Fijas (1989) provided good discussions on mechanical problems often encountered with heat exchangers. With exchanger performance, the priority issues are good knowledge of fouling resistances to avoid poor design, continuous monitoring of U value to maintain good performance, pressure drop survey for troubleshooting, managing of two-phase flow, fouling mitigation, and heat transfer enhancements. The general methodology for improving heat exchanger performance is to monitor performance trends, identify opportunity, and develop and implement solutions.

14.8.1 How to Identify Deteriorating Performance

14.8.1.1 Fouling Resistances Inappropriate estimate of fouling resistances results in either too much or too little overdesign. Although the TEMA fouling resistances were originally only considered to be rough guidelines for heat exchanger design, they are often treated as accurate values. This may cause considerable errors because the transient character of the fouling process is neglected. Conditions in initially overdesigned heat exchangers often promote fouling deposition, thus making fouling a self-fulfilling prophecy. Thus, one needs to be critical of the fouling resistances listed in the public domain and make proper adjustment based on historical fouling data. For existing services, obtain historical fouling trends and assess the characteristics of the system to determine the root causes for fouling and design accordingly.

14.8.1.2 U Value Monitoring Due to the fact that heat exchanger performance varies with flow rates, compositions and fouling conditions, heat exchanger assessment must be conducted on a regular basis so that a performance trend over time can be measured and problems can be detected at an early stage. A single rating of an exchanger is good for getting a baseline data on its performance, but it must be done on a regular basis to define trends. From a single point of rating, you can calculate a single U value, pressure drops of shell side and tube side, and a single value of heat duty. However, single point assessment cannot provide insights into fouling evolution over time and sudden changes in U value due to process variations. But a U value trend can help you with these operating issues.

The most important thing of a U value trend is its capability of showing the fouling behavior. The purpose of U-trend monitoring is to identify any abnormal fouling behavior. In general, fouling accumulation in heat exchanger depends on the type of fouling, the service (fluid compositions, temperature, and pressure), and the exchanger design, and so on. Under normal operation, a U trend should display gradual changes in U value. However, operation changes could affect fouling, which include feed rate variations, fluid composition change, by-pass operation, and hydraulic head change. If an operation change suddenly distorts the normal fouling behavior (e.g., U value reduces sharply), this change must be investigated and appropriate actions must be taken.

When an exchanger is new, a detailed performance evaluation is warranted and it should be repeated after 6 months or so. One should trend data in-between and afterward. Process temperatures, pressures, and flows around the heat exchanger are measured in daily averages. It is recommended to use distributed control trend logs for data collection. It is important to keep the data and the calculations for reference in future. The potential actions for fouling mitigation are discussed in Zhu (2014).

14.8.1.3 Pressure Drop Monitoring The importance of pressure drop calculations cannot be overemphasized as it can help with analyzing performance problems and troubleshooting of heat exchanger malfunction. Calculated pressure drop for single-phase flow can be reasonably close to measured pressure drop if there is no fouling. For two-phase flow, calculated pressure drop can also be reasonably close to measured pressure drop if pressure drop zones are used and flow patterns are considered. With these two assumptions as the basis, pressure drop calculations can be used as a tool for identifying problems.

If measured pressure drops are significantly lower than calculated drops, this might indicate fluid bypassing, which could occur either on tube side or shell side. On the other hand, if measure pressure drop is too high, this is often caused by severe plugging or fouling, or freezing or slug flow for two-phase flows.

14.8.1.4 Avoid Poor Design Chemical and petroleum industries have been plagued for decades with poorly operating and occasionally inoperable heat exchangers. One of common causes is usually traced to poor design, which should be avoided in the design stage by all means. Careful considerations of major design choices must be made in order to obtain an "optimal" heat exchanger design. The design issues include fouling considerations, tube-side design (tube counts, tube passes, tube length, tube pitch, and tube layout), shell-side design (shell diameter, shell types (including TEMA Types E/F/G/H/J/K/X, shell flow distribution), and baffle design (baffle types, segmental baffle including single/double/triple, baffle spacing, and baffle cut).

The essential design task is to optimize velocity in both tube and shell sides by the best use of allowable pressure drop available. For example, when the number of tube passes is increased from one to two passes, the velocity could be twice that of one-pass velocity as the travel distance is doubled. Then the heat tube-side transfer coefficient will increase according to the 0.8 power of velocity. At the same time, the tube-side pressure drop will increase according to the square of velocity and to the travel distance. Therefore, pressure drop will rise to the cubic of the increase in tube passes for a given tube counts and tube-side flow rate. When the pressure drop is higher than allowable one, reduction of tube length could reduce pressure drop. Besides the number of tube passes and tube length, other design choices include the tube outside diameter, tube pitch, tube counts, and layout.

Design choices are available to reduce shell-side pressure drop. The number of baffles (N_B) is proportional to the baffle spacing. Baffle spacing and baffle cut have a profound effect on shell-side pressure drop. In many cases, the shell-side pressure drop is still too high with single segmental baffles in a single-pass shell even after increasing the baffle spacing and baffle cut to the highest values recommended. These cases may accompany with very high shell-side flow rate. The next design choice is to consider double segmental baffles. When double segmental baffles at relatively high baffle spacing cannot satisfy shell-side allowable pressure drop, a divided-flow shell (TEMA J) with single segmental baffles could be considered. Since pressure drop is proportional to the square of velocity (u^2) and to the length of travel (L), a divided-flow shell could have one-eighth the pressure drop in an identical single pass exchanger. This discussion can go on as there are other design choices that are available to deal with high shell-side pressure drops, for which Mukherjee (1998) provided detailed explanations.

NOMENCLATURE

Variables

A	surface area (ft^2)
C, c	fluid heat capacity of hot and cold streams (Btu/(lb °F))
d_i	inner diameter of the tube (ft)
d_o	outer diameter of the tube (ft)
D_e	equivalent shell diameter (ft)
D_s	shell diameter (ft)
f_s	shell-side friction factor ((ft^2 °F h)/Btu)
f_t	tube-side friction factor ((ft^2 °F h)/Btu)
F_t	LMTD correction factor (fraction)
h_t	tube-side file coefficient (Btu/(ft^2 °F h))
h_s	shell-side file coefficient (Btu/(ft^2 °F h))
k	fluid thermal conductivity (Btu/(h ft °F))
L	tube length (ft)
LMTD	ΔT_{lm} (°F)
M, m	mass flow rate for hot (cold) streams (lb/h)
N_B	number of baffles
p	tube pitch (ft)
ΔP	pressure drop (psia)
Q	heat duty (MMBtu/h)
R	temperature ratio: $(T_1 - T_2)/(t_2 - t_1)$ (dimensionless)
Re	Reynolds number (dimensionless)
R_f	overall fouling resistance $= R_t + R_s$ ((ft^2 °F h)/Btu)
R_t	R_s = tube (shell) side fouling resistance, (ft^2 °F h)/Btu
P	temperature ratio: $(t_2 - t_1)/(T_1 - t_1)$ (dimensionless)
r_w	resistance of the inner tube referred to the tube outside diameter ((ft^2 °F h)/Btu)
T_1, t_1	supply temperature of hot (cold) stream (°F)
T_2, t_1	target temperature of hot (cold) stream (°F)
ΔT_1	hot end temperature approach (°F)
ΔT_2	cold end temperature approach (°F)
ΔT_{lm}	logarithmic mean temperature difference (LMTD) (°F)
u	shell-side cross-flow velocity or tube velocity (ft/h)

U overall heat transfer coefficient (Btu/(ft^2 °F h))
V_S superficial gas velocity (ft/s)

Greek letters

ρ fluid density (lb/ft^3)
μ fluid viscosity (lb/(ft h))
μ_w water viscosity (lb/(ft h))
τ_0 sheer stress (lb/ft^2)

Subscript and superscript

A actual
C clean
D design
e equivalent
f friction
G gas
i inside of tube or shell
L liquid
lm logarithmic mean
o outside of tube or shell
s shell side
t tube side

REFERENCES

Barletta AF (1998) Revamping crude units, *Hydrocarbon Processing*, 51–57, February.

Bell KL (1973) Thermal design of heat transfer equipment, in Perry RH and Chilton CE (eds) *Chemical Engineers Handbook*, 5th edition, p. 10, McGraw-Hill.

Bennett CA, Kistler RS, Lestina TG, King DC (2007) Improving heat exchanger designs, *Chemical Engineering Progress*, 40–45, April.

Dittus FW, Boelter LMK (1930) *Publications on Engineering*, University of California, Berkeley, Vol. **2**, p. 443.

Fijas DF (1989) Getting top performance from heat exchangers, *Chemical Engineering*, 141–145, December.

Hewitt GF, Shires GL, Bott, TR (1994) *Process Heat Transfer*, CRC Press, pp. 275–285.

Kern DQ (1950) Process Heat Transfer, *McGraw-Hill*, pp. **148**.

Melo LF, Bott TR, Bernardo CA (eds) (1988) *Advances in Fouling Science and Technology*, Kluwer Academic Publishers.

Mukherjee R (1998) Effectively design shell-and-tube heat exchangers, *Chemical Engineering Progress*, February Issue, AIChE.

Shah RK, Sekulić P (2003) *Fundamentals of Heat Exchanger Design*, John Wiley & Sons.

TEMA (1999) *Standards of TEMA*, 8th edition, Tubular Exchanger Manufacturers Association, New York.

TEMA (2007) *Standards of TEMA*, 9th edition, Tubular Exchangers Manufacturer Association, New York.

Zhu XX (2014) Energy and Process Optimization for the Process Industries. John Wiley & Sons.

15

DISTILLATION COLUMN ASSESSMENT

15.1 INTRODUCTION

Distillation is the core of a process unit for converting multicomponent streams into desirable products and accounts for the majority of energy consumptions. Improving energy utilization, reducing capital costs, and enhancing operational flexibility are spurring increasing attention to distillation column optimization during design and operation. A good understanding of distillation fundamentals, feasible operation, and equipment constraints will enable process engineers gain insights for the distillation performance.

15.2 DEFINE A BASE CASE

The first step for tower performance evaluation is to simulate the original tower design because it is uncommon that the original tower datasheets are unavailable or inaccurate. To do this, selection of a proper VLE (Vapor and Liquid Equilibrium) calculation package is critical. For hydrocarbon separation, the Peng–Robinson equation of state model is a common choice. By providing process data including feed and product flows and compositions, together with tower data including temperature and pressure, feed tray, the number of theoretical stages and reflux rate, the simulation will generate mass and composition balances as well as heat balances indicating reboiling and condensing duties.

Once the process simulation is developed, it is desirable to verify the simulation fidelity using different process conditions. The predicted product rates and purity

Hydroprocessing for Clean Energy: Design, Operation, and Optimization, First Edition.
Frank (Xin X.) Zhu, Richard Hoehn, Vasant Thakkar, and Edwin Yuh.

and compositions as well as key operating parameters such as reflux rate and reboiling/condensing duties can then be compared with measurement. In some cases, performance tests are required to gather key data to compare with simulation for the accuracy and reliability of the simulation. To do this, the performance tests must be conducted under steady and smooth conditions to mimic steady-state operations.

If the simulation fidelity is proven to be sufficient enough, it is ready to move to the next task, which is evaluation of the tower performance because the purpose of reproducing the original design data is to understand the tower hydraulic and thermal performances of the base case.

An important aspect of defining the base case is gathering all the important data for the material and heat balances in one single sheet for a tower of interest. It would be very informative to have important mass flows, temperature, pressure, and composition data in one table so that a snapshot of the tower performance can be seen at a glance. Such an example is a heat-pumped C3 Splitter shown in Figure 15.1 with required data shown in Table 15.1. In building such a table, it is a good practice to include tag number of the instrument for each parameter so that the data can be retrieved readily from historian to produce the table with snapshots of different time for evaluation of tower performance in future. The last column shows the high and low values of the corresponding parameters. The accuracy is determined by recording (or observing) operation during the steady-state period and noting the average high and low values of the various instruments during this period of time. This information could be very helpful when establishing heat and mass balances with indication of closure percentage. Typically, smaller flows than feed streams can have a higher inaccuracy than larger streams and not severely affect the material balance. Therefore, it is good to know which streams have the highest reliability when determining the material or heat balance. It is also important to record the date and the time period that the data was taken for future reference. A ready reference of what data is needed in a typical tower evaluation can be seen here.

Defining a base case is to determine the base case operation of the tower of interest. This requires extracting two kinds of data. One kind is process data in terms of

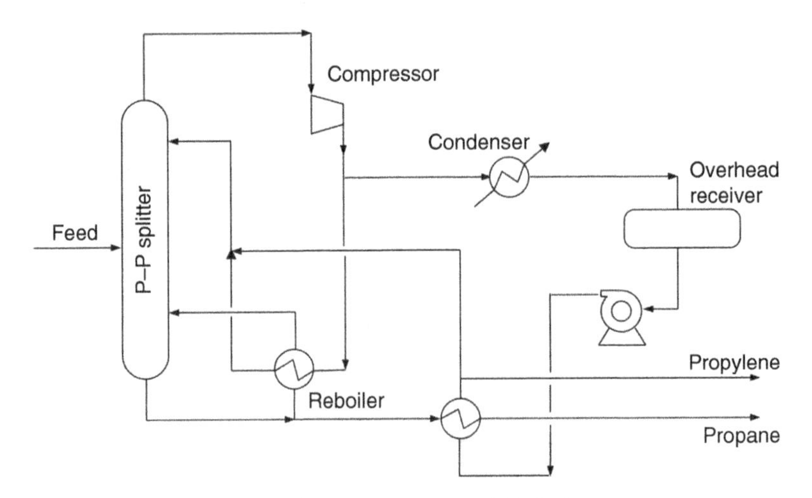

FIGURE 15.1. Heat-pumped C3 Splitter.

TABLE 15.1. Major Data Set for a Heat-Pumped C3 Splitter

Data	Units	Tag. No.	Value	Accuracy
Feed rate	BBL/D	FE-8854	4975	±50
Feed temperature	°F	Pyrometer	87	
Top pressure	psig	PI-8831	100	
Pressure drop	psi	PDI-8827	9.2	
Top temperature	°F	TI-8774	53	
Bottom temperature	°F	Pyrometer	73	
Compressor suction temperature	°F	Pyrometer	55	
Compressor discharge pressure	psig	PC-8832	230	
Compressor discharge temperature	°F	TI-8776	119	
Compressor discharge temperature	°F	Pyrometer	135	
Main reflux rate	MSCFD	FT-8858	34.55	Too low
Main reflux temperature	°F	Pyrometer	74	
Trim reflux rate	BBL/D	FT-8857	600	
Trim reflux temperature	°F	Pyrometer	99.5	
Bottom flow	BBL/D	FT-8864	1060	±50
Propylene product temperature	°F	Pyrometer	110	
Propylene product flow rate	BBL/D	FT-8860	3840	±100
Overhead composition	vol% C3=	AR 869-3	92.1	±0.5
Bottom composition	vol% C3 —	AR 869-2	97.1	±0.1

Source: Summers (2009). Reproduced with permission of AIChE.

feed and product conditions such as flows and compositions, while the other is tower operating data including temperature, pressure, and reflux rate. The former defines the mass and composition balances and the latter sets the heat balance around the tower with Table 15.2 giving such an example of C2 Splitter column.

Due to the importance of developing a reliable base case as the basis for evaluation, Summers (2009) gives excellent discussions for this topic. For understanding the difference between simulation and measurement, readers can refer to Kister (2006).

15.3 CALCULATIONS FOR MISSING AND INCOMPLETE DATA

Plant historian data is the best source, but they are usually incomplete. This is particularly true for old process units. In order to avoid wasted time and rework, you need to make sure critical meters are working properly. In most cases, design and operating data is of interest and the key is to understand the difference and reasons. For example, the knowledge about heat exchanger fouling can help evaluation of current operation performance considerably. Major consumption and any critical inputs must be verified carefully. The first stage of verification is to compare design data with operating data and perform some adjustments. This first pass verification can separate the important data from the trivial data so that the effort for chasing high precision and gathering miniature data and nit-gritty details can be avoided.

Since most correlations for heat exchangers are empirical based, the heat transfer calculations for exchangers are only accurate to about 85–90% when all the necessary

TABLE 15.2. Heat and Mass Balances for a C2 Splitter

Composition (wt%)	Feed	Vent	Ethylene Product	Dilute Ethylene Product	Ethane Bottoms
Hydrogen	0.0016%	0.26%	0.21 ppm	0	0
CO_2	0.0001%	0.0006%	0.0002%	0.61 ppm	0
Methane	0.091%	14.45%	0.007%	0.007%	0
Ethylene	77.77%	85.28%	99.98%	80.44%	1.55%
Ethane	21.66%	0.0002%	0.0109%	19.56%	96.17%
Propylene	0.291%	0	0	0.002%	1.37%
Propane	0.0071%	0	0	0	0.033%
Isobutane and heavier	0.187%	0	0	0	0.88%
Total (lb)	100,000	588	71,853	6,292	21,267
Phase	Vapor	Vapor	Liquid	Liquid	Liquid
Temperature (°C)	−13.0	−43.4	−29.8	−26.1	−7.0
Pressure (psig)	340	250	270.5	276.2	279.7

DA-2410 condenser pressure	250 psig
DA-2410 top pressure	251 psig
DA-2404 condenser pressure	269.9 psig
DA-2404 top pressure	269.9 psig
Vent condenser duty[a]	0.73 MMBtu/h
Condenser duty[a]	49.87 MMBtu/h
Reboiler duty[a]	23.18 MMBtu/h
Side reboiler duty[a]	13.17 MMBtu/h
Reflux rate to DA-2410[b]	4.73 lb/h
DA-2410 reflux temperature	−43.4 °C
DA-2410 top temperature	−36.1 °C
Vapor rate to DA-2410[b]	5,318 lb/h
DA-2404 reflux rate[b]	349,370 lb/h
DA-2404 reflux temperature	−33.7 °C
DA-2404 top temperature	−30.4 °C

Source: Summers (2009). Reproduced with permission of AIChE.
[a]All duties adjusted to a 100 klb feed basis.
[b]All flows adjusted to a 100 klb feed basis to make the true capacity of the unit.

data is known. When some data has to be estimated, the accuracy gets worse. However, this accuracy is sufficient to tell if a heat exchanger is functioning as expected or not.

Example 15.1 Obtain Missing Data

In many cases, short-cut calculations can fill in the gaps. An example used in Kenney's book (1984) gives good illustration for how to do it. This example is revised to reflect the reality. Consider the tower in Figure 15.2. As for many plants, cooling water rates are not measured and overhead product comes off on level control. However, since feed rate and composition and overhead product composition are known, much of the missing data can be inferred by energy and mass balances and the primary heat transfer equation.

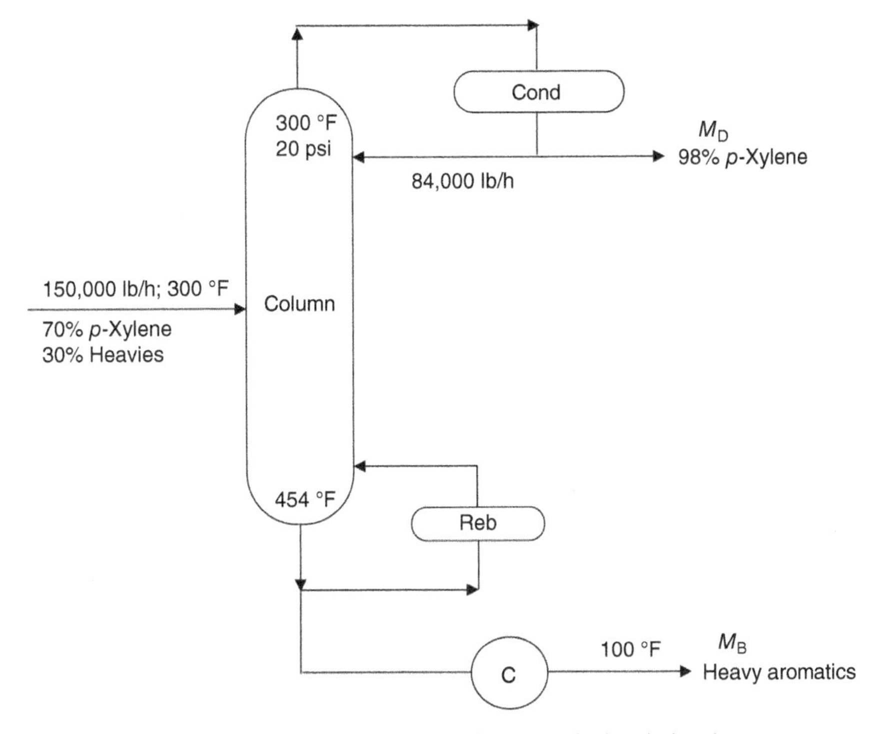

FIGURE 15.2. Use of heat/mass balances to obtain missing data.

In this problem, *p*-xylene is to be recovered from a stream containing heavier aromatics. Neither product rate is measured, but feed rate, reflux rates, and the *p*-xylene content of the overhead are. No heat exchanger duties are measured. With some data from a readily available source, the energy use for the tower can be estimated.

Given: For *p*-xylene: normal boiling point = 138.5 °C; latent heat of vaporization = 146.2 Btu/lb; specific heat = 0.38 Btu/°C lb at 0 °C and = 0.43 at 41 °C and = 0.55 extrapolating to 140 °C. For heavier aromatics: specific heat = 0.4 Btu/°C lb for naphthalene at 87 °C and = 0.5 for pentadecane at 50 °C = 0.8 extrapolating to 230 °C.

Calculate missing data: *p*-Xylene product rate; heat duty for the overhead condenser, bottom cooler and reboiler.

Solution:

(i) Calculate *p*-xylene product rate: Applying component balance on *p*-xylene and mass balance on the tower gives

$$70\% \times F = 98\% \times m_D; \quad m_D = \frac{0.70 \times 150{,}000}{0.98} = 107{,}143\,\text{lb/h},$$

$$m_B = F - m_D = 150{,}000 - 107{,}143 = 42{,}857\,\text{lb/h},$$

where F is feed rate.

(ii) Calculate the bottom cooler duty: The heat rejected in the bottom cooler is

$$Q_{\text{bottom cooler}} = m_B \times C_p \times \Delta T = 42{,}857 \times 0.8 \times (454 - 150)/10^6$$
$$= 10.4\,\text{MMBtu/h}.$$

If the heat capacity data is in error, the calculated duty would vary ±0.1 Btu/(lb °F). This is within the precision of other data.

(iii) Calculate the overhead cooler duty:

$$m_{\text{overhead}} = 107{,}143 + 84{,}000 = 191{,}143\,\text{lb/h}.$$

Case 1: No subcooling, the condenser duty is

$$Q_{\text{condenser}} = m_{\text{overhead}} \times q_{\text{latent}} = 191{,}143 \times 146.2/10^6 = 27.9\,\text{MMBtu/h}.$$

Case 2: Assuming 30 °F subcooling, the condenser duty is summation of latent heat duty and subcooling duty, which can be calculated as

$$Q_{\text{condenser}} = m_{\text{overhead}} \times q_{\text{latent}} + m_{\text{overhead}} \times C_p \times \Delta T$$
$$= 191{,}143 \times (146.2 + 0.55 \times 30)/10^6 = 31.1\,\text{MMBtu/h}.$$

(iv) Calculate the reboiler duty: Applying the energy balance around the tower indicates that the reboiler duty is the summation of the condenser duty and the heat required to raise the bottom from 300 to 454 °C.

Case 1: No subcooling in the overhead, the reboiler duty is

$$Q_{\text{reboiler}} = Q_{\text{condenser}} + m_B \times C_p \times \Delta T$$
$$= 27.9 + 42{,}857 \times 0.8 \times (454 - 300)/10^6 = 33.2\,\text{MMBtu/h}.$$

Case 2: 30 °F subcooling in the overhead, the reboiler duty is

$$Q_{\text{reboiler}} = Q_{\text{condenser}} + m_B \times C_p \times \Delta T$$
$$= 31.1 + 42{,}857 \times 0.8 \times (454 - 300)/10^6 = 36.4\,\text{MMBtu/h}.$$

With these approximations, the heat duties on condenser, reboiling, and cooler are established, which provides the basis for the process simulation.

15.4 BUILDING PROCESS SIMULATION

A separation column simulation is conducted in a process simulation tool based on tray-by-tray equilibrium calculations for mass and heat balances. Given the data for feed and products in terms of flow rates and compositions, as well as column operating conditions in terms of pressure and temperature, the column simulation can mimic the mass and heat balances for the current operation. Table 15.2 shows an example of the data required for conducting simulation of a C2 Splitter column.

For some processes that involve a process stream with many components, it could be too difficult to gather all the components for simulation. In this case, the concept of pseudocomponents is applied so that a group of components is lumped together into a pseudocomponent with similar physical properties. For oil refining processes, crude oils and refining products are mixture of many different chemical compounds. They cannot be evaluated based on chemical analysis alone. In order to characterize any crude oil and refining products, the petroleum industry applies a shorthand method of describing hydrocarbon compounds by number of carbon atoms and unsaturated bonds in the molecule and uses distillation temperatures and properties to define crude and products. For example, commercial jet fuel can be represented by an ASTM D-86 distillation temperature plot with the kerosene boiling range of 401 °F at 10% and 572 °F endpoint, while naphtha jet fuel, also called aviation gasoline, can be represented by a shorter distillation range of 122 °F at 10% and 338 °F endpoint.

The first step is feed simulation. If detailed feed analysis is available, which includes composition and conditions, a feed can be readily defined in simulation. Otherwise, the feed can be simulated from back-calculated as the summation of all products for which compositions and conditions are provided as part of the mass and heat balances data.

The second step is to determine feed tray position. Theoretical stages should be used in simulating a column. If tray efficiency is known, the feed tray in terms of theoretical stage can be determined from the actual feed tray and tray efficiency. However, tray efficiency is usually unknown. In this case, a sample lab test may be warranted. It is recommended to take a side sample one tray away from the feed tray. The feed point is one stage away from the theoretical stage, which matches the sample composition the best. Taking the sample from the feed tray would give compositions that are highly influenced by the feed and hence cannot truly represent the internal compositions inside the column.

The third step is to determine the number of theoretical stages required. With the feeds defined and product conditions given in the tabulated data, a column simulation can be established. For a simple column with two products, one from the overhead and from the bottom, the number of theoretical stages can be determined from the measured reflux rate. For a given reflux rate, the required number of theoretical stages is the one that can match the product specifications. As reflux rate defines the reboiler duty and hence the column heat balance for a simple column, the heat balance determines the product specifications for given column conditions. For a complex column involving side-draws and pumparounds, the column should be simulated section by section because the column heat balance is defined by reflux rate together with the column pumparounds. It is recommended to simulate a complex column from top to bottom. The top section has an overhead product, a side-draw and a pumparound next to the side-draw. For a given reflux rate and pumparound duty, the number of theoretical stages in the top section is determined by matching the given product specifications. The section next to the top section is then simulated similarly.

The simulation can provide the sound basis for conducting other assessment tasks, which are discussed as follows.

15.5 HEAT AND MATERIAL BALANCE ASSESSMENT

One of the early steps of assessing the fractionation system is to obtain good material and energy balances. Otherwise, it could be possible that assessment yields misleading conclusions.

The material and energy balances can be built based on the input of feeds and energy as well as outputs of products and energy in operation. The purpose of conducting a column material balance is to make sure feeds and products are measured accurately and desirable products are obtained. The energy balance is to verify if all major sources of energy input are accounted and if efficient use of energy is achieved.

Heat input is the driving force for fractionation. For a simple fractionation column, heat input comes from feed and bottom reboiling, while heat is removed from products and overhead condenser. For a complex fractionation column, multiple products are produced while pumparounds are located to remove excess heat in the column and recover this heat for process usage.

Both material and energy balances should be conducted on the basis of steady-state operation as this is a stable operation away from any transient excessive flooding or weeping operation. The steady-state operation can be viewed from historian when process data remains virtually the same within a very narrow band. On the other hand, steady-state operation can be obtained in operation after a minimum time from making operating adjustment to a tower. This minimum time can be expressed as follows:

$$t_{\text{min}} = \frac{M_{\text{hold}} R_{\text{f}}}{F}, \tag{15.1}$$

where M_{hold} is the summation of material holdup in sump and receiver drum. R_{f} is the reflux ratio and F is the tower feed rate.

For any fractionation column, there are overall mass balance and component mass balances, which follow mass conservation law:

For overall mass balance,

$$\text{Total mass input} = \text{Total mass output}. \tag{15.2}$$

For component balance,

$$\text{Total input of component } j = \text{Total output of component } j. \tag{15.3}$$

Similarly, heat balance follows energy conservation law:

$$\text{Total heat input} = \text{Total heat output}. \tag{15.4}$$

15.5.1 Material Balance Assessment

Good understanding of a material balance and key component balance can give insights for maximizing desirable product yields while minimizing undesirable products. Material flows are measured for feed and products, which are readily available online. Component measurement is usually obtained from lab tests for key components. Samples are taken daily on most towers and analyzed in the plant's local laboratory. However, these laboratories are typically set up to measure for certain key compounds that can contaminate the final product and do not have the capability of measuring the full range of multicomponents involved in feed. Therefore, unless the tower of interest has only a few components in the feed, a complete component balance will typically need special laboratory assistance, which more than likely will come from outside the local plant. In this case, be very careful with compositions and understand the units of measurements that are provided by the laboratory. However, the component balance could be difficult to obtain for hydrocarbon separation towers due to the fact that individual components cannot be fully characterized. But for most chemical and natural gas separation towers, it is possible to establish individual component balances.

When the material balances achieve at least ±10% offset ([total mass input – total mass output]/total mass input), it is acceptable for tower evaluation (Summers, 2009). For hydrocarbon separation processes, the closure could be as high as ±5%. It is common that a poor mass balance is caused by transmission error between pressure drop and flow rate for some of the material streams. Most flow meters are pressure drop-based devices and they could give wrong readings if physical properties are not used properly for converting pressure drop into flow rate. In some cases, wrong readings can be corrected by meter calibration including proper zeroing and spanning. Consult with instrument engineers and they can help resolve the meter-related issues.

Example 15.2 Overall Mass Balance

This example comes from the main fractionation tower in a hydrocracking process that is operated to make naphtha, kerosene, and diesel products. The bottom product is called unconverted oil, part of which is recycled back to reaction for further conversion and the rest sold as fuel oil to the market. Table 15.3 shows the material balance for the fractionator indicated by expected yields versus the actual yields as well as the flow rates for feed and products measured online.

Let us look at the overall material balance for the tower. The measured product rates in barrel per day are 18,130 of naphtha, 12,200 of kerosene, 9968 of diesel, and 9968 of unconverted oil, which gives a total of 50,266 barrel per day. The difference between the measured total of 50,266 and feed rate of 51,548 is 3 vol%. As the light

TABLE 15.3. Mass Balance Around a Fractionation Tower

	vol% of Feed (Yield Expected)	vol% of Feed (Yield Produced)		Barrel Per Day (Produced)		Barrel Per Day
	Feed	Distillate	Bottom	Distillate	Bottom	Feed
Naphtha	35.2	35.2		18,130		
Kerosene	22.1	23.7		12,200		
Diesel	23.6	19.3		9,968		
Unconverted oil	19.1		21.8		9,968	
Total	100	78.2	21.8	40,298	9,968	51,548

end products are not measured, it usually accounts for around 3 vol% of feed. Thus, the material balance for the tower is in a good closure at less than 1% of uncertainty.

What could we learn from this material balance? The first observation is that 1.6% extra kerosene is produced than expected. This has a simple explanation because the tower was operated to maximize kerosene production as kerosene was more valuable in the local market at the certain time. In contrast, 4.3% of less diesel was made, which was surprising. Two plausible causes were thought of by the process engineer responsible for the tower operation. One reason was that kerosene cuts 1.6% deep into diesel while the other was that 2.7% (4.3% – 1.6%) diesel slumps into the bottom unconverted oil. If the latter was true, it could be a significant yield loss and should be resolved.

This was only a hindsight, which could be wrong. The expected yield comes from yield estimates based on empirical correlations, which could give inaccurate estimates sometimes. Distillation temperature for the bottom product could provide the answer to this question. Thus, a lab test was conducted for the bottom unconverted oil, which shows that the 5% distillation temperature is 720 °F. This temperature cut should belong to the diesel range. It was clear that the unconverted oil contains a good portion of diesel that could have been sold to the market at a premium price. The diesel price was at 2.45/gallon, and thus $143,000/day was lost under this tower operation, or $50 MM/year could be lost if this problem was not resolved. This price tag alarmed large enough to secure swift actions for troubleshooting.

In summary, the investigation revealed two root causes. The first one was that the stripping steam at the bottom of the tower was insufficient as it was put on constant control based on the original design point. Although the throughput was increased by 10% over the years and feed became heavier, the stripping steam did not change. After adjusting it accordingly, the diesel recovery was improved. The reason is more stripping steam into the tower reduces hydrocarbon partial pressure, which helps to vaporize or lift more diesel components from the heavier in the bottom. The second cause was insufficient number of trays in the bottom section as the tower was designed for dealing with lighter feed than what it is handling now. A revamp project to add a few trays in the bottom section was scheduled for the turnaround.

A lesson learned: A simple mass balance identifies a major yield loss.

Example 15.3 Component Mass Balance

A stripper column in a naphtha hydrotreating process unit needs to remove H_2S, which is corrosive and could poison the catalyst in a downstream naphtha reforming unit. Another objective is to remove as much C_5 as possible from the stripper bottom, which is the feed to the naphtha reforming unit.

The stripper was operated with these two objectives in mind. However, the lab test showed the C_5 component distribution. The stripper bottom contained 2 mol% C_5 that exceeds the targeted C_5 removal from the bottom, which is undesirable. Two negative results were observed. As C_5 does not involve in the catalytic reforming reaction, C_5 material not only occupied the space in the reactor and hence reduced reaction throughput but also consumed extra heat in the feed heater for the reforming reactors.

Once the problem was identified, the process engineer discussed it with the control engineer who quickly changed the set point for the reboiling duty and reflux rate. With increased reboiling duty and consequently increased reflux rate, better fractionation in the top section and hence more C_5 was stripped out of the bottom. However, the reflux rate is controlled on minimum overflash so that no extra energy than necessary was consumed.

A lesson learned: Therefore, the component mass balance could help determine the desirable locations for key components to go from separation. Improper separation could cost not only the energy but also have negative effect on yields.

15.5.2 Heat Balance Assessment

A major part of tower heat balance is checking reflux rate and temperature, which determines both the condenser and reboiler duties. It is important to measure reflux temperature as it affects the heat balance significantly when the reflux is subcooled. Calculations in Figure 15.2 demonstrated the effect of reflux subcooling.

The common problem with measuring reflux rate is that reflux meters are typically set at startup and then never adjusted again. Therefore, the reflux flow rate is typically not reliable. The reflux ratio is checked and monitored as important operating parameter, but the absolute value of the reflux rate is rarely monitored. However, to have a correct heat balance, the reflux flow meter must be checked and calibrated in order to achieve at least ±5% closure of heat balance ([total heat input – total heat output]/total heat input). Only with this accuracy of heat balance, tray efficiency or packing HETP can be accurately determined (Summers, 2009).

15.6 TOWER EFFICIENCY ASSESSMENT

A benchmark efficiency for a tower should be established. By comparing the actual efficiency with the benchmark efficiency, it is important to obtain a trend of efficiency over time and see the sign of poor separation. A good efficiency indicates a healthy operation of the tower in general while a poor efficiency identifies signs of unstable operation, which warrants a tower rating assessment to reveal root causes of abnormality and thus determine actions for corrections.

Calculation of distillation efficiency requires process simulation. From the number of theoretical stages simulated for each section, column section and overall efficiency can be determined, respectively. For a simple tower with two products, one from top and one from bottom together with one condenser and one reboiler, the separation efficiency can be calculated via

$$\eta_o = \frac{N_{eq}}{N_{act}}. \tag{15.5}$$

As an example for illustration, McCabe–Thiele diagram (McCabe and Thiele, 1925) in Figure 15.3 indicates 12 actual stages required in comparison with 8 theoretical stages in the tower. Partial condenser and partial reboiler are counted in both the theoretical stages and actual stages. Thus, the overall tower efficiency is 72%.

For a complex column, equation (15.5) cannot give the right answer as it is only applied to each section. Instead, fractionation correlation plots by O'Connell (1946) (Figure 12.3) is widely used for overall tower efficiency, which is the standard of the

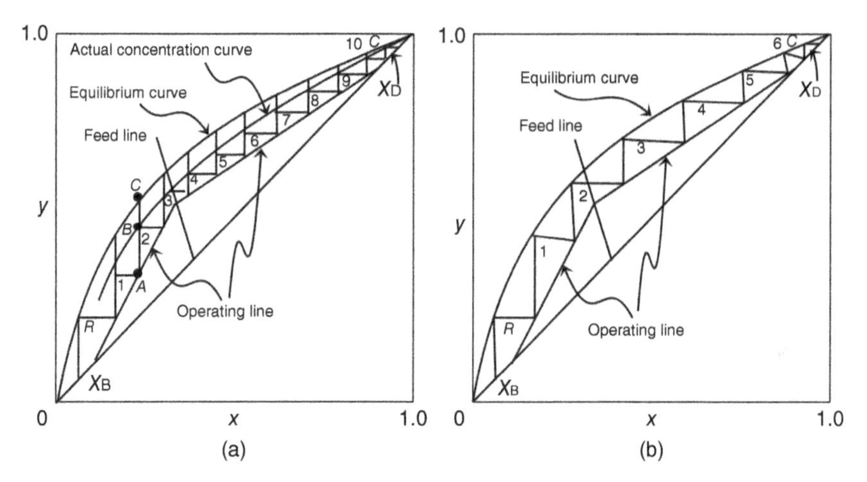

FIGURE 15.3. McCabe–Thiele diagram.

industry for industrial tower efficiency. Lockett converted the O'Connell's plots into an equation form as

$$\eta_o = 0.492(\mu\alpha)^{-0.245}, \tag{15.6}$$

where μ is the viscosity of liquid and α relative volatility, which are calculated based on average temperature and pressure between column top and bottom. Thus, O'Connell correlation states that higher viscosity leads to lower efficiency due to greater liquid phase resistance while higher relative volatility also reduces efficiency as it increases the significance of the liquid phase resistance.

However, O'Connell's correlation plots (Figure 12.3) and Lockett's equation (15.6) are developed based on efficiency data points for industrial towers and does not reveal fundamental reasons for what to do, why, and how in order to improve efficiency.

Thus, the natural question is: what parameters affect tower separation efficiency? Mainly, there are three kinds of parameters. The first is flow properties such as relative volatility and viscosity, which are intrinsic. The second one is tray layout such as tray deck type (sieve or valve), flow path length, tower diameter, tray spacing, and weir length, which affect the liquid and vapor distribution and flow regime and are determined by design. The third one is process conditions such as tower feed rate and reboiling duty. The common effect of these parameters is in impacting the balance between vapor and liquid loadings.

Efficiency varies very little in the region of stable operation or feasible operating window as discussed in Chapter 10 while efficiency falls off the cliff outside the feasible region. Figure 15.4 shows a typical trend of tower efficiency dependent on the balance of vapor and liquid rates. In the middle of the efficiency curve corresponding to stable operation, there is a relative flat region although with marginal variation. Trays with good turndown features such as valve tray compared with sieve tray have a wider flat or stable operating region. On the either side of the curve, efficiency drops off dramatically. Efficiency declines under low feed rate corresponding to turndown operation and falls off the cliff when dumping occurs. On the other hand, efficiency reduces at excessive entrainment and thus plummets when spray or flooding happens. Optimization in design and operation tends to push the tower toward the boundary

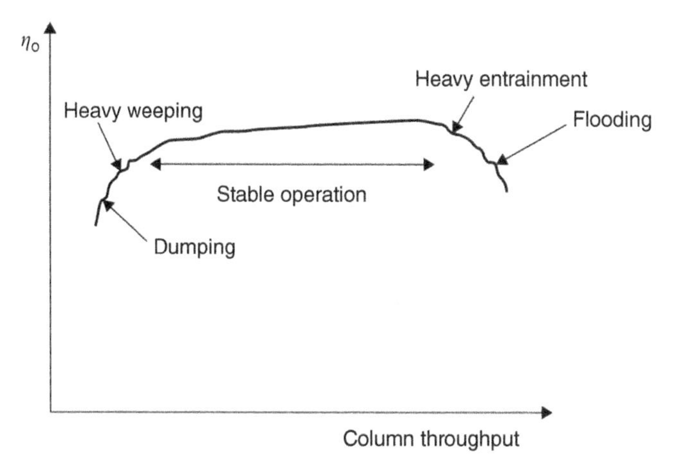

FIGURE 15.4. A typical trend of tower efficiency.

of stable operation. Understanding of these controlling mechanisms can shed insight into how to optimize tower design and operation while achieving stable operation.

Efficiency assessment can detect the section(s) with poor efficiency from which root causes can be found. A section or whole column could be flooded due to too high vapor or liquid loading. This could be caused by changes in the conditions of feed and products in terms of rates, compositions and product specifications. It could be also caused by too high reboiler duty, high feed temperature, and low column pressure or combination of these. For the case of changes in feed compositions, it could be traced back to processing issues in upstream.

From retrofit point of view when dealing with too high liquid loading, enhanced capacity trays could be used such as UOP MD/ECMD or Shell HiFi or Sulzer high capacity or Koch-Glitsch high-performance trays. For too low liquid loading, packing could be the solution. For too high vapor loading, valve trays could be considered. Tray damage could also cause malfunction of a column operation. In whichever case, identification of low fractionation efficiency triggers the search for the root causes and solutions.

Example 15.4 Overall Efficiency Estimate Using O'Connell's Correlation

This example (Wankat, 1988) was revised. A sieve tray distillation column is separating a feed that is 50 mol% *n*-hexane and 50 mol% *n*-heptane. The feed is a saturated liquid. Tray spacing is 24 in. The average column pressure is 114.7 psia. Distillate composition is 99.9% mol of *n*-hexane and 0.1% mol of *n*-heptane. Feed rate is 1000 lb mol/h. Internal reflux ratio *L/V* is 0.8. The column has a total reboiler and total condenser. Estimate the overall efficiency.

Solution:

To apply equation (15.6), we need to estimate α and μ at the average temperature and pressure of the column. The column temperature can be obtained from the modified DePriester chart (Dadyburjor, 1978) as shown below.

x_{C6}	0.000	0.341	0.398	0.500	1.000
y_{C6}	0.000	0.545	0.609	0.700	1.000
T (°C)	98.4	85.0	83.7	80.0	69.0

Relative volatility is $\alpha = (y/x)/[(1-y)/(1-x)]$. The average temperature can be estimated in several ways.

Average temperature $T = (98.4 + 69.0)/2 = 83.7$; x and y at $T = 83.7\,°C$ can be interpolated based on the table above. Thus, $\alpha = 2.36$ at $T = 83.7\,°C$. If average at $x = 0.5$, $T = 80$, $\alpha = 2.33$ at $T = 80\,°C$. Not much difference. Use $\alpha = 2.35$ corresponding to $T = 82.5\,°C$.

The liquid viscosity of the feed can be estimated (Reid et al., 1977) from

$$\ln \mu_{mix} = x_1 \ln \mu_1 + x_2 \ln \mu_2. \tag{15.7}$$

The pure component viscosities can be estimated from

$$\log_{10}\mu = A\left(\frac{1}{T} - \frac{1}{B}\right), \tag{15.8}$$

where μ is in cP and T in K (Reid et al., 1977).

$$nC_6 : A = 362.79; B = 207.08; \quad nC_7 : A = 436.73; B = 232.53.$$

The two equations above for μ_{mix} and μ give $\mu_{C6} = 0.186$, $\mu_{C7} = 0.224$, and $\mu_{mix} = 0.204$. Thus, $\alpha\mu_{mix} = 0.479$. Applying equation (15.6) gives $\eta_o = 58.9\%$, which agrees well with $\eta_o = 59.0\%$ obtained from O'Connell correlation plots. The lower value should be used for conservative purpose.

15.7 OPERATING PROFILE ASSESSMENT

Another simple assessment method is based on tower profiles generated from simulation, which include flow, temperature, pressure, and composition profiles. What can we learn from these profiles? In a nut shell, tower profiles can allow us to observe what is going on inside the tower, such as X-ray photos by vision.

The flow profile shows internal liquid and vapor flows across the column, which can vary from tray to tray with sudden change at feed stage and withdraw stages. In general, in the rectifying section above the feed stage up to condenser, the vapor flow is higher than the liquid flow while it is opposite in the stripping section below the feed stage. As part of the flow estimates, the feed is flashed at the feed tray conditions. The importance of flow estimates is not so much the absolute values but the ratio of L/V, which determines the internal reflux and the slope of the operating line. This behavior can be observed in Figure 15.5.

The temperature and pressure profiles show a general trend of monotonic reduction in both temperature and pressure from the reboiler to the condenser. Figure 15.6 shows an example temperature profile where the steep parts of the curve would be, where light and heavy keys are significantly separating. In some cases, temperature profiles feature plateaus in certain trays where little temperature change occurs. In these flat regions, it indicates virtually no separation taking place although nonkey components are being distributed. When there are a large number of stages, these plateaus can be more self-evident. These stages represent the pinch region where the

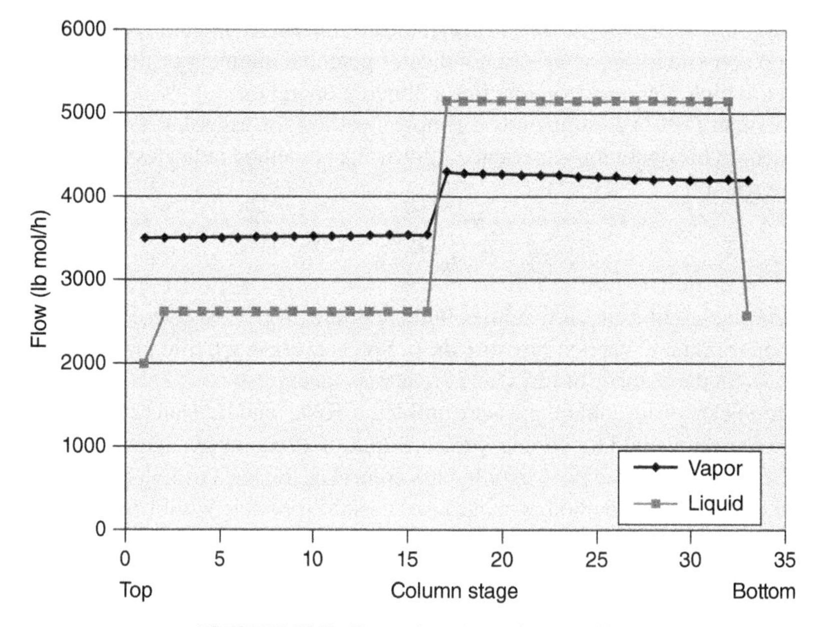

FIGURE 15.5. Example column flow profile.

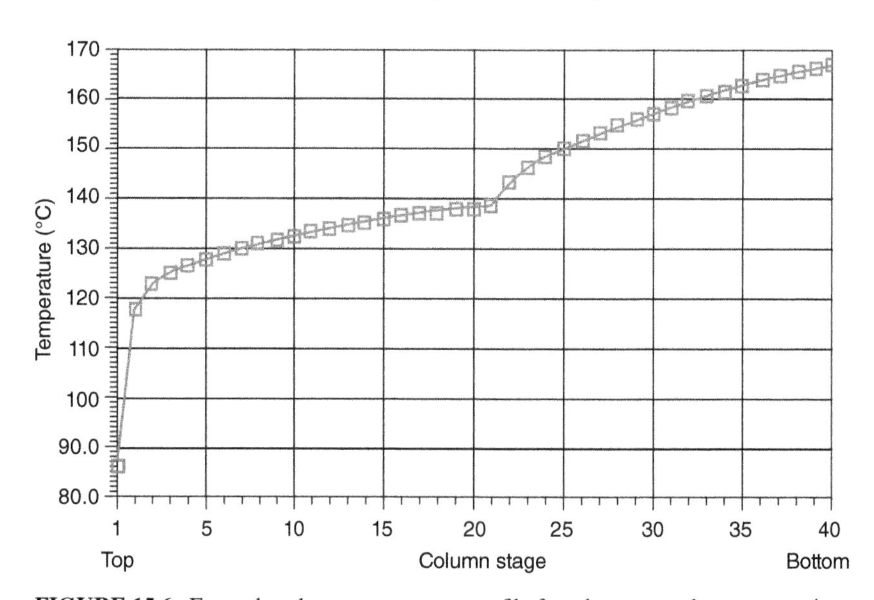

FIGURE 15.6. Example column temperature profile for a benzene–toluene separation.

operating line is very close to the equilibrium curve. In this pinch region, the ratio of relative volatility between key components is very small corresponding to a difficult fractionation.

Obviously, when a column or a section is flooded or dumping, a flat temperature profile can be obtained since there is no fractionation taking place.

A tower *pressure drop profile* can also indicate abnormal operation. A too low pressure drop across a tower or a section indicates potential dumping or flow channeling while a too high pressure drop manifests flooding operation.

Lieberman (1991) recommends a simpler method for assessing flooding condition based on his operation experience. Lieberman's method indicates occurrence of flooding when

$$\frac{\Delta P}{\mathrm{Sp}_{\mathrm{L}} N_{\mathrm{T}} H_{\mathrm{S}}} \geq 22 - 25\%, \tag{15.9}$$

where ΔP, inches of water, is overall column pressure drop between column overhead and reboiler outlet or section pressure drop. Sp_{L} is average specific gravity of liquid on tray, N_{T} is the number of trays, and H_{S} is tray spacing, in.

To troubleshooting column pressure problems, Kister and Hanson (2015) provided a simple retrofit method for column pressure control. A survey by Kister (2006) identified the poorly designed hot-vapor bypass control as the most troublesome pressure and condenser control method, which causes unstable pressure within column. Unstable pressure results in an unsteady column as pressure affects column vaporization, condensation, temperature, volatility, and so on.

The simple method that Kister and Hanson proposed is to add a throttle valve in the condensate outline. Ideally, the valve should have a pressure drop larger than 3–4 psi and should be installed more than 10 in. diameter away from the reflux drum liquid inlet to minimize turbulence at the drum inlet. Another practice is to install a horizontal baffle in front of the vapor nozzle to disperse the vapor flow and prevent it from impinging on the liquid surface upon intensification.

The *composition profile* can reveal the details of separation taking place inside the tower. For component balances, it is highly important to know the composition of the feed and product streams around the tower. Samples are taken daily on most towers and analyzed in the plant's local laboratory. However, these laboratories are typically set up to measure for certain key compounds that can contaminate the final product and do not have the capability of measuring the full spectrum of a multicomponent column's feed. Therefore, unless the tower of interest has only a few components in the feed, a full component balance will typically need special laboratory assistance, which will more than likely come from outside the local plant. One needs to understand the units of measurements for compositions by the laboratories. Frequently, they will not provide the units such as molar, volume, or weight percentages.

This example is about separation of toluene (light key) from ethyl benzene (heavy key). Figure 15.7 shows the liquid mole fraction for these four components. Stage 21 is the feed stage.

Let us follow the toluene mole fraction curve, which is more obvious. Concentration of LK toluene increases through the tower in a monotonous manner until it peaks at the top of tower but dips at the receiver. This is because benzene, the nonkey light component, is the most volatile component and peaks at the receiver but should not be in the bottom.

On the other hand, the concentration of the heavy key, ethyl benzene, enriches toward the bottom of the tower monotonously through the tower and peaks a few stages above the reboiler because it is the least volatile component.

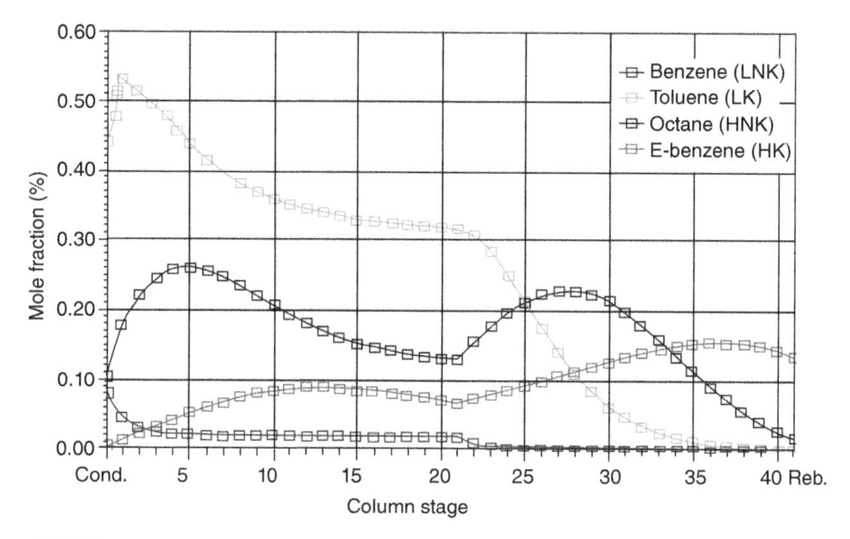

FIGURE 15.7. Example composition profile for toluene–ethyl benzene separation.

The concentration profile for the nonheavy key, octane, is the most confusing one as it goes up and down throughout the tower with two maxima because of the competition with other compounds on those trays and how they concentrate. From the feed stage 21 to stage 28, the separation takes place between the LK toluene versus HK ethyl benzene and HNK octane. As a result, the concentrations of both ethyl benzene and octane increase while toluene concentration reduces. From stage 29 to stage 40 where LK toluene concentration decreases to very low, the separation takes place mainly between HK ethyl benzene and HNK octane. In this section, octane concentration reduces steeply and ethyl benzene concentration increases. In summary, octane concentration increases from the feed stage until it peaks at stage 28 and then starts to decrease toward the bottom. This creates the first maxima in octane concentration.

From stage 15 to stage 5, LK toluene is separated from HK ethyl benzene. At the same time, the separation between HK ethyl benzene and HNK octane occurs where ethyl benzene concentration goes down as octane concentration steps up until octane peaks at stage 5. Above stage 5 toward the top of the tower, the separation takes place between LK toluene against both HK ethyl benzene and HNK octane where toluene concentration climbs and peaks at the top. In contrast, HNK octane concentration plummets and HK ethyl benzene concentration reduces to distinction. In summary, octane concentration increases from stage 15 until it peaks at stage 5 and then it reduces sharply. This creates the second maxima in octane concentration.

15.8 TOWER RATING ASSESSMENT

Tower simulation is only mimicking the current operation and can provide the vapor and liquid loadings as the basis for rating assessment while rating assessment will tell how the tower is operating under current conditions in relation to the feasible operating window.

Tower rating is applied to assess the effects for changing process conditions, in particular feed rate change, on tower performance. It can also be applied when an existing tower is considered to be used for a new service. Briefly, a tower rating can answer three questions:

(1) Can the tower operate with increased throughput or with changing process conditions within the feasible operating window? The calculations for the operating window were shown earlier in this chapter. With the internal *L*/*V* from simulation, we can determine the current operating point in relation to the operating window.
(2) What is the hydraulic performance of the tower under new conditions? The operating pressure drops can be calculated based on hydraulic calculations discussed earlier in this chapter.
(3) What are the limiting factors of the tower under new conditions? The limitations could come from the size of the tower, tray spacing, downcomer geometry, and so on.

The criteria can be established as necessary and sufficient conditions for the suitability of an existing tower for changing conditions or new services:

(i) The operating point must fall within the operating window. For example, the actual vapor and liquid rates should be less than the maximum limits.
(ii) Operating pressure drop must be less than allowable pressure drops.

When these two conditions are fulfilled, an existing tower is suitable for different conditions for which it is rated. When the process conditions undergo significant changes, the rating assessment should be performed to make sure the tower can perform the task satisfactorily under new conditions. Otherwise, either operating conditions should be altered or modifications to the existing tower need to be implemented.

Why would we want to use a tower for a service that it was not designed for? The main reason is that it is less expensive and quick to modify an existing tower than to purchase a new one. It is rare that the existing tower provides a perfect fit to a new service. But engineers are keen to take the challenge of modifying existing equipment as it is their second nature of seeking the most economic solution with quick turnaround.

Tower rating assessment can be conducted using tower evaluation software by vendors (e.g., Sulzer's Sulcol tool). A tower simulation provides basic data required for rating. In generating data from simulation to rating assessment, a tower is divided into sections and the stage with highest vapor loading in each section is selected to represent this section as this tray is the most constrained tray for the whole section. Thus, the data for this stage is entered into the tower rating software. The input data includes (1) vapor and liquid loadings and physical properties for both vapor and liquid, which are obtained from simulation; (2) tower geometry layout (e.g., tower diameter). Execution of the rating software will give percentages of tray flooding, downcomer backup and dumping, vapor maximum capacity, liquid maximum capacity, froth/spray transition, pressure drop, dry tray pressure drop, downcomer velocity, and weir loading.

The rating assessment software will indicate the current operating point in relation to operating window and thus reveal what operating limits the tower may have gone beyond, which are the root causes for sudden decrease of tower efficiency. Some commercial rating tools can generate operating window or performance diagrams (Summers, 2004). Performance diagrams, if plotted with vapor and liquid volume loadings, can represent tray performance independent of operating pressure and composition.

15.9 GUIDELINES

The guidelines discussed as follows are recommended by Wankat (1988) and the following things in order of increasing costs can be explored when the existing column cannot produce desired product purities:

- Find out whether the product specifications can be relaxed. A purity of 99.5% is much easier to obtain than 99.99%.
- Increase reflux rate and see if it can meet product specifications. Remember to check if column vapor capacity is sufficient as flooding could be an issue with an increased reflux rate. And also check if existing reboiler and condenser are large enough. If the tower can make purer products, usually reducing reflux rate can make product back to specification, which also reduces operating cost.
- Change the feed temperature. This change may require altering of feed stage and could result in an optimal feed location.
- Will a new feed stage at the optimal stage allow meeting product specification?
- Consider replacing the existing column internals with more efficient or tighter spaced trays or new packing. This is relatively expensive but is cheaper than a new column.
- Add a stub column to increase the total number of trays.

If the column vapor loading is more than the limit implying the existing column diameter is not large enough, engineers can consider the following:

- Operating at a reduced reflux ratio, which reduces vapor loading, but this could make it difficult to meet product specifications.
- Operating at a higher pressure, which increases vapor density. Need to check if the column can operate at the increased pressure.
- Using two columns in parallel.
- Replacing the existing downcomers with large ones.
- Replacing the trays or packing with higher capacity ones.

On the other hand, if the column diameter is too large, vapor velocities will be too low. Trays will operate at too lower efficiency and in severe cases they may not operate since liquid may dump through the holes. Engineers can consider the following:

- Decrease column pressure to decrease vapor density and hence vapor velocity.
- Increase reflux ratio.
- Recycle some distillate and bottom products.

Using existing columns for new services often requires innovative solutions. Thus, it can be both challenging and fun; they are also often assigned to engineers just out of school but under supervision of experienced engineers.

NOMENCLATURE

Variables

C_p specific heat
F feed rate
H_S tray spacing
M mass flow
N_T total number of trays
N_{eq} number of theoretical trays
N_{act} number of actual trays
ΔP pressure drop
q latent heat
Q heat content
R_f tower reflux ratio
Sp average specific gravity of liquid on tray
t time
T temperature
ΔT temperature difference

Greek letters

α relative volatility
μ liquid viscosity
η_o overall tower efficiency

REFERENCES

Dadyburjor DB (1978) SI units for distribution coefficients. *Chemical Engineering Progress (CEP)*, AIChE, P85, April Issue.

Kenny WF (1984) *Energy Conservation in the Process Industries*, Academic Press.

Kister HZ (2006) *Distillation Troubleshooting*, AIChE–John Wiley & Sons.

Kister HZ, Hanson DW (2015) Control column pressure via hot-vapor bypass, *Chemical Engineering Progress (CEP)*, AIChE, P35–45, February Issue.

Lieberman N (1991) *Troubleshooting Process Operation*, 3rd edition, Penn Well.

McCabe WL, Thiele EW (1925) Graphic design of fractionation columns, *Industrial and Engineering Chemistry*, **17**, 605–611, June.

O'Connell HE (1946) Plate efficiency of fractionating columns and absorbers, *Transactions of the AIChE*, **42**, P741.

Reid RC, Prausnitz JM, Sherwood TK (1977) *The Properties of Gases and Liquids*, 3rd edition, McGraw-Hill, New York.

Summers DR (2004) Performance diagrams – all your tray hydraulics in one place, AIChE Annual Meeting – Distillation Symposium, Austin, paper 228f.

Summers DR (2009) How to properly evaluate and document tower performance, AIChE Spring Meeting, April 27, Florida.

Wankat PC (1988) *Equilibrium Staged Separations*, PTR Prentice Hall, New Jersey.

PART 5

PROCESS SYSTEM EVALUATION

16

ENERGY BENCHMARKING

16.1 INTRODUCTION

When you are given a task to improve energy performance for the plant or process unit, your immediate response would be where I should start. The answer to this question is to know where the process unit stands in energy performance. In other words, you need to determine both current energy use and an energy consumption target. Only then it is possible to establish the baseline and to know how well the process unit is doing by comparing current performance against a target. We call the exercise of establishing a baseline as benchmarking.

The most important result of energy benchmarking is the indication of energy intensity for individual processes. If a performance target can be defined based on a corporate target, industrial peer performance, or the best technology performance for each process, then the benchmarking audit can determine the process energy performance in overall in comparison with targeted performance. In general, benchmarking assessment can give several indications:

- The need of having an overall energy optimization effort: If large gaps are available for majority of the process units, this could imply there are many opportunities available and require consorted effort across the plant. A dedicated energy team may need to be established to coordinate the overall effort in identifying and capturing the opportunities.
- Areas for focus: Some process units are identified with large performance gaps, and these processes can be selected as focus areas. This allows us to effectively

Hydroprocessing for Clean Energy: Design, Operation, and Optimization, First Edition.
Frank (Xin X.) Zhu, Richard Hoehn, Vasant Thakkar, and Edwin Yuh.

concentrate efforts on the areas with the greatest potential for improvement. Specialists may need to be assembled to form a project team for individual process units.

- Update targets: If all major process units are under good performance relative to the targets, the plant may concentrate the effort on continuous improvements via monitoring and control.

16.2 DEFINITION OF ENERGY INTENSITY FOR A PROCESS

Let us start with the specific question: how to define energy performance for a process? People might think of energy efficiency first. Although energy efficiency is a good measure as everyone knows what it is about, it does not relate energy use to process feed rate and yields and thus it is hard to connect the concept of energy efficiency to plant managers and engineers.

To overcome this shortcoming, the concept of energy intensity is adopted, which connects process energy use and production activity. The energy intensity was originated from Schipper et al. (1992) who attempted to address intensity of energy use by coupling between energy use and economic activity through the energy use history in five nations: the United States, Norway, Denmark, West Germany, and Japan. The concept of energy intensity allows them to better examine the trends that prevailed in both cases of increasing and decreasing energy prices.

By definition, energy intensity (I) is described by

$$I = \frac{\text{Energy use}}{\text{Activity}} = \frac{E}{A} \tag{16.1}$$

Total energy use (E) becomes the numerator and common measure of activity (A) is the denominator. For example, commonly used measures of activity are vehicle-miles for passenger cars in transportation, kW h of electricity produced in power industry, and unit of production for the process industry.

Physical unit of production can be tons/h or m^3/h of total feed (or product). Thus, industrial energy intensity can be defined as

$$I = \frac{\text{Quantity of energy}}{\text{Quantity of feed or product}} \tag{16.2}$$

Energy intensity defined in equation (16.2) directly connects energy use to production as it puts production as the basis (denominator). In this way, energy use is measured on the basis of production, which is in the right direction of thought: a process is meant to produce products supported by energy. For a given process, energy intensity has a strong correlation with energy efficiency. Directionally, efficiency improvements in processes and equipment can contribute to observed changes in energy intensity.

Therefore, we can agree that energy intensity is a more general concept for measuring process energy efficiency indirectly.

Before adopting the concept of energy intensity, you may ask the question: which one, feed rate or product rate, to be used as the measure of activity? For plants with single most desirable product, the measure of activity should be product. For plants making multiple products, it is better to use feed rate as the measure of activity. The explanation is that a process may produce multiple products and some products are more desirable than other in terms of market value. Furthermore, some products require more energy to make than other. Thus, it could be very difficult to differentiate products for energy use. If we simply add all products together for the sum to appear in the denominator in equation (16.2), we encounter with a problem, which is the dissimilarity in product as discussed. However, if feed is used in the denominator, the dissimilarity problem is nonexistent for cases with single feed and the dissimilarity is much less a concern for multiple-feed cases than for multiple products because, in general, feeds are much similar in compositions than products.

The above discussions lead us to define the process energy intensity on the feed basis as

$$I_{\text{process}} = \frac{\text{Quantity of energy}}{\text{Quantity of feed}} = \frac{E}{F} \tag{16.3}$$

It is straightforward to calculate the energy intensity for a process using equation 16.3, where E is the total net energy use and F is the total fresh feed entering the process. Net energy use is the difference of total energy use and total energy generation. Process energy use mainly includes fuel fired in furnaces, steam consumed in column stripping and reboiling as well as steam turbines as process drivers and electricity for motors. Process energy generation mainly comes from process steam generation, power generation from process pressure reduction. In many cases, a process makes fuel gas and/or fuel oil, which is exported to other processes for firing or sold to markets. This type of fuel is not counted as energy generation as it is regarded as a part of product slates.

16.3 THE CONCEPT OF FUEL EQUIVALENT FOR STEAM AND POWER (FE)

There is an issue yet to be resolved for the energy intensity defined in equation (16.3). The energy use (E) for a process consists of fuel, steam, and electricity. They are not additive because they are different in energy forms and quality. However, if these energy forms can be traced back to fuel fired at the source of generation, which is the meaning of fuel equivalent, they can be compared on the same basis, which is fuel. In other words, they can be added or subtracted after converted into fuel equivalent. For simplicity of discussions, definitions of fuel equivalent (FE) for different energy forms are given here while examples of FE calculations are provided as follows.

In general, fuel equivalent (FE) can be defined as amount of fuel fired (Q_{fuel}) at the source to make a certain amount of utility (G_i):

$$\text{FE}_i = \frac{Q_{\text{fuel}}}{G_i} \quad i \in (\text{fuel, steam, power}) \tag{16.4}$$

In most cases, Q_{fuel} is calculated based on the lower heating value of fuel. G_i is quantified in different units according to specifications in the market place, namely Btu/h for fuel, lb/h for steam, and kW h for power. Thus, specific FE factors can be developed as follows based on this general definition of fuel equivalent.

16.3.1 FE Factors for Fuel

By default, fuel is the energy source. No matter what different fuels are used, tracing back to itself makes "fuel equivalent for fuel" equal to unity, that is,

$$\mathrm{FE_{fuel}} = \frac{Q_{\text{fuel@source}}}{G_{\text{fuel}}} \equiv 1\ \text{Btu/Btu} \tag{16.5}$$

16.3.2 FE Factors for Steam

A typical process plant has multiple steam headers, typically designated as high pressure, medium pressure, and low pressure. In some cases, very high pressure steam is generated in boilers, which is mainly used for power generation. For calculating fuel equivalent of steam, a top-down approach is adopted starting from steam generators. The total FE for each steam header is the summation of all FEs entering the steam header via different steam flow paths, which include steam generated from on-purpose boilers and waste heat boilers, steam from turbine exhaust, and steam from pressure letdown valves. The FE for each steam header is the total FE divided by the amount of steam generated from this header, that is,

$$\mathrm{FE}_i = \frac{Q_{\text{fuel}}}{G_i} = \left.\frac{\text{Total FE consumed}}{\text{Total steam generated}}\right|_i \quad \text{header } i \in (\text{HP}, \text{MP}, \text{LP}) \quad (\text{kBtu/lb}) \tag{16.6}$$

16.3.3 FE Factors for Power

For power, $\mathrm{FE_{power}}$ is expressed as

$$\mathrm{FE_{power}} = \frac{Q_{\text{fuel}}(\text{Btu/h})}{Q_{\text{power}}(\text{Btu/h})} = \frac{1}{\eta_{\text{cycle}}} \quad (\text{Btu/Btu}) \tag{16.7}$$

where η_{cycle} is the cycle efficiency of power generation and Q_{power} represents the amount of energy associated with power in the unit of Btu/h.

By using the conversion factor of 1 kW = 3414 Btu/h, equation 16.7 is converted to

$$\mathrm{FE_{power}} = \frac{1}{\eta_{\text{cycle}}}(\text{Btu/Btu}) \times 3414(\text{Btu/kWh}) = \frac{3414}{\eta_{\text{cycle}}}(\text{Btu/kWh}) \tag{16.8}$$

Equation (16.8) can be generally applied to different scenarios for power supply such as power import, on-site power generation from backpressure, and condensing steam turbines as well as from gas turbines.

16.3.4 Energy Intensity Based on FE

By converting different energy forms to fuel equivalent, process energy intensity in equation (16.3) can be revised to give

$$I_{\text{process}} = \frac{\text{FE}}{F} \quad \text{Btu/unit of feed} \tag{16.9}$$

where FE is the total fuel equivalent as the summation of individual fuel equivalent for different energy forms across the process battery limit.

Let us walk through calculation of process energy intensity via an example.

16.4 DATA EXTRACTION

For the purpose of energy benchmarking of a process unit, the important thing is to identify main energy consumers and give reasonable estimate for missing data. Going overboard to collect miniature details and chase utmost precision should be avoided. Doing so may actually waste the effort because such fine details are most likely not needed in the benchmarking calculations and will not make reasonable impact on energy optimization.

Table 16.1 gives an example for the relevant data needed at this stage for establishing the process energy balance and calculating energy performance. Although all the data look familiar in the table, you may question the need of including the fuel generated in the unit as part of energy generation. As a general guideline, the fuel produced from a process unit, in the forms of fuel gas, LPG, and fuel oil, is treated as part of products from the process and thus should not be included in the energy balance for the unit. However, it is accounted for in the overall fuel balance for the overall site, which has a central fuel pool.

In order to have a clear view of energy flows into and out of the process, we can obtain Figure 16.1 based on the data in Table 16.1. The left-hand side of the figure shows the energy needs by the process in the forms of electricity, high pressure steam, and fuel. At the same time, exothermic reaction provides additional heat to the process. In the right-hand side of the figure, energy leaves the process, which includes heat exported and lost. In addition, the raw feed and boiler feed water carry a certain amount of energy into the process based on the assumed reference temperature of 100 °F. A different reference temperature could be used and fuel equivalent calculations should conduct based on the chosen reference temperature. For example, if BFW temperature at 250 °F is selected as the reference temperature, the BFW does not carry heat into the process anymore. A reference temperature is selected based on the consideration that any heat below the reference temperature is not economically viable to recover.

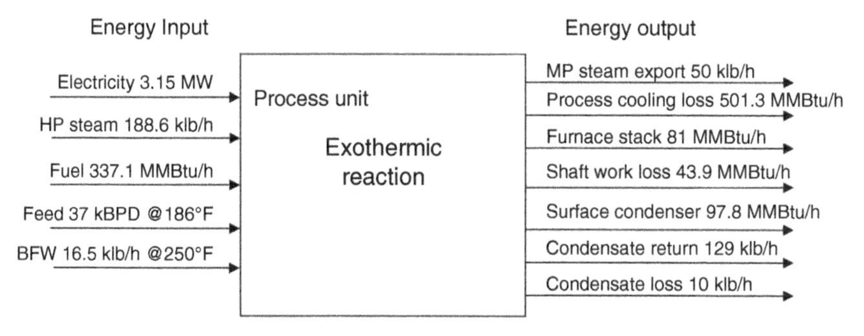

FIGURE 16.1. Energy flows into and out of the process unit.

TABLE 16.1. Example Data Set for Energy Use and Generation

Items	Normal Duty (MMBtu/h)	Normal Load (kW)	Electric Power (kW)	HP Steam at 610 psig (klb/h)	MP Steam at 150 psig (klb/h)	LP Steam at 60 psig(klb/h)	Condensate at 3 psia (klb/h)	Fuel Fired Duty (MMBtul/h)
Steam import from steam headers				−188.6	0	0		
Pumps and drivers								
Charge pumps			−2101.0					
Lean amine pumps			−53.3					
Rich amine pumps		−798.8			−18.9		18.9	
Stripper bottoms pumps			−101.8					
Stripper overhead pumps			−68.4					
Product fractionator bottoms pumps		−512.2			−12.1		12.1	
Diesel product pumps			−43.0					
Kerosene product pumps			−28.5					
Product fractionator overhead pumps			−89.6					
Wash water pumps			−80.4					
RXAIR cooler			−175.2					
Separation coolers			−405.8					
Water coolers								
Heavy naphtha product	6.0							
Kerosene product	4.5							
Diesel product	11.4							
Reaction effluent	27.5							
Fractionation column kerosene pump around	39.4							
H_2 make-up gas	30.0							
H_2 make-up compressor intercoolers	6.4							
Lean amine	1.4							

Air coolers							
Reaction effluent	192.0						
Debutanizer column overhead	21.3						
Fractionation column overhead	73.3						
Naphtha splitter overhead	12.9						
Compressors							
Recycle gas compressor		−2610.0		−86.6	86.6		
Make-up H_2 compressor		−5936.0		−70.5		70.5	
Fired heaters							
Charge heater for train 1	−29.5						−53.5
Charge heater for train 2	−29.5						−55.9
Product fractionator feed heater	−95.0						−111.8
Debutanizer column reboiler	−80.0						−94.1
Diesel stripper reboiler	−12.0						−21.8
Steam generation							
Product fractionator bottom steam generation					16.50		
Energy export							
Condensate return to boiler house						129.4	
Condensate lost						−10.0	
MP steam goes to steam header					50.0		
Energy entering into the process			3147.0	188.6			337.1
Energy exporting outside the process					50.0	129.4	

Note 1: A positive value indicates quantity produced. A negative value (−) indicates quantity consumed.

16.5 CONVERT ALL ENERGY USAGE TO FUEL EQUIVALENT

In industry, steam is measured in mass flow while fuel in volumetric flow and electricity in electrical current. To compare them on the same basis, all the energy use and generation need to be traced back to fuel fired at the source of energy generation in order to obtain fuel equivalent, which is a cardinal rule for energy balance calculations. The following illustrates how to conduct FE calculations based on Figure 16.1.

Assumptions: First, assumptions for related fuel equivalent factors need to be made and the basis for deriving these assumptions is explained later. Assumed fuel equivalent (FE) factors are

- FE for purchased power = 9.09 MMBtu/MW
- FE for HP steam = 1550 Btu/lb
- FE for MP steam = 1310 Btu/lb
- FE for condensate = 94.6 Btu/lb
- FE for BFW @221 °F = 177 Btu/lb

Convert energy inputs and outputs to fuel equivalent:

- FE for power = 3.15 MW × 9.09 MMBtu/MW h = 28.6 MMBtu/h
- FE for HP steam = 188.6 klb/h × 1.55 MMBtu/klb = 292.3 MMBtu/h
- FE for fuel fired = 337.1 MMBtu/h
- FE for MP steam export = 50 klb/h × 1.31 MMBtu/klb = 65.5 MMBtu/h
- FE for condensate return = 129.4 klb/h × 94.6 Btu/lb × 10^3 lb/klb × 1MMBtu/10^6 Btu = = 12.2 MMBtu/h
- FE for condensate loss = 10 klb/h × 94.6 Btu/lb × 10^3 lb/klb × 1 MMBtu/10^6 Btu = 0.9 MMBtu/h

To reveal the significance of FE calculations, let us assume a process receives 20 klb/h of HP steam, of which 10 klb/h comes from a boiler with an efficiency of 75% and another 10 klb/h from a boiler with an efficiency of 85%. Obviously, the fuel required or fuel equivalent for the same amount of HP steam, that is, 10 klb/h, by the two boilers is very different: the fuel equivalent from the boiler with 85% efficiency is 15.35 MMBtu/h, while the fuel equivalent for the boiler with 75% efficiency is 16.38 MMBtu/h. We can think of another example of power generation on site by a combined cycle (Gas and Steam turbines) cogeneration facility versus a coal-fired steam turbine power plant. The fuel equivalent for the same amount of power from these two sources is very different. Therefore, we cannot overstate the importance for tracing any energy back to fuel equivalent.

16.6 ENERGY BALANCE

After converting all energy forms to fuel equivalent, these energy forms are leveled on the equal basis and thus we are ready to conduct energy balance. For a chemical process, energy balance is defined as

$$\text{Energy supply} + \text{Energy generation} = \text{Energy export} + \text{Energy loss} \quad (16.10)$$

The sum of energy supply and energy generation makes total energy input, while both energy export and energy loss forms total energy output. Energy supply implies the energy coming into the process battery limit. Energy generation for a chemical process implies heat of reaction. If a reaction is exothermic, the term of energy generation takes a positive sign as it contributes to total energy input. An endothermic reaction takes a negative sign as it takes energy away from energy supply and needs energy input to make up the difference. Energy export denotes the energy leaving out of the process, which is used by other processes. Energy loss indicates the energy flows leaving out of the process but lost the environment.

After obtaining fuel equivalent values for all energy flows, we can convert Figure 16.1 to Figure 16.2, which gives a visualized energy balance around the process unit including energy supply, energy generation by heat of reaction as well as energy export and losses. The heat of exothermic reaction is calculated as 141 MMBtu/h for this example based on the feed composition and reaction conditions. Heat content of the feed and boiler feed water above 100 °F are treated as energy input. At the same time, the figure shows energy output including energy export and energy losses. It can be observed that only energy flows entering and leaving the process battery limit are addressed in the energy balance described in Figure 16.2.

The detailed energy balance is given in Table 16.2. The total energy input is 819.2 MMBtu/h for the process unit currently operated. The heat of exothermic (endothermic) reaction contributes positively (negatively) to the total energy input. Fuel fired in process heaters is 337.1 MMBtu/h, which is the most dominant accounting for about 40% of total energy input. The second most dominant energy use is the process power demand. HP steam of 292.3 MMBtu/h is used for steam turbines as process drivers while purchased electricity of 28.6 MMBtu/h is for running motors. The total fuel equivalent for meeting the process power demand is 321 MMBtu/h (28.6 + 292.3), which accounts for another 40% of total energy input. Heat of

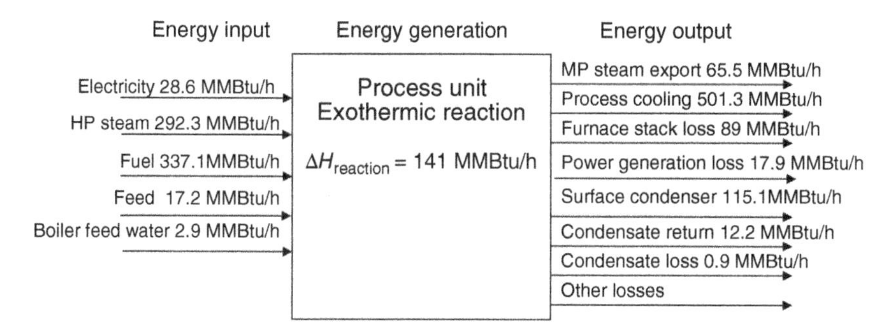

FIGURE 16.2. Energy balance in a visualized form.

TABLE 16.2. Tabulated Energy Balance for the Example

Energy Input	FE (MMBtu/h)	Energy Output	FE (MMBtu/h)
	(above 100 °F)	Energy export	(above 100 °F)
Power	28.6	MP export	65.5
Fuel	337.1	Condensate return @141 °F	12.2
HP steam	292.3	Total	77.7
Heat of reaction	141.0	Energy lost	FE MMBtu/h (above 100 °F)
Feed @170 °F	17.2	Power gen losses	17.9
Boiler feed water @250 °F	2.9	Air coolers	352.4
Total	819.2	Water coolers	148.9
		Furnaces stack loss	89
	FE MMBtu/h	Pumps and motors (mechanical loss)	2.86
Net energy input	741.4	Surface condensers	115.1
	FE kBtu/bbl	Condensate loss	0.9
Specific energy use	480.9	Unaccounted losses	14.6
		Total	741.5
Balance check: Energy input – Energy output = 819.2 – (77.7 + 741.5) = 0			

reaction contributes a significant portion of the energy input at 17%. The remaining minor contributions to the energy input come from feed and boiler feed water.

Energy output are grouped in two categories, namely energy export and energy losses. Energy export includes any energy flows going out of the process and being used for a meaningful purpose. In the example, the energy export is 77.7 MMBtu/h, which includes MP steam to the steam header and condensate return to the boilers. It could also include hot feed directly sent to downstream processes, which does not present in this example.

Energy losses are mainly caused from process water and air cooling. To derive fuel equivalent, a process cooling duty is divided by the boiler efficiency (85% for this example) assuming low temperature heat available in process cooling could be used for boiler feed water preheating. Total cooling duty accounts for 68% of total energy losses. Therefore, one critical area for improving process energy efficiency is to identify opportunities to reduce heat losses in process cooling although the heat is usually available at low temperature.

Fuel equivalent for purchased power is assumed at 9090 Btu/kW h comparing the normal conversation factor of 3414 Btu/kW h. This assumption implies power generation loss of 5676 (=9090 – 3414) Btu for each kW h imported. Thus, power generation loss is 17.9 MMBtu/h for 3.15 MW h purchased. The rationale for this assumption is discussed later with the FE calculation given in equation (16.16).

Furnace stack loss is calculated based on actual heater efficiency. For this example, 55% furnace efficiency is assumed for the charge heater and the diesel stripper heater, which have radiant section only. Eighty-five percent furnace efficiency is used for the product fractionator heater and the debutanizer reboiler heater, which have both radiant and convection sections.

The mechanical losses for pumps and motors are calculated based on motor efficiency, which is assumed to be 90% for this example.

The net energy input is expressed as

$$\text{Net energy input} = \text{Energy input} - \text{Energy export} \tag{16.11}$$

For the example in question, net energy input = 819.2 − 77.7 = 741.5 MMBtu/h.

Let us define specific energy use the same as energy intensity

$$\text{Specific energy} = \frac{\text{Net energy input}}{\text{Feed rate}} \tag{16.12}$$

Applying equation (16.12) yields

$$\begin{aligned}\text{Specific energy} &= 741.5\,\text{MMBtu/h} \times 1000\,\text{kBtu/MMBtu}/37{,}000\,\text{bbl/day}\\ &\times 24\,\text{h/day} = 480.9\,\text{kBtu/bbl}\end{aligned}$$

where 37,000 bbl/day is the process feed rate.

Specific energy use is a very insightful concept as it represents the *energy intensity of production* indicated by the amount of energy required for processing one unit of feed.

16.7 FUEL EQUIVALENT FOR STEAM AND POWER

In previous discussions, some assumptions of fuel equivalent factors were made for power and steam. You may ask: what is the basis for making these assumptions? How to determine fuel equivalent values for power and steam in your plant? Let us first consider the calculation of fuel equivalent for power.

16.7.1 FE Factors for Power (FE_{power})

FE_{power} is expressed as

$$FE_{power} = \frac{Q_{fuel}}{Q_{power}} = \frac{1}{\eta_{cycle}} \quad (\text{Btu/Btu}) \tag{16.13}$$

where η_{cycle} is the cycle efficiency of power generation and thus $\eta_{cycle} = Q_{power}/Q_{fuel}$ with Q_{power} (in Btu/h) representing the amount of heat associated with power with a conversion factor of 3414 Btu/kW h.

By using the conversion factor of 1 kW = 3414 Btu/h, equation (16.13) can be converted into

$$FE_{power} = \frac{1}{\eta_{cycle}} \quad (\text{Btu/Btu}) \times 3414 \quad (\text{Btu/kWh}) = \frac{3414}{\eta_{cycle}} \quad (\text{Btu/kWh}) \tag{16.14}$$

Rearranging equation (16.14) leads to

$$FE_{power} = \frac{3414}{\eta_{cycle}} = \frac{3414}{\frac{Q_{power}}{Q_{fuel}}} = \frac{Q_{fuel}}{\frac{Q_{power}}{3414}} = \frac{Q_{fuel}}{W} \quad (\text{Btu/kWh}) \tag{16.15}$$

where $W = (Q_{power}/3414)$ and W (in kW) represents the amount of power. By converting the unit of FE_{power} from Btu/Btu in equation (16.14) to Btu/kWh in equation (16.15), the expression of FE_{power} in equation (16.15) becomes exactly the same as that of heat rate for power generation. Let us look at three cases for applying equation (16.15) as follows.

Case 1 Importing power from coal power plants

Average efficiency for today's coal-fired plants is 33% globally, while pulverized coal combustion can reach an efficiency of 45% (LHV, net) (IEA, 2012). Thus, fuel equivalent factors (FE^{ST}_{power}, MMBtu/MW) for purchased coal power are in the range of 7.58 (45% of power efficiency) and 10.34 (33%). For example, if assuming steam cycle efficiency is 37.56%, applying equation (16.14) yields

$$FE_{power} = \frac{3414}{\eta_{cycle}} = \frac{3414}{0.3756} = 9090 \quad (\text{Btu/kWh}) \tag{16.16}$$

Note that 9090 Btu/kWh is the FE_{power} factor used in the previous assumption for power.

Case 2 On-site power generation from steam turbines

For on-site power generation, usually heat rate is known and it should be used as FE_{power}. If unknown, a typical condensing steam turbine cycle efficiency of 30% could be used to yield

$$FE_{power} = \frac{3414}{\eta_{cycle}} = \frac{3414}{0.3} = 11{,}380 \quad (\text{Btu/kWh}) \tag{16.17}$$

FE_{power} factors for backpressure steam turbines could be much higher than 11,380 Btu/kWh. What is the interpretation of a higher FE_{power} from on-site power generation than that of purchased power? The implication is that a commercial power plant can make power more efficient than a process plant if cogeneration is not involved. Does it mean that use of motor is more efficient than using on-site condensing turbine for process drivers? The answer is Yes. You may stretch out to think: the backpressure turbines could be even worse as process drivers. Is it true? The answer for this question relies on the steam balances. If the exhaust steam from the backpressure turbines is used for processes, the backpressure turbines have much high cogeneration efficiency (power plus steam).

Case 3 On-site power generation from combined gas and steam turbines

When power is generated by a gas turbine (GT), gas turbine exhaust is usually sent to heat recovery steam generator (HRSG) for steam generation. Steam is

then used for further power generation via steam turbines. A configuration such as this is known as a gas turbine–steam combined cycle.
The combined cycle efficiency can be expressed as

$$\eta_{CC} = \eta_{GT} + \eta_{ST} - \eta_{GT} \times \eta_{ST} \tag{16.18}$$

By applying equation (16.14), fuel equivalent factor for power generated from a combined cycle would be

$$FE_{power}^{CC} = \frac{3414}{\eta_{CC}} \tag{16.19}$$

Suppose that a gas turbine cycle has an efficiency of 42%, which is a representative value for gas turbines, and the steam turbine has an efficiency of 30%. The combined cycle efficiency (η_{CC}) is 59.4% based on equation (16.18) and FE factor is 5747 MMBtu/MW based on equation (16.19). In general, the combined cycle is much efficient in power generation than the steam cycle alone.

16.7.2 FE Factors for Steam, Condensate, and Water

Steam headers are the central collection points where steam enters each header from different sources and distributes to different sinks. The total FE for each steam header is the summation of all FEs entering the steam header via different flow paths. The FE for each steam header is the total FE divided by the amount of steam generated from this header, that is,

$$FE_{Header\,i} = \left. \frac{\sum \text{FE consumed}}{\sum \text{steam generated}} \right|_{Header\,i} \quad i = (\text{HP}, \text{MP}, \text{LP})\text{MMBtu/klb} \tag{16.20}$$

A top-down approach is adopted for FE calculations. First, FE for HP steam is calculated and then cascading down in the order of pressure levels, FEs for other steam headers are determined. Let us look at the following example.

Example 16.1

Calculate the fuel equivalent values for the steam headers in Figure 16.3.

Solution

To determine the fuel equivalent for steam headers, the actual ways of producing steam must be identified, which could have influence on the fuel equivalent for the steam.

(a) FE for HP steam

There are two paths for making HP steam, namely boiler 1 with 75% thermal efficiency and boiler 2 with 85% thermal efficiency. The FE factors for both HP

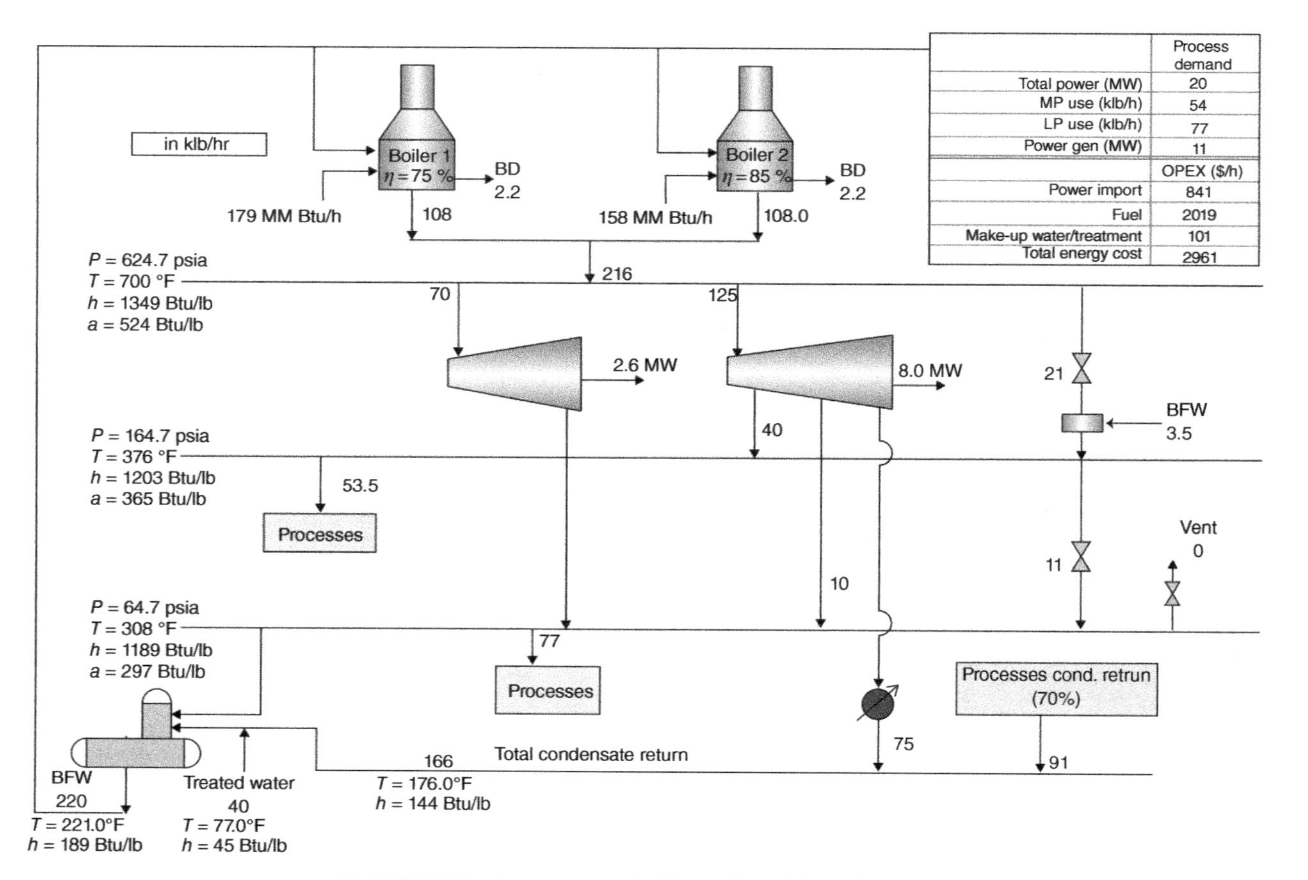

	Process demand
Total power (MW)	20
MP use (klb/h)	54
LP use (klb/h)	77
Power gen (MW)	11
	OPEX ($/h)
Power import	841
Fuel	2019
Make-up water/treatment	101
Total energy cost	2961

FIGURE 16.3. Steam system for Example problem 16.1.

generation sources can be calculated as

$$FE_{HP,boiler\ 1} = \frac{Q_{B1}}{M_{B1}} = \frac{179}{108} = 1.66\,\text{MMBtu/klb}$$

$$FE_{HP,boiler\ 2} = \frac{Q_{B2}}{M_{B2}} = \frac{156}{108} = 1.44\,\text{MMBtu/klb}$$

The average FE for HP steam can be calculated as

$$FE_{HP} = \frac{Q_{B1} + Q_{B2}}{M_{B1} + M_{B2}} = \frac{179 + 156}{108 + 108} = 1.55\,\text{MMBtu/klb}$$

For evaluating a base case scenario, the average FE factor for HP steam should be used. In the case when opportunities for steam saving or extra steam use are explored, the generation-source-based FE factors must be considered. For this example, when capturing the steam saving opportunity, steam generation should be reduced from boiler 1, the less efficient boiler. On the other hand, when extra HP steam is required from processes, it should be generated from boiler 2, the more efficient boiler.

In general, high-pressure steam is defined as steam produced from steam generators, mainly boilers. If using boiler feed water as the reference point, the fuel equivalent of high pressure steam can be derived as

$$FE_{HP} = \left.\frac{Q_{fuel}}{M_{HP}}\right|_{boiler\ i} = \frac{h_{HP} - h_{BFW}}{\eta_{boiler\ i}}\ \text{kBtu/lb} \tag{16.21}$$

where h_{HP} *and* h_{BFW} are specific enthalpies for high-pressure steam and boiler feed water while $\eta_{boiler\,i}$ is the boiler efficiency and M_{HP} is the amount of HP steam generated from the boiler.

In most cases, multiple boilers are used. In this case, equation (16.20) can be applied to derive the weighted average of fuel equivalent for combined HP steam going to the HP header as

$$FE_{HP} = \frac{\sum_i^{Boilers} M_{i,HP} FE_{i,HP}}{\sum_i^{Boilers} M_{i,HP}} = \frac{\sum_i^{Boilers} M_{i,HP} \times (h_{HP} - h_{BFW})/\eta_{Boiler\ i}}{\sum_i^{Boilers} M_{i,HP}}\ \text{kBtu/lb} \tag{16.22}$$

(b) FE for MP steam

Three paths of MP generation are identified as follows:

- Path 1: 40 klb/h of MP extraction from TG-1001 with specific steam rate $m_{HP\text{–}MP}$ at 35.6 klb/MW h. The fuel equivalent for the MP steam exhaust can be calculated via

$$FE_{MP\text{-}steam} = FE_{HP\text{-}steam} - \frac{FE_{power}^{import}}{m_{HP\text{–}MP}} \tag{16.23}$$

The reason why FE factor for power import is used in equation (16.23) is that power import is the marginal power source. In other words, if a steam turbine is replaced by motor, purchased power will be used.

Assume fuel equivalent factor for purchased power as 9.09 MMBtu/MW h, thus

$$\mathrm{FE}^{\mathrm{P1}}_{\text{MP-steam}} = 1.55 - \frac{9.09}{35.6} = 1.29\,\mathrm{MMBtu/klb\,MP}$$

- Path 2: 21 klb/h of the letdown valve, $\mathrm{FE}^{\mathrm{P2}}_{\text{MP-steam}} = \mathrm{FE}_{\mathrm{HP}} = 1.55\,\mathrm{MMBtu/klb}$ because a letdown is an adiabatic process and thus FE does not change through the letdown valve.
- Path 3: 3.5 klb/h of BFW addition for desuperheating, $\mathrm{FE}^{\mathrm{P3}}_{\text{MP-steam}} = \mathrm{FE}_{\mathrm{BFW}} = 177\ \mathrm{Btu/lb}$. The FE factor for BFW is calculated based on equation (16.24), which assumes that LP steam is used for BFW preheat. It must be pointed out that the FE of the pumping power is ignored as it is very low (~10 Btu/lb of BFW), even with the very high pump ΔP.

$$\mathrm{FE}_{\mathrm{BFW}} = \mathrm{FE}_{\text{LP steam}} \times \frac{h_{\mathrm{BFW}} - h_{\text{Ambient water}}}{h_{\text{LP steam}} - h_{\text{Ambient water}}} \tag{16.24}$$

Thus, the average FE for the mixed MP steam can be calculated based on equation (16.20) as

$$\mathrm{FE}^{\mathrm{av}}_{\mathrm{MP}} = \frac{M^{\mathrm{P1}}_{\mathrm{MP}} \times \mathrm{FE}^{\mathrm{P1}}_{\mathrm{MP}} + M^{\mathrm{P2}}_{\mathrm{MP}} \times \mathrm{FE}^{\mathrm{P2}}_{\mathrm{MP}} + M^{\mathrm{P3}}_{\mathrm{MP}} \times \mathrm{FE}^{\mathrm{P3}}_{\mathrm{MP}}}{M^{\mathrm{P1}}_{\mathrm{MP}} + M^{\mathrm{P2}}_{\mathrm{MP}} + M^{\mathrm{P3}}_{\mathrm{MP}}}$$
$$= \frac{40 \times 1.29 + 21 \times 1.55 + 3.5 \times 0.177}{40 + 21 + 3.5} = 1.31\,\mathrm{MMBtu/klb}$$

(c) FE for LP steam

There are three paths for making LP steam:

- Path 1: 70 klb/h from the TG-1002 turbine with a specific steam rate of 26.8 lb/kW h. The fuel equivalent for the MP steam exhaust can be calculated via

$$\mathrm{FE}_{\text{LP-steam}} = \mathrm{FE}_{\text{HP-steam}} - \frac{\mathrm{FE}^{\mathrm{import}}_{\mathrm{power}}}{m_{\mathrm{HP\text{-}LP}}} \tag{16.25}$$

Assume fuel equivalent factor for purchased power as 9.09 MMBtu/MW h, thus

$$\mathrm{FE}^{\mathrm{P1}}_{\text{LP-steam}} = 1.55 - \frac{9.09}{26.8} = 1.21\,\mathrm{MMBtu/klb\,LP}$$

- Path 2: 10 klb/h from the TG-1001 LP extraction with a specific steam rate of 22.9 lb/kW h.

$$\mathrm{FE}^{\mathrm{P2}}_{\text{LP-steam}} = 1.55 - \frac{9.09}{22.9} = 1.15\,\mathrm{MMBtu/klb\,LP}$$

- Path 3: 11 klb/h of the letdown valve, $FE^{P3}_{LP\text{-}steam} = FE_{MP} = 1.31\,MMBtu/klb$ because letdown is an adiabatic process. BFW desuperheating is not needed for the LP steam in this case because the superheated fraction in LP steam is very small.

Thus, the average FE for the mixed LP steam is calculated based on equation (16.20):

$$FE^{av}_{LP} = \frac{70FE^{P1}_{LP} + 10FE^{P2}_{LP} + 11FE^{P3}_{LP}}{(70 + 10 + 11)}$$

$$= \frac{(70 \times 1.21 + 10 \times 1.15 + 11 \times 1.31)}{91} = 1.21\,MMBtu/klb$$

What about the FE for vented LP steam? In this case, FE_{LP} should also be calculated based on the path from which this vented LP steam is generated. This is because a certain amount of fuel equivalent is consumed to make the LP steam no matter it is used or vented or not. For vented LP steam, the value is zero but FE is not.

(d) FE for condensate

Condensate temperature is similar to the deaerator temperature typically around 200 °F. The condensate FE_{Cond} can be determined by the difference of condensate temperature and raw water temperature (ambient). FE condensate is usually in the range of 100–150 Btu/lb of condensate. Although FEBFW is small relative to steam, accumulated loss could be significant for a large amount of condensate loss. Also condensate loss is costing due to extra chemicals required to treat make-up water.

(e) *FE for BFW water*

The energy required for providing boiler make-up water includes the heat content and the pump power used to elevate its pressure. The BFW heat content is the major portion of the BEW FE factor, which is determined by the difference of BFW temperature (typically around 250 °F) and raw water temperature (ambient). The FE_{BFW} is in the range of 150–200 Btu/lb of BFW.

(f) FE for cooling water

Energy for providing cooling water includes pump power and the fan power in running the cooling tower fans. The FE_{cw} is in the range of 7–15 Btu/gallon of cooling water.

16.8 ENERGY PERFORMANCE INDEX (EPI) METHOD FOR ENERGY BENCHMARKING

Naturally one would think that the first and second laws of thermodynamics should be used as methods for assessing energy efficiency for industrial processes. If we apply the first law of thermodynamics to Table 16.2, we have

$$\eta = \frac{\text{Useful energy}}{\text{Energy input}} = 1 - \frac{\text{Energy loss}}{\text{Energy input}} = 1 - \frac{741.5}{819.2} = 10\% \qquad (16.26)$$

Clearly, this efficiency is not very insightful as energy input is provided at very high quality in terms of temperature, pressure, and composition, while the energy losses are at much lower quality; but they are compared on the same basis.

In contrast, an efficiency based on the second law of thermodynamics could make more sense as it takes energy quality into account. Exergy analysis is the method developed for industrial applications based on the second law of thermodynamics. Exergy could be done in reality, but the effort in data requirement could be prohibitive. More importantly, the second law of thermodynamics is a difficult concept to grasp for many process engineers, which is not common for applications in the process industry.

Instead, a much simpler yet effective method is presented and discussed here. This method is built on the concept of guideline energy performance (GEP), which is used as a benchmark against which actual energy performance (AEP) is compared. The rationale of using this concept as the basis for assessing process energy efficiency is revealed in the following discussion.

Let a ratio of AEP and GEP be defined as follows. This ratio shall be labeled the energy performance index (EPI):

$$\text{EPI} = \frac{\text{Actual energy performance}}{\text{Guideline energy performance}} = \frac{\text{AEP}}{\text{GEP}} \tag{16.27}$$

By definition, EPI represents the energy efficiency for the process unit on the basis of GEP. In this way, any improvements in operation, design, equipment, and technology upgrade can be measured using EPI. Application of the EIP method is discussed as follows.

Generally speaking, an EPI gap of less than 5% between AEP and GEP belongs to an operational gap. In other words, better operating practices and control could close this gap. An EPI gap of larger than 10% may require small energy retrofit projects, which can feature a quick payback, for the gap to be closed. If the EPI gap is in the order of 20%, it may require significant energy and process retrofit projects to close the gap.

16.8.1 Benchmarking: Based on the Best-in-Operation Energy Performance (OEP)

By applying the method for calculating specific energy use, you can obtain a plot of specific energy versus time based on the historic data. This plot can pinpoint the best-in-operation energy performance (OEP) that your process unit has achieved at a time when there was institutionally dedicated effort for operation performance and with technical know-how available. You could confirm this by checking with engineers and operators who have worked in the plant during this period. As a result, you will be able to determine the OEP as the GEP representing the best-in-plant performance. Assume the specific energy use on the basis of OEP is 438.7 kBtu/bbl feed, which can be obtained from historian. With the actual energy use of 480.9 kBtu/bbl calculated as above, EPI for the process unit can be calculated by

$$\text{EPI} = \frac{\text{AEP}}{\text{OEP}} \times 100\% = \frac{480.9}{438.7} \times 100\% = 109.6\% \tag{16.28}$$

Equation (16.28) indicates a 9.6% deviation of AEP from the OEP. Such a gap is significant, which should alert you to initiate investigations for root causes. The mere factor of determining EPI gives you an immediate indicator as to where your process unit stands in energy performance, so that you can quickly spot problematic areas.

At this point, we have a very good starting point. You know three essential facts: the energy intensity for your process unit, the performance target, and the gap against the target. Your mind may be racing with questions such as the following: What has gone wrong with my process unit? How can the AEP be reduced to OEP for the process unit? These questions are answered in Chapter 17.

16.8.2 Benchmarking: Based on Industrial Peers' Energy Performance (PEP)

In industry, there are peer survey groups organized based on industrial sectors and process technology. Organizers for the survey groups send questionnaires to survey members to gather sample data on yearly and conduct performance calculations. Consequently, the peer performance results are shared among survey members. If your plant belongs to the survey group, you could obtain the best peer energy performance (PEP) via the representative in your organization. For certain large companies, there are CoP (Community of Practice) networks based on process technology. You should seek out the best PEP for your process unit via the CoP in your company.

Assume the specific energy for PEP is 430 kBtu/bbl for the example. Based on the actual energy use of 480.9 kBtu/bbl, we can calculate EPI for the process unit as

$$\text{EPI} = \frac{\text{AEP}}{\text{PEP}} \times 100\% = \frac{480.9}{430} \times 100\% = 111.8\% \tag{16.29}$$

Usually, the survey group is divided into tiered performance structure such as first, second, third, and fourth quartiles. Based on the EPI calculated above, you can find out which performance quartile your process unit belongs to. This indicates where your process unit stands among your peers.

16.8.3 Benchmarking: Based on the Best Technology Energy Performance (TEP)

With technology advancement in catalyst, equipment, process design, and control, process energy efficiency could improve. It is not difficult to gather the performance data for state-of-the-art technology. In some cases, the data is published in public by government offices and you could find them via web search. If not available in public, you can contact technology companies – they are often eager to provide the data to customers.

Assume the operation is improved for the example process and the energy use is reduced from 480.9 to 438.7 kBtu/bbl. The plant management is interested to know the scope of further energy improvement by applying better process technology and design. Assume the TEP is 380 kBtu/bbl. Thus, the EPI for the process unit can be calculated by

$$\text{EPI} = \frac{\text{AEP}}{\text{TEP}} \times 100\% = \frac{438.7}{380} \times 100\% = 115\% \tag{16.30}$$

Technology updates can make big step changes in both production and energy performance but usually with high capital costs and long implementation periods. Therefore, it should be applied very selectively.

16.9 CONCLUDING REMARKS

There are three fundamental concepts discussed in this chapter. The first one is the concept of converting all energy back to fuel equivalent. This concept makes all forms of energy on the same basis, that is, fuel fired or fuel equivalent. The second one is specific net energy, which describes the energy intensity for production. The third concept is GEP as the best alternative for comparison with actual performance.

In combination, these three concepts make it a much simpler yet effective task for assessing the energy performance for a process unit and require minimal data. Therefore, the EPI method is designed for practical applications.

The strategy for achieving the target can involve changes to operating practice, new control strategy, process equipment modifications, technology upgrade, or combinations of the above. In general, closing the gap between the average and the best potential performance of an individual unit involves operational and maintenance improvements. Eliminating the gap between an existing unit and its peers in industry often involves retrofit with modifications to operation and the process design. Reaching the state-of-the-art performance usually involves technology upgrade.

You may have questions during data extraction. Which data periods should be used as the basis for energy benchmarking? What data is more representative than other? How to prevent inefficient usage of time from data collection? Although the general guideline is to collect data that represents the most common operation, specific guidelines are provided as follows:

16.9.1 Criteria for Data Extraction

- Near maximal feed rate or the most commonly used feed rate
- Use middle-of-the-run historian data
- Use 24-h rolling average based on hourly average data to smooth out fluctuation
- One year of data could be a good representation; get rid of bad data by all means.

The reason for using middle-of-the-run historic data is because it represents an "average" operation performance. In contrast, both SOR (start-of-the-run) and EOR (end-of-the-run) represent two extreme operation modes and hence the data would give biased indications of energy use.

Annual data can cover changes in season and operation modes while monthly data can zoom into focus on a particular operation mode.

16.9.2 Calculations Precision for Energy Benchmarking

At this stage, you need to start focusing on important data. Make quick estimates for small consumption users as they usually do not have meters. Do not chase decimal point of precision as the key is getting the order of magnitude right. Some of the following guidelines could be helpful to you:

- Ask instrumentation engineers to recheck critical meters to make sure they are functioning properly.
- Major consumptions need to be verified. Use design data for small consumptions if meters are not available. Corrections may be necessary to reflect the difference in temperature, pressure, and mass flow.
- Fill missing data by heat and mass balances.
- All forms of energy must be converted to fuel equivalent. Adjustments may be necessary in order for the actual energy use to be on the same basis as guideline energy use.
- Specific energy can be on feed volume or mass basis depending on the norm used in the industry. Specific energy can also be on a product basis, which is the ratio of total net energy usage to a desirable product rate on either volume or mass basis. For a process involving both reaction and separation, use feed as the basis for calculating specific energy use. If a process only involves separation and makes single product, use product as the basis for calculating specific energy use.

16.10 NOMENCLATURE

Variables

AEP actual energy performance
BFW boiler feed water
EPI energy performance index
FE fuel equivalent; amount of fuel at the source to make a unit of energy utility (power, steam)
GEP guideline energy performance
h enthalpy
m, M mass flow
OEP best-in-operating energy performance
PEP peer's energy performance
Q heat content

Greek letters

η_{cycle} power generation efficiency; the amount of fuel (energy input) required to make a unit of power (energy output)

Subscript

CC combined gas and steam cycle
GT gas turbine
ST steam turbine

REFERENCES

IEA (2012) Technology Roadmap: High-Efficiency, Low-Emissions Coal-Fired Power Generation, December 4.

Schipper L, Howarth RB, Carlassare, E (1992) *Energy Intensity, Sectoral Activity and Structural Change in the Norwegian Economy*. Energy: The International Journal, **17**, P215–233.

17

KEY INDICATORS AND TARGETS

17.1 INTRODUCTION

If you ask operators and engineers how their plant is doing, they would tell you that the plant is under good control. Although true, the process performance could become much better economically and efficiency wise. The root cause is the lack of monitoring key indicators and no process optimization capability available to operators and engineers in the face of many variables and strong interactions, which are typical in operation.

What really needs to happen is the indication of key parameters to monitor, and the target values to achieve for these parameters in order to achieve better energy performance. Although process energy benchmarking in Chapter 16 gives measure of process energy intensity for process units, the energy intensity does not provide indications of the root causes and the operating parameters to turn in order to improve energy performance. To determine how well a process unit is doing, a system of performance metrics should be developed so that actual energy usage can be compared with a consumption target. Only then is it possible to conduct root cause analysis and take appropriate remedial actions. To accomplish this goal, the concept of key energy indicator is introduced (Zhu and Martindale, 2007), which is the foundation for systematic performance assessment and optimization.

The rationale of introducing key energy indicators is to seek answers to this critical question: "How can engineers characterize energy use in a process unit with an emphasis on *major energy users* in terms of *their needs and the reasons and practical ways to minimize energy use for the needs*?" Application of the key energy indicators in reality follows a methodology based on three steps: defining key indicators,

Hydroprocessing for Clean Energy: Design, Operation, and Optimization, First Edition.
Frank (Xin X.) Zhu, Richard Hoehn, Vasant Thakkar, and Edwin Yuh.

setting targets, and identifying actions to close gaps. Using this methodology, a process unit can be described by a small number of key indicators to measure energy performance, which can be developed based on process knowledge and experience. Application of key indicators will allow focus on important issues and avoid falling into trap of details.

17.2 KEY INDICATORS REPRESENT OPERATION OPPORTUNITIES

The intention of defining key indicators is to describe the process and energy performance with a small number of operating parameters. A key indicator can be simply an operation parameter. Some examples of key indicators are product rates, column overhead reflux ratio, column overflash, spillback of a pump, heat exchanger U value, etc. The parameter identified as a key indicator is important due to its significant effect on process and energy performance.

In defining key indicators, one needs to understand the strong interactions between process throughput, yields, and energy use. In the traditional view, energy use is regarded as a supporting role. Any amount of energy use requested from processes is supposed to be satisfied without question and challenging. This philosophy loses sight of synergetic opportunities available for optimizing energy use for more throughput and better yields. The following discussions will provide insights into these kinds of opportunities, which form the basis for defining key indicators for your process units to capture specific opportunities.

17.2.1 Reaction and Separation Optimization

Optimizing energy use in reaction and separation systems could lead to significant energy saving because both reaction and product separation consume the majority of overall energy use. Much effort is commonly put into reducing energy losses incurred in heat exchangers, furnaces, steam leaks, insulations, etc.; but little effort is spent on minimizing energy use for reactions and separations, which are the heart of the processes. Very often, energy demands in these systems are considered as "must meet" with expectation of no challenges from engineers and operators. In reality, however, there is a large scope in minimizing energy use in these areas.

Reaction condition optimization considers reaction severity in terms of temperature and pressure profiles in accordance with catalysts performance in the entire run length. Optimizing reaction conditions, selecting better catalysts, and maintaining catalyst performance in operation have significant effects on both yields and energy efficiency. Consider reaction temperature as an example. In the catalyst cycle, the catalyst performance deteriorates, which affects the reaction conversion. To compensate, the reaction temperature may be increased. However, more severe reaction conditions require more heat from hot utilities such as fired heaters, while severe conditions also produce more desirable products as well as undesirable by-products. The question is how to determine the optimal reaction temperature, which is a function of reaction conversion, production rate, and energy use.

Separation optimization is to achieve product recovery and quality with minimum energy use. Consider overflash for a given separation. Overflash is the feed vaporized in excess of the products drawn above a flash zone. Overflash is typically expressed as a percentage of a distillation column feed. Higher overflash leads to more excess heat in the distillation column, which is rejected in overhead cooling leading to higher reflux rates. Overflash is typically provided by heat input from a hot utility on the feed stream to the distillation column. In a reboiled column, excess heat in the distillation column is provided by a reboiler. A higher overflash can achieve better separation but at the expense of extra hot utility duty such as a reboiler or feed heater. On the other hand, a too low overflash will make poor separation and product quality suffers. Yet, in another circumstance, a lower overflash can allow higher throughput with distillation corrections taking place further downstream. Achieving the optimal overflash or optimal vapor to liquid ratio for a separation process is a critical operation optimization issue.

Minimizing recycle is another optimization example in separation. Another common observation in a process plant is that unconverted streams or off-specification streams are recycled back to the front of a process. Recycle streams undo separation. Minimizing recycle presents a significant opportunity for process plants to smartly reduce energy consumption. However, minimizing recycle requires proper operation of separation columns across whole plant from upstream to downstream.

17.2.2 Heat Exchanger Fouling Mitigation

Fouling mitigation represents large opportunity for increased heat recovery in operation. Fouling occurs in many heat exchange services and fouling reduces heat transfer duty significantly. Effective fouling mitigation can save substantial amount of energy. The overall heat transfer coefficient, or the U value, is the single most important parameter for fouling monitoring for a stand-alone heat exchanger. However, for a complex exchanger network involving many exchangers, determining which exchangers should be selected for cleaning and the frequency of cleaning is not trivial. Selection of the most fouled heat exchanger for cleaning could lead to a suboptimal solution. The primary objective of heat exchanger fouling mitigation should be to minimize energy use. It is possible to select the exchanger for cleaning, which may not be the most fouled, but it could yield the greatest reduction in energy use than cleaning the most fouled one.

17.2.3 Furnace Operation

Fired heaters provide major part of heat sources for reaction and separation. Reliability is the major concern for furnace operation with heat flux for large heaters and tube wall temperature (TWT) for small heaters as the most important reliability parameters. Increasing either heat flux or TWT could increase furnace efficiency. When operating heat flux is much lower than the maximum limit; this is an indication of the furnace being underutilized and thus presents an opportunity for increased feed rate. Increased feed rate is a win–win operation since more feed also results in reduced energy intensity.

Besides reliability, efficient furnace operation is a major part of an energy management system. Oxygen content (O_2%) in the flue gas and furnace stack temperature are the two key operating parameters. Correct measurement of O_2% in the flue gas is the first step while good control of air intake is the necessary action to achieve low O_2%. Maintenance is essential for burners to function properly and to eliminate air leaks into combustion zones. However, too low O_2 content could promote uneven distribution of combustion flame, damage furnace tubes, and cause reliability incidents. Typically, 3% of oxygen is industrial average although furnaces with state-of-the-art burners and control systems could achieve less than 3%. When reducing air intake to minimize oxygen content in operation, it is imperative to minimize air leaks and have proper O_2% measurement.

Reducing stack temperature could yield greater reduction in combustion fuel and the limit for the stack temperature is usually set by sulfur dew point. For furnaces with a convection section, the reduction of stack temperature could be achieved by adding an "economizer" service into the existing convection section. Economizer services are typically water heating services for boiler feed water preheat or saturated steam generation. For furnaces without a convection section, installing a convection section could be the answer. However, these options need to go through a thorough feasibility evaluation of foundation strength, thermodynamic and hydraulic considerations. Preheating combustion air is another opportunity in reducing stack temperature. Low pressure steam could be used for air preheating.

17.2.4 Rotating Equipment Operation

Rotating equipment used in process industry includes pumps and compressors. For pumps and compressors with fixed speed motors, minimizing spill backs could result in significant power saving. Optimizing steam turbine operation and maintenance could save steam. Selection of process drivers, namely motors versus steam turbines, could save energy cost.

17.2.5 Minimizing Steam Letdown Flows

Steam letdown valves are used to give steam supply flexibility and temperature control. However, letdown steam represents lost opportunity for power generation. Steam balance optimization could minimize the letdown steam flow and hence reduce the loss in power-generating potential.

17.2.6 Turndown Operation

Poor turndown operation implies that when feed rate reduces, energy use does not reduce accordingly. For example, air intake for furnaces should reduce accordingly when feed rate drops. Stripping steam for separation columns should be reduced based on steam-to-feed ratio instead of fixed steam flows. Some of the hand valves for steam turbines can be partially closed when feed rate drops significantly so that the governor could be maintained wide open in order to minimize steam rate. Proper turndown operation could generate significant energy saving for plants operating under large feed variations.

17.3 DEFINE KEY INDICATORS

The above discussions reveal the fact that major improvement opportunities can be captured by key operating parameters, which is the foundation of introducing key indicators. The method for determining key indicators follows a thinking process: understand the process objectives → understand energy needs → develop the measures for the needs → define the key indicators representing the measures. As an example of how to define key indicators for process, let us consider a single-stage hydrocracking process in oil refining industry.

The single-stage flow scheme as shown in Figure 17.1 is a commonly used hydrocracking unit in the past as it allows greater than 95% conversion of a wide variety of feedstocks to high-quality products such as gasoline, jet fuel (kerosene), and diesel. To reduce the charge heater duty, the reactor effluent stream is heat-exchanged against both feed and recycle gas to recover process heat. Afterward, reaction effluent is sent to high-pressure separator where the gas containing high concentration of hydrogen is recycled back to mix with raw feed for reaction. The conversion products as the liquid of the high-pressure separator go to the flash drum first and then join with the hot separator bottom liquid going to the stripper where H_2S and lighter components are separated while the liquid products go to the main fractionator to obtain kerosene and diesel. The fractionator is typically total condensing. Offgases are generated at a stripper. The overhead goes to the debutanizer column where LPG and naphtha are made off the column top and bottom, respectively. Majority of the main fractionation column bottom is recycled back to the reactor to reduce the conversion per pass.

17.3.1 Simplifying the Problem

Clearly, hydrocracking is a relatively complex process. To simplify the overall task of defining key indicators for the whole unit, we can divide it into three sections, namely reaction and product fractionation and naphtha stabilizer. The naphtha

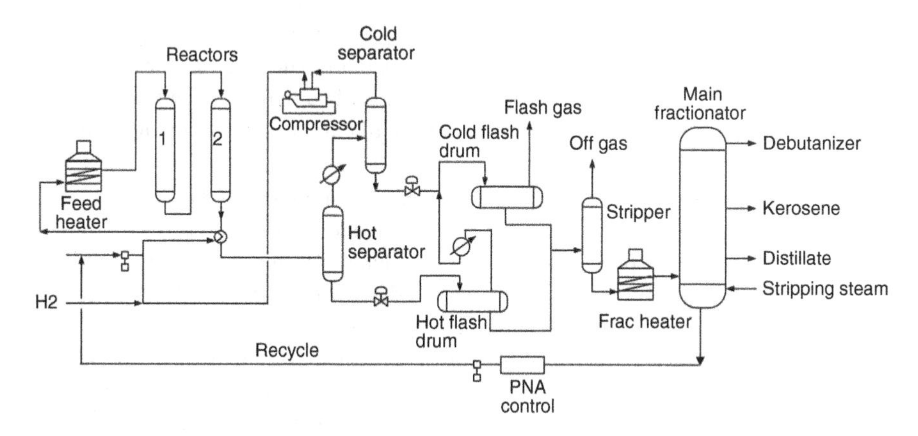

FIGURE 17.1. Typical single-stage hydrocracking unit.

stabilizer section is not shown in Figure 17.1. The goal is to define a set of key indicators for each section, which can be used for monitoring and optimization.

17.3.2 Developing Key Indicators in the Reaction Section

(i) *Understanding the process*
In this example, the process objective is to achieve 95% conversion of feedstocks to high-quality products such as naphtha, diesel, and jet fuel. The reactor contains catalysts that allow maximum production of desirable products. Hydrocracking reactions are highly exothermic and thus requires cold hydrogen quench injection to control reactor temperature. The energy efficiency for this unit will largely depend on how effectively the reaction effluent heat is recovered.

(ii) *Understanding the energy needs*
The following items are identified as major energy users in the reaction circuit and their distinct roles and significances are discussed as follows:

- The feed heater is used to increase the feed temperature and control the reactor inlet temperature. Although the heater efficiency depends on how it is designed and operated, the heater duty is determined by feed preheating. The feed heater inlet temperature can also be a function of the reaction temperature profile – that is, ascending temperature profile will lower the heater outlet temperature. For a given feed preheat, the heater duty is mainly a function of the heat of reaction and heat recovery. A process engineer can determine the ways to maintain process heat recovery, heater efficiency, and heat flux.
- The compressor for recycle gas is a large power user. The role of a recycle gas compressor is to provide the required amount of hydrogen to the reaction and to provide quench for heat release. The recycle compression work depends on the gas flow and its molecular weight. The gas flow depends much on the hydrogen purity in the recycle gas, while the molecular weight of the recycle gas depends on the amount of the light ends brought into the recycle gas from the high pressure separator. The role of a process engineer is to optimize gas compression ratio. The optimization may indicate to operate the compressor at maximum rate to provide the highest catalyst on-stream efficiency.
- Stripping steam is injected to the stripper column for the purpose of removing H_2S and noncondensable light components from the bottom product. The role of a process engineer is to determine the dew point approach, which is the difference of stripper overhead temperature and the water dew point. The dew point of water is a function of the amount of stripping steam and stripper overhead compositions. The dew point of water is important because the hydrogen sulfide in liquid water is corrosive to the stripper internals. There are handles for the stripper operation. One is to maintain a certain feed temperature such that there is enough enthalpy in the feed to generate sufficient reflux to keep the overhead vapor above the water dew point.

The second handle is to maintain a certain amount of stripping steam to remove H_2S and light ends from the bottom product. If the feed temperature is below a limit, the overhead vapor temperature could be below the water dew point and corrosive water could accumulate in the top tray. On the other hand, if too much stripping steam is injected, particularly in turndown operation, the water dew point could increase and reach a point where water in vapor condenses out.

(iii) *Effective measures for the energy needs*

Based on the above understanding of the energy needs, we can go one step further to develop efficiency measures in providing these needs.

- Heat of reaction: A higher heat of reaction occurs when dealing with feeds containing more aromatics, higher sulfur concentrations, or higher olefin concentrations. In these cases, the hydrocracking reaction becomes more severe, resulting in shorter catalyst life, higher H_2 consumption, and compression work. Thus, specific heat of reaction is the parameter that connects reaction severity, catalyst life, H_2, and power consumption based on feed compositions and products.
- Heater reliability: Heat flux (for large heaters) or TWT (for small heaters) is the key reliability parameter for a heater. When heater operation is higher than the limit of heat flux or TWT, a heater is under risk of reliability because the tube life is shortened for higher fluxes. On the other hand, operating a heater at much lower than the limit makes the heater underutilized, which represents an opportunity for increased feed rate or higher process severity. Besides these two operating limits, flame impingement is another reliability measure, which is usually caused by too low O_2 content. Combustion flame becomes longer with too little O_2 and could reach the tube and pose a serious reliability risk.
- Heater efficiency can be affected by the excess O_2 content or extra air for combustion and a high stack temperature. Inappropriate O_2 content could be caused by lack of control, air leaks, and poor burner performance, while a high stack temperature corresponds to high heat loss in flue gas. A heater approach temperature, defined as the temperature difference between flue gas to the stack and heater feed inlet, could be caused by heater fouling in operation and by heater design.
- The feed is mainly heated by the reaction effluent in feed exchangers before the charge heater. The hot end approach temperature on feed exchangers is a good indication of heat recovery performance by the feed preheating system.
- Reactor effluent air cooler (REAC) inlet temperature: After the reaction effluent transfers its heat to feed and recycle gas, the reactor effluent goes to a REAC. Thus, the REAC inlet temperature on the reaction effluent side is a good indication of how effective the reaction effluent heat is recovered.
- Hydrogen to hydrocarbon (H_2/HC) ratio: The recycle gas containing high percent of hydrogen from the high pressure separator is recycled back to the reactor. The recycle gas rate is determined by desirable hydrogen partial

pressure for the reaction purpose, which affects compression, yields, and catalyst life time. Too high H_2/HC ratio could cause greater power usage due to increased recycle rate while too low ratio could impact on yield and shorten the catalyst life. Thus, hydrogen ratio is a parameter affecting energy and yields.

(iv) *Developing key indicators for the energy needs in the reaction section*
Through the above exercise of simplifying the problem and developing understanding of major energy needs and measures of efficiency in providing the needs, we can define the following key indicators for the reaction section:

- Specific heat of reaction
- H_2/HC ratio
- Combined feed exchanger (CFE) hot end approach temperature
- REAC inlet temperature

Specific indicators for heaters could include the following:

- Heater O_2 content
- Heater stack temperature
- Heat flux
- Flame impingement.

17.3.3 Developing Key Indicators for the Product Fractionation Section

(i) *Understand process characteristics*
H_2, H_2S, NH_3, and light ends are removed from reaction effluents through a series of separation and flashes, resulting in the reaction products in liquid form, which goes to the stripper, the feed heater, and then to the main product fractionator. The task of the main fractionation is to separate different products based on their product specifications such as distillation endpoint, ASTM D-86 T90%, or T95% point. Side draws from the column go to the product strippers where kerosene and diesel products are made. The net draw from the column bottom is called unconverted oil (UCO), while the recycled oil also produced from the main column bottoms is recycled back to the reaction section for nearly complete conversion. There are two pumparounds, namely kerosene and diesel pumparounds, as a main feature of heat recovery from the main fractionation.

It is essential to avoid flooding and dumping, which could severely affect fractionation and thus energy efficiency. Fractionation efficiency can be monitored by column internal vapor to liquid (V/L) ratios. Desired V/L can be achieved jointly by optimizing feed heater outlet temperature, fractionation stripping steam, and overhead reflux rate together with pumparound heat duties, which are used to control excess heat in the column.

(ii) *Understand the energy needs*

- Main fractionator feed heater provides the driving force for the fractionator by vaporizing feed partially to generate sufficient vapor traffic within the column. The heater duty and stripping steam in the column bottom determine the level of product recovery from the column bottom, which can be monitored by the bottom 5% boiling point. For a given product recovery, the heater duty is a function of feed rate, temperature and compositions, feed preheat, and the enthalpy of vaporized products.

- Fractionation side-stripper steam: Stripping steam is used to remove the lighter materials in diesel in order to control the flash point. This steam can be minimized and, in some cases, even eliminated until the flash point requirement is met.
- Column pumparounds: Heat is recovered by the pumparounds and transferred to other process streams. The pumparound duty affects the downcomer flow and temperature profile in the pumparound section. Thus, a pumparound not only affects the heat recovery but also the fractionation efficiency below the pumparound.

(iii) *Effective measure for the energy needs*

- Feed heater efficiency: The discussions for heaters are similar to the reaction charge heater given previously.
- Heater outlet temperature: This temperature can affect the lift of diesel out of UCO. A too low heater outlet temperature would cause slump of diesel into UCO, which degrades the value of diesel into fuel oil. Too high of a heater outlet temperature will cause unnecessarily high reflux rate at the expense of extra heater duty. The heater outlet temperature is mainly a function of the hydrocarbon partial pressure in the flash zone because the TBP cut point is a function of the diesel distillation specification and fractionation efficiency. An optimal heater outlet temperature could be determined by the fractionation overflash, which measures the internal reflux rate.
- Fractionation efficiency can be determined by the gap of diesel 95% cut point and UCO 5% cut point

(iv) *Developing key indicators for the energy needs in the main fractionation system*

Based on the understanding of major energy needs and measures of efficiency in providing the needs, we can define the following key indicators for the fractionation section:

- Recycle oil combined feed ratio
- Fractionator overhead pressure
- Fractionator stripping steam
- Fractionator column reflux ratio
- Fractionator column overflash
- Diesel stripping steam to diesel ratio
- The TBP gap of diesel 95% cut point and UCO 5% cut point
- Heater O_2 content
- Heater stack temperature
- Heat flux
- Flame impingement.

17.3.4 Developing Key Indicators in the Naphtha Stabilizer Section

The main equipment in the naphtha stabilizer section is debutanizer. Similar to the above discussions, we can identify key indicators for the debutanizer column as follows:

- Reflux ratio: affecting separation in the column and can control C5 in the LPG (overhead product) and C4% in naphtha (the bottom product).
- C5% in LPG: too much C5 in LPG is a liquid loss as C5 that could be blending stock for gasoline.
- C4% in naphtha: too much C4 could cause higher Reid Vapor Pressure (RVP) than specification, which is an indication of combustion instability in car engines.

17.3.5 Remarks for the Key Indicators Developed

- For a typical hydrocracking unit, there could be a couple of thousands of operating data measured. The key indicators identified above only accounts for very small fraction of the overall data but capture the key performance that contributes to the major portion of operating costs. If these indicators can be monitored and optimized, the unit can operate close to optimal performance.
- In most cases, parameters related to feed and product yields are measured and controlled using basic control systems or APC (Advanced Process Control) systems. However, traditionally, the process indicators are not integrated with energy use. Furthermore, many energy parameters are not measured.
- By identifying the key process and energy indicators and optimizing them together, the optimization does not only reduce energy cost but may also allow increasing throughput when needed, improve product quality, and minimize product specification give-away.

17.4 SET UP TARGETS FOR KEY INDICATORS

To improve from current performance, targets must be established for the key indicators and these targets provide standard against which existing facilities are measured and equipment improvement are evaluated. The difference between a target and the current performance for each key indicator defines the performance gap. Each performance gap should be associated with dollar value, which represents opportunity to be captured. Each indicator is correlated to a number of parameters including process and equipment conditions together with equipment limits. In this way, energy optimization is connected with process conditions and constraints.

How would one make the concept of key energy indicators working for a process unit or the company? Let us use the example of the debutanizer for the aforementioned hydrocracking unit.

Problem: The debutanizer tower sketch is given in Figure 17.2. The reboiling at the bottom of the tower is to provide sufficient vapor flow on the trays for separating C4 from C5 and heavier components in the feed. C4 and lighter components

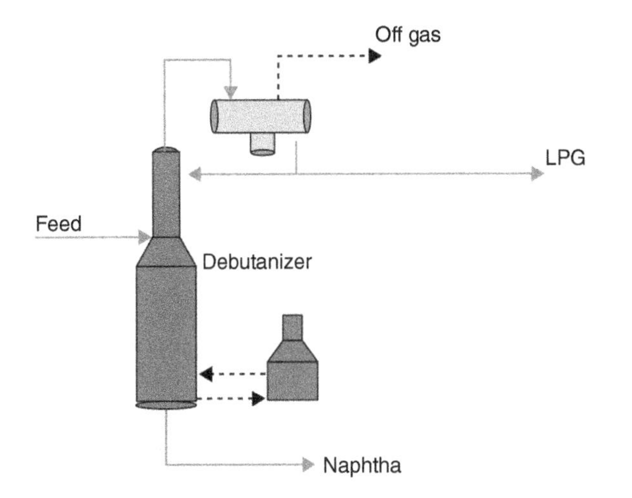

FIGURE 17.2. Debutanizer column in hydrocracking unit.

will be withdrawn at the overhead, while the C5 and heavier components leave at the bottom. A certain amount of C4 in the bottom product is allowed based on the maximum RVP specification for gasoline. Reboiling duty is the main variable in controlling the C4 amount in the bottom product. In other words, the reboiling duty must increase when C4 amount in the bottom exceeds the specification. However, it is undesirable to reduce C4 amount to lower than the specification as this would be product specification give-away at the expense of additional utilities. The operating objective is to minimize the reboiling duty while achieving the maximum RVP specification at all time.

Rationale: The task at hand is to develop a relationship between reboiling duty and C4% in the Debutanizer bottom product so that C4% can be controlled by adjusting the reboiling duty. However, other operating parameters also affect the reboiling duty, which include feed conditions (rate and composition), feed preheating (feed temperature), tower conditions (overhead temperature and pressure). If a correlation of reboiling duty against the above influencing parameters could be generated, reboiling duty can be adjusted according to any of the changes in the related parameters and thus avoid the need of trial and error.

Solution: There are a couple of ways to develop such a correlation. The simplest way is use of a data historian. This method can be applied if three conditions are met: (1) the related parameters are measured and data available in the historian; (2) the measured data must reflect the operation at the time the butane content was measured, and (3) the historian data cover all possible operating scenarios. After all, online data is the true representation of real simulation!

Development of a correlation using the historian data can be conducted readily in a spreadsheet using regression techniques. After gathering the data from historian, multiple-variable regression can be applied to develop such a correlation. The overall correlation coefficient must be higher than 85% for sufficient regression fidelity.

The second option is to use the step-test method usually for developing parametric relation for control systems. By making a small step change to the manipulatable (independent) variable of interest, a response from the control variable (dependent variable; reboiling duty in this case) can be recorded after reaching the steady-state condition. This response can be called an energy response. Finally, the regression method is applied to derive the correlation of reboiling duty against all related variables.

However, in many cases, the conditions above for using data historian are difficult to satisfy. It could be also labor intensive and inconvenient in operation to adopt the step-test method. Thus, the most common method is to use the simulation method for developing relationship correlations. To do this, a simulation model for the tower can be developed readily based on the feed conditions (rate and compositions), tower conditions (temperature, pressure, and theoretical trays) with product specifications (C4% in the bottom and C5% in overhead) established as set points in simulation. Operating parameters such as reflux rate and reboiling duty can be adjusted to meet product specifications. The simulation model is verified and revised against high-quality performance test data.

For evaluating the effect of individual parameters, simulation cases can be developed by prespecifying the values for independent variables of interest; the energy response (reboiling duty) will be recorded automatically. For example, to evaluate the effect of feed preheating, UA value of the feed preheating exchanger is varied with prespecified values. During simulation runs, the feed temperature before the tower will change according to the UA values, which cause the reboiling duty to vary automatically in the simulation. A set of four curves for such one-to-one relationships can be obtained as shown in Figure 17.3a–d. Brief explanations are given for each figure.

Reducing reflux drum pressure will reduce reboiling duty with the trend as shown in Figure 17.3a. C4% in the bottom is the specification that the gasoline product must meet. But, too low C4% is not necessary as it does not generate commercial benefit but at the cost of extra reboiling duty due to the steeper part of the curve (Figure 17.3b). This product specification give-away operation must be avoided by all means. C5% in overhead product (Figure 17.3c) is the indication of gasoline blending component lost in LPG, which should be avoided as well. Feed preheat also reduces reboiling duty but increases condensing duty (Figure 17.3d).

One must be aware of the capacity limit at the existing condenser when increasing feed preheating, which requires extra condensing. On the other hand, when the feed quality in terms of the butane concentration, the reboiling duty is affected for the separation of C4 from the C5+ materials. There is no control for the feed quality for the debutanizer operation because the feed quality is the consequence of different raw feeds processed in the hydrocracking unit and the processing severity. For the other four parameters above, operators can make changes to reflux drum pressure, C4% in the bottom product, C5% in the overhead product and feed preheating. Optimizing these parameters could give around 5% reduction in reboiling duty than operated based on experience only, which is very significant.

Reboiling and condensing duty can be described based on the relationship with individual parameters as shown in Figure 17.3a–d. If assuming a polynomial form of correlations with order of three is used, then we have

$$R_i = a + bx_i + cx_i^2 + dx_i^3$$
$$x_i = (\text{C4\% in bottom}, \text{C5\% in overhead}, \text{Preheat}, \text{Drum } P), \qquad (17.1)$$

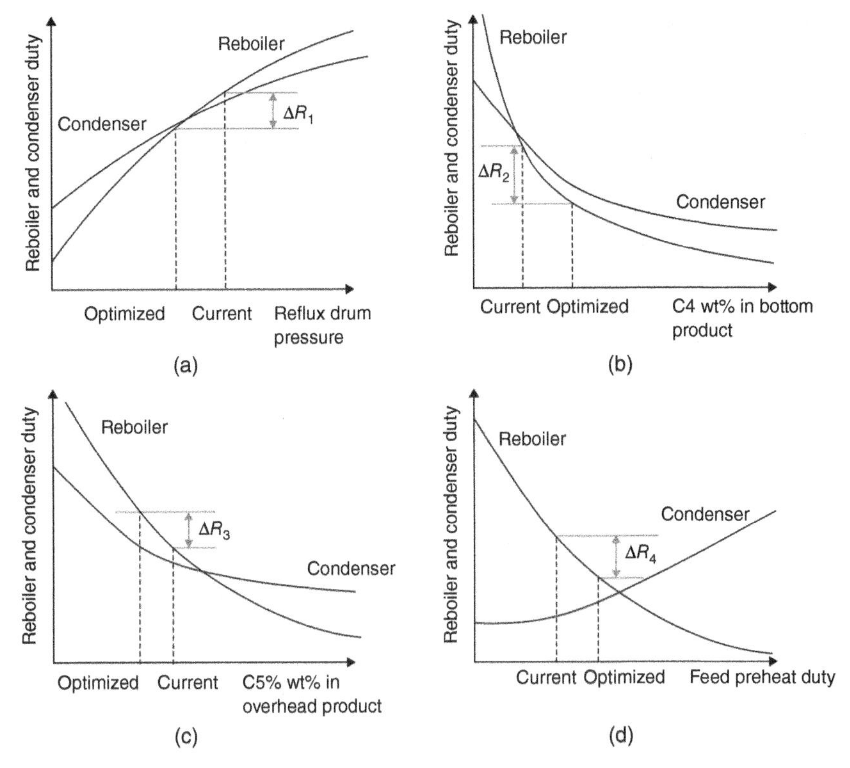

FIGURE 17.3. Correlations of debutanizer reboiler duty and other parameters.

where x_i is one of the four operating parameters and equation (17.1) describes the relationship of reboiling duty (R_i) and x_i.

The incremental effect (ΔR_i) from individual parameter (Δx_i) can be determined as

$$\Delta R_i = R_{i,\text{new}} - R_{i,\text{base}} = a + b\left(x_{i,\text{new}} - x_{i,\text{base}}\right) + c\left(x_{i,\text{new}} - x_{i,\text{base}}\right)^2 + d\left(x_{i,\text{new}} - x_{i,\text{base}}\right)^3. \tag{17.2}$$

If there are no interactions among these four operating parameters, the total effect of changes in these parameters on reboiling duty would be the simple summation of individual effects, that is,

$$\Delta R = R_{\text{new}} - R_{\text{base}} = \sum_{i=1}^{4} \Delta R_i. \tag{17.3}$$

However, in many cases, there could be strong interactions among operating parameters. In this case, two or more parameters could appear together in one term and the bilinear ($x_1 \cdot x_2$) is the simplest form of interaction. To develop relationship of parameters with interactions, several parameters need to vary at the same time in plant test or simulation and the effect on reboiling duty can be seen as the statistically significant result of the interaction parameter. A set of data with changes to the operating parameters and the energy response can be obtained and regression is subsequently applied to derive correlation involving interactions.

When dealing with correlations involving multiple variables, economic sensitivity analysis is essential to determine the most influential parameters. For example, feed preheat and reflux drum pressure are very sensitive to reboiling duty more than other operating parameters for the debutanizer. Getting the most sensitive parameters right in operation can get the greatest economic and technical response.

The correlation developed can be implemented into the control system so that reboiling duty can be controlled automatically to achieve the minimum at all time. On the other hand, the correlation can be used as a supervisorial tool. Whenever a variation is expected, adjustments to operating parameters need to be made to optimize the reboiling duty. This reboiling duty is the minimum with all things considered and it is the target for the conditions at hand. This target and dollar value for closing the gap must be communicated with board operators in each shift so that actions will be taken for achieving targets while dollar value saved could give operators a sense of pride as a recognition of their actions.

17.5 ECONOMIC EVALUATION FOR KEY INDICATORS

Operation variability is the major cause of operation inefficiency. In general, there are two kinds of variability, which can be observed in reality, namely inconsistent operation and consistent but nonoptimal operation. Figure 17.4 represents the operating data of a stripping steam rate in the main fractionator in a hydrocracking unit. In Figure 17.4a, the stripping steam rates appear to be randomly scattered showing an example of inconsistent operation. This is usually caused by either poor control strategy or different operating policy used by operators for running the tower. In contrast,

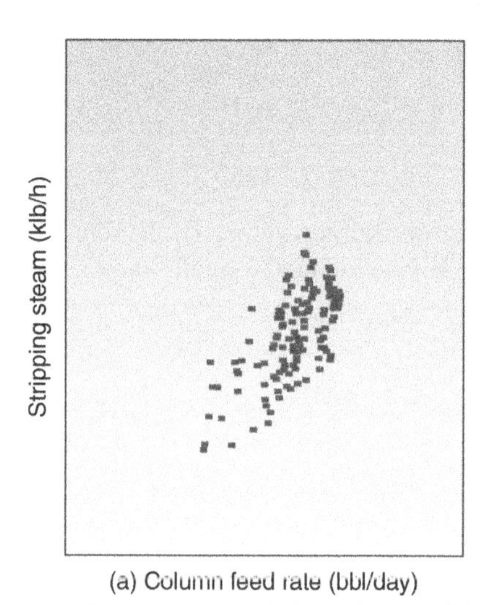

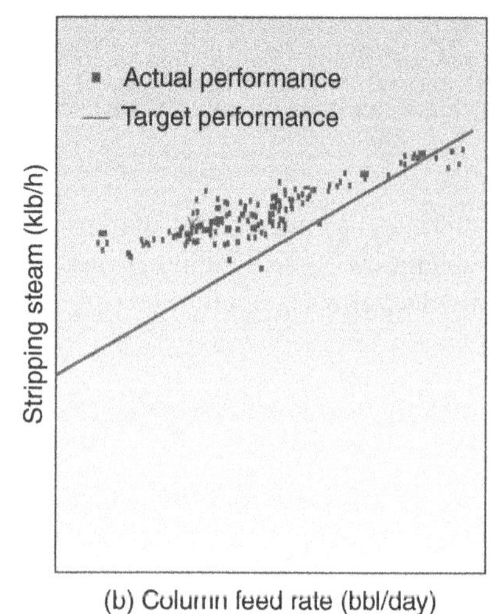

FIGURE 17.4. Two common operating patterns: (a) inconsistent operation; (b) consistent operation but nonoptimal.

Figure 17.4b shows a consistent operation but nonoptimal. In this case, a consistent operating strategy was adopted, but it was far away from the target for adjusting reboiling duty against column feed rate. The target operation represents the minimum reboiling duty to achieve product specification.

The variability of any operating parameter occurs due to various reasons. The question is how to identify operation variability and the economic value of minimizing variability.

Variability assessment starts with simple statistical analysis of operating data. For example, the operating data for C5% in the debutanizer column overhead product under normal conditions can be extracted from historian as shown in Figure 17.5a with specification limit provided. To understand the variability, data in Figure 17.5a is converted into a normal distribution curve, which represents frequency of observations as shown in Figure 17.5b. In many cases, the operating data mimics the normal distributions.

Two parameters describe the normal distribution, namely mean or average (μ) and variance or variability (σ). σ defines the shape of normal distribution. The larger (smaller) the σ value, the thicker (thinner) the curve. μ and σ can be calculated as follows:

$$\mu = \frac{\sum x}{N}, \tag{17.4}$$

$$\sigma = \frac{\sum (x-\mu)^2}{N}, \tag{17.5}$$

where x is the value of key indicator obtained from historian while N is the number of sample data points for the key indicator.

Referring to the example discussed in White (2012) as shown in Figure 17.5, there are two shortcomings in the operation performance. The first one is the large variability while the second shortcoming is too conservative in reaching the specification limit. If the operation or control strategy improves, the variability could be minimized to achieve more consistent operation (Figure 17.6b) but still far away from the limit. The limit is usually set by a physical limit such as product purity specification, maximum temperature or pressure, maximum valve opening, maximum vapor loading in

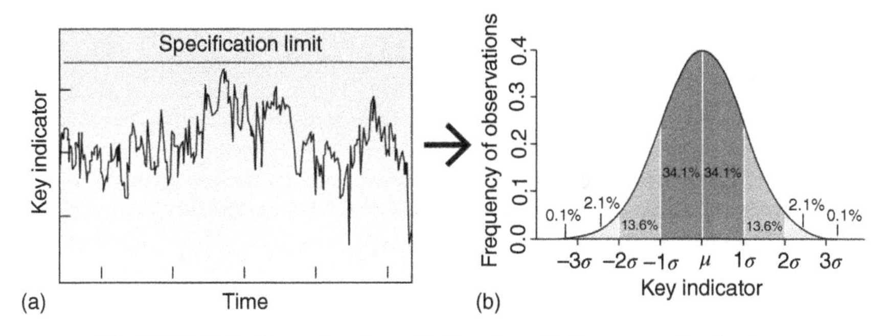

FIGURE 17.5. Operating data: (a) historian; (b) frequency distribution.

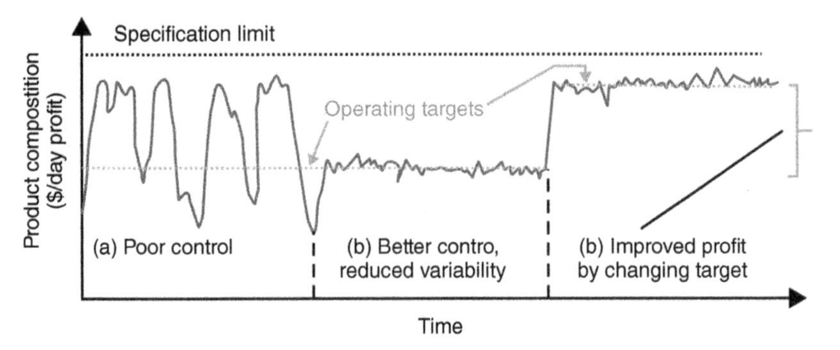

FIGURE 17.6. Operation performance: (a) current operation; (b) reduced variability; (c) increased profit. Source: White (2012). Reproduced with permission of AIChE.

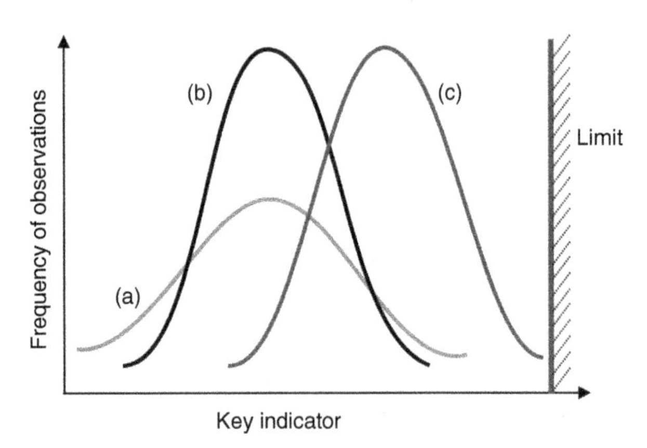

FIGURE 17.7. Convert time series data in Figure 17.6 into normal distribution curves: (a) current; (b) reduced variability; (c) increased profit.

a separation column, maximum space velocity in a reactor, and so on. The operation can be improved further (Figure 17.6c) by moving the average closer to the limit by adopting a better control strategy. Time series data in Figure 17.6 can be converted into normal distribution curves as shown in Figure 17.7.

If the price for the component i of interest is known as C_i, and the loss of this component is x_i with a frequency of observations as f_i, then the money loss V_i can be calculated as follows:

$$V_i = C_i \cdot x_i \cdot f_i \tag{17.6}$$

For example, C_5% in the debutanizer overhead product represents the high-value component C_5 lost in LPG. The key indicator could be defined the difference of actual C_5% in LPG and C_5 specification. If LPG produced from the column is 1000 barrel per day with C_5% in LPG at 1% higher than the specification or $x_i = 1\%$ with frequency of occurrence as $f_i = 30\%$, and C_5 is valued at \$75/barrel, the economic value

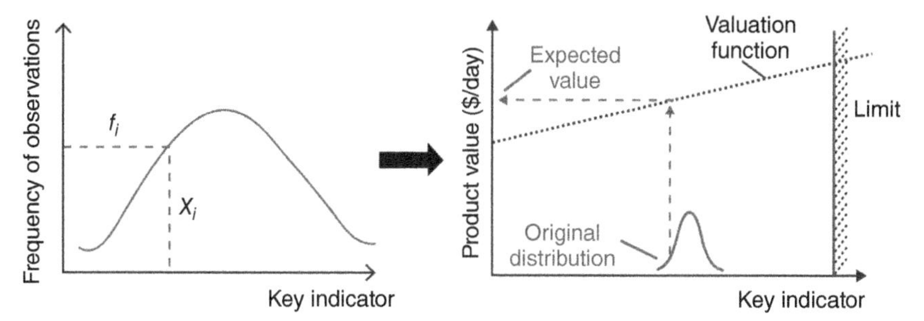

FIGURE 17.8. Converting the normal distribution curve to economic curve.

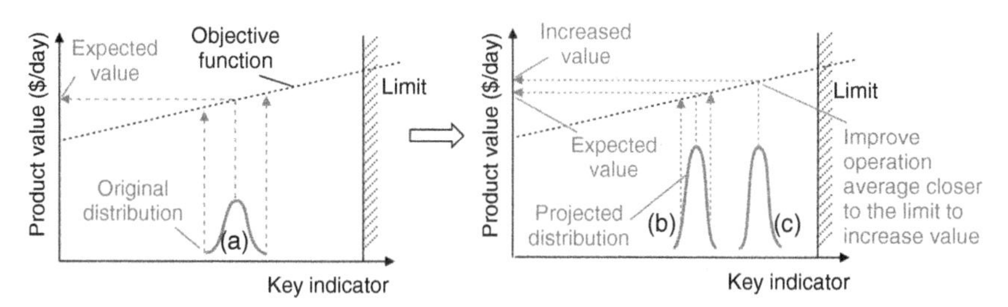

FIGURE 17.9. Economic curves generated based on normal distributions. Source: White (2012). Reproduced with permission of AIChE.

to avoid this occurrence is

$$V = C_i \cdot x_i \cdot f_i = \$75/\text{bbl} \times 1\% \times 1000\,\text{bbl/day} \times 30\% = 225\,\$/\text{day}.$$

Similarly for other occurrences, C_5% in LPG could be lower or higher than 1%, and the economic values can be calculated accordingly.

In this way, the normal distribution curve can be converted into economical curve as shown in Figure 17.8. The conversion is calculated to economic difference.

With the statistic-based economic evaluation method mentioned, improved operation can be quantified with economic values based on statistical distribution of operating data. The current operation with large variance (case a in Figure 17.9) is improved by more consistent operation and/or control strategy to reduce variability (case b in Figure 17.9) while optimized operation (case c in Figure 17.9) utilizes the potential capability available in the process and equipment and pushes the economic value even higher.

17.6 APPLICATION 1: IMPLEMENTING KEY INDICATORS INTO AN "ENERGY DASHBOARD"

The concept of key energy indicators (KEIs) and targets can be readily implemented into an energy dashboard, which can quickly show the performance gaps between current and targets on the computer screen. The level of a gap indicates the severity of deviations and forms the basis to assign a "traffic light" for each KEI – that is, a green light implying the current performance is acceptable as it is within the target range; a yellow light, a warning sign indicating that a gap occurs and requires attentions; or a red light, an alarm sign urging to take actions at the earliest time possible. An example tool of monitoring key indicators is Honeywell's Energy Dashboard (Sheehan and Zhu, 2009). This tool could be tremendously valuable to operators and engineers as to what to watch, what to focus, and which knobs to turn and when.

As a system, KEIs can be defined in a hierarchical structure, from overall site to each process unit down to major equipment and individual operating parameters. The sum of all incentives (opportunity gap between current and targets) from all pieces of equipment represents the total opportunity for the entire process and the overall site. This hierarchical structure allows engineers to drill down from overall performance to specific parameters and thus identify specific actions.

- *Overall site view* shows the site-wide energy consumption and greenhouse gas (GHG) emission versus overall targets. At the same screen, the overall site view shows the relative size of energy consumptions and GHG emissions in each process unit. Traffic light color is assigned to indicate which processes are furthest away from the targets.
- *Process unit view* indicates the process performance that can be measured by 10–20 key energy indicators. These key energy indicators are developed from a combination of design, process simulation, and historical data. These predicted energy targets are automatically adjusted to reflect current operating conditions such as feed rate and compositions, operating mode, product yields, and quality. Color coding is assigned to each KEI, which could indicate the need for drill down in the next level of key indicators for identifying root causes and actions.
- *Equipment view* describes equipment performance via several key operating parameters with indications of current values versus corresponding targets. The operators may decide to perform a more detailed investigation for the root causes if the gap is large.
- *Deviation trends view* allows operators to review over the time periods when the KEI deviate significantly from the targets and to determine the major causes of the deviation. By building up a history of causes, operators are able to look back over time and see the most common causes of deviations. This can lead to recommendations about remediable actions for improving equipment performance and, hence, overall process performance.

For each key indicator, a target is established as the basis to compare with current performance. The difference between the target and the current performance for each key indicator defines the performance gap. Different gap levels indicate the severity and level of urgency for actions.

Gap analysis is then used to identify root causes – potential causes include inefficient process operation, insufficient maintenance, inadequate or lack of operating practices, procedures and control, inefficient energy system design, and outdated technology. Gap analyses are translated into specific corrective actions to achieve targets via either manual adjustments or by automatic control systems. Finally, the results are tracked in order to measure the improvements and benefits achieved.

17.7 APPLICATION 2: IMPLEMENTING KEY INDICATORS TO CONTROLLERS

Many opportunities for energy improvement can be achieved directly changing the plant conditions by adjusting the set point of key variables. In some cases, these opportunities for energy improvement may be possible by incorporating these key variables into an APC if the investment for such an APC is justified by the value to be captured.

Multivariable, predictive control, and optimization applications have been commonly applied in the process industry. The ability to take models derived from process data and simulations and configure the models in a highly flexible manner allows the engineers to design controllers that can be suitable for multiple purposes. The same controller can be used to maximize throughput, maximize yields, and/or minimize energy use just by changing the cost factors in the objective function. This APC environment is suitable for incorporating energy strategies into overall operating objectives. In fact, adding energy operating costs into existing objective function and inserting related KEIs with corresponding correlations and operating limits is generally advisable. In this manner, minimizing energy cost will not be accomplished at the expense of the most valuable product yields.

There are many energy-saving opportunities that can be incorporated into APC applications, such as

- Furnace pass balancing and excess O_2 control
- Distillation column controls combined with pressure minimization and flash zone temperature control to maintain yields of the most valuable products while minimizing energy use in reboiling or feed heater
- Reaction conversion control
- Feed preheating maximization
- Separation column reboiler duty control
- Recycle minimization
- Water dew point control for steam strippers.

One example of a single variable control strategy is applied to a stripper in a hydrocracking unit. The main purpose of the stripper is to remove H_2S and noncondensable components from the bottom product. One of the key indicators identified earlier is the water dew point at the top of the stripper column. As a matter of fact, the dew point is a function of column overhead vapor composition and the amount of water. There was no monitoring capability available for the dew point temperature. If the column top temperature is lower than the dew point, the hydrogen sulfide will dissolve in the condensed water and cause corrosion to the column overhead system. To avoid this, operators usually run the column with a high dew point approach to make sure the top temperature is sufficiently higher than the dew point at the expense of extra amount of stripping steam. By applying a new control device recently developed for dew point for this type of stripper columns, the actual dew point can be calculated accurately. With measured column top temperature, a tight dew point approach can be maintained, which results in 10% reduction of the stripping steam.

Another example is a large multivariable control strategy, which was applied to an ethylene complex (Sheehan and Zhu, 2009). This involves 17 multivariable controls that were linked together by an overarching optimization strategy that included the use of nonlinear cracking model to predict product yields. The result of the APC applications enabled the operators to increase the feed rate by 3% over the previous best rate by being able to operate the process up against multiple constraints simultaneously. In addition, the application is able to reduce energy consumption by 3.3% by reducing steam consumption in the fractionators and minimizing excess O_2 in the furnaces. This results in a payback of less than 5 months for the APC investment.

17.8 IT IS WORTH THE EFFORT

As demonstrated, the concept of key indicators and targets can play important roles for process and energy integrated optimization. Process optimization without taking energy use into account will lead to high energy costs while energy optimization without fully addressing process needs will cause penalty in processing capacity, product quality, and yields. With appropriate work process fitting into the existing technical management system, the concept of key indicators and optimization can become the cornerstone of energy management.

Due to this significance, developing key indicators and targets should become a corporate concerted effort as management's support is critical. First of all, significant effort is required in developing technical targets for key indicators. Setting up targets needs modeling of major equipment. Once a base case of operation is defined and simulated, operation variations can be simulated and correlations can be developed for key indicators in relation with other operating parameters. Using the debutanizer as an example, the reboiling duty is the key indicator, which can be affected by feed rate and compositions, reflux drum pressure, preheat, and C4% in the bottom and C5% in the overhead products. Regression analysis of simulation results may be required to develop the correlation of the reboiling duty and these operating parameters. Whenever operating conditions change, the correlation could be applied to determine the minimum reboiling duty under the new conditions.

The engineering effort is appreciable in developing a system of simulations for setting up technical targets for key indicators at different operating conditions. It may take one to two man-years to develop a system of key indicators and targets for processes and major equipment. In one example of a large refinery complex, two man-years were required to develop such a target system. Implementing this target system in day-to-day operation is essential and integrating the target system into existing technical management system. The confidence will grow when people observe the operating cost saving of applying such a target system. The benefits will pay for the investment alongside the development. The lessons learned and experience gained will be disseminated to other process units.

NOMENCLATURE

Variables

C economic value for a key indicator
x measured value of key indicator
f frequency of observations
R reboiling duty
N the number of sample data points
V economical value

Greek letters

μ mean value in normal distribution
σ variance in normal distribution

REFERENCES

Sheehan B, Zhu XX (2009) The first step in energy optimization, *Hydrocarbon Engineering Journal*, pp. 25–29, April,

White DC (2012) Optimize energy use in distillation, *Chemical Engineering progress (CEP)*, pp. 35–41 AIChE, March.

Zhu, XX, Martindale, D (2007) Energy optimization to enhance both new projects and return on existing assets, 105th NPRA (National Petroleum Refining Association) Annual Meeting, March, Texas, USA.

18

DISTILLATION SYSTEM OPTIMIZATION

18.1 INTRODUCTION

Distillation is the core of a process unit and also the major energy users. Design and operation of distillation and separation columns involves a trade-off between energy use and product recovery. When energy usage is less than design, product recovery and quality may suffer. On the other hand, when energy is more than design, product quality is better than specification, which is called product spec giveaway. In abnormal operations, little product recovery can be achieved regardless of how much energy is used.

However, reducing energy usage in a distillation system is not straightforward because a distillation system involves many operating parameters including those within and outside the process battery limit. In particular, variations in conditions of feed and products as well as prices of feeds and products add much complexity to the economic operation of the process. This feature leads to strong dynamic behaviors of operating parameters. Furthermore, most of the parameters interact in a nonlinear manner and have numerous constraints on their operation, which further complicate the task of energy optimization. If some of the constraints can be relaxed, this could improve operating margin significantly.

The concept of tower feasible operating region is discussed in Chapter 10 and performance assessment in Chapter 15. This chapter focuses on economic operation within the feasible operation region.

Hydroprocessing for Clean Energy: Design, Operation, and Optimization, First Edition.
Frank (Xin X.) Zhu, Richard Hoehn, Vasant Thakkar, and Edwin Yuh.

18.2 TOWER OPTIMIZATION BASICS

Tower optimization is a difficult task as product pricing and unit constraints often change daily or weekly, but changing unit operating philosophy and addressing hardware constraints can take months to accomplish. Even after the steps to improving optimal performance have been identified and implemented, if the desire to improve is removed, operation tends to return to the older, more comfortable routine, or constraints in other areas often prevent operation in the most profitable mode. Thus, it is highly recommended that for a complex system, performance optimization should be implemented in APC, which can improve tower operation to the most economic mode on a regular and consistent basis and in an automatic manner.

To establish tower optimization, key operating parameters must be defined and correlations must be developed to understand the relationships between key parameters. Finally, optimization objective function must be developed to determine the optimal set points for the key parameters. The optimization can be conducted in two ways: one is semimanual based, while the other is APC based. In the semimanual-based approach, operating parameters are manually adjusted, while optimization is done in an off-line (online) manner. In contrast, in the APC approach, operating parameters are automatically adjusted, while optimization is done online. But both methods adopt the common ground of optimization: using an objective function to derive the optimal set points for key parameters based on economic trade-offs; correlations to represent relationships between key parameters, and constraints to define process and equipment limits. Noticeably, the optimization pushes operating limits in obtaining the optimal solution, and relaxation of sensitive constraints or limits could generate significant benefits.

18.2.1 What to Watch: Key Operating Parameters

As discussed extensively in Chapter 17, it is important to define major operating parameters or key indicators as they can describe the process and energy performance. A key indicator can be simply an operation parameter such as desirable product rate, column overhead reflux ratio, column overflash, column temperature, and pressure. By the name of key indicator, the parameter identified is important and has significant effect on process and energy performance.

Although primary operating parameters affecting both fractionation and energy use are tower specific, common operating parameters can be identified, which are discussed in the following sections.

18.2.1.1 Reflux Ratio Reflux ratio is defined as the ratio of reflux rate to distillate rate (R/D) or the reflux rate to feed rate (R/F). In essence, a reflux rate is to set a tower top temperature required for making the distillate (overhead product) to meet specification. Reflux is generated by energy via either tower reboiler or feed heater. Lower reflux rate saves energy, but too low reflux rate could affect product quality. On the other hand, too high reflux rate could be wasting of energy if product quality is better than specifications already. In this case, the quality that is better than the specification is given away for free because there is no credit in pricing for the extra better quality.

Optimal reflux rate in operation depends on the operating margin, which is defined as the difference of product sales and feed cost and energy cost. When energy cost is too high, it could drive the operation toward lower reflux rate and vice versa for lower energy cost.

In tower design, the reflux ratio is determined based on the trade-off between operating cost in reboiler and capital cost for the tower. In other words, use of more separation stages requires less reflux rate and in turn less reboiling energy but at the expense of additional capital cost. The minimum reflux ratio is calculated based on Underwood (1948). A tower requires an infinite number of stages to achieve the minimum reflux ratio. To make tower feasible in operation and affordable in cost, a reflux ratio larger than the minimum is used. Typical reflux ratio is 1.1–1.3 of the minimum reflux ratio. With a high reflux ratio, the number of theoretical stages is lower, resulting in lower capital cost for a tower but at the expense of higher reboiler duty; and vice versa. The optimal reflux ratio in tower design is determined based on the minimum total cost.

18.2.1.2 ***Overflash*** Overflash is defined as the ratio of internal reflux at the feed vaporization zone and the feed rate. By definition, overflash represents the percentage of feed vaporized more than the amount of products drawn from above the feed tray. Overflash is a function of reflux rate, feed temperature, and tower pressure.

Overflash is generated from the overhead reflux rate. Thus, it can be said that overflash is generated by reboiler or feed heater. Overflash is an indication of reflux rate sufficiency for proper separation throughout the tower. A small overflash implies less reboiling duty and thus saves energy, but it could negatively affect the fractionation efficiency and hence product quality; and vice versa. Therefore, overflash connects fractionation efficiency and energy efficiency for a tower. A tower could be making poor product quality even with high overflash when the tower is operated under abnormal operations such as flooding or dumping.

Overflash is typically controlled between 2% and 3%. An operation policy focusing on throughput would operate a tower at very low reflux rate; it is not uncommon to observe that a tower is operated at close to 1% overflash. This low reflux operation could be beneficial if the tower produces intermediate products that will be processed further via downstream reaction and separation processes. In this case, this operation could lead to energy efficiency as well as high economic margin.

18.2.1.3 ***Pressure*** Lower pressure typically saves energy. This is because the lower the tower pressure, the less heat required for liquid to vaporize and thus less energy required. This results in better fractionation as it is easier for vapor to penetrate into liquid on the tray deck.

The condenser pressure controls the tower pressure and thus the feed tray pressure. There is a pressure valve in the overhead that can be used to control tower pressure. The lower limit of the tower pressure is defined by the column overhead condensing duty, net gas compressor capacity, and column flood condition. During extended turndown periods, reducing pressure up against an equipment limit can improve efficiency. Many of the new APC systems use pressure control to save energy.

Heat exchanger fouling in overhead condensers could cause higher pressure drop or lower heat transfer and thus result in high tower pressure. On the other hand, higher reflux rate could lead to high pressure drop in the overhead loop causing high tower pressure.

18.2.1.4 ***Feed Temperature*** A hotter feed can increase feed vaporization and thus reduce reboiling duty. However, higher temperature feed could cause too much vapor, resulting in rectification section flooding. For a given tower, the optimal feed temperature corresponds to the lowest reboiling duty while the tower can meet product specifications.

18.2.1.5 ***Stripping Steam*** Some towers may have stripping steam in the feed zone. Stripping steam reduces flash zone pressure and provides partial heat to the feed and thus helps increase the lift of light components from the bottom product. Stripping steam for a fractionation tower is controlled based on the lift, while stripping steam for a stripper is controlled based on stripper product specification. Be aware of too much stripping steam as it could lead to high energy cost and also cause vapor loading limitations in the overhead system.

18.2.1.6 ***Pumparound*** Many fractionation towers have pumparounds to remove excess heat in the key sections of the tower. The effect of increasing pumparound rate is reduced internal reflux rate in the trays above the pumparound but increased internal reflux rate below the pumparound. Thus, change in pumparound duty affects fractionation. On the other hand, pump around rates and return temperature affect heat recovery via the heat exchanger network. It is not straightforward in optimizing pumparound duties and temperatures since the effects on both fractionation and heat recovery can only be assessed in a simulation model. An APC application incorporated with process simulation should be able to handle this optimization.

18.2.1.7 ***Overhead Temperature*** In hot weather, tower overhead fin fan condenser could be limited and thus the tower top temperature can go up. As a result, valuable components could be vaporized into overhead vapor, leading to yield loss. There are a number of ways to reduce the overhead temperature such as increasing cooling water rate; turning on spare overhead fan for air cooler, and increasing reflux rate. On the other hand, when overhead temperature is too cold, salt condensation in the condenser could occur and cause corrosion.

18.2.2 What Effects to Know: Parameter Relationship

How do operating parameters relate to each other? Which parameters are more sensitive to fractionation and energy use? What is the impact of changing one parameter to another? Understanding these could provide insights and guidelines for operational improvements. The objective of developing key indicators is to understand the strong interactions between process throughput, yields, and energy use so that the trade-off among them can be optimized with the objective of maximizing operating margin. In the traditional view, energy use is regarded as a supporting role. Any amount of energy use requested from processes is supposed to be satisfied without question and challenging. This philosophy loses sight of synergetic opportunities available for optimizing energy use for more throughput and better yields.

In developing correlations, one needs to connect energy with product yields and quality. One such example is discussed in detail in Chapter 17. The correlations can be applied for operation optimization. For automatic control, the correlations can be implemented into an APC system, which determines the set points for primary or independent operating parameters. For manual control, operating targets for primary parameters can be obtained based on the correlations.

A process simulation could be a very good vehicle in developing correlations of primary parameters. To do this, a simulation model for the tower can be developed readily based on the feed conditions (rate and compositions), tower conditions (temperature, pressure, and theoretical trays) with product specifications established as set points in simulation. Operating parameters such as reflux rate and reboiling duty can be adjusted to meet product specifications. The simulation model is verified and revised against performance test data based on clean conditions. Different operating cases can be generated in simulation and simulation results can be transferred to a spreadsheet with relationship between dependent and independent variables. Then the regression method is applied to derive the correlations.

When dealing with correlations involving multiple variables, an economic sensitivity analysis is essential to determine the most influential parameters on process economics. For example, in a debutanizer, feed preheat and reflux drum pressure are very sensitive to reboiling duty more than any other operating parameters. Getting the most sensitive parameters right in operation can get the greatest bang.

18.2.3 What to Change: Parameter Optimization

A tower is built to make separation of products. Therefore, tower optimization is to maximize operating margin and minimize energy usage. This processing goal can be described mathematically in an objective function with the parameters in the objective function connected to other processing parameters. All these parameters are defined as constraints in two forms: inequality equations (larger and smaller than), which are used for describing operating minimum and maximum operating limits. Therefore, these constraints form a feasible operating region, which the objective function is constrained within during optimization. The objective function plus the set of constraints form an optimization model. The results of solving this model yield the values for a set of operating parameters, which can be adjusted in operation to achieve the maximal operating margin defined in the objective function.

A generic form of a process optimization model are provided as

$$\text{Objective function: Maximize } Z = \sum_{i}^{m} c_i P_i - \sum_{j}^{n} c_j F_j - \sum_{z}^{p} c_z Q_z$$

$$\begin{aligned}\text{Subject to}: \quad & f_k(X_c, X_m, P, F, Q) = 0; \quad k = 1, 2, \ldots, q \\ & X_{c,\min} \le f_l(X_c) \le X_{c,\max}; \quad l = 1, 2, \ldots, u \\ & X_{m,\min} \le f_p(X_m) \le X_{m,\max}; \quad p = 1, 2, \ldots, w\end{aligned}$$

where c_i are the unit prices of products, c_j are the unit prices of feeds while c_z are the unit costs of energy including steam, fuel, and power. F's and P's are the mass flows of feed and products while Q's are the amount of energy. Essentially, the objective function Z represents the upgrade value from feed to products at the expense of energy in terms of fuel, steam, and power. X's are the operating parameters, X_c's are the independent or control variables, and X_m's are the dependent or manipulated variables for the tower. $f(X_c, X_m, P, F, Q)$ are the relationship constraints.

Maximizing the objective function Z under these correlations with proper limits of the operating parameters will change the related parameters to the values that economical value Z will be maximal. In this case, the operating parameters achieve optimal values that can be used as set points for the close-loop or open-loop control.

In building this optimization model, the most important thing is to include all the major operating parameters that affect operating margin. Then, correlations are developed to describe the relationship among these major parameters and between these major parameters and other operating parameters. When defining operating limits, it is very important to distinguish soft and hard constraints. Hard constraints refer to mechanical performance limits, for example, the tower tray flood limit, the compressor flow rate limit, or the furnace heat flux limit. While making sure that hard constraints must be satisfied, relaxing soft constraints could play a significant role in improving operating margin.

18.2.4 Relax Soft Constraints to Improve Margin

Finding ways to relax plant limitations is one of the most important tasks in improving operating margin. Mathematically, relaxing constraints will lead to a large operating region and push the objective function toward the edge of the enlarged region. But the bottom line is what can be done in reality for relaxing plant limitations. Equipment rating analysis could be a very effective way to identify equipment spare capacity and limitations. Utilization of spare capacity can allow capacity expansion up to 15–20% in general and accommodate improvement projects without or with little capital cost. The important part of a feasibility study is to find ways to overcome soft constraints, at some times, hard constraints if they can be overcome using cheap options.

Three general limitations for equipment are pressure, temperature, and metallurgy as each equipment is designed with the limits for these parameters. If it is identified that equipment will operate at a higher pressure than the design limit, operating pressure needs to be reduced if possible. Otherwise, it is necessary to replace it, which comes with a cost, and it is usually expensive. This is similar to the case when operating temperature will exceed the design limit. These constraints could be resolved if process conditions could be changed. However, if metallurgy is found less than required, this could be a major hard constraint and it needs to be flagged out for the metallurgist's attention as earlier as possible. In some cases, the equipment could still be usable if it is agreed by the metallurgist and the plant takes actions for routing inspection.

For fired heaters, the limitations could be heat flux or TWT (tube wall temperature). The former is applied to heaters with pressure drop larger than 20 psi, which is the most common. The latter is for low pressure drop heaters. When the heater duty must be increased to handle duty much larger than design, the existing heater may be insufficient in meeting the limits. Installing a new heater could be very expensive.

The most effective way to avoid this is to increase feed preheating via process heat recovery, adding a feed preheater or adding tube surface area into the heater or installing new one.

For separation/fractionation columns, the major limitation is tray loadings. Typically, the column is designed with 80–85% jet flooding. In operation, the column can run harder and flooding limit can be relaxed to 90% or even 95%. It is possible to push extra throughput at the column flooding limit if product cut point specifications can be relaxed. If higher throughput is desired in revamp, more efficient and higher capacity trays could be used to replace the old ones partially or completely. Although it is costly, it is cheaper than installing a new column.

For compressors, when it reaches the capacity limit in flow or head, the direct solution is reducing the gas flow. The alternative solutions include increasing speed by gear change or adding wheels to rotor or adding a booster compressor. Similarly for pumps, the major constraint is insufficient capacity in flow or head. The low-cost solution is replacing impeller with a larger size. The alternative solution is to add a booster pump.

For heat exchangers, the constraint is insufficient surface area. In resolving this constraint, adding surface area via more tube counts or tube enhancement to the existing exchanger could help; but area addition could only be up to typically 20%. Beyond this, adding a new shell in parallel may be required, which could reduce ΔP, but it can cause problems in flow distribution.

The above constraints occur in the ISBL (Internal System Battery Limit). However, it is also important to identify OSBL (Outside System Battery Limit) limitations such as offsite utility, layout, piping, and substations. For example, a revamp project requires additional HP steam for process use, but existing steam system may reach the boiler capacity limit of HP steam generation. Installing a new boiler could kill the economics of a revamp project. Another revamp case could require installing a new exchanger, which is well justified from ISBL conditions; but the piping to connect two process streams in the exchanger is too long in distance, which becomes cost prohibitive. A recent revamp project determined the great benefit of installing a new motor of 5 MW in replacement of the steam turbine to run recycle gas compressor in a reforming unit. The benefit of this project was due to the fact that the turbine has a low power generation efficiency and the electricity price is very low. But the project was deemed infeasible because the electricity requirement of the new motor is beyond the capacity of the nearby substation. Installing a new substation could cost multiple million dollars. Numerous ISBL and OSBL limitations could occur, which are not listed here in detail.

18.3 ENERGY OPTIMIZATION FOR DISTILLATION SYSTEM

Operation of separation columns involves a trade-off between energy use and product recovery or purity. When energy usage is lower than the requirement, product recovery and purity suffer. On the other hand, when energy usage is more than the requirement, product purity and yields are improved. The optimal trade-off determines the operating target. The operating parameters involved in the trade-off include reflux ratio, feed temperature, column temperature, reboiling duty, column pressure,

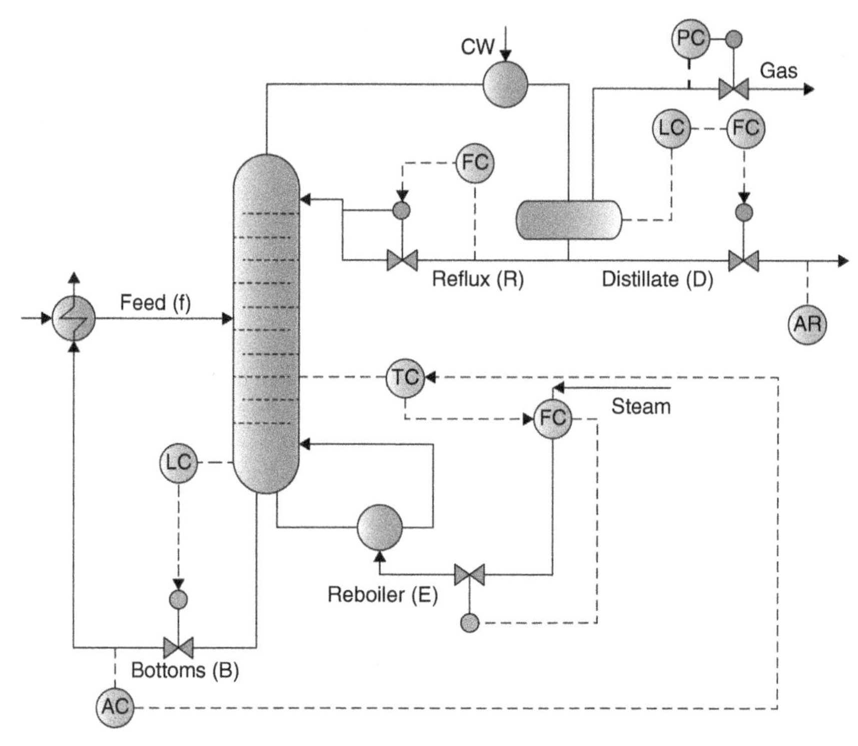

FIGURE 18.1. Debutanizer example: energy optimization based on reflux ratio. Source: White (2012). Reproduced with permission of AIChE.

and so on. The questions are: how to obtain the energy target for a fractionation column and how to achieve the most economic operation of the column?

Let us look at an example that consists of a debutanizer column (Figure 18.1) for which White (2012) gave excellent explanations. This example gives perspective and guidelines in principles for optimizing a separation column, and this example is reproduced here with permission from AIChE.

The feed and product specifications and prices for this example are listed in Table 18.1. Both products have tiered prices: on-specification product is priced much higher than the one that does not achieve specification. If the top product, butane, achieves the specification, that is, less than 3% C_5, it is sent to downstream unit for further processing leading to eventual sales. Off-spec butane will be used as fuel, which has a low value than selling as a product. Similarly, the bottom product, pentane, if achieving specification, is used for making high-value products. Otherwise, it will be sent to tank for reprocessing. In operation, the operator changes column temperature and reflux to feed ratio to achieve the product specification.

18.3.1 Develop Economic Value Function

The objective function (economic value) representing the economic operating margin for the column is shown in Figure 18.2, which is defined as the difference of the

TABLE 18.1. Product Specifications and Prices

Stream	Composition/ Specification	Value
Feed, 20,000 bbl/d	25% C_3	\$60/bbl
	25% nC_4	
	25% nC_5	
	25% nC_6	
Bottoms Product = C_5	$\leq$5% C_4	\$80/bbl
	>5% C_4	\$60/bbl
Top Product = C_4	$\leq$3% C_5	\$60/bbl
	>3% C_5	\$40/bbl
Steam		\$15/MBtu

Source: White (2012). Reproduced with permission of AIChE.

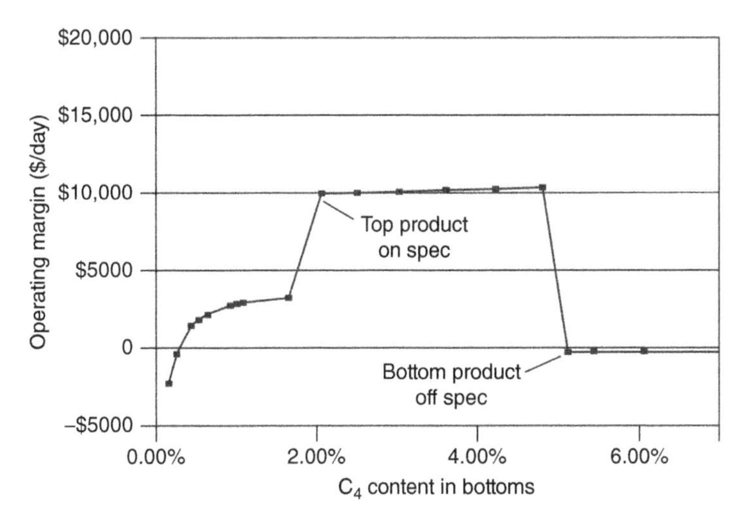

FIGURE 18.2. Operating margin as a function of the bottom composition. Source: White (2012). Reproduced with permission of AIChE.

value of products (top butane product and pentane bottom pentane product) and costs of feed and energy. The value function features two discontinuities. The first, which occurs when the composition of the bottom product is about 1% butane, corresponds to a change in the top product from off-spec to on-spec. The second discontinuity occurs when the bottom product becomes off-spec at 5% butane.

18.3.2 Setting Operating Targets with Column Bottom Temperature

To choose the bottom temperature target, first assume that the reflux rate is fixed and that the bottom product is on-spec but the top product is off-spec because of its high pentane content. This would correspond to a very hot bottom temperature. When the bottom temperature is slowly reduced, the amount of bottom product increases, but the percentage of butane in the bottom also increases simultaneously. As the amount

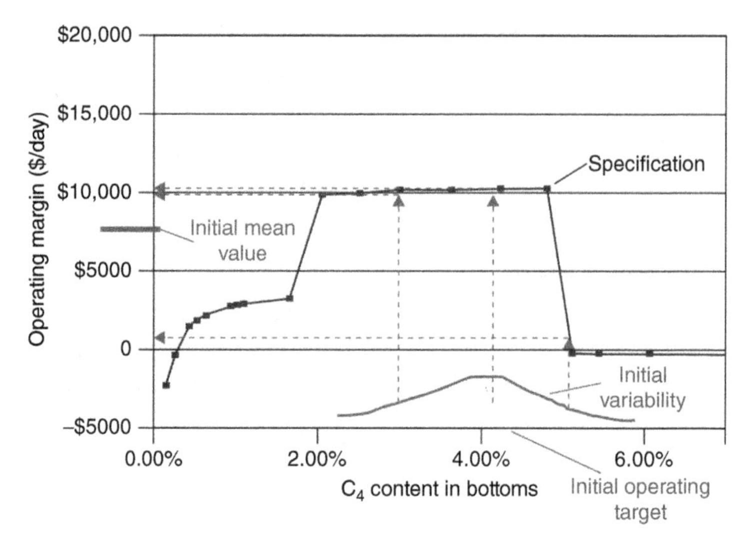

FIGURE 18.3. Observed composition normal distribution versus operating margin. Source: White (2012). Reproduced with permission of AIChE.

of pentane product increases, the total product value improves. The middle line in Figure 18.2 represents the above-mentioned operation.

Normally one would select a temperature target such that the bottom composition is as close to the specification limit as possible. There will always be some variability in the control performance due to external disturbances and limitations on loop control action. If composition control is poor and has a high variance, the observed composition probability distribution could look like a normal distribution in relation to operating margin as shown in Figure 18.3. A more detailed explanation for using normal distribution to represent operating data can be found in Chapter 17.

The product composition target is the mean value of observed composition distribution as shown in Figure 18.3. The mean value of the operating margin in Figure 18.3 is calculated based on the weighted average composition of the observed distribution – that is, the percentage at each composition is multiplied by the margin value at that composition to determine the overall value.

Figure 18.3 shows that part of the column operation is the bottom product being off-spec. The mean product value does not correspond to the value at the mean of product composition, which is also the operating target. This is because of the non-symmetrical nature of the operating margin and low value of off-spec products.

After reducing the variability through improved control-valve performance and reduced measurement error, the new mean value of the operating margin increases at the same operating target or bottom composition target (Figure 18.4). It can be seen that reduced variability results in increase in the mean value of operating margin.

18.3.3 Setting Operating Targets with Column Reflux Ratio

The above discussions involved constant reflux ratio. Next, consider the situation where the reflux rate is varied and bottom temperature is constant. Fundamentally,

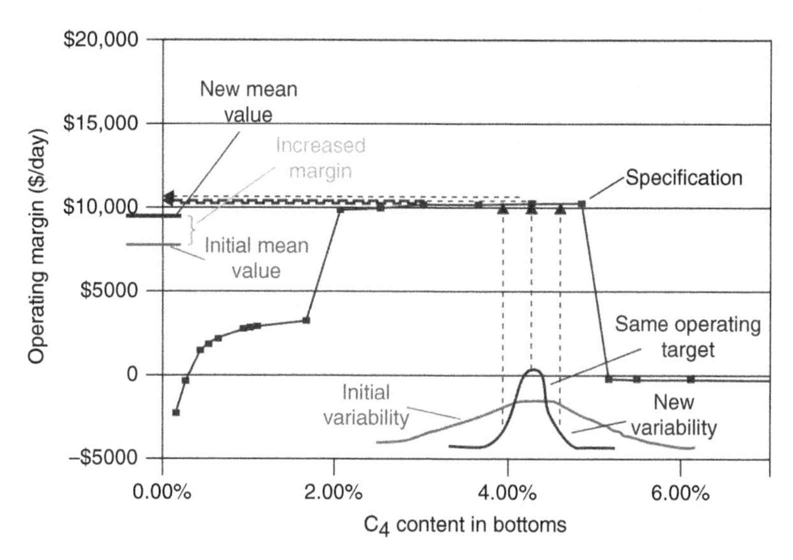

FIGURE 18.4. Improved composition normal distribution versus operating margin. Source: White (2012). Reproduced with permission of AIChE.

reflux rate provides internal reflux needed for separation in the tower and it is generated by either feed heater or reboiler. Thus, lower reflux rate saves energy, but too low reflux rate could affect product quality. On the other hand, high reflux rate could improve production of more valued products.

In tower design, the reflux ratio is determined based on the trade-off between operating cost in reboiler and capital cost for the tower. In other words, use of more separation stages requires less reflux rate and thus less reboiling energy but at the expense of additional capital cost. The minimum reflux ratio is calculated based on Underwood (1948). A tower requires an infinite number of stages to achieve the minimum reflux ratio. To make separation feasible and at the same time a tower affordable, a reflux ratio larger than the minimum is used. Typical design reflux ratios are 1.1–1.3 of the minimum reflux ratio. With a high reflux ratio, the number of theoretical stages is lower, resulting in lower capital cost for a tower but at the expense of higher reboiler duty; and vice versa. The optimal reflux ratio in tower design is determined based on the minimum total cost.

However, in operation, optimal reflux ratio is determined based on the trade-off product value and energy cost. When the reflux ratio increases, the separation improves at the expense of increased reboiling duty (Figure 18.5). As a result, top product rate decreases while the bottom product rate increases. As shown in Figure 18.5, the cost of reboiling duty presents a linear relationship with reflux ratio, but the product rate is nonlinear and presents a different trend as reboiling duty.

Figure 18.6 shows the operating margin for different energy prices, assuming constant product prices. The optimum reflux rate depends on the price of energy. At a high energy price, the optimum reflux rate is at the minimum value, which allows the column to maintain the top product in specification. At the lower energy prices, the optimum reflux rate increases.

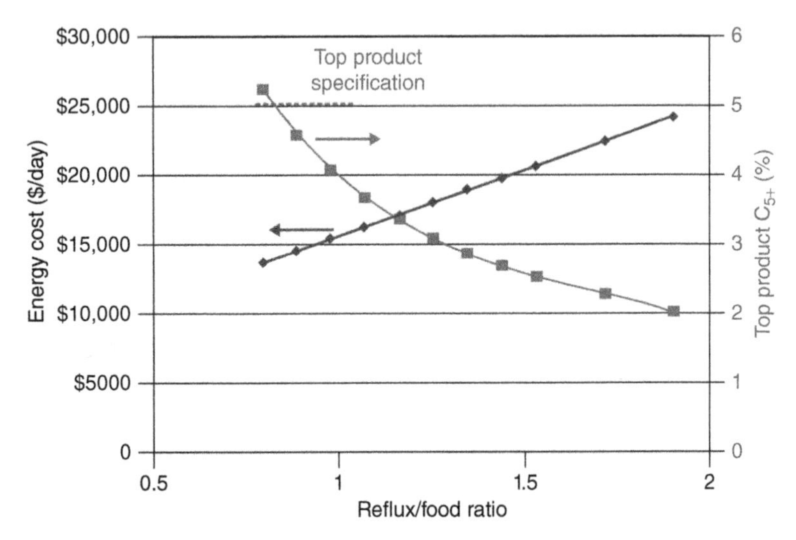

FIGURE 18.5. Energy-separation trade-off: energy cost increases linearly as reflux rate while the top product quality improves. Source: White (2012). Reproduced with permission of AIChE.

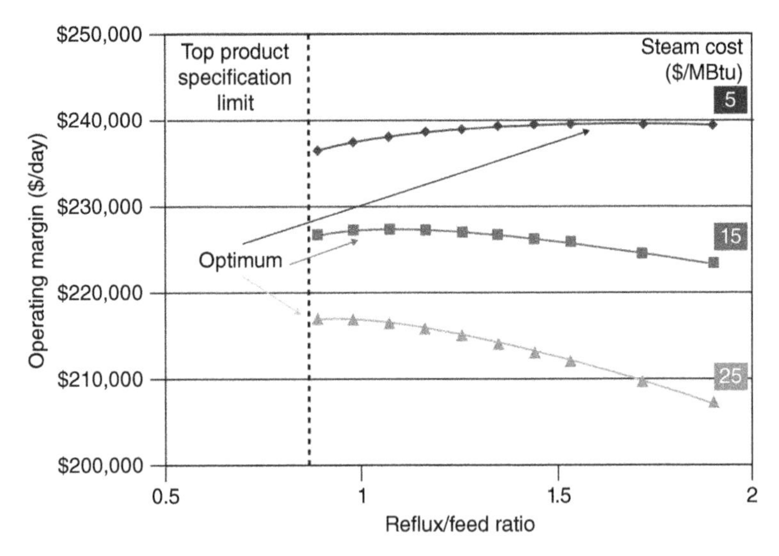

FIGURE 18.6. Optimum reflux rate depends on energy price. Source: White (2012). Reproduced with permission of AIChE.

The conclusion is that operating targets should be a function of energy costs rather than a fixed number even with fixed composition limits. It is common to observe separation columns operating at reflux rate that are 50% higher than the optimum. For the debutanizer column operation discussed here, such an operation could cost operating margin in excess of $500,000 per annum.

18.3.4 Setting Operating Pressure

It is generally known that reducing the operating pressure of separation columns reduces energy consumption. This is because the lower the tower pressure, the less heat required for liquid to vaporize (thus less energy required) and the easier for vapor to penetrate into liquid on the tray deck (thus better separation). Yet, many columns are operated well above their potential minimum pressure. One may ask: if benefit of reducing pressure is well known, why is it not widely implemented? There appear to be three primary reasons for this.

First, changing column pressure requires simultaneously changing the bottom temperature set point to hold the product composition at their targets. This is difficult to do manually – advance composition control is required.

Second, changes in column pressure have other impacts such as changes in the offgas rate, the amount of reboiler duty, and hydraulic profile of the plant. In the case of partial condensation, pressure control can interact with the overhead receiver level. While these effects are real, their magnitudes are sometimes exaggerated and cited as reasons for not making any changes.

Finally, plant personnel frequently do not agree on the amount of operating margin required to handle major disturbances. For instance, questions often arise about the dynamic response of an air-cooled condenser to a rainstorm and the ability of the overall control system to handle such conditions. A well-designed overall control system for the column can compensate for such disturbances.

The condenser pressure controls the tower pressure and thus the feed tray pressure. There is a pressure valve in the overhead that can be used to control tower pressure. The lower limit of the tower pressure is column overhead condensing duty, net gas compressor capacity, and column flood condition. Many of the new APC systems are using pressure control to save energy. During extended turndown periods, reducing pressure up against an equipment limit can improve efficiency (Figure 18.7).

18.4 OVERALL PROCESS OPTIMIZATION

This example comes from Loe and Pults (2001) and is reproduced, with permission from AIChE, for the purpose of explaining how a tower could be optimized.

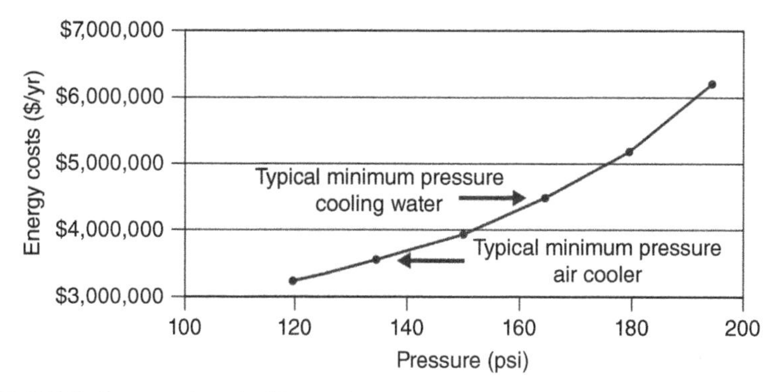

FIGURE 18.7. Pressure has significant effect on energy cost. Source: White (2012). Reproduced with permission of AIChE.

This example provides a case history of how operational improvements for a single deisopentanizer fractionation tower are identified, implemented, and sustained. To do so, the current operation is simulated and assessed. Then improvement opportunities are identified and the limiting factors are determined. Optimal solutions are obtained by optimizing tower ISBL (Inside System Battery Limit) conditions and OSBL conditions. The improvements on this single tower have generated over $500,000 compared with historical operation over the first 6 months.

18.4.1 Basis

The Deisopentanizer (DIP) tower, shown in Figure 18.8, processes light straight-run naphtha from two sources. The LPG fractionation unit debutanizer bottoms consists of primarily iso and normal pentane (iC_5 and nC_5), with small fractions of butane and C_6+ components. The overhead from the naphtha fractionator tower contains mostly C_5 and C_6 paraffin compounds, with some benzene and C_6 naphthenes and a small amount of butane. The combined feed to the DIP is typically in the range of 9000–15,000 bbl/day.

The DIP overhead product is normally rich in isopentane and is routed to gasoline blending along with other high-octane, low Reid Vapor Pressure (RVP) gasoline components. The DIP bottoms, which is rich in nC_5 and C_6 paraffins, is routed to the light naphtha isomerization unit, along with light raffinate from the aromatics extraction unit.

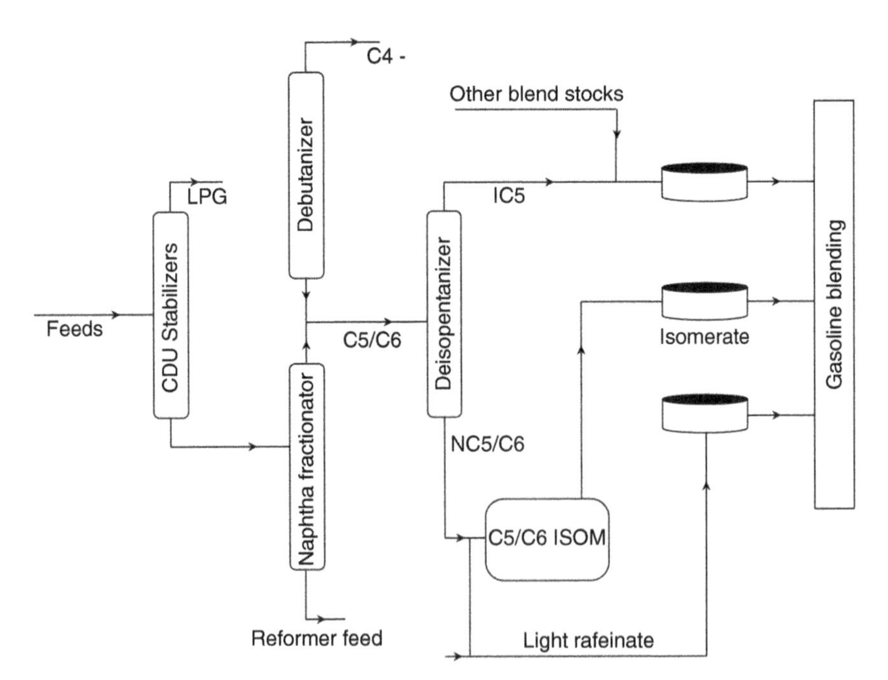

FIGURE 18.8. Deisopentanizer flow scheme. Source: Loe (2001). Reproduced with permission of AIChE.

The DIP tower has 50 trays in comparison with 70+ trays used in a typical deisopentanizer. As a consequence, the DIP tower often has a difficult time making a good split between isopentane and normal pentane components.

18.4.2 Current Operation Assessment

Historically, the tower had been operated with a target of 10% nC_5 in the overhead and 20–30% iC_5 in the bottom product. The tower was reported to be limited by reboiler or condenser duty. One of the two steam reboilers had been out of service for some time, and 20–30% of the condenser fin fan motors were not operating and in need of repair. The DIP equipment had not been a maintenance priority, in part because no economic penalty had been calculated for having a reboiler or condenser out of service. There was also a concern that the tower could flood if both reboilers were placed in service.

The DIP process control was accomplished with a DCS system equipped with an advanced process control (APC) algorithm. The controller was set to target 10% nC_5 in the overhead and 10% iC_5 in the product and would increase reboiler steam and reflux rate until reaching the maximum limits for these flows. The tower pressure was also controlled within a specified range by the APC, and this could indirectly limit the reboiler duty as well, if the tower pressure increased beyond its set maximum. Inferential estimates for product iC_5 and nC_5 qualities were calculated based on tower temperatures and pressures, and a bias for these values was continually updated based on daily lab data.

To establish the historic performance, the operating and laboratory data were collected from the past year, eliminating periods of known equipment failure or poor unit volume balance. The data showed an average of 19% nC_5 in the DIP overhead, and 27% iC_5 in the bottoms product, with a wide variation in the product quality, as shown in Figure 18.9. An average of 18% of normal butane was observed in the overhead product, indicating poor debutanization in the upstream fractionation towers.

The process design for the tower showed an available reboiler duty of 51,000 pounds per hour of steam, but the average for the data collected showed only 27,000 pounds per hour. The data shows an erratic variation in reboiler duty.

18.4.3 Simulation

Using the averaged process and lab data, a simulation model for the DIP was developed. The feed rate and composition to the tower were fixed, as well as the reflux rate, tower pressures, and overhead rate. The model results for reboiler duty, tower temperatures, and compositions compared favorably with the unit operating data, as shown in Table 18.2.

The calibrated model was used to simulate DIP performance at the design reboiler duty, to determine if available condenser duty and tower tray capacity would be adequate for this operation. As shown in Table 18.3, the predicted condenser duty for this operation was only slightly above the design value, and tower tray parameters indicated that flooding was unlikely. Also, the separation of iC_5 and nC_5 improved

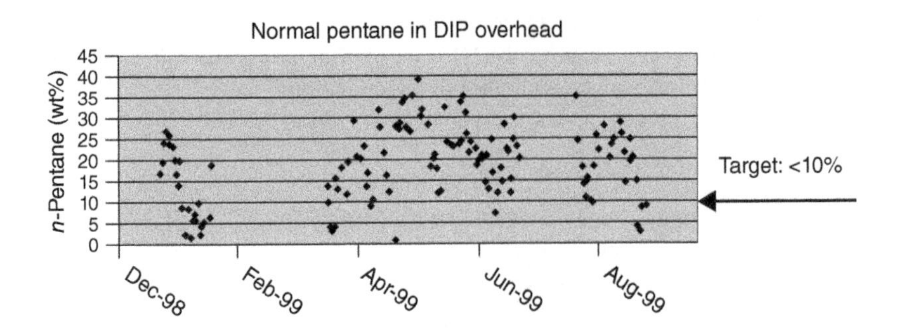

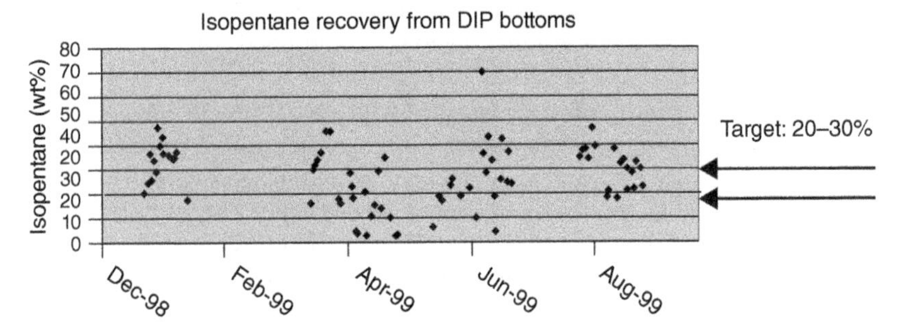

FIGURE 18.9. Variation of DIP performance. Source: Loe (2001). Reproduced with permission of AIChE.

TABLE 18.2. Simulation Results Versus DIP Operating Data

Result	Data	Model
Reboiler duty (MMBtu/hr)	23.6	26.7
Top temperature (F)	143	153
Bottom temperature (F)	179	183
NC_5 in overhead (wt%)	18%	16%
IC_5 in bottoms (wt%)	31%	30%

Source: Loe 2001. Reproduced with permission of AIChE.

TABLE 18.3. Simulation Results at Designed Reboiler Duty

Result	Design	Model
Reboiler duty (MMBtu/hr)	48.8	48.8
Condenser duty (MMBtu/hr)	45.0	48.1
Max jet flood	85%	74%
Max downcomer backup	50%	37%
NC_5 in overhead (wt%)	NA	5%
IC_5 in bottoms (wt%)	NA	18%

Source: Loe (2001). Reproduced with permission of AIChE.

dramatically versus the historical operation as would be expected with the increased tower traffic.

18.4.4 Define the Objective Function

The profitability of the DIP column was determined based on the value of separating iC_5 for direct blending to gasoline and nC_5 to be used as feed for the C_5/C_6 isomerization unit, less the utility and downstream unit opportunity costs incurred to do so. Lighter feed components, such as *n*-butane, were assumed to always be fractionated into the DIP overhead, and components heavier than nC_5 were assumed to always be found in the DIP bottoms stream. Thus, only the disposition of iC_5 and nC_5 components was considered in the profitability calculation.

Therefore, the objective function for optimizing the DIP tower is defined as follows:

$$\text{DIP Upgrade Value} = \text{Overhead Value} + \text{Bottoms Value–Feed Value}$$
$$\text{–Reboiler Steam Cost–Isom Operating Cost–Isom Capacity Penalty.}$$

The DIP feed and overhead values were calculated as the gasoline blending value of the iC_5 and nC_5 in this stream, with corrections for road octane and RVP of these components versus those of conventional regular gasoline. The DIP bottoms stream was normally processed at the isomerization unit, where 75% of the exiting C_5's were assumed to be iC_5. After this equilibrium conversion, the value of the resulting iC_5/nC_5 stream was calculated at gasoline blending value as described for feed and overhead above. The reboiler steam cost was calculated assuming a 70% generation efficiency from refinery fuel gas, and the isomerization unit operating cost (for fuel, power, and catalyst) was taken to be the same value per barrel as used in the refinery planning model.

During some periods, the refinery isomerization unit had more feed available than can be processed. If additional DIP bottoms was produced, less capacity was available to process light raffinate from the aromatics extraction unit. The Isom capacity penalty, or the opportunity cost for processing additional DIP tower bottoms at this unit, was therefore estimated by evaluating the octane upgrade of light raffinate.

18.4.5 Off-Line Optimization Results

Once the economic evaluation criterion was determined, it was implemented in the simulation model. Numerous case studies were conducted via process simulation to determine the optimum operating point for the DIP tower under different scenarios. It quickly became apparent that in nearly all economic and operating situations, maximizing the DIP reboiler duty up to the maximum limit gave the highest profitability.

For subsequent case studies, the simulation was completed with maximum reboiler duty, and tower pressure and nC_5 content of the overhead product were also fixed. These constraints completely specified the tower operating conditions.

The DIP profitability was first examined for scenarios where the isomerization unit has available capacity. The DIP feed rate and overhead nC_5 content were varied, and profitability calculated, as shown in Figure 18.10. This analysis showed that

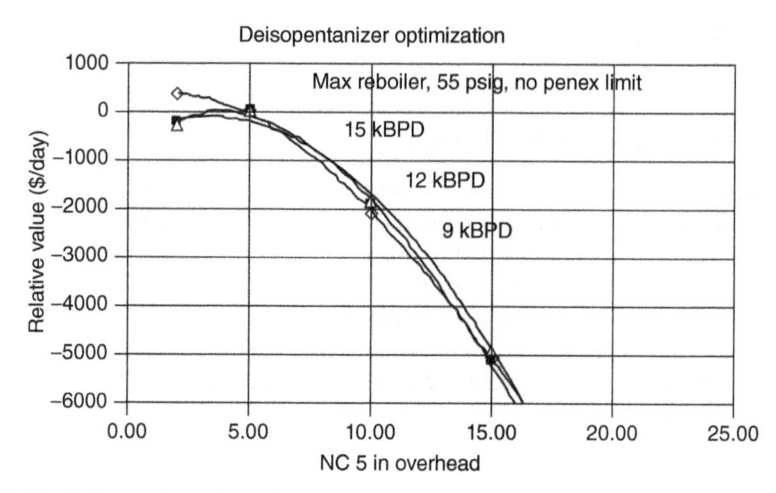

FIGURE 18.10. Optimization without Isom capacity constraint. Source: Loe (2001). Reproduced with permission of AIChE.

the optimum target was around 5% of nC_5 in the overhead, regardless of the tower feed rate.

Profitability was then examined assuming the isomerization unit was at its maximum charge rate, and additional production of DIP bottoms would result in bypassing of light raffinate around the Isom, direct to gasoline blending. The cost of losing the light raffinate octane upgrade can vary between $2 and $5 per barrel, and so simulation cases were completed for both of these scenarios as shown in Figure 18.11. In these cases, the optimum nC_5 in DIP overhead target is dependent on the charge rate to the tower. At low charge rates, the available reboiler duty is sufficient to obtain good separation between iC_5 and nC_5 components, so that minimal iC_5 is lost into the DIP bottoms when targeting 5% nC_5 in the overhead. At higher charge rates, more iC_5 is lost to the bottoms stream, and it is more profitable to increase the overhead nC_5 target, reducing the DIP bottoms rate to the Isom unit and allowing additional raffinate upgrading. Thus, the optimal nC_5 in overhead target varies between 5% and 20%, depending on DIP charge rate and the value of light raffinate upgrading.

Based on a comparison of the optimal tower operation as determined above, and the historical performance, an incentive of around $1.5 million per year was identified to improve DIP fractionation.

18.4.6 Optimization Implementation

In order to realize the benefit indicated by the optimization results above, several unit hardware, process control, and operating philosophy changes were needed. First, it was clear that both reboilers would be required, so the spare bundle was leak-tested and returned to service by operations. Several fin fan motors were also quickly repaired to ensure that design condenser duty was available and overpressurization of the tower would not limit the reboiler duty that could be applied.

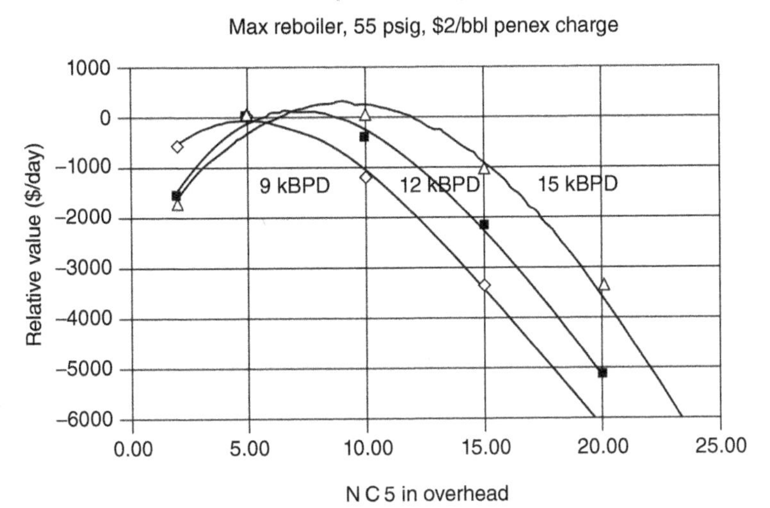

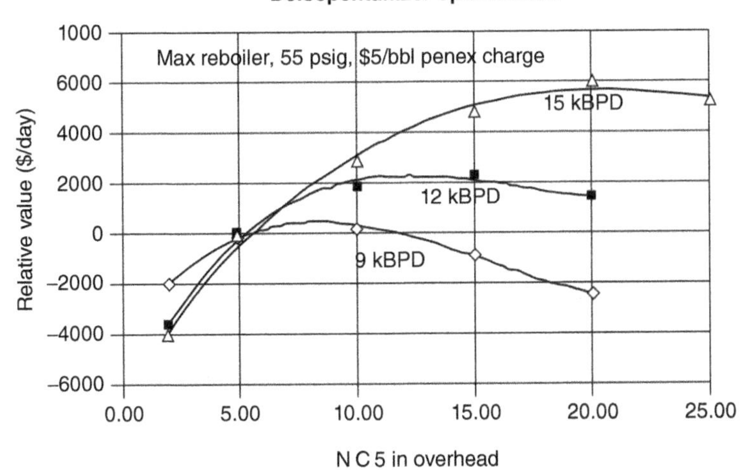

FIGURE 18.11. Optimization with Isom capacity constraint. Source: Loe (2001). Reproduced with permission of AIChE.

Second, the optimum target for the DIP overhead nC_5 was implemented into APC system. The APC controller on the DIP DCS system was reconfigured to operate at this nC_5 target, while maximizing the reboiler duty as limited by the high limit on tower pressure. This APC system allowed the DIP operation to be maintained at an economic optimum, accounting for the isomerization unit capacity and the economics of the day.

Third, communication of the new operating philosophy was also critical in improving DIP performance. Operators and unit supervisors were trained on the importance of always maximizing the reboiler duty and setting the nC_5 in overhead target. The iC_5 content of the bottoms stream was still measured by lab and inferred analysis, but this was no longer a tower control variable. The reboiler duty and overhead nC_5 were tracked on a daily basis, and performance for these Key Performance Indicators (KPIs) were discussed at weekly operations and planning meetings.

18.4.7 Online Optimization Results

The economic benefit of improving DIP fractionation was tracked on a monthly average basis shortly after implementation of the new optimization strategy. Economic performance versus the baseline operation is shown in Figure 18.12, using actual monthly averaged economics and unit operating and lab data. Monthly benefits of over $100,000 were achieved in several cases during the summer months, when octane values were at their highest level. As octane values dipped during spring and fall months, benefits from the improved DIP fractionation dropped off as well.

A significant drop in benefits can be seen for July. This was due to poor operation of the DIP tower, caused by a high butane content in the tower feed from the crude unit stabilizers. The C_4's caused the DIP tower pressure to increase up to its safe operating limit, and the reboiler duty was cut back to avoid overpressuring the tower. This resulted in a reduction in fractionation efficiency and profitability during part of July and represented one of the challenges encountered in sustaining the improved DIP performance.

18.4.8 Sustaining Benefits

Without ongoing attention, optimization improvements from initiatives such as that on the Deisopentanizer tower tend to fade over time, for a multitude of reasons. Challenges to the new level of performance must be tackled as they arise, whether they

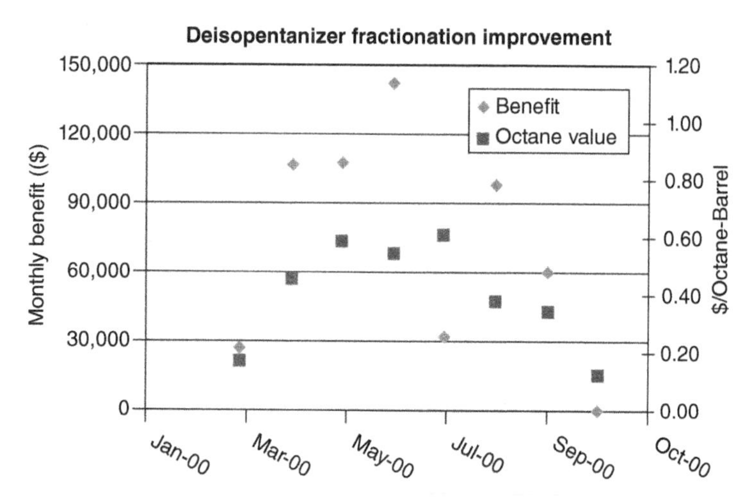

FIGURE 18.12. DIP economic improvements. Source: Loe 2001. Reproduced with permission of AIChE.

result from hardware or control problems, misunderstandings, or operating changes in other parts of the refinery. Tracking the economic benefits of the initiative is a critical element of sustaining the change, since knowing the lost profits associated with a loss in performance helps to set work priorities within the refinery. Several problems occurred during the first 6 months of improved DIP operation, which reduced the profit derived from this tower, and these issues were quickly addressed to sustain the improvement.

During the summer months, a change in routing of a portion of the refinery condensate resulted in a hydraulic constraint on the amount of condensate that could be removed from the DIP reboilers into the condensate header. This caused reboiler duty reduction leading to fractionation efficiency drop. The economic calculations clearly showed that the benefit derived from the additional reboiler duty was much higher than the value of recovering the condensate. For this reason, condensate was safely spilled into the sewer until normal condensate header operation was restored, and DIP fractionation was maintained.

A second challenge occurred when it was noticed that the DIP reflux drum temperature had increased above 140 °F, which was higher than the recommended rundown limit to tankage. Initially, reboiler duty was decreased to ensure safe operation. However, in meeting with the tank farm operators about the problem, it was found that the DIP overhead mixed with several other much larger streams before entering a gasoline blending tank. Calculations showed that the effect of the higher DIP rundown temperature on the tank temperature was minimal, and DIP reboiler duty was again increased.

Another problem in maintaining reboiler duty was identified when the DIP tower pressure increased due to butanes in the feed, as mentioned above. It was found, however, that the maximum tower pressure was set well below the vessel design pressure, and so the safe operating limit of the tower could be increased after an appropriate Management of Change (MOC) review. Simulation modeling showed that operation at the higher pressure did not significantly impact the tower profitability if the reboiler duty could be maintained near the maximum. The pressure limit was increased and performance of the tower again improved.

A fourth reduction in DIP profitability was noticed when the APC controller began cutting reboiler duty, for no apparent reason. Further investigation showed that the inferred nC_5 content of the overhead stream had been deviating from the lab value for a few days, because of a computer glitch. This was quickly corrected and DIP operation returned to normal.

Although all of these obstacles reduced profitability for a short period, timely identification and resolution of the constraints averted potentially long periods of underperformance.

18.5 CONCLUDING REMARKS

The profit improvement process should pass through several phases. First, current performance must be assessed, evaluating upstream and downstream constraints and unit equipment limitations. Understanding of how the unit operation can be optimized can then be gained by use of an appropriate process model and refinery economics, and a new operating strategy is then developed to improve profitability. Implementation of this strategy requires good communication of its benefits throughout the plant. Upgrading or repair of process equipment may be needed to allow operation under desired conditions. Tracking of key operating parameters as well as economic performance of the unit and distributing these results within the plant is essential to sustaining the improvement, as this process flags deterioration of profitability. The profit improvement process requires input from technical, operations, safety, environmental, and economics groups within the plant.

REFERENCES

Loe B, Pults J (2001) Implementing and sustaining process optimization improvements on a de-isopentanizer tower, AIChE Spring Meeting, April, Houston.

White DC (2012) Optimizing energy use in distillation, *Chemical Engineering Process (CEP)*, pp. 35–41 AIChE, March

Underwood AJV (1948) Fractional distillation of multi-component mixtures, *Chemical Engineering Progress (CEP)*, **44**, 603.

PART 6

OPERATIONAL GUIDELINES AND TROUBLESHOOTING

19

COMMON OPERATING ISSUES

19.1 INTRODUCTION

Hydroprocessing units in the refinery are designed to upgrade different boiling range feedstocks derived from crude oil having various contaminants under catalytic and hydrogen-rich high-pressure environment. Successful and reliable hydrotreating and hydrocracking operation can add substantial upgrading values to refinery economics. These units are key to producing clean fuels meeting environmental regulations or providing downstream petrochemical feedstock. An unplanned shutdown due to operating issues is undesirable that can create stressful situations affecting fuel supplies to consumers and opportunity loss affecting corporate profits.

Successful implementation of hydroprocessing projects starts with careful licensor selection, then strong project execution with quality design, construction, and commissioning. It is especially critical at the inception of hydrocracking projects to choose the optimal process conditions with matching catalyst combinations in order to achieve the project goals and meeting market flexibility goals due to feedstock and product demand variations.

Operating personnel will require extensive training and repeated refresher courses to keep their practices sharp. An experienced hydroprocessing engineer or unit operator should have a clear grasp of the reaction chemistry fundamentals. A good understanding of the design and metallurgical selection practices and control philosophy is also critical. If the unit undergoes a HAZOP analysis, the operating staff should participate. Furthermore, appreciation of the unit economic potential coupled with a good understanding of the unit's integration in the refinery allow

Hydroprocessing for Clean Energy: Design, Operation, and Optimization, First Edition.
Frank (Xin X.) Zhu, Richard Hoehn, Vasant Thakkar, and Edwin Yuh.

the operating staff to use integrative thinking behavior to make well-considered operating decision.

Detailed step-by-step startup, shutdown, normal operating, and emergency procedures should be drafted. Some units include very specific operating cards covering each valve position and line-up. These procedures should be reviewed at regular periods and continuously improved for clarity. Use management-of-change methodology to document revision reasons. Once a procedure is agreed upon, the operating staff should follow without any deviation with full empowerment by management of their mitigation actions. The emergency procedure should be readily available and checklists created for quick reference at the time of emergency in the control room to avoid confusions.

Before startup, the unit should pass through a thorough checkout, punch listing, and corrections to prevent future shutdown due to errors. Deviations are assigned severity levels, with highest level items being those affecting safety and ability to meet performance, and should be corrected before startup never accepting shortcuts.

Follow licensor's specifications and recommendations to install key equipment such as, for example, reactor internals. Incorrect reactor internal installation will cause extensive shutdown to unload and reload. Incorrect distillation column tray installation can affect product separation and prevent the unit from meeting specifications.

During startup, leak check rigorously and hot bolt unit carefully until meeting specifications. History has taught us that a simple oversight could lead to catastrophic failures creating unsafe conditions resulting in equipment damages that could lead to injuries and casualties.

Careful and thoroughly planning catalyst loading tasks including selecting qualified loading contractors for the job. Work with catalyst supplier or licensor's engineers to ensure the good loading to achieve best loading density. Catalyst support materials are important too and should be purchased from reputable suppliers with proper specification to avoid operating problems. Record loading observations, loading speed, and any loading adjustments in detail in a loading report including planned and actual loading diagrams with every loading dimension. Since catalyst loading in infrequent, the more details recorded in a report, the more useful the report will be as a good reference for future reload operations.

During normal operation, attentive monitoring and anticipating abnormal conditions allow effective troubleshooting and resolution. Operating data and regular lab data are imperative to help engineer analyze unit status and project future operation plan meeting the production objectives. Without such data and regular performance assessment with licensor or catalyst supplier, it will be difficult to optimize the unit performance.

A combination of the aforementioned mentioned practices will certainly make hydroprocessing unit more reliable and safe. However, sometimes problems are unavoidable. This section discusses various common issues facing hydrocracking operation.

19.2 CATALYST ACTIVATION PROBLEMS

Proper startup is to ensure best catalyst activity following licensor's and catalyst supplier's procedure. When heating up catalyst beds before sulfiding, be cautious of the heat-up rate to remove water from catalyst and avoid high temperature conditions with prolonged exposure of catalyst to hydrogen in the absence of hydrogen sulfide that may risk reducing catalyst. During sulfiding, conduct sulfiding with the bulk of the sulfiding, for example, >75%, at about 230–250 °C then conduct 290–315 °C high temperature step to finish the sulfiding. The high temperature step will depend on catalyst type and methods of sulfiding, either gas phase or liquid phase.

When transitioning to normal feedstock, operating the unit initially with a less severe and lighter feedstock is preferred. Heavy feedstock laden with contaminants or processing cracked stock should be avoided the first week, if not longer, to prevent catalyst damage that can lead to high catalyst deactivation.

Should the unit be idled when encountering problems during catalyst activation, follow preagreed procedure and parking conditions to cool catalyst bed and protect equipment. Deviation and shortcuts often result in mishaps that compromise long-term catalyst performance.

19.3 FEEDSTOCK VARIATIONS AND CONTAMINANTS

Frequently, catalyst deactivation is often related to more severe feedstock or frequent feed variations. It is rare that a hydrocracking unit will be processing only single feedstock deriving from one crude source. Frequently, the feed is a blend of many streams from different upstream units. Feed having higher sulfur and nitrogen and other contaminants are harder to process. Cracked feedstock derived from Coking Unit and Visbreaking Unit contains fragments from residual cracking. Feedstock derived from Solvent Deasphaltene Unit (SDA) can also be a potential feed. These feeds have more trace metals and coke precursors with multiple aromatic rings, leading to increased polynuclear aromatic formations and catalyst coking.

Sometimes, it is unavoidable to process difficult feeds, but good feed quality control in the upstream units can still alleviate downstream processing severity stress. The worst case is unanticipated feed quality change that leads to premature deactivation. A single combined feed analysis is insufficient and frequent individual feed stream analysis is a key to avoid surprises and mitigate the uncertainty.

When poorer quality feed is processed, pretreating catalyst beds should protect the cracking catalyst by maintaining adequate level of denitrification. Some cracking catalysts are more nitrogen tolerant. If pretreat catalyst activity is tight, these nitrogen-tolerant cracking catalysts can pick up the denitrification shortage. However, for long-term operation, both catalysts will suffer and result in higher catalyst deactivation rate.

Many refineries purchase crudes in the spot market based on economics. The crude slate processed can be variable and complex. Crude oil production and oil movement can also introduce extraneous components such as surfactant, which can poison catalysts. The refinery procurement department can proactively work with operating team to mutually optimize operating by selecting appropriate crudes to be processed. Cheapest crude may be profitable but shorter run length and frequent turnaround are the price.

However, controlling feed quality is more easily said than done because the operating engineers often do not know the crude source and tend to take the approach that they will process any crude that the refinery decides to process, thus limiting their ability to work proactively. Treating crude processing as a molecular management with a full understanding of each molecule's impact on performance will certainly eliminate some of the unanticipated feed-related deactivation. If the refinery can afford shorter turnaround frequency, then processing lower grade crudes can provide more value. However, this is an economic optimization that requires participation of refinery planner and the operating engineers.

As demand for ultra-low-sulfur diesel increases, many hydrocracking units are coprocessing diesel boiling range LCO stream from FCC Units. These streams are converted streams lighter from endpoint perspective, but they contain much olefins and multiring aromatics. Depending on the sulfur and nitrogen level, these streams create a different issue, that is, temperature control. Olefins are easily saturated and generate plenty of heat from heat of reaction. Aromatic saturation also generates a large amount of heat. While this stream may have less coking impact compared to heavier streams such as HCGO (Heavy Coker Gas Oil) or HCO (Heavy Cycle Oil), LCO processing nevertheless impacts on the pretreat catalyst performance of the hydrocracking unit. The heat release also can lead to high temperature rises in the pretreat catalyst beds requiring lowering lead bed inlet temperature to distribute the exothermic bed rise. Since quench rate can be limited, catalyst bed temperature cannot be kept at equal peak outlet temperature. When this happens, an ascending temperature profile is maintained leading to faster deactivation of the lower bed catalysts than the upper catalyst beds. The result is the unit may operate, but it is not in optimal conditions.

19.4 OPERATION UPSETS

Operation upsets can be of a variety of reasons. It is difficult to list all the upset possibility, but a few examples are given. It is important to maintain operating pressure and hydrogen purity because hydrogen partial pressure has a significant impact on catalyst deactivation. These variables are normally not an issue since operating pressure should not fluctuate much day-to-day. Sometimes, hydrogen supply can be an issue that requires supplementing with lower grade makeup hydrogen. This lowers recycle gas hydrogen purity and may limit feed processing capacity and result in higher catalyst deactivation.

Maintain a steady recycle gas flow with sufficient compressor capacity. The compressor also supplies emergency quench requirements. Gas-to-oil ratios should be held at design levels to protect catalyst from deactivation. At lower throughput, maintaining higher gas-to-oil ratio is acceptable.

Any operation changes should be adjusted in small increments. Each unit has different time lags that depend on capacity and flow scheme. Changes in operating temperature will generally take a period of time to appear as changes in product rates or product quality. Caution is advised while making changes in reactor temperatures – it is better to make small rather than large moves. During transient operation, it is always safest to reduce severity first. Always put increasing temperature the last when increasing feed rate or increasing operating severity. Vice versa – reduce temperature first before reducing operating severity.

Operation problems obviously are not desirable. When the unit is upset, be prepared to take actions. Common issues such as feed pump trip, while it may seem easily managed, could lead to bigger problems if not handled correctly. Avoiding frequent feed pump stoppage will keep reactor vapor/liquid distribution steady with feed flow maldistribution. When feed is reintroduced for restart, judgment is required to ensure bed temperature has sufficiently decreased for feed reintroduction. When feed is introduced at high temperature, it could lead to temperature excursion that can damage catalyst, resulting in higher deactivation.

Maintain steady makeup compressor operation to hold operating pressure. If the makeup gas compressor experiences an upset resulting in system pressure decrease, lower operating severity to protect catalyst from deactivation operating at lower hydrogen partial pressure. When pressure has dropped too low for sustained operation, feed can be stopped until makeup gas compressor return to normal operation.

Reliable wash water injection to the reactor effluent cooler is a key practice to ensure unit on-stream efficiency. Sustained operation is not possible when wash water is completely lost. Partial loss of wash water would require reducing operating severity in feed rate or conversion because recycle gas ammonia concentration will increase suppressing catalyst activity. While partial loss of wash water does not directly impact on catalyst deactivation, improper handling of wash water practice can lead to not only reactor effluent condenser fouling but also increase in operating severity if reactor temperature is instead increased to compensate for activity loss. The first corrective action is fix water injection as soon as possible. For a short duration partial wash water loss, reactor temperature should be lower as wash water is reduced. Wash water should be reintroduced as soon as possible. Without wash water, the unit will need to be shut down.

Any temperature excursion in the catalyst beds can result in coke formation, which affects flow distribution aggravating catalyst performance. When a coke ball is formed in the catalyst bed, low flow regime with longer residence time can lead to hot spots. If the hot spot is near a bed thermocouple, then the hot spot can be detected and monitored. If not, it is very risky that the unstable hot spot can lead to further temperature excursion. Operating unit at such a risky condition is abnormal. When such a situation exists, the catalyst deactivation will not be as ideal requiring operating adjustments to control the hot spot by quenching the problematic catalyst bed. The catalyst utilization will not be optimal. In most cases, the only way to remove the hot spot is unload and reload the catalyst. Unplanned turnaround is costly that may not be readily feasible due to production requirements. As a result, the unit may continue to suffer operating at lowering severity such as minimum conversion or reducing feed rate. Avoiding upset situations that create the hot spot is the first and best mitigation.

19.5 TREATING/CRACKING CATALYST DEACTIVATION IMBALANCE

When the unit is designed, it is based on a set of design feedstock with certain design objectives. The catalysts are selected to meet these objectives. Once the unit is in operation, every attempt is made to balance the deactivation of pretreat and cracking catalysts to ensure equal deactivation to reach the end of run at the same time. Such long-term operation requires good operating data. Many refineries do not lack operating data but lack laboratory data. Without any operating and laboratory data or with partial data set, it is nearly impossible to monitor catalyst deactivation and optimize operation. Any operating adjustment recommendations under such circumstances are not quantitative but qualitative.

For vigilant operation monitoring, the process engineer should plot data rather than relying on spot data analysis. Track unit operation with plots of corrected flows, stream temperatures and pressures, and lab data. Licensors can also use the data to normalize operating temperatures to a standard reference condition. The normalization basis could be the design conditions, current average operating conditions, or past cycle operating conditions. Normalization eliminates daily operation fluctuations by placing the evaluation basis on a constant basis.

A kinetic adjustment is applied to reactor bed temperatures to derive at normalized reactor temperature and then a deactivation trend is determined as shown in Figure 19.1 for pretreat catalyst from a UOP Unicracking unit.

The reactor temperature is tracked as catalyst ages. The deactivation rate is monitored. Licensors and catalyst suppliers generally provide such assessments periodically in an operating report that include comments on catalyst life projection, production yields, and product quality trends. The operating engineer can use the assessment report with his own judgments to determine if the catalyst deactivation rate will meet production requirements or catalyst activity can be fully utilized.

It is highly recommended to work with the licensor or catalyst supplier to mutually optimize unit performance tapping into their resources in research data or other unit operating references. The licensor should also provide refiners with operating correlations so that the refinery engineer can use these correlation and actual deactivation assessments to project various scenarios to explore the best catalyst utilization and operation case.

For simplicity, the operating correlation curves normally cover one single variable change in feed quality or process variable change showing its effect on temperature adjustment requirement and deactivation impact on each main catalyst. One has to

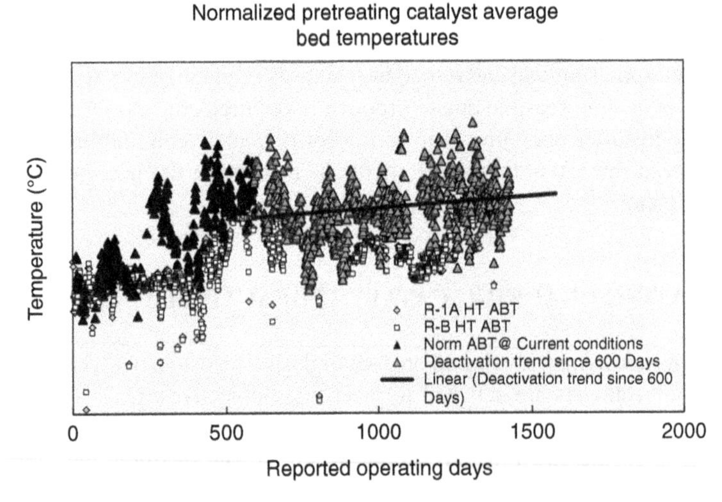

FIGURE 19.1. Example of catalyst NABT and ABT.

apply judgment using these operating correlation curves because the synergy of the operating variables cannot be separated into discrete individual impacts. Nevertheless, these correlation curves do provide a quick way to assess performance impact and project potential operation scenarios.

An application of the correlation curves are shown for balancing nitrogen slip out of pretreat reactor and its impact on cracking catalyst performance in Figures 19.2 and 19.3, respectively. The vertical axis shows temperature requirement and impact on catalyst cycle as nitrogen slip is varied on the horizontal axis.

If a unit is constantly going through large feed and process condition fluctuations, obtaining a consistent data analysis is more challenging even if attempts were made to normalize the data. In this case, if feasible, it is recommended to periodically conduct a unit mini-test run at comparable operating conditions to determine the deactivation rate.

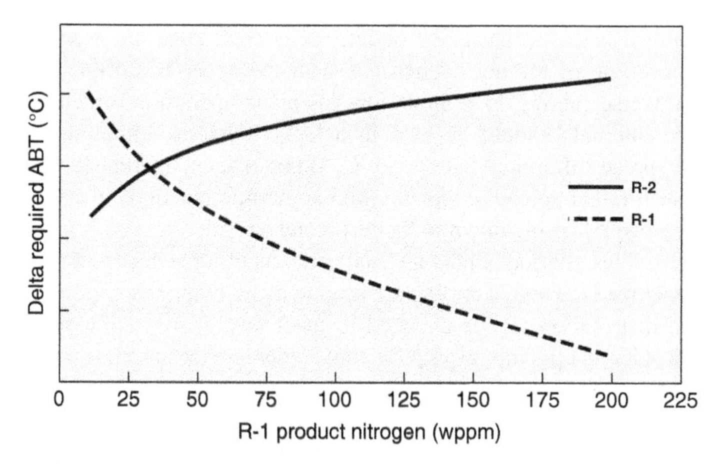

FIGURE 19.2. Pretreat nitrogen slip on pretreat (R1) and cracking (R2) ABT.

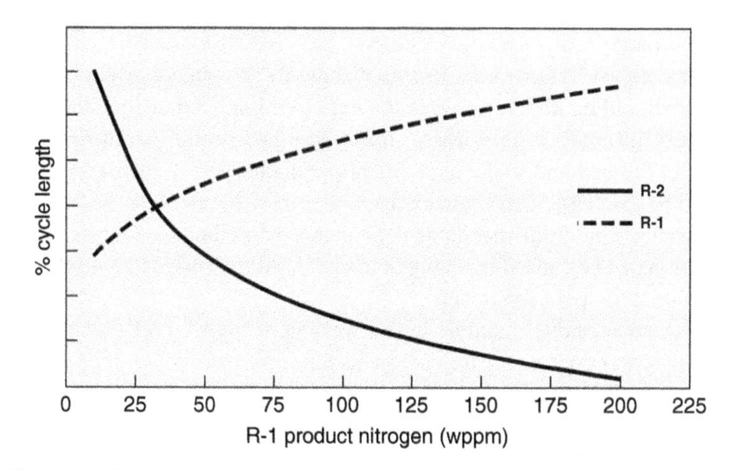

FIGURE 19.3. Effect of pretreat nitrogen slip on pretreat (R1) and cracking (R2) catalyst life.

As catalyst ages, product selectivity can shift. Also, when feed changes, product yield will also change. Using as-produced product yields for yield selectivity assessment can be misleading. More rigorously the mass balanced product yields from each stream should be blended to recut to constant TBP cut points. Such yield trends are monitored along with catalyst activity to gain full understanding of catalyst performance. Some units track operating data very closely that multiple cycle data is available in order that it is used in future catalyst selection. These data can also be benchmark with lab testing data to further optimize product yields. The conclusion is that a properly operating unit relies on quality data collection and frequent assessment of operating data.

19.6 FLOW MALDISTRIBUTION

In hydroprocessing units, mixed gaseous and liquid reactants flow downward through multiple catalyst beds. At each bed outlet, the quench zone has reactor internals serving the purpose of mixing quench gas with the upper bed effluent to achieve two-phase traverse mixing to remove any lateral temperature variance. Effective quench zone internals should be able to achieve a lateral temperature or radial temperature spread difference within ±3 °C. When reactor internals are inadequate to achieve such radial spread, coupled with exothermic reactions, the radial spread can quickly worsen flowing down to the next catalyst bed.

During initial installation, all the internal components are checked against specifications. Proper packing and clamping are used to avoid bypass or catalyst migration. Vapor/liquid distribution tray is checked to meet levelness specification from the licensor. Consider an upgrade revamp for some older reactor internals that are unable to achieve the radial spread target.

The internals can experience fouling during a cycle. Reactor internals should be cleaned during turnaround to ensure they are clean before they are returned to operation. The design of reactor internal has to consider ease of cleaning. Catalyst support grids and outlet collectors can be cleaned by wire brushing and vacuuming. If fouled areas of the profile wire or mesh screen remain, it can create channeling causing poor flow distribution.

Improper catalyst loading can also contribute to flow maldistribution. The catalyst supplier should be able to provide the expected loaded density. During loading, frequent check of catalyst levelness to make sure bed profile is flat not concave or convex. Level the bed and make adjustment parameter and if necessary the catalyst may need to be discharged and reloaded.

Besides reactor internal mechanical or catalyst loading problems, reactor flow maldistribution can be caused by many other different reasons. Many of the problems can be attributed to operation upsets.

During startup, conduct catalyst liquid wetting at high liquid flux to avoid flow channelizing. Never operate the unit below its minimum turndown, if necessary, recycle unconverted oil liquid from the fractionator bottom to ensure good reactor flux. Avoid unit upset especially frequent feed interruptions. Layout sketch of location of the reactor thermocouples at each reactor level should be readily available on the control room panel. These sketches help flow distribution troubleshooting. Reactor bed temperatures, skin temperatures, and inlet and outlet readings are also tracked over the run to monitor the trend of temperature distribution.

19.7 TEMPERATURE EXCURSION

During normal steady operation with constant reactor feed, reactor conversion, and quench control, reactor temperatures should not fluctuate and rise unexpectedly. But unexpected problem situations always exist that emergencies should be recognized and acted upon immediately.

Hydrocracking reaction is exothermic. If the emergency response is not immediate, the temperature in the catalyst beds can increase rapidly called temperature excursion and, in some uncontrollable cases, very high temperature is reached in a temperature runaway.

For understanding of why temperature excursion occurs, think of the Boy Scout fire triangle analogy applied to the hydroprocessing reaction triangle shown in Figure 19.4.

Making a fire requires oxygen, fuel, and an ignition source. For hydroprocessing reaction triangle requires hydrogen, hydrocarbon feed, and catalyst. Breaking the link in the triangle should stop fire or stop reaction. For hydroprocessing, it is not possible to remove catalyst easily, so it should be removing either hydrogen or hydrocarbon. Stopping the exothermic reactions requires quick removal of these reactants.

New hydrocracking units normally have two-rate depressuring systems. Low-rate depressuring can be activated manually or automatically upon recycle gas compressor failure. High-rate depressuring can either be activated manually or automatic upon reactor reaching a preset temperature. In some hydrocracking units, design metallurgical temperature is used to trigger automatic depressuring. The number of thermocouples that should be configured in the automatic depressuring is frequently debated. The most conservative safe guard is every thermocouple around the reactor envelope should be entered into the configuration because a reactor hot spot can be localized and undetected. Hydrocracking unit also has temperature rate change alarm to warn of abnormal condition. The depressuring shutdown interlock should be checked during startup to ensure it functions and never be bypassed during normal operation.

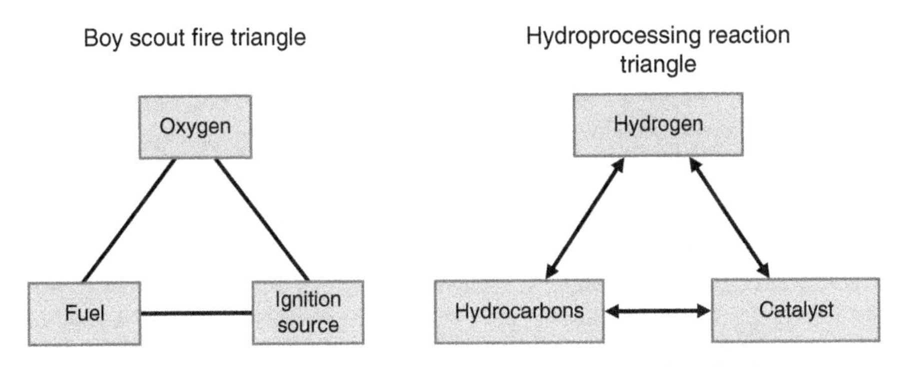

FIGURE 19.4. Boy scouts fire and hydroprocessing reaction triangles.

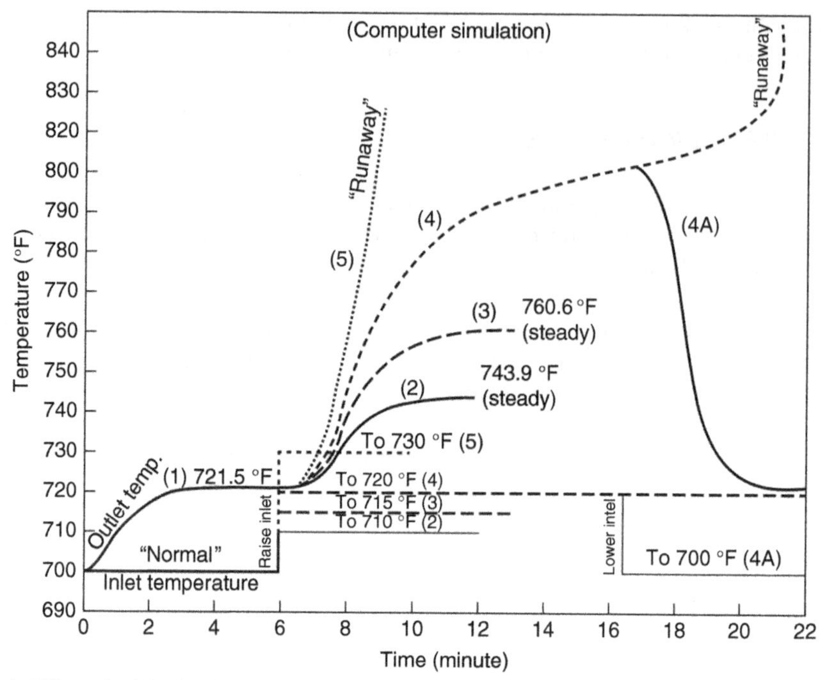

FIGURE 19.5. Temperature excursion simulation.

Figure 19.5 is an example illustrating temperature response across a hydrocracking bed from simulation that starts with a hypothetical "normal" reactor operating condition when inlet temperature is at 700 °F with a bed rise of 21.5 °F and outlet temperature of 721.5 °F.

Curve 2 at 6 min point the bed inlet temperature is increased suddenly by 10 °F that the bed outlet temperature will increase and steady out. Correspondingly, for Curve 3, if the bed inlet temperature is increased suddenly by 15 °F, then the bed outlet temperature will increase and level out at 760.6 °F with 45.6 °F bed rise. These two curves illustrate that under steady reactor operating conditions, for example, constant feed quality and feed rate, the reactor temperature will respond quickly to 15 °F reactor inlet temperature change. Furthermore, a 45.6 °F bed rise would have resulted in substantial change in reactor conversion of 40% plus and high hydrogen consumption that other areas in the unit would have been affected.

Continuing with the illustration, when the bed inlet temperature is increased 20 °F in Curve 4, an unsteady condition is created that until about 11 min point the reactor temperature increases rapidly and then seems to be leveling off between 11 and 17 minute points, that is, 6 minute duration, reactor temperature increases from 788 to 800 °F at ~2.3 °F/minute rate. At 17 minute point, total bed rise is 60 °F. This simulation implies that when the bed inlet temperature is increased 20 °F, it is above the increment necessary to exceed the threshold of instability. Beyond 17 minute point, if there is not any emergency response, the speed of temperature increase picks up rapidly, again creating a runaway situation at 21 minute point that the outlet temperature increase rate is uncontrollable. The simulation showed that it is possible that emergency activation depressuring before 17 minute point can avert an excursion, implying a well-trained operating staff can save the unit without a total shutdown.

Curve 5 shows when the bed inlet temperature is increased 30 °F, the bed outlet temperature immediately takes off and shows no signs of leveling off and immediately in less than 3 minute time starts a runaway condition at more than 60 °F/minute and accelerating continuously. When temperature is rising this fast, it would be futile to try to turn around the "runaway" by decreasing the bed inlet temperature back to normal, for example, by reestablishing quench gas flow at the outlet of the preceding bed. Emergency depressuring at maximum rate is the recommended procedure. From commercial experience, runaway reaching 800–900 °C is possible. Such an incident can not only create coke balls in the catalyst bed but also cause equipment damage and, in extreme cases, hydrocarbon release.

The example is from hypothetical simulation but in practical situations any sudden large reactor temperature fluctuation is definitely hazardous and has to be avoided.

On many occasions when depressuring is activated, the depressuring is manually stopped at intermediate pressure too soon with the wish of holding reactor pressure trying to save restart time. This is absolutely a wrong approach that can result in another emergency when restarting the unit with a hot and insufficiently cooled catalyst bed with feed remnants remaining in the catalyst pores.

One needs to appreciate that in any unit depressuring, high driving force only limits to the initial minutes and then the rate gradually slows. Stopping depressuring at intermediate pressure level lowers the driving forces to remove the reactants. It is critical to insist on not deviating from depressuring the unit completely to near slightly positive pressure removing reactant as much as possible cooling catalyst beds in temperature excursion. Deviating from such operating procedures is dangerous.

After a temperature excursion, nitrogen should be used to pressure the unit. Nitrogen is used because it is an inert gas unlike hydrogen. Failure to do so and instead pressuring the unit with hydrogen has resulted in repeated excursions immediately after another in many units. This is quite a common mistake operators make with the wish to restart and turnaround unit too quickly.

Following is an example in Figure 19.6 from a hydrocracking unit excursion incident. Reactor pressure is the dotted line and Bed 5 and Bed 6 temperatures are black solid and grey solid lines, respectively.

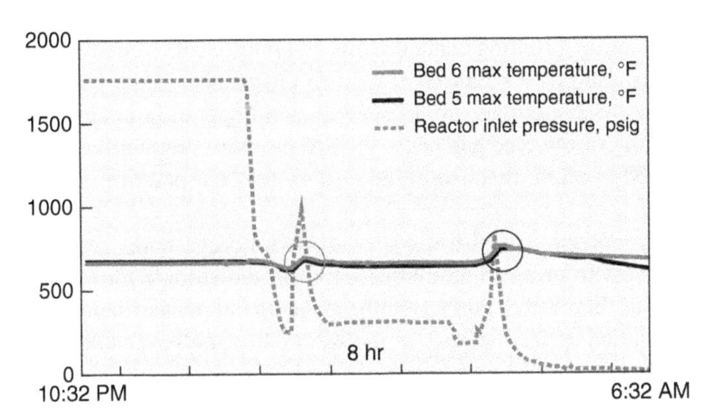

FIGURE 19.6. Two temperature spike excursion.

In early morning, that recycle gas compressor tripped and then unit depressuring system initiated automatically as shown in the dotted line that the reactor pressure dropped from 1800 psig to about 250 psig. The reactor temperature decreased and then the unit was repressured with hydrogen to 1000 psig about 45 min later, but reactor temperature in lower two beds increased quickly as shown in black solid lines. The unit was depressured again to 300 psig and ran for about 2 hours. Another attempt was made to pressure the unit with hydrogen and reactor temperature in lower two beds again increased as shown in grey solid circle. At this point, the unit was depressured completely as shown in the dotted line and purging once through with nitrogen in order to bring reactor temperatures down to 300 °F. This temperature excursion is the not worst example but it illustrates that until reactor temperature is low enough and unit is totally depressured bringing in hydrogen can result in temperature instability.

Every year there are temperature excursion incidents in hydrocracking units. Recycle gas compressor failure is a common cause of unit depressuring. UOP has conducted an internal survey and found about 50% of survey respondents reported to have had 1–2 temperature excursions, with only around 40% having none. This is an alarming trend because the survey does not include those who did not respond. Most of the reasons for the excursion can be attributed to operation errors that related to not following emergency procedures. Every temperature excursion emergency should be analyzed as a lesson learned.

19.8 REACTOR PRESSURE DROP

The operating staff should know reactor bed maximum allowable pressure drop in order that the limit is not exceeded during normal, transient, and emergency situations. The pressure drop is normally tracked over the operating cycle. Excessive reactor pressure drop and premature pressure drop increase are undesirable, which limit the unit throughput and shorten the run length. When the reactor bed has high pressure drop, the only way to correct the problem is catalyst unloading or skimming. Excessive reactor pressure drop can occur in different locations and be caused by various reasons.

Most of the reactor pressure drop is related to feed handling. Both cold feed from storage and hot feed from upstream units can be processed in the unit. Hot feed is preferred to minimize fouling contaminants and improve unit energy efficiency. Cold feed is usually used as a balanced stream to ensure that when hot feed is lost, its flow can be increased to make up the feed to avoid operation interrupting. Improper storage handling of the feed can cause reactor pressure drop and exchanger fouling. Since cold feed is often included in the design, the best practice is ensuring feed is stored properly.

Many hydrocracking units use gas blanketed storage tanks for their feedstock, which is the best to ensure minimum fouling in the storage by reducing exposure to oxygen. Gas blanketed tanks require more maintenance and are more costly. The second choice of feed storage is the use of floating-roof tank. Although many floating-roof tanks do not adequately exclude air, many units have successfully stored feed in well-sealed, floating-roof tanks. A floating-roof tank not perfectly round from the top to the bottom of the tank will permit air ingress or blanket nitrogen to leak away. A good tank seal is very important to keep out oxygen. Toroidal seals of rubber-type material are effective in preventing air contact.

Cracked feedstocks, such as coker gas oil, are extremely reactive with oxygen in storage and require the highest level of protection possible. Avoid processing shipped-in cracked feedstock or rerunning this material through the vacuum distillation unit to eliminate fouling contaminants. Cracked feeds are normally not stored. If storage is absolutely necessary, then use gas blanketed storage tanks.

Feed filters, cartridges, or auto backwash type, designed with proper flux and particulate filtering, represent another level of protection. Filters need to be maintained by periodically replacing broken filter elements or filters with excessive gaps. Never bypass feed filters and checking whether the bypass line is cold, indicating it is without any flow is a quick way to ensure no bypass.

Another common cause of reactor pressure drop is corrosion products from acidic crude processing from upstream crude/vacuum columns. Industry experience showed that the biggest naphthenic acid problem occurs with concentration of the organic acids at the vacuum distillation unit in the 290–345 °C range. Most crude/vacuum distillation units are not designed for high naphthenic acid crude, that is, >0.5 mg KOH/g acid numbers. For example, Nigerian crudes especially Forcados in some units have resulted in plugging of catalyst bed. The best way to control naphthenic acid corrosion is using an appropriate metallurgy and avoid high velocity and turbulent in crude/vacuum distillation unit design.

Feed-related pressure drop problems commonly limit to only the lead reactor bed. However, corrosion-induced pressure drop problems can pass the lead bed into lower beds especially iron sulfide that can break down easily. Following is an example that corrosion has passed through the first bed.

Corrosion in the fractionation section such as Fractionator column can cause reactor plugging from recycle oil. Proper H_2S stripping in the upstream Stripper can minimize corrosion in the Fractionator. Proper Fractionator heater metallurgy is also important to minimize corrosion.

Graded catalysts with different activities, trace metal capacities, sizes, and shapes started appearing in the 1990s. The application has helped to reduce reactor pressure drop, and it is now common practice having the lead reactor top bed to include several layers of different graded catalysts. In more recent years, the traditional ceramic ball hold-down material on top of catalyst bed has also been replaced by newer materials that increase the void fraction for particulate trapping. This material has proven effective to mitigate pressure drop problems. The example in Figure 19.7 after 300 days on-stream included a layer of such materials despite the fact that corrosion material continue to enter the reactor.

Catalyst migration is another possible cause of excessive reactor pressure drop. The leakage can lead to a flow constriction. While uncommon, flow reversal during emergency situation can also lead to catalyst migration. During commissioning, reactor distributor trays and outlet collectors should be inspected to avoid any gaps or

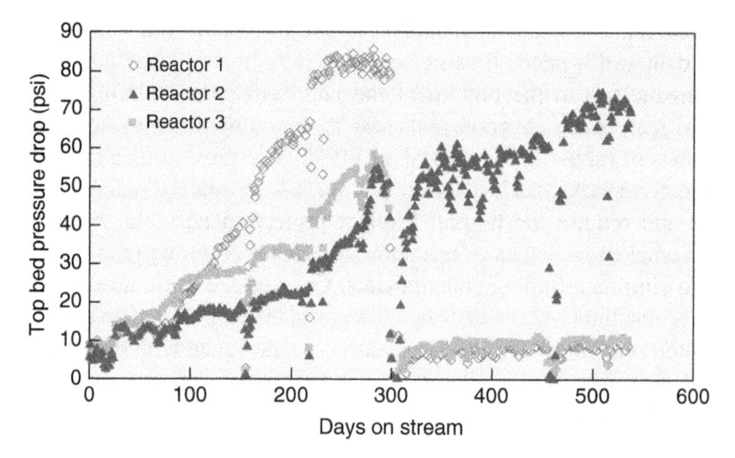

FIGURE 19.7. Iron sulfide fine pressure drop problem.

any mechanical problems, which may cause catalyst leakage. As mentioned earlier, these reactor internal components could have scales or coke accumulated over long operation that should be cleaned during turnaround. An improperly cleaned reactor is just a formula for later pressure drop problems.

19.9 CORROSION

Corrosion of hydroprocessing units is a common concern in the refinery industry. The intrinsic nature of the process operating at elevated temperature and hydrogen partial pressure with feedstock laden with contaminants, such as sulfur, nitrogen, and chloride, makes it challenging regarding proper material selection, fabrication, and maintenance. The unit can be subjected to various corrosion phenomena requiring proper corrosion control to make the unit safe to operate.

The unit is normally divided into high-pressure reactor section and low-pressure fractionation section. Do not exceed design temperature and pressure of any piece of equipment. Excessive temperature can reduce life of the equipment and degrade the material properties.

Material selection in hydroprocessing unit can vary from killed carbon steel, low and high chrome alloys to various grades of stainless steels. While each material upgrade gives increasing degradation resistance to corrosion, each material requires special consideration in its use.

Reactors in hydrocracking unit are thick-walled pressure vessels. With continuous improvement of material and fabrication, the reactor diameter can be ~6 m and wall thickness >300 mm. The base material for the reactor is normally made of 2 ¼ Cr or 3 Cr low alloy steels for strength and an austenitic stainless steel overlay lining for corrosion resistance. The thick-walled vessel can be subjected to a phenomenon called temper embrittlement that needs to be considered during operation. Temper embrittlement is defined as brittleness that results when susceptible alloys are held within or cooled through a temperature range of about 340–540 °C. The degree of embrittlement is manifested by an increase in the ductile-to-brittle fracture transition temperature. When the reactor metal temperature is below the ductile-to-brittle transition temperature, the operating pressure must not exceed more than 20% of the reactor metal yield strength which produces stresses in the reactor.

Obtaining a precise Minimum Pressurization Temperature (MPT) is not easy requiring working with reactor fabricator. As a practical guideline, the unit should not be pressured up above 30% of the normal operating pressure at Cold High Pressure Separator until the equipment temperature is above the transition temperature called MPT. In general, for vintage 2 1/4 Cr low alloy reactors, 93 °C is used as MPT. For newer modern reactor made of 2 1/4 or 3 Cr low alloy, 66 °C is used as MPT. The lower the MPT, the shorter the startup time. Operating staff should depressure during shutdown if MPT is reached or not pressure up until MPT is reached.

Austenitic stainless steel is used extensively in reactor section for its tolerance to high-temperature corrosion, but it has issues with chloride and polythionic acid stress corrosion cracking (SCC) at low temperature. Chloride coming in with the feedstock or with makeup hydrogen under wet condition can cause chloride SCC. Chloride can also cause NH_4Cl deposition in Feed/Effluent exchanger. When controlling chloride ingress is not effective, the next choice is upgrade material to alloy 625 or alloy 825 to provide a better protection. Low point drains in the Reactor Feed/Reactor Effluent Exchangers are particularly prone to chloride SCC. It is worthwhile to upgrade the metallurgy especially at these drain points where water could accumulate during unit idle.

Polythionic acid stress corrosion (PTASCC) is created in wet condition with sulfide in an oxygen atmosphere. Avoiding wet condition is the best way to protect against PTASCC. However, keeping reactor or heater coils warm is not always practical during shutdown or turnaround. An acceptable alternative is to neutralize the equipment with soda ash solution according to NACE recommendations.

The Reactor Effluent Air Cooler (REAC) is vital piece of equipment. It cools reactor effluent before entering the product separator. The pipe arrangement is critical to ensure even vapor/liquid distribution and control fluid velocity to minimize erosion/corrosion. A balanced design is considered the best practice as shown in Figure 19.8. Potential problem areas for REAC are shown in Figure 19.9.

Reliable wash water injection practice is essential in hydrocracking operation. Wash water is injected to minimize ammonium salt fouling before REAC and before the last Reactor Feed/Reactor Effluent Exchanger. Control good wash water quality by using condensate and minimizing stripper sour water reuse. If necessary, limit stripped sour water to <50% of total wash water need from a Sour Water Stripper not processing Coking or FCC unit's sour water to avoid cyanides. At the water injection point, use a quill to distribute water and make sure water is not injected to a boiled dry condition as the initial condensed water will be too corrosive. If water injection quantity is significant, then upgrade REAC to alloy 825 or 625. API RP 932 is a good reference for design, fabrication, operation, and inspection for REAC system.

Fractionation section equipment is typically carbon steel. Complete H_2S rejection should be achieved in the Stripper or Debutanizer column upstream of the main Fractionator to avoid high-temperature sulfidic corrosion. The Stripper top section should also be operated to avoid having water dew point occurring in the column. But as an extra protection, the top section of the Stripper is lined with TP316L stainless steel.

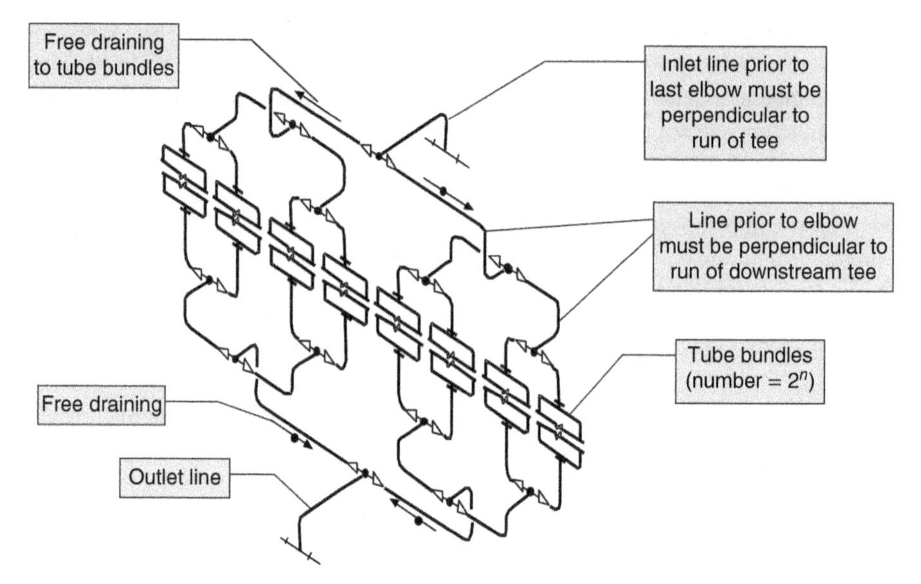

FIGURE 19.8. REAC system – balanced design.

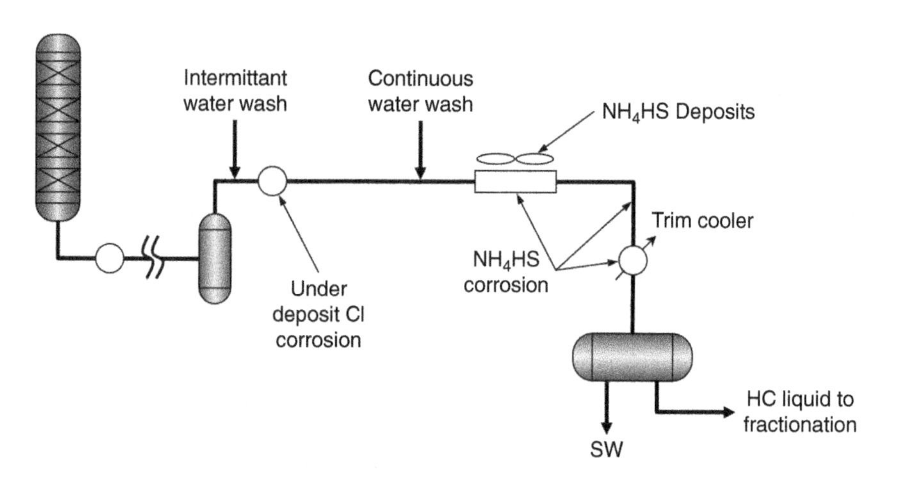

FIGURE 19.9. REAC system problem areas.

Licensor and design contractor should review material selection and handle deviation request carefully, especially related pipe class breaks between different metallurgies and services. The design methodology for new unit is more rigorous typically following project specification process but for later operation changes or revamps, failure to repeat these careful practices can lead to catastrophic damages. For example, several failure analyses from large catastrophic failures that resulted in ruptures can be attributed to improper pipe class break between carbon and alloy steels for hydrogen service or inferior equipment and construction. It is prudent that every deviation request should consider a metallurgical impact review.

19.10 HPNA

Hydrocarbon compounds containing more than two aromatic rings are polynuclear aromatics, that is, PNA or PCA (polycyclic aromatics) or PAH (polyaromatic hydrocarbons) as shown in Figure 19.10.

Heavy Polynuclear Aromatics (HPNAs) are aromatic molecules consisting of 7+ rings and in the industry nicknamed "Red Death" because of its infamous color. HPNA compounds do not exist in hydrocracking feedstock. Crude oil normally has 2–6 rings PNAs. HPNAs are formed from compounds containing lower aromatic rings through condensation reactions.

HPNA molecules can lead to increased coking or cause equipment fouling. As reactor effluent is cooled as shown in Figure 19.11, REAC is particularly susceptible to fouling due to low HPNA solubility. When this occurs, the unit has to be shut down and REAC tubes are individually hydroblasted. Therefore, a hydrocracking operator needs to pay attention in order to avoid HPNA-created problems.

Formation of HPNAs depends on several factors. Lower pressure units and high conversion operations are unfavorable conditions, which increase HPNA formation. Heavier feed with high endpoint bring in more PNA precursors. Controlling feed endpoint is one way to control HPNA formation. However, this requires lowering feed endpoint reducing feed availability, which is not the best way to operate a hydrocracking unit. To control HPNAs, a small slip stream can be removed from the Fractionator bottom to purge HPNA out of the unit or unconverted oil can be reprocessed in a vacuum distillation unit to prevent HPNA accumulation in reactor system. Purging a small stream of unconverted oil reduces overall conversion and negatively impacts operating economics, but this is an economic trade-off against downtime and lost opportunities.

As shown in Figure 19.12, UOP has patented two better ways of physically removing HPNA or concentrating HPNA. Unconverted oil can be processed in UOP HPNA RM® over lead-lag adsorbent beds. This technique has been commercialized and

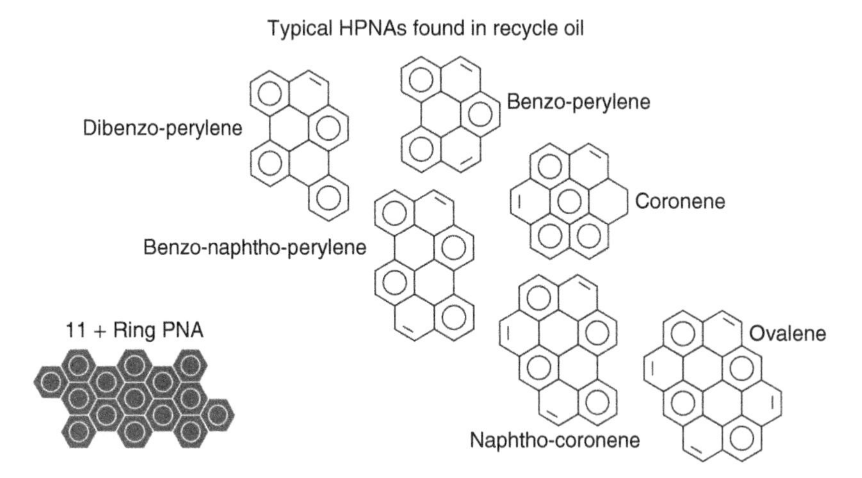

FIGURE 19.10. PNAs and HPNAs.

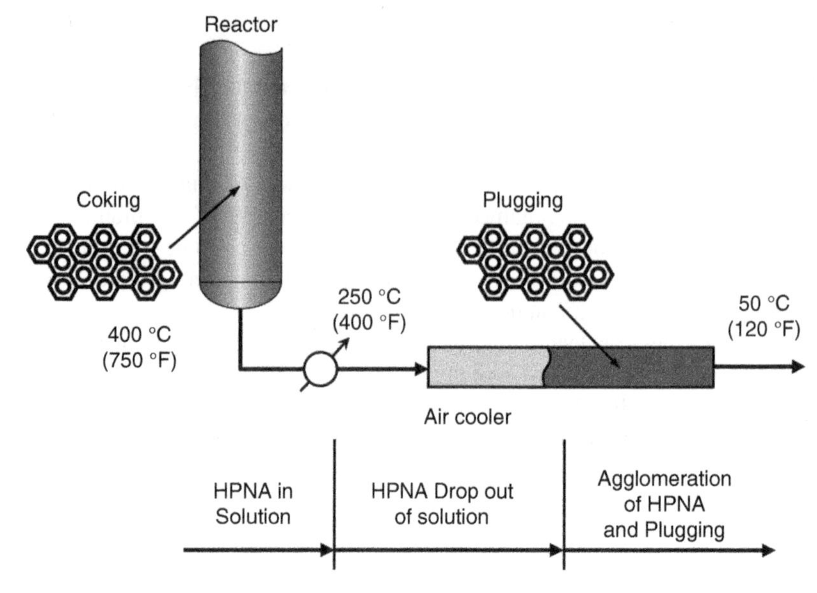

FIGURE 19.11. Mechanism of HPNA fouling.

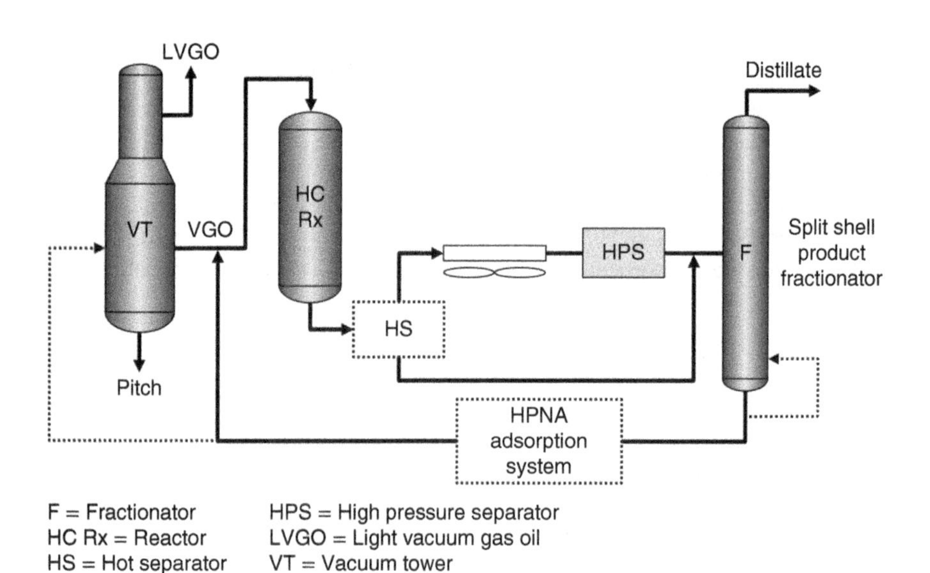

FIGURE 19.12. HPNA management.

allows the unit to operate at 100% conversions processing a heavy feedstock without any unconverted oil purge. As shown in Figure 19.13, a unit processing a heavy feedstock has resulted in significant HPNA formation. The left bottle is the dark red unconverted oil with HPNA. The middle and right bottles are lead adsorption bed outlet and lag adsorption bed outlet, respectively, with HPNA removed having reduced

FIGURE 19.13. Unconverted oil color.

fluorescence. The adsorbent HPNA RM® is replaced periodically. Another technique is using a patented split-shell fractionator (SSF) in which a portion of unconverted oil is reprocessed in the SSF concentrating HPNA to the small 0.5 volume% purge out of unit. The SSF option does not require any adsorbent and is a fractionation solution.

Successful hydrocracking operation requires effective HPNA management. HPNA RM® and SSF control techniques are commercially proven to effectively control HPNA achieving very high ~100% conversion. Comparing to early days when only controlling feed endpoint or purging fractionator bottoms were mainly the only solutions, these new techniques while requiring extra capital investment represent a more favorable solution and pay back quickly.

19.11 CONCLUSION

This chapter has quickly reviewed common operation problems. Hydroprocessing technology has been around in refinery nearly a century. The units operate in a severe environment. Design advancement and catalyst innovation have come a long way, but safe and reliable operation remained fundamentally dependent on good design, diligent operation, and maintenance.

20

TROUBLESHOOTING CASE ANALYSIS

20.1 INTRODUCTION

Influenced by global economics, refiners often need to consider volatile issues such as crude and feed quality changes and product demand fluctuations. Adapting to such refining situations becomes challenging, especially while operating at optimized conditions to reduce refining costs while meeting environmental regulations and maintenance constraints.

Hydrocracking is a critical unit in the refinery that requires good planning and catalyst selections. Because it has a great impact on overall refining profitability, it is crucial to avoid startup delays or unplanned shutdown.

Hydrocracking units are typically expected to have run lengths of between 2 and 6 years. When a conversion unit such as the hydrocracking unit is not operating well, the refinery can undergo a state of stress and economic loss. Once unit conditions are selected and catalysts are chosen, its operating window is fixed, making it difficult to accommodate large swings and changes. A complete understanding of the operation is critical to meet the unit performance targets. Figure 20.1 includes examples of operation concerns.

Often, there are many solutions to a problem. There are three key qualities that operating engineers must have to be proficient to operate hydrocracking units and be effective in troubleshooting.

(1) Having an inquisitive (I) mind-set to diligently examine the issues and ask every possible question, leaving no stone unturned.
(2) Upholding a sharing (S) attitude by exchanging findings and experiences with peers to learn from each other.

Hydroprocessing for Clean Energy: Design, Operation, and Optimization, First Edition.
Frank (Xin X.) Zhu, Richard Hoehn, Vasant Thakkar, and Edwin Yuh.

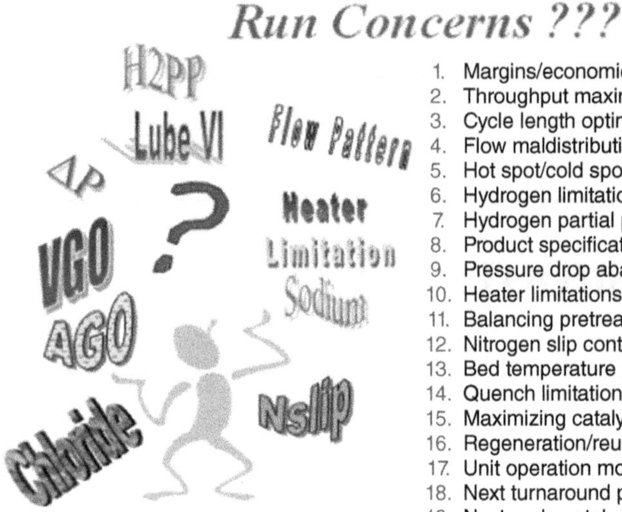

FIGURE 20.1. Hydrocracking operating concerns.

(3) Effective communication (C) is crucial and requires documenting troubleshooting and takeaways to avoid similar problems being repeated. A good forum for such practice is participating in industry's community of practice so that everyone benefits.

This section illustrates a few troubleshooting case studies related to operating hydrocracking units. These are realistic and relevant cases drawn from commercial operations without disclosing the specific client information.

20.2 CASE STUDY I – PRODUCT SELECTIVITY CHANGES

A high-pressure hydrocracking unit is designed and loaded with distillate selective catalyst from initial startup with a very successful first run. Another distillate selective type catalyst was used for second run. After 2 years of operation, the market demands more naphtha to fill downstream catalytic reforming unit, while the kerosene and diesel produced continued to meet specification requirements. Several options can be explored. Refer to Table 20.1.

20.2.1 Naphtha Cut Point Change

Without changing overall conversion and major operating conditions, one can explore increasing the naphtha cut point to as highest possible value that the downstream catalytic reforming unit can accept. As seen in Table 20.1, increasing heavy naphtha cut point from 170 to 180 °C allowed squeezing +2.4 wt%. More naphtha can be made if cut point is raised further >180 °C; however, the refiner chooses not to do so due to the coking burning ability in the continuous catalytic reforming unit.

TABLE 20.1. Options to Increase Naphtha Yield

Case	1	2	3	4	5	6	7
Catalyst type		Distillate			Flexible		Naphtha
Heavy naphtha endpoint (°C)	170	180	180	180	180	180	180
Combined feed ratio (CFR)	1.5	1.5	1.3	1.3	1.3	1.3	1.3
Conversion cut point (°C)	350	350	350	320	320	320	320
Heavy naphtha yield (wt%)	19.0	21.4	21.9	22.9	27.9	29.9	34.9
Δ Naphtha, +wt%	Base	2.4	0.5	1.0	5.0	2.0	5.0
Net naphtha gain (wt%)	Base	2.4	2.9	3.9	8.9	10.9	15.9
Feed changes						[a]	[a]

[a]Remove diesel in feed.

20.2.2 Combined Feed Ratio Change

The combined feed ratio (CFR) is inversely proportional to the conversion per pass as

$$\text{CFR} = \frac{1}{\text{Conversion per pass}}$$

Product selectivity changes with conversion per pass so that a lower conversion per pass favors heavier products. As CFR is lowered from 1.5 to 1.3 at constant conversion cut point, heavy naphtha increased +0.5 wt%.

When conversion cut point is reduced, more reactor quench will be needed as conversion increases. Saturated recycle liquid to the reactor is a good heat sink and as it is reduced, quench needs to be increased. A well-designed hydrocracking unit should have quench excess that should be able to withstand the CFR changes. CFR can be further reduced to 1.1 in extreme situations. As CFR is reduced, catalyst deactivation will increase with operating severity.

CFR is a variable to change to increase naphtha yield when reactor quench is not limited or the catalyst life is in excess. From the operating perspective, CFR may not be able to change significantly because hot recycle liquid is often used for unit heat integration in fractionation section.

20.2.3 Conversion Cut Point Change

The next step is to consider changing conversion cut point. Changing diesel endpoint from 350 to 320 °C increases naphtha yield +1.0 wt%. Continuing to lower the conversion cut point makes a lighter diesel stream. Similar to lowering CFR, as conversion cut point is lowered, operating severity increases and requires the examination of quench limitations and catalyst life impact. These three changes bring the net gain on the heavy naphtha yield +3.9 wt%; however, this is still not sufficient.

20.2.4 Catalyst Type Change

A refiner can manipulate the three preceeding operating variable changes to ascertain unit limitations within a catalyst system selected. It is recommended that the catalyst supplier and the unit licensor are next involved in the endeavor to increase naphtha yield.

In this case, the licensor was contacted to benchmark the operating yield model. After some deliberation, the licensor suggested changing the catalysts. Two catalyst choices were evaluated. It was preferred to use a flexible catalyst that favors increased production of naphtha while still maintaining the ability to produce good quality distillate. Integrating the flexible catalyst and operating at lower CFR, lower conversion cut point, and higher naphtha endpoint allowed a larger step change (+5.0 wt%) in heavy naphtha yield (Case 4 vs Case 5).

During the investigation, the licensor also suggested removing feed front end, that is, product boiling range materials in feed, so that overall feed is heavier in the front-end portion to reduce the unconverted material riding through the unit. The feed change is estimated to gain +2.0 wt% more naphtha on top of a flexible catalyst (Case 5 vs Case 6). For the extreme case, a further change from flexible catalyst to naphtha selective catalyst is estimated to increase naphtha yield further +5.0 wt% (Case 6 vs Case 7).

The unit was originally designed for distillate operations with a high bed rise limitation. When more active catalyst was used, though, the operating temperature was lowered, the bed rise increased beyond the original design. The licensor evaluated that the naphtha selective catalyst will require a short loading to limit the temperature rise. The less active naphtha selective catalyst quantity, even when short loaded, can meet catalyst life requirements.

The refiner then took the evaluations in Table 20.1 to judge the direction and check the economics of switching the catalysts. Next they confirmed that the unit will work without significant changes to the reactor section so as to limit the need to redesign the reactor. As net naphtha yield increases to +15.9 wt%, the lower pressure fractionation section will also need to be studied to ensure it is hydraulically capable in separating the products. As naphtha yield increases, light ends and hydrogen consumption also increase, requiring further evaluation on the capacity to handle these changes.

20.3 CASE STUDY II – FEEDSTOCK CHANGES

Hydrocracking units rarely process single feedstock from single crude. The amount of heavy sour crude supplied to the market is always increasing. Consequently, to cope with such a supply trend, refiners often needed to modernize these processes with new investments to process heavier crudes. Hydrocracking units originally designed for a narrow range of feeds several decades ago may be required to process feeds derived from different crude and other units such as the Coker Unit and Solvent Deasphaltene (SDA) Unit. Furthermore, increasing ULSD demand requires coprocessing of products from FCC Unit. A hydrocracking unit is very well suited for handling such new dimensions in the overall scheme of clean fuels production.

These heavier feeds contain higher nitrogen and sulfur contaminants with higher Conradson Carbon Residue and trace metals. The heavier feed is also more aromatic and hydrogen deficient, evident by lower °API and UOP K factor. Figure 20.2 illustrates the product cut distribution of various crudes. Compared to typical crude such as WTI (West Texas Intermediates) or Arabian Light, heavier crudes have significantly

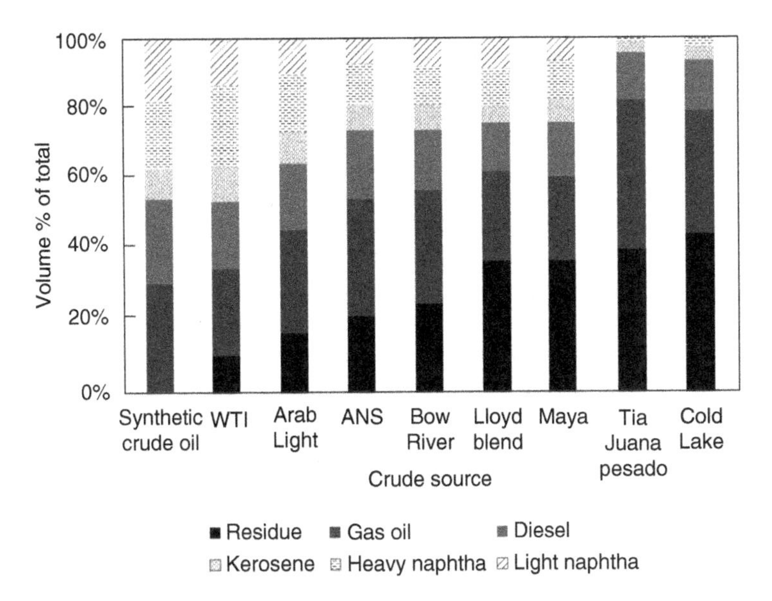

FIGURE 20.2. Product distribution of different crudes.

more residue and VGO materials. The VGO feed is considerably more refractive, making it challenging to process because PNA (polynuclear aromatics) precursors are difficult to crack and can condense in the hydrocracking unit to form HPNA.

A hydrocracking unit operation troubleshooting case related to feed processing is as follows.

This hydrocracking unit is a two-stage unit designed from the 1970s. The unit normally processes paraffinic HVGO to produce 360°C cut point diesel with a target run length of 48 months. The catalyst loading was conducted normally, achieving consistent and expected loaded density.

The unit experienced a higher-than-expected first stage cracking catalyst temperature requirement and deactivation after 1-month induction period. The product quality was negatively affected and HPNA level increased in recycle oil. Several causes are examined in the troubleshooting.

20.3.1 Catalyst Issues

The first question to address in troubleshooting is whether the catalyst is working properly, and the very first step in this analysis is to examine the startup. This hydrocracking unit historically used the gas phase catalyst sulfiding. In the new catalyst reload, Type II pretreat catalyst is selected to meet activity requirements. But with a gas phase startup, it is necessary to treat Type II catalyst with a preparatory treatment technique to avoid catalyst damages due to exposure to high bed temperatures before full wetting.

The initial hypothesis was that the pretreat catalyst treatment material prematurely migrated from the pretreat catalyst to the cracking catalyst and coked the cracking catalyst during startup, resulting in lower catalyst activity. The refiner consulted a licensor/catalyst supplier for assistance. Prior startup experiences and testing of treated and untreated pretreat catalyst with cracking catalyst did not show any performance loss on the cracking catalyst. In addition, the material used to treat the pretreat catalyst is not a high-coking-tendency hydrocarbon material. Furthermore, sulfiding conditions are mild from a coking tendency perspective because operating temperatures are not sufficiently high to promote coking. Thus, migration of the pretreat catalyst treatment material to the cracking catalyst and subsequent coking was ruled out as a contributing factor for activity problem.

The licensor also reviewed cracking catalyst manufacturing certification data to reexamine both physical and chemical analyses. These analyses showed the catalyst shipped met manufacturing specifications. Although the initial investigation indicated the cracking catalyst should not be affected by treatment of pretreat catalyst, the retained cracking catalyst samples were assembled from manufacturing composites and tested in the pilot plant. The results confirmed the loaded cracking catalyst had reference catalyst activity and yields.

If catalyst is on-spec, then is there any possible catalyst damage between loading and sulfiding? There was a 1-month idle operation period between loading the catalyst and the catalyst activation step. Catalyst handling conditions during this 1-month period between loading and catalyst sulfiding was reviewed for any bed exotherm or abnormal situation. The review revealed there were no unusual deviations to compromise catalyst performance while waiting for startup. Therefore, any contribution of catalyst quality issues to the observed activity debit was ruled out.

20.3.2 Startup Issues

The troubleshooting focus then switched to startup events once catalyst quality and handling issues were ruled out. Material balance calculations confirmed the gas phase sulfiding laid down expected quantity sulfur on the catalyst eliminating that as a cause.

Startup oil introduction and bringing catalyst to conversion practices were reviewed. After the initial prewet for the first stage, mechanical issues caused the recycle gas compressor to trip. The feed was stopped then when compressor issues were resolved, wetting was continued. The second stage catalyst prewet was conducted without problems. The only interruption was in first stage prewet. One could therefore question whether the compressor disruption during catalyst prewetting contributed to the problem with first stage cracking catalyst.

Further analysis of the startup revealed an additional deviation. During startup oil introduction, licensor recommendation was to utilize startup oil consisting of straight-run distillate and having endpoint between 315 and 370 °C, and nitrogen content <200 ppm. Instead, the refiner had used an HVGO containing 1040 ppm nitrogen as the startup oil.

This is not the best practice for a fresh catalyst using excessively heavy oil instead of a lighter diesel. Would this cause activity debits? The licensor then drew upon their experience in the troubleshooting logics including examining pilot plant experiences from licensor's database for impacts of organic nitrogen on fresh catalyst.

While it is not a best practice, pilot plant work showed that prewetting the catalyst with VGO is not a major issue compared to using straight-run diesel. Therefore, this pilot plant work supported the standard applied for sulfided catalyst startup procedure where restart does not undergo a diesel wetting step, saving startup time and slop oil reprocessing. However, taking precautions are necessary when using a heavier feed, adjusting catalyst operating conditions to compensate for the impact of organic nitrogen, since heavy feed has more nitrogen than straight-run diesel. Review of operating data from multiple hydrocracking units indicates at least one instance where the cracking catalyst lost 4 °C activity and the pretreat catalyst also suffered with higher nitrogen slip from 2 to 6 ppm at constant pretreat temperature. It showed that when using a heavy feed, pretreat catalyst has to be at the appropriate severity level to convert feed nitrogen in order to protect cracking catalyst. Pilot plant testing confirmed uncontrolled denitrification will impact on catalyst activity with minimum impact on hydrogen consumption and yields.

The licensor then searched for troubleshooting cases and found two additional commercial operation cases to support the investigation on the impacts of introducing excessive organic nitrogen during startup or during transition operation.

20.3.3 Case A: A Two-Stage Unit Processing VGO/CGO with ~1300 ppm Feed Nitrogen

Two similar incidents occurred in this unit within the first 100 days of initial operation. VGO rate was reduced significantly and replaced with coker kerosene and coker diesel. Pretreat and cracking catalyst temperatures were reduced to avoid olefin polymerization. The pretreat temperature was reduced due to lower feed rate and lighter feed. The cracking temperature was also reduced, responding to lower feed rate and low conversion. During this interim period, large quantities of unconverted organic nitrogen slipped into the cracking catalyst. After this event, the cracking catalyst was raised excessively to reach conversion goal. The organic nitrogen adsorbed on the catalyst did not have the chance to be removed and turned into coke. Both incidences resulted in activity debits as shown in Figure 20.3, commercial data compared to expectation from pilot plant data.

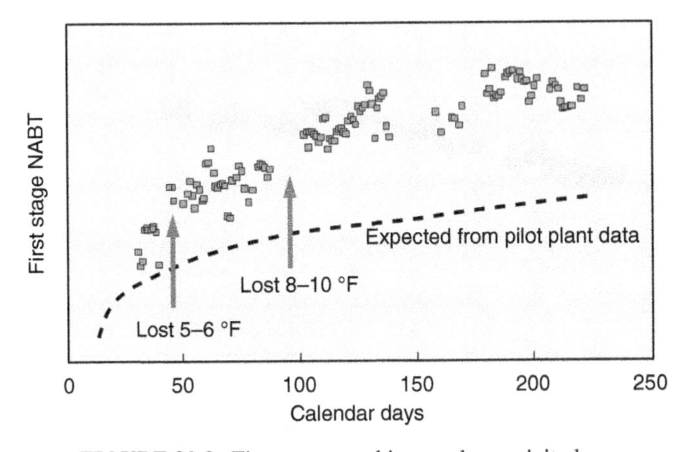

FIGURE 20.3. First-stage cracking catalyst activity loss.

20.3.4 Case B: A Single-Stage Hydrocracking Unit Processing VGO

This unit experienced feed nitrogen poisoning at 466 days into operation. While the catalyst is no longer considered fresh catalyst, nitrogen poisoning can lead to activity loss. The unit experienced a problem with the oil/water interface control at Cold High Pressure Separator, leading to hydrocarbon contamination at Sour Water Stripper. This disruption at the Sulfur Plant leads to an upset in the hydrocracking area. Hydrocracker feed was stopped and reactor temperature cooled as per standard practice. The problem occurred when the hydrocracking unit feed was introduced before downstream processing units were ready. Right after feed introduction, feed rate was purposely kept at very low pretreat temperatures from 285 to 330 °C while feed was brought in from storage. Unfortunately, this feed contained some high endpoint material. The combination of operating at a pretreat temperature incapable of converting feed nitrogen, and increased feed nitrogen content led to the cracking catalyst activity illustrated in Figure 20.4, which shows a step change in reactor temperature at 466 days on stream in Cycle 11.

Troubleshooting analysis examples in Case A and Case B, as well as pilot plant tests, support that feedstock may be the cause of higher-than-expected initial cracking catalyst temperature requirement and deactivation. Back to the two-stage unit case study, when compressor had issues during prewetting, the catalyst was subjected to high feed nitrogen level for over 2 days at 160–190 °C low temperatures. It is suspected that the adsorption of heavy nitrogen compound and heavy aromatics on freshly sulfided acid sites coupled with insufficient time for nitrogen to desorb resulted in more coking on the fresh catalyst and contributed to the decreased catalyst performance.

Since deactivation rate remained higher after initial catalyst induction period, the troubleshooting was further extended to examine the feed processed after startup. Operation reviews between the refiner and licensor showed that the initial 70 days had

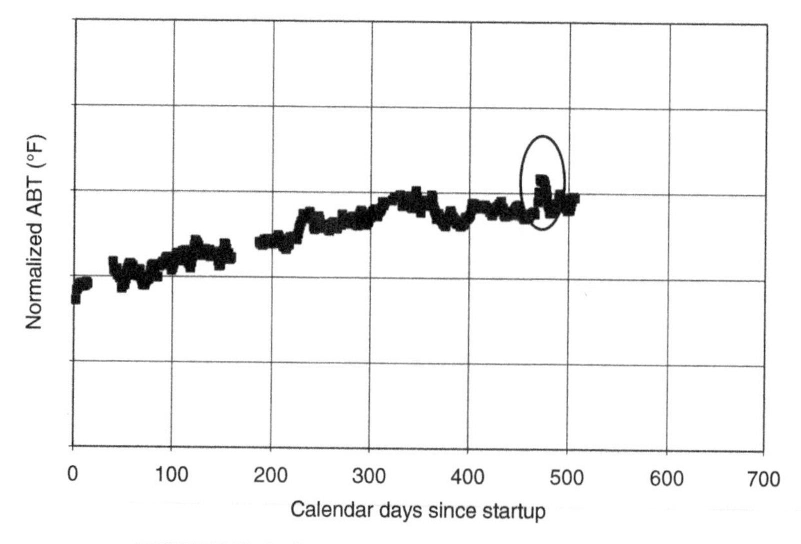

FIGURE 20.4. Step change in cracking catalyst activity.

large increases in West African and heavy Russian crudes, requiring the processing of less low-density feed streams. At 110 days into the run, feed samples were sent to licensor for examination. Normal D-2887 simulated distillation did not show anything unusual. Investigating further, high temperature simulated distillation (ASTM D-6352) was applied to new samples and showed very heavy tails in both samples, with endpoints 75–90 °C higher compared to design feed with a 525 °C cut point (Figure 20.5).

Two-dimensional GCXGC was used to examine the aromatic types in Figure 20.6. While the feed nitrogen is not as high, it contains a higher concentration of 4+ aromatic rings compared to the reference sample taken earlier. This illustrates a switch of feed quality that is more refractive in nature.

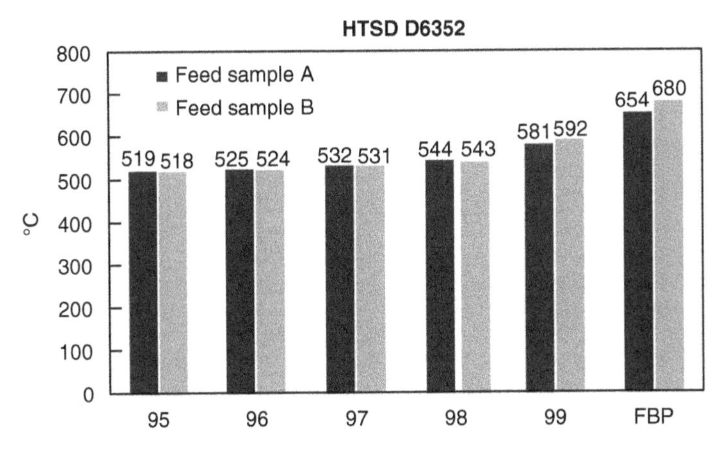

FIGURE 20.5. High-temperature simulated distillation of feedstock.

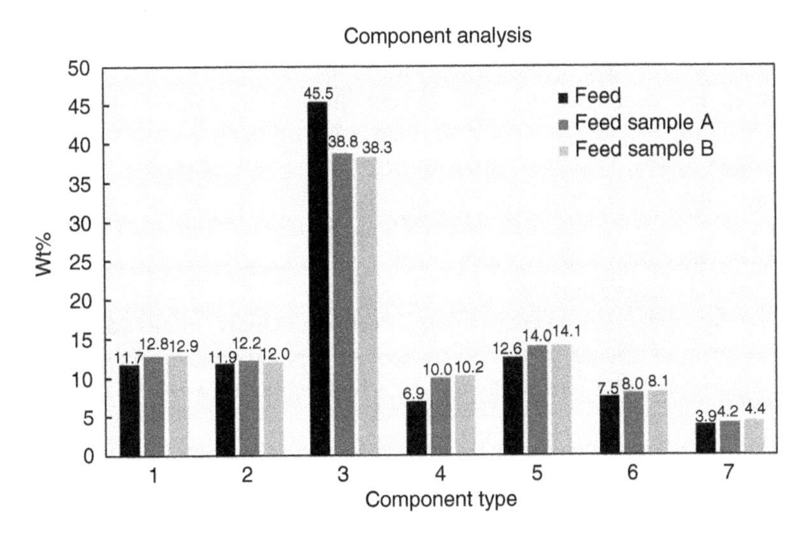

FIGURE 20.6. Component analysis.

TABLE 20.2. Feed Analysis – Aromatics Distribution

DBE	Hydrocarbon type, PPM	Feed A	Feed B	Feed
1	Monocycloparaffins	0.0	0.0	0.0
2	Dicycloparaffins	885.5	952.3	500.8
3	Tricycloparaffins	4,489.8	4,327.5	5,230.7
4	Benzenes	26,064.7	25,325.0	36,440.2
5	Indans/Tetralins	39,640.2	38,234.5	46,162.0
6	Indenes/Dinaphthene Benzenes	37,654.2	36,525.4	44,934.7
7	Naphthalenes	41,037.6	40,141.2	44,664.0
8	Acenaphthenes/Biphenyls	43,472.4	42,490.9	46,758.8
9	Fluorenes	41,311.1	40,456.6	46,021.0
10	Phenanthrenes/Anthracenes	35,183.7	34,588.0	38,908.4
11	Aceanthrenes	28,203.3	27,798.1	29,537.3
12	Pyrenes	23,727.4	23,428.8	25,499.8
13	Chrysenes	18,562.7	18,080.8	19,988.8
14	Cholanthrenes	13,272.2	12,902.0	12,808.1
15	Benzopyrenes	9,635.8	9,304.8	8,837.0
16	Dibenzathracenes	6,441.6	6,096.4	5,446.4
17	Benzoperylenes	3,969.4	3,751.5	3,167.8
18	Dibenzopyrenes	2,394.4	2,229.5	476.9
19	Coronenes	652.4	765.0	176.1
20	Naphthcoronenes	187.9	142.5	115.1
21	Dibenzoperylenes	68.7	45.7	24.6
22	Benzocoronenes	18.9	0.0	16.7
23	Dinaphthopyrenes	0.0	0.0	0.0
24	Naphthocoronenes	0.0	0.0	0.0

Furthermore, component analysis of feed aromatics analysis showed higher concentrations of dibenzopyrenes, coronenes, naphthcoronenes, and dibenzoperylenes, all of which are HPNA precursors (Table 20.2).

The troubleshooting analysis concluded that neither the catalyst quality nor the startup treatment material on the pretreat catalyst for gas phase sulfiding was the issue. Sulfiding procedures were also ruled out. Two likely causes were identified. First, some portion of activity loss was attributed to the exposure of active fresh catalyst to heavy feed and nitrogen poisoning, resulting in increased coking. Second, the analysis indicated that unit feed shifted to one that contains more refractive aromatics with more PNA precursors. When combined, these factors resulted in compromised catalyst activity.

Steps can sometimes be taken to repair catalyst activity. In this case, the refiner later had an opportunity to shut down the unit for compressor maintenance. The licensor recommended that a diesel flush and a hot hydrogen strip could be beneficial during shutdown.

The refinery flushed catalyst with straight-run diesel and ultra-low-sulfur diesel in the attempt to remove HPNA. For reference, one could also utilize LCO from an FCC as it could also be beneficial. Following diesel flush, the catalyst went through a hot hydrogen strip at 400 °C. The subsequent startup had much better controlled wetting

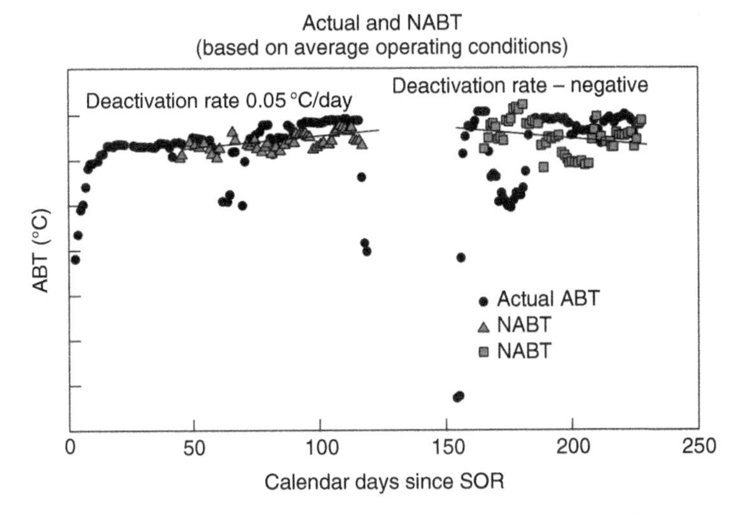

FIGURE 20.7. Catalyst performance after diesel flush and hot hydrogen strip.

and the overall catalyst activity had recovered significantly. Catalyst performance is shown in Figure 20.7. The conclusion in this scenario is that avoiding exposure to fresh active catalyst with heavy nitrogen compounds is the best practice.

20.4 CASE STUDY III – CATALYST DEACTIVATION BALANCE

A single-stage recycle hydrocracking unit is designed for distillate production. Over the years, the unit has switched to a catalyst system to enable distillate and lube base stock production processing straight-run VGO derived from heavy Iranian crudes. The reactors (Figure 20.8) are configured with one two-bed pretreat reactor (R1) and one three-bed cracking reactor (R2).

However, because of limitations of the pretreat catalyst, one of the cracking beds in Reactor 2 (R2B1) was replaced with a bed of pretreat catalyst. The refiner's engineer contacted licensors at 630 days on-stream to inquire the following situations and hope to receive advice on how to achieve a 3-year cycle length.

(1) Is it acceptable to decrease the temperature of the pretreating bed ABT 4 °C and increase the temperature of the cracking bed ABT 2 °C? The refiner questions the optimum way to operate pretreat and cracking catalysts because their data trends estimated that the cracking catalyst has more run life and lower deactivation compared to the pretreat catalyst deactivation. Less firing in heater is necessary as pretreat temperature is lowered. The high cracking temperature also gave more preheat duty at the reactor effluent. These changes will alleviate recycle gas heater duty, lowering maximum heater tube wall temperature.
(2) The unit also has a radial temperature spread in R1B1. The refiner believes lowering the pretreat temperature can help lower the radial spread.
(3) R1B1 is being poisoned by iron. The refiner wants to know the capacity of graded bed materials and considers skimming the lead pretreating bed.

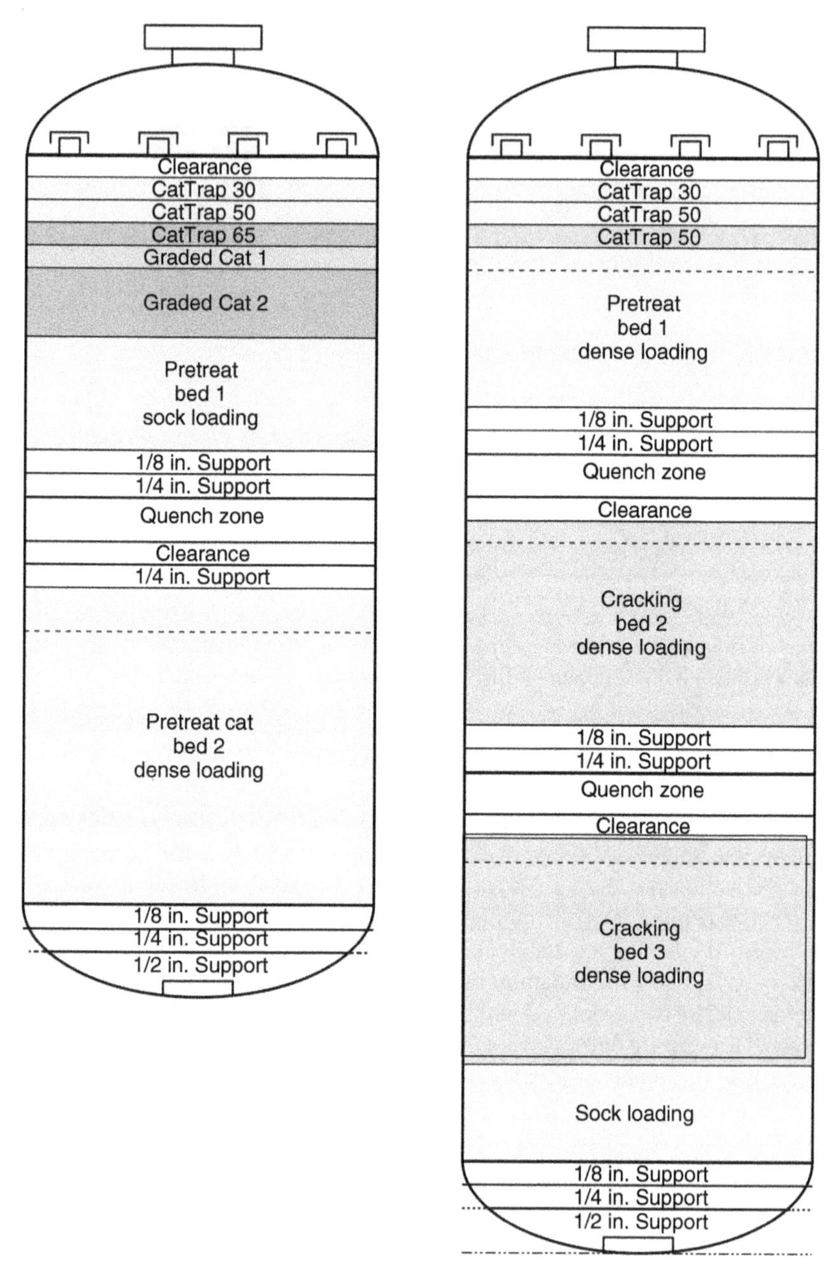

FIGURE 20.8. Reactor bed configuration.

The licensor examined the operating data and provided the following assessments and recommendations. Using the data as a reference, when an ascending temperature profile was applied, the last pretreat bed R2B1 peak temperature would be too high. It would be very close to the inlet temperature of R2B2 first cracking catalyst bed. Such R2B1 bed temperatures will result in excessive deactivation. Instead of equally lowering all pretreat bed temperatures, varying the temperature at each of the three pretreat beds was considered.

- Reduce the inlet temperature to R1B1 by 9 °C and WABT by about 10 °C.
- Maintain R1B2 inlet temperature to keep WABT in this bed about the same. It may require decreasing the inlet slightly because R1B2 will want to pick up some of the reactions passed on from R1B1.
- Decrease the inlet temperature of R2B1 by 1 °C. This will decrease the WABT in this bed by approximately 1 °C, and lower peak temperature by 1 °C.
- The combined effect of temperature adjustments in R1B1 and R2B1 is calculated to reduce overall pretreat catalyst WABT by about 2 °C in the overall HDN WABT.
- Using the correlation, R1 outlet nitrogen increased to 100 ppm, but R2B1 is 5–10 ppm at R2B1 outlet.
- The higher nitrogen slip out of the pretreat catalyst would result in about a 0.6 °C increase in R2 WABT. This is required to maintain conversion.

Examining the reactor temperature adjustment is only half of the work and evaluation of deactivation balance is still needed.

- The R1B2 pretreat catalyst was estimated to deactivate at about 0.025 °C/day. Its remaining life would be ~31 months using 425 °C end-of-run peak temperature.
- The R2B1 pretreat catalyst would deactivate at about 0.024 °C/day as a result of the reduction in temperature.
- Cracking catalyst deactivation as a result of increased nitrogen slip would increase by ~2.4% to 0.015 °C/day. The remaining life would be ~31 months using a 424 °C end-of-run peak temperature.

The net effect of these changes on deactivation is summarized as follows:

- Overall pretreat deactivation would decrease by 7% to 0.023 °C/day. The remaining life would be about 33 months.
- Overall cracking deactivation would increase by 2% to 0.015 °C/day. The remaining life would be about 30 months.

The postmortem complete run temperature plots are shown in Figures 20.9 and 20.10. The conclusion is that these adjustments at 630 days on-stream achieved a slight improvement over the remaining life estimates based on the temperature profile, but by not much. The run achieved 36 months.

The metallurgical temperature limit for the reactor is 454 °C. It was suggested to adopt a higher end-of-run peak temperature above those used in past cycles. For example, using 6 °C above past end-of-run peak is not unreasonable and will still maintain a good temperature margin below metallurgical limit. This was not adopted due to concerns with product quality, especially the unconverted oil quality for the end of run.

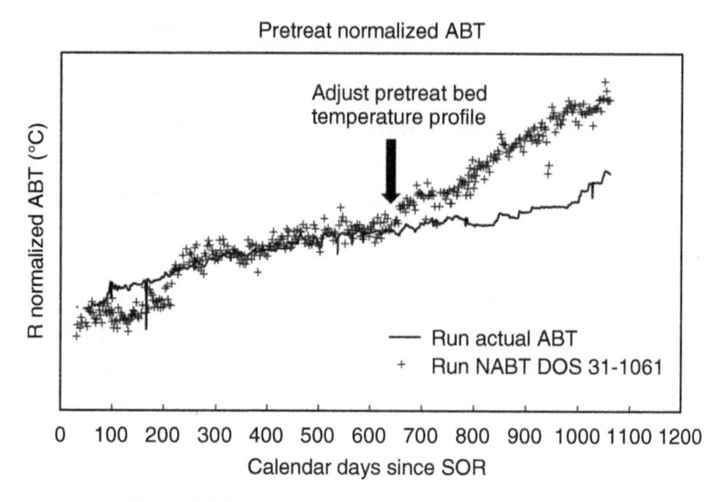

FIGURE 20.9. Pretreat temperature performance.

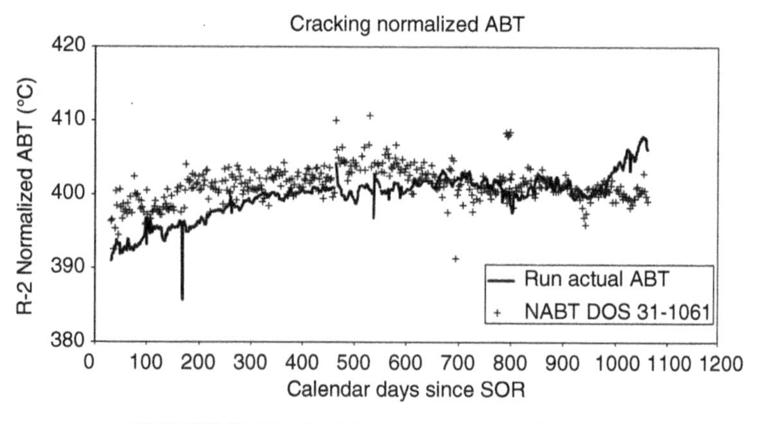

FIGURE 20.10. Cracking temperature performance.

20.5 CASE STUDY IV – CATALYST MIGRATION

Switching from catalyst performance issues demonstrated by the previous case studies, fixed bed process units such as hydrocracking occasionally experience catalyst migration incidents. In these cases, they could also experience temperature maldistribution or a pressure drop in the catalyst bed. In some cases, the catalyst appears in downstream equipment. Often, such a problem requires unit shutdown followed by unloading the catalyst to identify the problem area and correcting the mechanical issues. Circumstances are not always the same. Generally, more migration occurs soon after startup compared to during an operation that has already run for a period of time. The best practice to prevent such an issue is proper inspection before catalyst loading and proper startup to prevent inadvertent incidents.

If the operating severity is increased, a unit revamp may be required to fit the new operating modes. If the reactor internals are modified, special attention is necessary to ensure all internal components are revamped and that they fit properly. Design records in older units may not always be as complete; therefore, contingency plans for the unexpected problems during the retrofit should also be considered.

Following is a case of a new unit that developed catalyst migration problems during startup commissioning. This is a two-stage hydrocracking unit originally designed to produce naphtha. However, prior to startup, the catalyst selection was changed to produce distillate. The unit consists of two reactors, that is, one six-bed reactor (R101) in the first stage and one two-bed reactor (R102) in the second stage.

20.5.1 Catalyst Support Issues

Both reactors were loaded in parallel. The R101 bottom bed inert catalyst balls (ICB) layers were loaded and began to load posttreat catalyst. The R102 bottom bed ICB layers, cracking catalyst and posttreat catalyst were loaded and the vapor/liquid distribution tray installation was underway. At this point, the operator reported that there were ceramic balls appearing in the R101 outlet pipe, suspecting that the support material has migrated. Loading in R101 was stopped. Examination of the fracture pieces indicated that inferior support quality was the cause (Figures 20.11 and 20.12). The ICB material seemed to be layered and was "spalling" off.

ICB sacks brought to loading site showed that some of the bags indeed had broken pieces. It was suspected that R102 could have similar issues; therefore, loading in R102 was also stopped. Borescope examination also showed that both reactors had experienced migration of support material as shown in Figures 20.13 and 20.14.

The refiner decided to unload both reactors. But, should gravity dump or vacuum unloading be used? Logistics are easier and less time consuming with a gravity dump. However, vacuum unloading allows a better segregation of the material already loaded. The refiner decided to use gravity dump that screens the catalysts and ICBs to segregate the sizes.

To prevent future migration, the refiner considered keeping a wire mesh wrapping around the outlet collector. After consulting the licensor's on-site engineers, licensor explained that adding a wire mesh has been implemented in older units but in newer

FIGURE 20.11. New ceramic support.

FIGURE 20.12. Broken ceramic support.

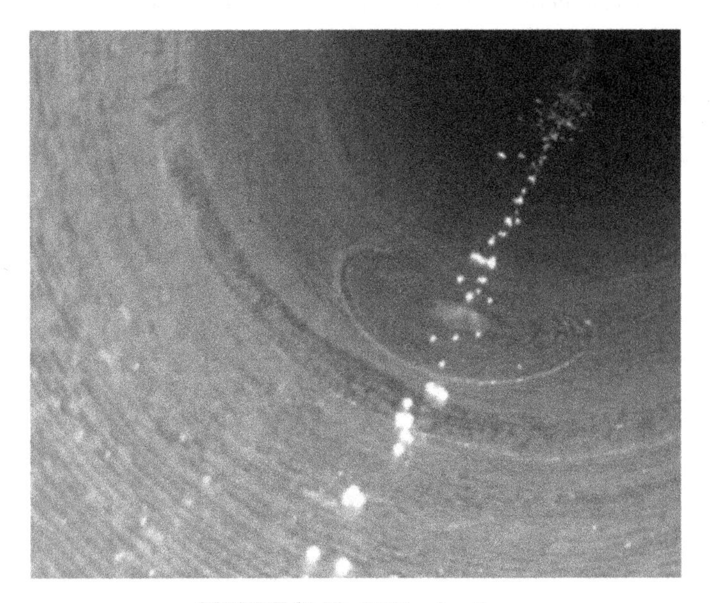

FIGURE 20.13. R101 migration.

units having proper size slots on the outlet collector and ICB meeting specifications, wire mesh is not required. Although adding wire mesh is a good way to prevent migration, not having proper welding of wire mesh can create problems. The wire mesh size can create other problems as well. For example, coke fines and catalyst chips could be lodged on the screen restricting flow and creating a pressure drop and can be hard to remove, leading to delays in turnaround time.

The refinery also debated adding a fourth layer of 15 mm ICB instead of three layers of 19, 13, and 6 mm sizes. Industry experience showed three layers are adequate.

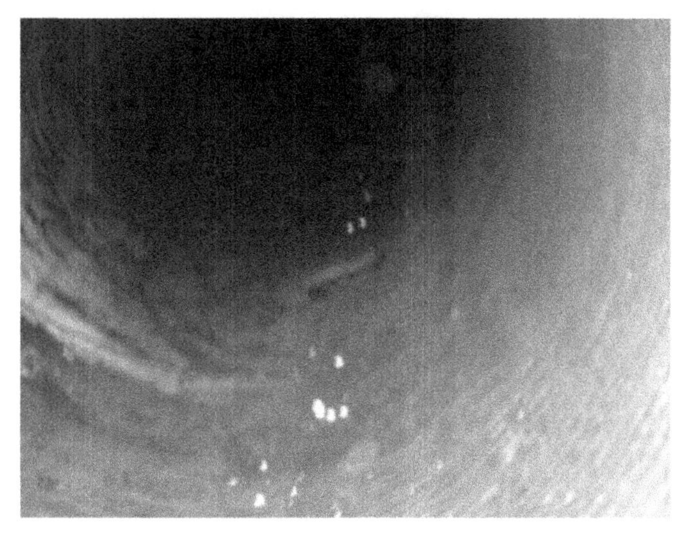

FIGURE 20.14. R201 migration.

But, it is important to ensure the largest size ICB is loaded typically 150 mm above the top of outlet collector top surface. To be conservative, the refiner added a fourth layer of 15 mm ICB as extra protection.

The refiner decided the following:

(1) Fine mesh on the outlet collector was removed and left as shown in Figure 20.15.
(2) New 19 mm ICB was purchased, since the broken ones tended to "spall" off after exerting crushing pressure.
(3) An additional layer of 15 mm ICB layer was added to the loading.
(4) ICBs were loaded gently using buckets instead of free falling from a loading hose, as that would fracture the balls.
(5) Screening cannot fully separate posttreat catalyst, cracking catalyst, and 3 mm ICB because all having a similar size. Therefore, some highly mixed catalysts with ICB were not reused. The rest was reloaded.
(6) The reactor outlet nozzle was vacuum-cleaned to remove damaged ICBs and particulates.

The reactor loading started again and was completed without any further incidents. This example showed that using proper catalyst support material is just as important as the quality of catalyst. It is crucial to pay attention to small details. Over the years, many ICB incidents occurred that caused unnecessary loading delays. Good loading preparation should always include confirming the quality of ICB. Procuring ICB from reputable vendors instead of choosing low cost options is the best choice. If there is any doubt, adhering to licensor's design standard and avoid using recycled ICBs.

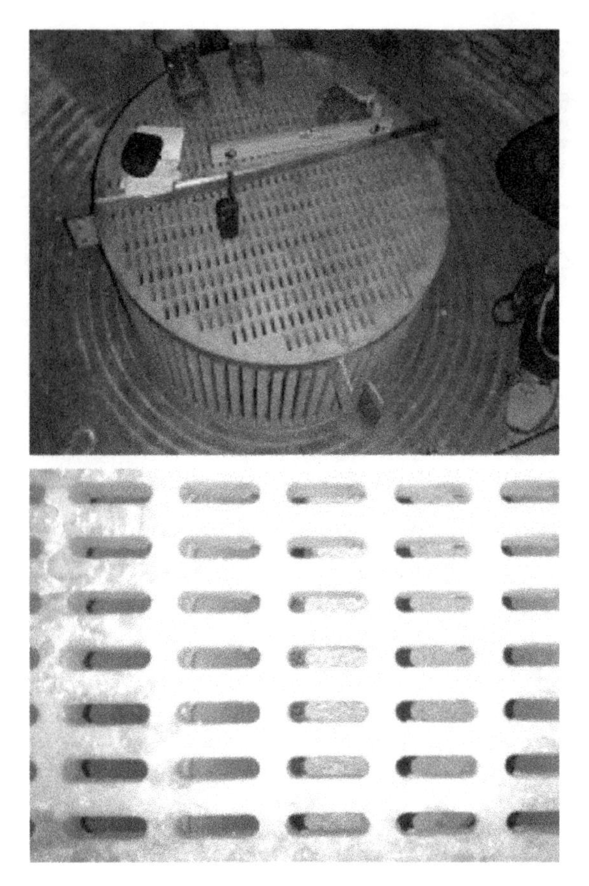

FIGURE 20.15. Reactor outlet collector.

20.5.2 Reactor Internal Issues

The problem of the catalyst migration did not end at the first ICB migration incident. During unit pressure testing at 120 MPa(g), the unit developed a leak at the first stage reactor feed/effluent exchanger. Upon opening the exchanger, catalyst and ICB were found in the exchanger again as shown in Figure 20.16.

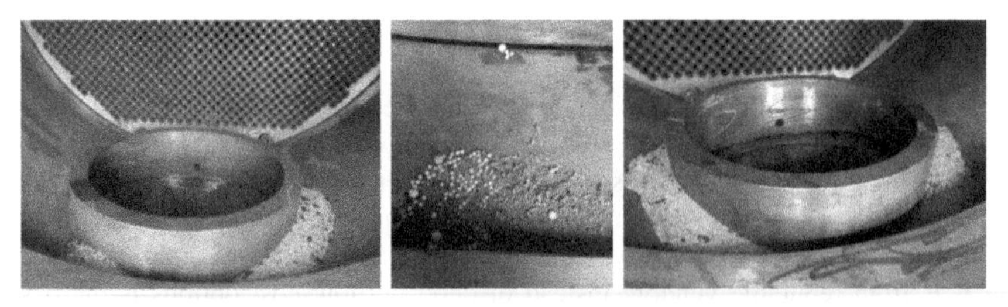

FIGURE 20.16. Catalyst migration into feed/effluent exchanger.

Naturally, with another startup delay, the refiner asked why catalyst migrated again and felt that the mesh screen should not have been removed at the outlet collector to start with to avoid such catalyst migration. The loading contractor was also puzzled by repeated finding and confirmed that they had done all they could to load ICB carefully.

The troubleshooting took the following path:

(1) Records were reviewed to confirm loading was according to the loading plan. The refiner actually loaded an additional 150 mm depth total 13 and 15 mm layers ICB at the bottom bed between the 19 and 6 mm ICB, further reducing the chance of the catalyst migration. Also, 19 mm ICB loaded on top of the outlet collector was procured from a different and reputable supplier.
(2) Pictures were taken at exchangers to confirm the quantity and type of catalyst/ICB migrated to confirm that it is a new leakage not left over from the first migration.
(3) Borescope examination of the first stage and second stage reactor outlet collectors was performed again.
(4) Reactor bottom elbows at Stage 1 and Stage 2 reactors were removed to see how much catalyst accumulated there.
(5) Nitrogen was purged in from the reactor top and it was checked if the gas flow would cause more catalyst coming through the bottom elbows.
(6) The system pressure and pressure drop trends were checked during the recent pressure tests to see if there were any abnormalities.
(7) Opening additional exchangers to ascertain extent of migration. However, the refiner did not open these high-pressure breech lock exchangers, worrying that if not handled properly, it could leak.

Borescope examination showed that the first stage outlet had catalyst but not the second stage. The problem so far seems to limit to only the first stage. The refiner questioned if the catalyst was those collected from the second loading or the first loading due to possible improper handling. It was suspected that the catalyst could be coming from dead space under the outlet collector and blown out when the recycle gas compressor was started. However, this possibility was ruled out.

The next is opening R101 outlet 24 in. elbow to allow videotaping, but the reactor wall was still hot at 70 °C from skin temperature heat up for pressure tests. This was too hot for inspection. Thus, the refiner considered using dry ice to cool the reactor.

While the refiner was making plan to unload the catalyst as the worst case, they discovered that the Bed 4 pretreat catalyst bed temperature started increasing while the unit was sitting idle. The temperature shot up to 190 °C at the bed bottom, while other bed temperatures showed high levels (Figure 20.17). Nitrogen was introduced through the quench gas lines to cool the catalyst bed, but the temperature dropped slowly, creating a concern about minimizing exposing pretreat catalyst to high temperatures before being fully wetted, as required in the catalyst activation. Instead of dealing with a catalyst migration issue, the refinery now had to deal with bed temperature increase issue.

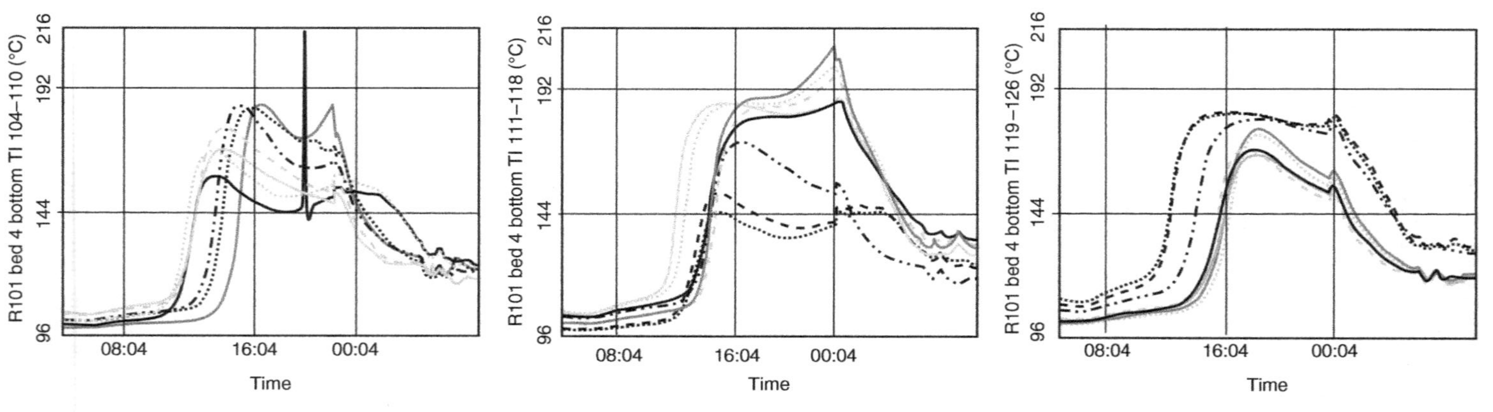

FIGURE 20.17. R101 bed 4 bottoms temperatures (24 points).

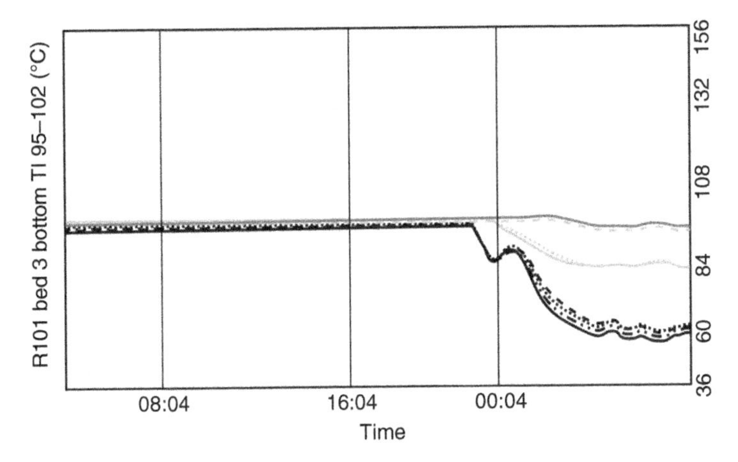

FIGURE 20.18. R101 bed 4 mid and top temperatures (8 points).

Data analysis showed that the TI trends at the center of the bed first started increasing followed by the peripheral TIs. The time at which the temperatures started increasing was around 2 am when the refiner opened the outlet line neutralization nozzle for borescopy work then the refiner opened next the reactor inlet neutralization nozzle and left both nozzles in an open condition throughout. This is an issue because nozzles open at both top and bottom of the 30-m-tall reactor create a slow natural draft of air across the reactor. Before shutting down the recycle gas compressor, the catalyst beds were slightly over 100 °C. The natural draft of humid air flowing over the dry pretreat catalyst could explain such a temperature rise pattern along the direction of air flow. It is somewhat encouraging to observe that the mid and top bed TIs did not see any similar temperature increases (Figure 20.18). After nitrogen purge, catalyst bed temperature subsided. The lesson learned is that steady and sufficient nitrogen purging is necessary to protect the catalyst to avoid air ingress.

After solving unexpected bed temperature increase problem, the troubleshooting focus returned to catalyst migration. Inspection staff opened one additional exchanger in the first stage reactor outlet, which revealed migration of a mix of catalyst and ICB as shown in Figure 20.19. The total amount was slightly more than those found in the exchanger that had seen catalyst migration and was mostly composed of ICB of various sizes and some 19 mm ICB fragments. Reviewing loading record, 19 mm size ICB was loaded to 150 mm depth above outlet collector top, then 15 mm ICB at 80 mm depth, then 13 mm ICB at 70 mm depth, and then finally 6 mm ICB at 100 mm depth and 3 mm ICB at 100 mm depth before the catalyst.

Pictures from the bottom of outlet collector are shown in Figure 20.20. It was observed that there were some 6 mm size ICB stuck on the outlet collector slots, which are 6–7 mm wide.

It took much deliberation about where those 6 mm ICB came from and whether they were coming from the loss of containment. The leading thought is that 6 mm ICB is difficult to travel through the 19, 15, and 13 mm three ICB layers to end up stuck on the side of the collector slots. It was found that most likely insufficient cleaning may have been the cause.

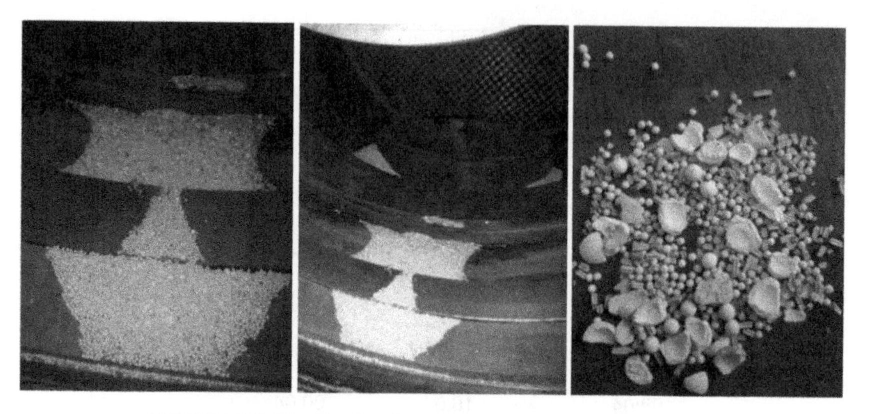

FIGURE 20.19. Catalyst in another feed/effluent exchanger.

FIGURE 20.20. Deformed first stage outlet collector.

There was also 6 mm ICB on the outlet collector top plate looking up from the bottom of outlet collector. If the surface was witnessed to be clean before loading, how could the 6 mm ICB end up there? Two possibilities were postulated that one reason could be that 6 mm balls remained in the loading chute and was transferred and mixed with 19 mm ICB when it was loaded. Another possibility is that the catalyst unloading nozzle on the side of reactor next to outlet collector was to be filled with 6 mm ICB before loading 19 mm ICB. However, for personnel standing convenience, 19 mm ICB was loaded up to the top of the outlet collector level to create a "standing platform" so that the unloading nozzle can be filled by pouring into unloading chute rather than lifted high for pouring. It is also possible that some 6 mm ICB bags spilled while loading the unloading chute and possibly not documented.

From examining the photographs of outlet collector, one odd item was that three locations on the side vertical runs had deformed slots on the outlet collector. ICB of 19 mm seemed to be lodged at the slots but not breaking through. This is definitely an abnormal situation with bent and deformed slots.

Reviewing the design drawing, it was confirmed that the first stage outlet collector slot width is indeed smaller than the second stage outlet collector because the original hydrocracker objective was for naphtha production requiring active support material be loaded around the outlet collector instead of the inert 19 mm ICB. Normally, posttreat catalyst is loaded at the bottom of the reactor for minimizing mercaptan formation. But posttreat catalyst takes up reactor volume. Instead of the inert ICB loaded at the bottom of reactor, the inert material can be substituted with the active support material. The active support size is a smaller size requiring reducing standard outlet collector slot width of 19 mm to contain the active support. The second stage outlet did not have any active support material, which had inert ICB, and the slot width was standard size as shown in Figure 20.15.

The deformed outlet collector slot was a major concern that perhaps the design calculation is incorrect. The licensor retrieved engineering calculation files and reviewed if the small slot specification for active support was valid. The design was found to be correct. Detailed engineering contractor also confirmed that the outlet collector was designed according to the mechanical load specifications. The first stage outlet collector arrived with the reactor shell and the second stage outlet collector was assembled on-site. Deformation was not present before loading at both outlet collectors. It appeared the first stage outlet collector was damaged after startup.

The refiner discussed possible solutions and questioned the possibilities of reinforcing the slots to prevent it from becoming bigger without unloading catalyst. Any welding would have to be done from bottom up and would not be safe.

Both reactors were not yet subjected to normal operation pressure drops since only recycle hydrogen gas was circulated and unit emergency depressuring was not used yet. Under such conditions, it is hard to imagine any mechanical force that could inflict such deformation to the outlet collector. The suspicion leads to poor fabrication.

Closer examination of the first stage outlet collector showed that the slots were poorly constructed and cut in irregular shapes as circled in Figure 20.21. It is possible

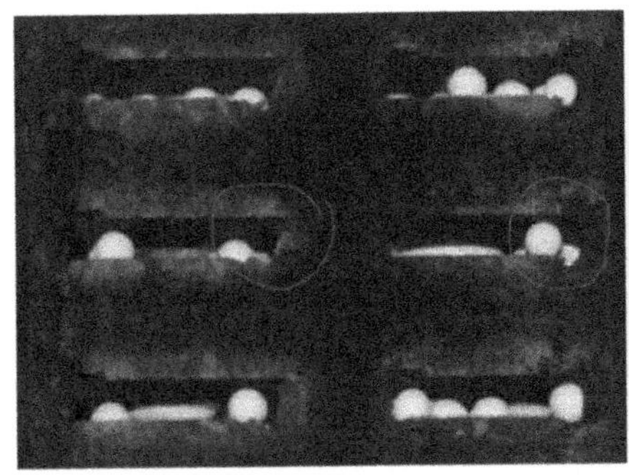
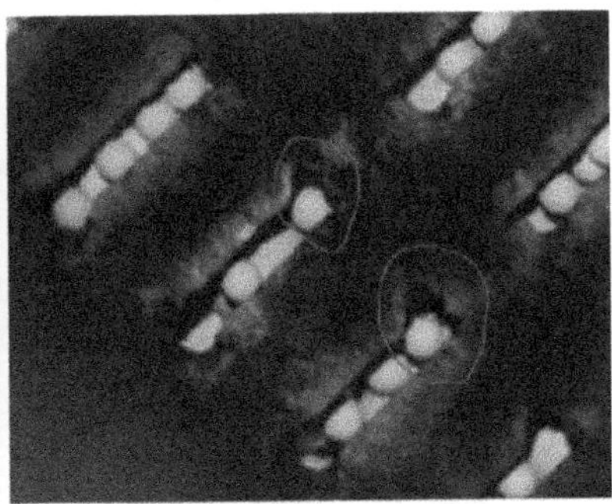

FIGURE 20.21. Outlet collector – chunks of metal missing near the junction.

that the outlet collector fabricator did not do a good heat treatment and upon heat up under load, the vertical slots bent releasing heat stress and deformed the collector.

The licensor recommended unloading and installing a new outlet collector. Upon more discussions on the need to start up the unit, the refiner decided not to unload the catalyst in R101 to fix the outlet collector. As long as the ICB is not leaking anymore, the damaged outlet collector will be replaced in the next turnaround.

The conclusion of this example is every piece of equipment and every step in the practice is critical from design, fabrication, installation, and proper operation. Also while dealing with a problem situation that every step involved in the troubleshooting requires careful consideration of what-ifs since one solution may lead to another problem. Attentiveness to details, diligent quality check-ups, considering all the what-ifs, and good planning are necessary to minimize problems.

20.6 CONCLUSION

Several examples were exhibited in this chapter. In nearly a century since hydrocracking technology is used commercially, many operation issues and incidents have occurred that it would be impossible to describe in detail every troubleshooting case. Use the "I," "S," and "C" mind-set suggested earlier in troubleshooting/problem solving and always learn from mistakes independently if this is from your own or from other units.

Troubleshooting is not always black and white and sometime the cause is not absolutely clear. As mentioned at the start of this chapter, the solution search should follow the approach of leaving no stones unturned. Design of hydrocracking units has also evolved becoming more reliable; but complacency is absolutely not tolerated needing careful/proper engineering, methodic project implementation, and day-to-day operation vigilance to achieve successful reliable performance.

INDEX

Hydroprocessing for Clean Energy: Design, Operation, and Optimization, First Edition.
Frank (Xin X.) Zhu, Richard Hoehn, Vasant Thakkar, and Edwin Yuh.